THE PESTICIC

GW00703385

For all those involved with pesticides, whether as manufacturers, users or advisers, **The Pesticides Monitor** is the only official monthly listing of UK approvals and other official announcements on pesticides and covers new developments in pesticides law. It includes:

— Details of new approvals, both full and provisional, making key information widely available for the first time; reproduces key points in full, e.g. statutory conditions of sale, supply, storage, use and advertisement;

— A separate section for off-label approvals. In particular this will benefit growers who are required by law to read the approvals before using the products.

— Official notices listing revocations, suspensions and amendments to the conditions of approval.

— Announcements on UK pesticides policy.

— News of routine reviews of older pesticides.

— Details on where to obtain information on pesticides.

Approved details of the following:

Agricultural, horticultural, forestry and home garden products;

Wood preservatives, public hygiene and household insecticides;

Surface biocides and antifouling products.

Commodity substances and adjuvants

Published by **The Stationery Office** for the **Pesticides Safety Directorate** (an executive agency of the **Ministry of Agriculture, Fisheries and Food**) and the **Health and Safety Executive**. Please order from The Stationery Office bookshops listed on the back cover, or by using the form overleaf.

Prices:
Annual Subscription £85 (Twelve Issues plus index)
Single Copy £7.95

These rates are subject to change and revision during 2000.

Order Form

☐ Please set up an annual subscription to The Pesticides Register @ £85
OR
☐ Please send me a single copy of The Pesticides Register (latest issue) @ £7.95

☐ I enclose a cheque for £ _____ made payable to The Stationery Office Ltd

☐ Please debit my

Mastercard/Visa/Amex/Connect /Diners Account

☐☐☐☐☐☐☐☐☐☐☐☐☐☐☐☐☐☐

Signature _____ Expiry date _____

☐ Please charge to my Stationery Office Account No. _____

Handling charge per order: £2.50

Total enclosed: £

PLEASE COMPLETE IN BLOCK CAPITALS

Name _____

Address _____

_____ Postcode _____

When completed, send to: AHN

   The Stationery Office
   PO Box 29
   Norwich
   NR3 1GN

PESTICIDES SAFETY DIRECTORATE
An Executive Agency of the Ministry of Agriculture, Fisheries and Food
HEALTH AND SAFETY EXECUTIVE

Pesticides
2000

Pesticides approved under:
The Control of Pesticides Regulations 1986

and

The Plant Protection Products Regulations 1995

London: The Stationery Office

First published 2000

ISBN 0 11 243051 1

CONTENTS

INTRODUCTION

1. Pesticides 2000 contains lists of agricultural and non-agricultural pesticide products whose uses held either full or provisional approval in the UK as at **15 October 1999** for PSD products and **28 October 1999** for HSE products. The exception are:

– PSD products for which approvals were known on **15 October 1999** to be due to expire on or before **31 December 1999** and which therefore do not appear

– HSE products for which approvals were known on **27 October 1999** to be due to expire on or before **31 December 1999** and which therefore appear with the appropriate symbol (see page vi).

Unless otherwise specified products listed in this book may be sold, supplied, stored, used or advertised *subject to the requirements of the pesticides legislation*, as explained in Chapter 2, and the conditions of their approvals, which are explained in Chapter 3.

2. Additional information on the products listed is available on product labels and literature; or from manufacturers and suppliers. You should always read the product label and use pesticides safely. *You have a duty to check that you know how to use a product safely in accordance with the statutory conditions of approval.* These approval conditions are set out on the product label and in the Schedules to the Control Of Pesticides Regulations (as amended) 1986 given by Ministers.

3. Further information on the approval status of pesticides may be obtained, in the case of PSD registered products listed in Part B, from the Pesticides Safety Directorate, External Relations Branch, Mallard House, Kings Pool, 3 Peasholme Green, York, YO1 7PX (Telephone No. 01904 455775) (e-mail: p.s.d.information@psd.maff.gsi.gov.uk). In the case of HSE registered products listed in Part C, from the Health and Safety Executive, Pesticides Registration Section, Magdalen House, Stanley Precinct, Bootle, Merseyside, L20 3QZ (Telephone No. 0151 951 3535 fax: 0151 951 3317, e-mail: info.prs@hse.gsi.gov.uk).

1. HOW TO USE THIS BOOK

PART A

The introductory chapters give general guidance on the use of pesticides and the relevant legislative controls. The remainder of the text is divided into lists of approved products, which are indexed.

PART B

This lists products registered with the Approvals Group of the Pesticides Safety Directorate, an Executive Agency of the Ministry of Agriculture, Fisheries and Food (MAFF). It is divided into four sections:

Section 1: Lists products approved for professional use in agricultural, horticultural, forestry and other pest control areas. This is sub-divided into the following classification groups: herbicides, fungicides, insecticides and vertebrate control products, biological pesticides and miscellaneous.

Within each of these sub-sections, approved products are listed by trade name under the appropriate alphabetically listed active ingredient along with details of the marketing company and the MAFF registration number. **In previous years, we have used a symbol to identify products under phased revocation in the 'Pesticides' publication. In order to provide more useful information, we have dispensed with the symbol in favour of the final date on which anyone may store, and persons other than the approval holder or their agents may sell, supply, advertise or use these products.**

For products with two or more active ingredients, the entry is repeated in respect of each, e.g.:
  2,4-D+Mecoprop and
  Mecoprop+2,4-D

Where there are three or more active ingredients, the secondary active ingredients in each entry are in alphabetical order, e.g.:
  Bentazone+Dichlorprop+Isoproturon
  Dichlorprop+Bentazone+Isoproturon
  Isoproturon+Bentazone+Dichlorprop

Products which have more than one area of use e.g. herbicide and fungicide, appear under all the appropriate activity headings.

Section 2: Products approved for amateur use feature in this section, although these products may also be used by professionals. The areas of use are detailed as part of the active ingredient heading.

Section 3: An alphabetical index of product trade names with a cross reference to the first entry of the product in each section by means of a code number.

Section 4: An alphabetical index of all active ingredient combinations with a cross-reference to the relevant part of each section by means of a code number.

Key to symbols
C – some or all uses of this product are "approved for agricultural use" as defined by the Schedules to the Control Of Pesticides Regulations (as amended) 1986 given by Ministers under Regulation 6 of the Control of Pesticides Regulations 1986, (for further information see Chapter 2.1, paragraph 10).

A – these products are approved for use in or near water as defined by the Schedules to the Control Of Pesticides Regulations (as amended) 1986 given by Ministers under Regulation 6 of the Control of Pesticides Regulations 1986, (for further information see Chapter 3 paragraph 17).

PART C

This covers products registered with the Pesticides Registration Section of the Health and Safety Executive (HSE) as antifouling products, biocidal paints, insect repellents, insecticides (including insecticidal paints), surface biocides, wood preservatives and wood treatments.

Within sections 1–7, relating to their area of use, approved products are listed below their active ingredients (which are given a code number and are listed alphabetically). In addition details of the marketing company, who can use the product and the HSE registration number are provided.

For products with two or more active ingredients, the entry is repeated for each active ingredient eg;

(1) 2-Phenylphenol and Cypermethrin
(2) Cypermethrin and 2-Phenylphenol

Products which have more than one area of use, e.g. wood preservation and surface biocide use, appear under each of the headings.

Who can use the product is given by the terms amateur, professional and industrial, which mean:

Amateur use – means that the product can be used by the general public.

Professional use – means that the product can only be used by people who are required to use pesticides as part of their work and who have received appropriate information, instruction and training.

Industrial use – means that the product is a wood preservative that can only be used for industrial pre-treatment of wood by operatives who have received appropriate information, instruction and training.

Key to symbols

W these products have either been voluntarily withdrawn for commercial reasons or revoked and hold approval for storage, supply and use but not for advertisement and sale under The Control of Pesticides Regulations 1986.

D these products have either been voluntarily withdrawn for commercial reasons or revoked and hold approval for the purposes of disposal only. HSE should be contacted for further information on individual products.

Active Ingredient Names

A number of active ingredients have several possible names. In order to compile the lists for sections 1–6, it has been necessary to standardise these names. Thus the active ingredient under which a product is listed may not be the same as that which appears on its label. These changes have been made for the purpose of Pesticides 2000 only and do not affect the name that should appear on the product label, which should be in accordance with the notice and schedule of approval.

The names that have been standardised are as follows:

Alkyldimethylbenzyl Ammonium Chloride   appears as **Benzalkonium chloride**

Ammonium Alkyldimethylbenzyl Chloride   appears as **Benzalkonium chloride**

Chromic Acetate   appears as **Chromium acetate**

Copper (I)   appears as **Cuprous**

Copper (II)   appears as **Copper**

Didecyldimethylammonium Chloride   appears as **Dialkyldimethyl ammonium Chloride**

Diethoxy Cetostearyl Alcohol   appears as **Ceto-stearyl diethoxylate**

Dioctyldimethylammonium Chloride   appears as **Dialkyldimethyl ammonium Chloride**

Disodium Tetraborate   appears as **Sodium tetraborate**

Gamma-HCH   appears as **Lindane**

Mono-ethoxy Oleic Acid   appears as **Oleyl monoethoxylate**

Parachlorometacresol   appears as **4-Chloro-meta-cresol**

Paradichlorobenzene   appears as **1, 4-Dichlorobenzene**

Pyrethrum/Pyrethrins/Pyrethrum extract   appears as **Pyrethrins**

Tributyltinmethacrylate Copolymer   appears as **Tributyltin methacrylate**

Hydrates – Hydrated and anhydrous substances are listed as the anhydrous form.

Part C is divided into eight sections;

SECTION 1 : Antifouling Products

SECTION 2: Biocidal Paints

SECTION 3: Insect Repellents

SECTION 4: Insecticides (including Insecticidal Paints)

SECTION 5: Surface Biocides

PART A

UK Regulation of Pesticides

2. THE LEGISLATION

2.1 Food and Environment Protection Act 1985 (FEPA) and The Control of Pesticides Regulations 1986 (as amended) (COPR)

Aims

1. Statutory powers to control pesticides are contained within Part III of The Food and Environment Protection Act 1985 (FEPA). Section 16 of the Act describes the aims of the controls as being to:

   i. protect the health of human beings, creatures and plants;

   ii. safeguard the environment;

   iii. secure safe, efficient and humane methods of controlling pests; and

   iv. make information about pesticides available to the public.

2. The mechanism by which these aims are to be achieved is set out in regulations made under the Act. The Control of Pesticides Regulations 1986 (SI 1986/1510) (as amended) (COPR) define in detail those types of pesticides which are subject to control and those which are excluded; prescribe the approvals required before any pesticide may be sold, stored, supplied, used or advertised; and allow for general conditions on sale, supply, storage, advertisement, and use, including aerial application, of pesticides. The 1986 Regulations were updated by the Control of Pesticides (Amendment) Regulations 1997 (SI 1997/188) and the text of the Schedules that set out these general conditions are at Annex A. Similar provisions apply in respect of the Plant Protection Product Regulations 1995 (as amended) (PPPR) (see chapter 2.2). Similar legislation exists in Northern Ireland and the majority of products approved for use in Great Britain are subsequently approved for use in Northern Ireland.

Definition of pesticides

3. Under FEPA, a pesticide is any substance, preparation or organism prepared or used, among other uses, to protect plants or wood or other plant products from harmful organisms; to regulate the growth of plants; to give protection against harmful creatures; or to render such creatures harmless.

The term pesticides therefore has a very broad definition which embraces herbicides, fungicides, insecticides, rodenticides, soil-sterilants, wood preservatives and surface biocides among others. A more complete definition and details of pesticides which fall outside the scope of the legislation is given in Regulation 3 of COPR.

What does COPR require?

4. The law requires that only pesticides approved by Ministers in the responsible Departments shall be sold, supplied, used, stored, or advertised. Post-devolution, the Minister of Agriculture, Fisheries and Food and the UK Environmental Minister, the UK Minister for Health and Ministers for the Scottish Executive, the Welsh Assembly and (for N.I. approvals) the Department of Agriculture and Rural Development in Northern Ireland are together responsible for the approval of all pesticides. There are three categories of approvals;

   i.   an experimental approval, to enable testing and development to be carried out with a view to providing the Ministers with safety and other data;

   ii. a provisional approval, for a stipulated period; or

   iii. a full approval, for an unstipulated period.

5. Products granted only an experimental approval cannot be advertised or sold. Such products do not appear in this book.

6. A data submission deadline, as the name suggests, is the deadline by which the Registration Authorities must be in receipt of any requested data required to secure continued approval. Failure to meet this deadline will result in the revocation of approval for products concerned. For PSD products, approval for advertisement, sale, supply and use by the approval holder and their agents is immediate, whilst approval for advertisement, sale, supply, storage and use by other parties, and storage by the approval holder and their agents remains for a limited period of up to two years.

7. Phased revocation of approval also occurs where the approval holder requests that a product be commercially withdrawn. This allows set periods of time to allow existing stocks to be marketed and used. For PSD products this allows up to two years for any persons other than the approval holder or their agents to market and use existing stocks. For HSE products this allows up to

three years for the advertisement and sale of the product, and up to five years for the supply, storage and use of the product.

8. Products granted provisional or full approval are normally allowed to be advertised, sold, supplied, stored and used.

9. New approvals are set out in the "Pesticides Register" available monthly from The Stationery Office book shops. For more details see inside the back cover.

10. Products listed in Part B and approved for agricultural use (as defined in the Schedules to the Control Of Pesticides Regulations 1986)(as amended) are indicated by the symbol "C". This formal classification is important for deciding whether the person storing, selling, supplying or using the pesticide needs to hold a recognised certificate of competence. The certification requirements are set out in Schedules 23 to the Control Of Pesticides Regulations 1986 (as amended) reproduced at Annex A at the end of chapter 3.

11. Products approved for amateur use (as listed in Part B, Section 2) and those registered with the HSE (as listed in Part C) do not attract storekeeper, salesperson or operator certification under COPR.

12. The supply of methyl bromide and chloropicrin is restricted to those who can prove that they are competent and have the necessary experience to use them safely.

The British Pest Control Association (BPCA) provide training and their own certification which is available to both members and non members. Competence can be proved by the possession of the relevant BPCA Certificate.

For further guidance on the use of methyl bromide, chloropicrin or other fumigants please refer to COP30 "The Control of Substances Hazardous to Health in Fumigation Operations".

13. Controls over the use of adjuvants and the tank mixing of pesticides are set out in the Schedules to the Regulations, which should be consulted. The list of authorised adjuvants is at Annex I (at the end of chapter 3) and is periodically updated and published in the "Pesticides Register".

14. Anyone who advertises, sells, supplies, stores or uses a pesticide is affected by the legislation, including people who use pesticides in their own homes, gardens and allotments.

Access to Information on Pesticides

15. A wide range of information about pesticides is published or made available by the Government. The Pesticides Act 1998 will widen public access to information on pesticides by enabling Ministers to provide the public with information on all pesticides, by way of new regulations. Previously, Ministers were able to provide information to the public only on those chemicals approved or reviewed since the introduction of statutory controls in 1986. PSD plan to have the regulations in place by early 2000.

The independent Advisory Committee on Pesticides publishes an annual report on its activities, a summary of each meeting and detailed evaluations of new or reviewed pesticide active ingredients, (a full list of published evaluations, as at 31 October 1999, is at Annex G). The data underlying published evaluations are available for inspection at PSD York, and HSE Bootle, on application (for addresses see page v).

Full results of pesticide residues monitoring in food stuffs and water are now published annually, as are usage surveys and reports of suspected incidents of human and wildlife poisoning. News of policy and procedural matters, product approvals and revocations are published monthly in the "Pesticides Register". Details of how to order the "Pesticides Register" can be found inside the back cover.

2.2 The Plant Protection Products Directive (91/414/EEC), The Plant Protection Products Regulations 1995 (as amended)(PPPR); The Plant Protection Products (Basic Conditions) Regulations 1997 (BCR)

The Directive

1. Council Directive 91/414/EEC harmonises the arrangements for authorisation of plant protection products within the Community, although product authorisation remains the responsibility of individual Member States. A list of plant protection products active substances will be assembled in Annex I to the Directive. This will be built up over a period of time as existing active substances are reviewed and new ones authorised. The requirements of this

Directive have been implemented in Great Britain under the Plant Protection Products Regulations 1995 (PPPR) (see Sections 10 – 14 below). The Pesticides Safety Directorate is the lead authority in Great Britain for the authorisation of products under this Directive.

Review Programme

2. Directive 91/414 provides for the review of active substances on the market on or before 25 July 1993. The details of the first tranche of the review programme are set out in Commission Regulations 3600/92, 933/94, 491/95, 2230/95, 1199/97 and 1972/99. Subsequent phases of the review programme will be drawn up in Commission Regulations in due course.

3. Following the evaluation of the submitted data, Member States are required to send their report to the Commission with a recommendation as to whether the active substance should be listed in Annex I of the Directive, refused Annex I listing, suspended from the market pending the provision of further data, or whether a decision on listing should be postponed pending the provision of further data.

New Active Substances

4. Companies with active substances new to the EU apply to a Member State of their choice for assessment on behalf of the Commission. The Member State initially ensures that the application dossier is compliant with the requirements of the Directive and then, following detailed evaluation of the data, submits a review report to the Commission with recommendations regarding whether or not the active substance should be included in Annex I of the Directive.

5. The recommendation of the Member State will be considered by all Member States in the framework of the Standing Committee on Plant Health and a decision taken by the Commission. The Commission has agreed the inclusion of a number of active substances (new and existing) on Annex I.

Uniform Principles, Transitional arrangements, Mutual Recognition

6. On completion of an EC review of an active substance Member States will review each of the products that contain it. "Uniform Principles" establishing common criteria for evaluating products at a national level were agreed in

September 1997 to enable harmonisation of the evaluation of applications for pesticide products by Member States.

7. Under transitional arrangements Member States may, prior to the listing of active substances in Annex I:

    a. Provisionally authorise products containing active substances not on the market by 25 July 1993 for periods of up to three years in advance of Annex I listing;

    b. maintain existing national approvals (in the case of Great Britain under the Control of Pesticides Regulations) until 25 July 2003, or until completion of the review of the active substance, whichever is earlier; and

    c. authorise new products containing active substances on the Community market on 25 July 1993 under existing national rules.

8. The Directive provides for "mutual recognition" for plant protection products with Annex I listed active substances which are authorised by a Member State in accordance with the provisions of the Directive. Subject to certain conditions, such plant protection products must be authorised by other Member States. Authorisation may be subject to conditions set to take account of differences such as climate and agricultural practice.

9. Detailed consideration of the impact of the Directive on approval procedures in the United Kingdom continues. Revised procedures are published in the "Pesticides Register" and in the Registration Handbook.

The Plant Protection Products Regulations 1995 (as amended); The Plant Protection Products (Basic Conditions) Regulations 1997

Aim

10. The Plant Protection Products Regulations 1995 (SI 1995/887) (as amended) (PPPR) implement the Plant Protection Products Directive, 91/414/EEC, in Great Britain. They define those products which are subject to regulation and prescribe the approvals required before any plant protection product (mainly agricultural pesticides) may be placed on the market or used. The Plant Protection Products (Basic Conditions) Regulations 1997 (SI 1997/189) (BCR) apply control and enforcement provisions similar to those of COPR to plant protection products, and thus ensure that enforcement powers for

products authorised under PPPR are equivalent to those under COPR. The general conditions relating to the sale, supply, storage, advertisement and use of plant protection products are set out in Schedules 1 to 4 of BCR. As with COPR, similar legislation exists in Northern Ireland.

Definition of a plant protection product

11. A plant protection product is an active substance or a preparation containing one or more active substances, put up in the form in which it is supplied to the user, intended to:

    a. protect plants or plant products against all harmful organisms or prevent the action of such organisms;

    b. influence the life processes of plants, other than as a nutrient (for example, as a growth regulator);

    c. preserve plant products, in so far as such substances or products are not subject to provisions of Community law on preservatives;

    d. destroy undesired plants; or

    e. destroy parts of plants or check or prevent the undesired growth of plants.

12. As active substances are placed on Annex I of the Directive, plant protection products containing only those active substances are required to be regulated under PPPR. Products based on active substances on the market after July 1993 (and not yet included in Annex I) are also required to be regulated, on a provisional basis, under PPPR.

What does PPPR require?

13. As with COPR, the Regulations require that only plant protection products approved by Ministers in the responsible Departments (see Section 2.1, para 4) shall be used or placed on the market. There are three main categories of approvals;

i.   Standard approval, to place on the market and use any plant protection product for a period not exceeding ten years. A standard approval is only appropriate where the active ingredient is on Annex I to Directive 91/414.

ii. Provisional approval, to place on the market and use any plant protection product for a period not exceeding three years. A provisional approval is appropriate where the Commission decision on the Annex I listing of the active substance is awaited.

iii. Approval for research and development, to enable experiments or tests for research and development to be carried out where such tests involve the release into the environment of a product which has not been otherwise approved.

14. Approvals may be renewed by Ministers for standard approvals (if requirements for approval continue to be satisfied) and for provisional approvals where the Commission's decision on the Annex I listing of the active substance continues to be awaited.

The Pesticides (Maximum Residue Levels In Crops, Food and Feeding Stuffs) (England and Wales) Regulations 1999 (as amended).

15. Residues of pesticides in food and feeding stuffs are controlled by a system of statutory maximum residue levels (MRLs). To date these have been set for certain key pesticides used on the more important components of the average national diet. MRLs are intended to check that pesticides are being used correctly, and also to provide additional protection for the consumer.

Codes of Practice

16. Government Departments have produced guidance to help those working with pesticides meet the requirements of the legislation. These are:

i) The MAFF "Code of Practice for Suppliers of Pesticides to Agriculture, Horticulture and Forestry" (PB3259), which gives practical guidance on the storage and transport of pesticides and the obligations on those who sell, supply and store for sale and supply. Copies are available free from MAFF Publications, ADMAIL 6000, London SW1A 2XX and;

ii) the MAFF "Code of Practice for the Safe Use of Pesticides on Farms and Holdings" (PB3528) which promotes the safe use of pesticides. It covers the requirements of the Control of Pesticides Regulations 1986 (COPR) and the

Control of Substances Hazardous to Health Regulations 1994 (COSHH). Copies are available free from MAFF Publications ADMAIL 6000, London SW1A 2XX.

iii) the HSC "The Safe Use of Pesticides for Non-Agricultural Purposes: Control of Substances Hazardous to Health Regulations 1994. Approved Code of Practice" which provides advice to those in the non-agricultural sector on compliance with COSHH priced £6.95 (ISBN 0-7176-0542-6)'.

iv) the HSC "Recommendations for Training Users of Non-Agricultural Pesticides" priced £2.00 (ISBN 0-11-885548-4).

v) the HSE/DOE "Remedial Timber Treatment in Buildings A Guide to Good Practice and the Safe Use of Wood Preservatives". Copies are available from The Stationery Office book shops and can be ordered from The Stationery Office Publications Centre (0171 873 9090) priced £4.00 (ISBN 0-11-885-9870).

vi) the HSC "Approved Code of Practice for the Control of Substances Hazardous to Health in Fumigation Operations" provides practical guidance on the COSHH Regulations 1999 priced at £8.50 (ISBN 0-7176-1195-7).

vii) The HSC "General COSHH (Control of Substances Hazardous to Health) ACoP, Carcinogens (Control of carcinogenic substances) ACoP and Biological Agents (Control of biological agents) ACoP" which provides guidance on the COSHH Regulations 1999 priced at £8.50 (ISBN 0-7176-1670-3).

17. The above HSC and HSE publications are now available by mail order from:
HSE Books
PO Box 1999
SUDBURY
Suffolk
CO10 6FS

Tel. (01787) 881165 Fax. (01787) 313995.

They can also be purchased from Dillons Bookstores and ordered from Ryman the Stationer and Ryman Computer Centres (0171-434-3000), but are no longer available from The Stationery Office.

2.3 *The Control of Substances Hazardous to Health Regulations 1999 (COSHH)*

1. Pesticides should only be used when necessary, and if the benefit from using them significantly outweighs the risks to human health and the environment. If a decision is reached to use a pesticide, the product selected should be one that poses the least risk to people, livestock and the environment yet is still effective in controlling the pest, disease or weed problem that has been identified.

2. The COSHH Regulations apply to a wide range of pesticides used at work. They lay down essential requirements and a step by step approach for the assessment and control of exposure to hazardous substances. Failure to comply with COSHH, in addition to possibly exposing people to risk, constitutes an offence and is subject to penalties under the Health and Safety at Work etc. Act 1974.

3. Substances (including pesticides and other chemicals) that are "hazardous to health" include those labelled very toxic, toxic, harmful, irritant or corrosive, substances with occupational exposure limits, harmful micro-organisms, or substantial quantities of dust. They also include fumes, gases and other materials which could cause comparable harm to health.

4. Employers and the self-employed are required by the COSHH Regulations 1999 to carry out an assessment of the risks to human health. The assessment may confirm that the pesticide chosen is the most appropriate, but if it becomes apparent that using another pesticide may involve less risk, then the choice will need to be reconsidered. Always make sure that the product is approved for the intended use and situation.

Making a COSHH Assessment

5. Before carrying out an assessment it should be remembered that the harm from pesticides depends mainly on the active ingredient, its form, e.g. granules, liquid etc. and how the product is to be applied. Hazardous substances can get into the body through the skin, the mouth or by inhalation. The greatest risk of exposure is often through skin absorption.

6. Information supplied by the manufacturer will provide the basis for an assessment but don't forget other sources of advice, such as Codes of Practice and industry guidance.

7. The COSHH assessment needs to be suitable and sufficient. It should include the following steps:

Step 1 – Identify the hazards

- Read the pesticide label and any other literature available from the manufacturer.
- What form is the pesticide in?

Step 2 – Who might be harmed and how?

- Who might be exposed – workers, members of the public?
- Consider both normal and accidental exposure

Step 3 – Decide what needs to be done.

- Consider all aspects of the job, e.g. storage, transport, application, disposal, cleaning and maintenance of both application equipment and personal protective equipment (PPE).
- Apply the hierarchy of control:–

  1. Could the product be substituted for a less hazardous pesticide

  2. Can engineering/technical controls reduce exposure, e.g. induction hoppers, closed transfer systems and in-cab controls?

  3. Are operational controls in use, e.g. warning signs and keeping people out of treated areas, notification of adjacent occupiers and others who need to know?

  4. PPE should only be used where substitution, engineering/technical controls and operational controls are not feasible. The PPE must be suitable for the job. If there is any doubt, the manufacturer or supplier of the pesticide should be consulted.

- Are those involved adequately trained and informed about items such as the importance of equipment calibration, weather conditions, timing of application and harvest intervals?
- Has the most appropriate application equipment been selected?
- Is health or medical surveillance required?

Step 4 – Record the findings if it is useful to do so.

Step 5 – Review the assessment from time to time.

2.4 The Biocidal Products Directive

1. This Directive (98/8/EC) is intended to remove barriers to trade by introducing a community wide scheme for authorising the placing of biocidal products on the market. Biocidal products include: all products currently regulated under CoPR where HSE takes the lead and, in addition, industrial preservatives, disinfectants, water biocides, certain cleaners claiming germicidal properties and pesticides other than for plant protection purposes.

The authorisation scheme will be similar to that introduced by the Plant Protection Products Directive described in 2.2 above. A two-stage process will be introduced whereby active substances are assessed at community level for possible inclusion on Annex I of the Directive and then products containing Annex I-listed active substances are authorised by Member States. Biocidal products containing active substances on the market in May 2000 will be able to stay on the market, subject to national legislation, if any, until the active substance has been reviewed and listing on Annex I has been agreed. The Directive envisages a 10-year programme for the review of these existing active substances. Biocidal products containing a new active substance cannot be placed on the market unless authorised following the Annex I listing of the active substance.

2. The Directive will be introduced in all Member States by May 2000. The European Commission, together with the Member States and Industry, continue to work hard to ensure that the necessary systems are in place by May 2000.

EU-wide Technical Guidance on data requirements should be finalised before May 2000. That on procedures for listing active substances on Annex I and for authorisation of products will be finalised later. There is active progress on the proposals for a Review Regulation and practical procedures for reviewing existing active substances. Again, this should be finalised before May 2000. Applications for Annex I listing of active substances or authorisation of products require the submission of dossiers. EU-wide guidance on the structure of such dossiers is being produced. Whilst much is being done, there are many areas where activity is unlikely to proceed quickly. This is especially the case in areas where the procedures are not likely to be needed for several years.

3. HSE is the lead authority in Great Britain for evaluation work and granting authorisations under this Directive. Draft Biocidal Products Regulations and accompanying guidance have been subject to formal consultation procedures

within the UK. The final Regulations will come into force in May 2000 and guidance will published around the same time.

Procedures for receiving and processing applications for Annex I listing and authorisation of products are being developed. An 'Applicant's Handbook', giving guidance to applicants is being drafted and should be available in 2000. It will draw heavily on the EU guidance. Consequently, for the reasons mentioned earlier, there will be many gaps in the guidance which can only be filled when they become a priority at community level.

4. The Directive requires Member States to charge for work done under the Directive. HSE is currently examining how it will recover costs of the regime from May 2000, including, where appropriate, the level of individual fees.

2.5 *The Campaign Against Illegal Poisoning*

The Campaign Against Illegal Poisoning of Wildlife, aimed at protecting some of Britain's rarest birds of prey and wildlife whilst also safeguarding domestic animals, was launched in March 1991 by the Ministry of Agriculture, Fisheries and Food and the Department of the Environment, Transport and the Regions. It is strongly supported by a range of organisations associated with animal welfare, nature preservation, field sports and game keeping including the RSPB, English Nature, the British Field Sports Society and the Game Conservancy Trust.

The three objectives of the Campaign are;
- To advise farmers, gamekeepers and other land managers on legal ways of controlling pests;
- To advise the public on how to report illegal poisoning incidents and to respect the need for legal alternatives;
- To investigate incidents and prosecute offenders.

All these contribute to the main objective which is to deter those who may be considering using pesticides illegally from doing so.

During the nine years in which the Campaign has been in existence, much work has been done to achieve these objectives. A freephone number (0800 321 600) has been established to make it easier for the public to report incidents and numerous leaflets, posters, postcards and stickers have been created and distributed in order to publicise the existence of the Campaign. A video has also been produced and this is used to illustrate the many talks, demonstrations and exhibitions which are regularly presented by ADAS Consulting Ltd. on behalf of MAFF in England and Wales.

The Campaign arose from the results of a MAFF scheme for the investigation of possible cases of illegal poisoning, the Wildlife Incident Investigation Scheme. Under this scheme, all reported cases are considered and thoroughly investigated where appropriate. Enforcement action is taken wherever sufficient evidence of an offence can be obtained. Since the launch of the Campaign there have been over 45 prosecutions.

The message is getting through that this dangerous and illegal poisoning will not be tolerated. Your help in reporting possible incidents and spreading the word will be much appreciated.

Further information about the Campaign is available from:

Pesticides Safety Directorate
Research Co-ordination and Environmental Policy Branch
Room 308
Mallard House
Kings Pool
3 Peasholme Green
York
YO1 7PX

Web-site address is: http://www.maff.gov.uk/aboutmaf/agency/psd/caip/ caip.htm

3. THE APPROVALS PROCESS

1. All approvals are granted by Ministers in response to an application from a manufacturer, formulator, importer or distributor (or in certain circumstances a user) supported by the necessary data on safety, efficacy and, where relevant, humaneness. Approvals are normally granted only in relation to individual products and only for specified uses.

Statutory Conditions of Use

2. It is an offence not to follow the statutory conditions of use of a pesticide: these are detailed in the relevant notice of approval. Approvals are published in the "Pesticides Register". Additionally copies of approvals may be obtained from PSD, Approvals Administration, Room 208, Mallard House, Kings Pool, 3 Peasholme Green, York, Y01 7PX (Telephone No. 01904 455715). Additionally COPR require that the product label be consistent with the statutory conditions of use. For "off-label" approvals and commodity substances – see below. Statutory conditions of use will differ from product to product and use to use: they cover:

(i)   Field of use restrictions, e.g. agriculture, wood preservative.

(ii)   User restrictions

(iii)   The crop or situation which may be treated.

(iv)   Maximum individual dose/application rate.

(v)   Maximum number of treatments or maximum total dose.

(vi)   Maximum area or quantity which may be treated;

(vii)   Latest time of application, harvest or re-entry interval.

(viii) Operator protection or training requirements.

(ix)   Environmental protection requirements.

(x)   Any other specific restrictions which Ministers may require.

It is therefore important to *read the label (including any accompanying leaflet) carefully, if possible before purchase and certainly before use.*

3. In order to highlight the statutory conditions to the user the following steps have been taken:

Since 31 December 1994 all approved pesticides must display the statutory conditions of use in a designated area on the label known as the "statutory box".

Approvals for (Extensions of Use) Off-Label Uses

4. Users of agricultural pesticides may apply to have the approval of specific pesticides extended to cover uses additional to those approved and shown on the manufacturer's product label. Any such specific off-label approvals granted may have additional conditions of use attached to them. Use in these cases is undertaken at the user's choosing, and the commercial risk is entirely theirs. Users are required to be in possession of the relevant notice of approval. Approvals are published in the "Pesticides Register" and are available from ADAS offices. For further details contact ADAS Kirton, 22 Willington Rd, Kirton, Boston, Lincs, PE20 1EJ Tel (01205) 723300 Fax (01205) 722922 or the National Farmers Union Offices via the NFU "order line" service (Telephone No. 0345 58324). Additionally copies of approvals may be obtained from PSD, Approvals Administration, Room 208, Mallard House, Kings Pool, 3 Peasholme Green, York, YO1 7PX (Telephone No. 01904 455715).

5. Since 1 January 1990 arrangements have been in place which permit many pesticide products to be used for additional specific minor uses, subject to adherence to various conditions. These arrangements were updated in December 1999 to current standards and additional guidance notes provided to improve clarity. Full details are given in Annex C at the end of this section. These arrangements are valid until 31 December 2005. Use in these cases is undertaken at the user's choosing, and the commercial risk is entirely theirs.

Approval of Commodity Substances

6. Commodity substances are chemicals which have a variety of non-pesticidal uses and also have minor uses as pesticides. If such a substance is to be used as a pesticide, it requires approval under COPR, and is granted approval for use only.

7. *Sale, supply, storage and advertisement* of a commodity substance as a pesticide is an offence unless specific approval has been granted for the substance to be marketed as a pesticide under an approved label. Anyone wishing to sell, supply, store or advertise any of these substances specifically as pesticides therefore has to seek specific approval under COPR.

8. The commodity substances that can be used as a pesticide are listed below. For some substances separate approval is given for different fields of use. The approvals are listed at Annex D and detail the conditions which a user must follow when the commodity substance is used as a pesticide;

4-chloro-*m*-cresol
Camphor
Carbon dioxide
Ethanol
Ethyl acetate
Formaldehyde
Isopropanol
Methyl bromide
Liquid nitrogen
Paraffin oil
Sodium chloride
Strychnine hydrochloride
Sulphuric acid
Tetrachloroethylene
Thymol
Urea
White spirit

List of Products Approved for Aerial Application

9. Products approved for use by aerial application, together with details of the crops on which they are approved for use are listed in Annex B at the end of this chapter. Only products approved for aerial application may be used for that purpose, and regular returns by any person who undertakes the aerial application will have to be made to the Pesticides Usage Survey Group, MAFF, Central Science Laboratory, Sand Hutton, York, YO4 1LZ.

Amendment, Expiry, Suspension and Revocation of Approvals

10. Pesticide approvals may at any time be subject to review, amendment, suspension or revocation. Revocation of approval may occur for a number of reasons, for example the identification of safety concerns or an approval holder's failure to meet a data submission deadline. Where possible a "phased revocation" will be implemented, but when safety considerations make it necessary immediate revocation may take effect. On expiry or revocation of approvals it becomes unlawful to advertise, sell, supply, store or use the products for the uses concerned.

11. Suspension of approval may take place when it is anticipated that the action is temporary, e.g. where approval might be reinstated by the provision of required and satisfactory data.

12. A list of pesticide active ingredients banned or severely restricted in the UK can be found at Annex E.

Authorised Adjuvants

Adjuvants

13. An adjuvant is a substance other than water which is not in itself a pesticide but which enhances or is intended to enhance the effectiveness of the pesticide with which it is used. Adjuvants for use with agricultural pesticides have been categorised as extenders, wetting agents, sticking agents and fogging agents.

14. Schedule 3, paragraph 5(1) of COPR (as amended) permits the use of an adjuvant with a pesticide only where:

i.   the use is in accordance with the conditions of approval of the pesticide;
ii. the adjuvant appears on a list of adjuvants published by Ministers; and
iii. the use is in accordance with any conditions to which use of that adjuvant with that pesticide is subject.

15. The List of adjuvants includes details of all the conditions to which use of particular adjuvants are subject. The List is published annually in full in The Pesticides Register, as are subsequent additions and amendments. Copies of the List or of individual list entries are available on request from PSD.

List of Products Approved for Use in or Near Water

16. Pesticides must not be used in or near water unless the conditions of approval, specifically allows such use. Such pesticides may be subject to additional statutory conditions which will be detailed on the product label. Revised "Guidelines for the use of Herbicides on Weeds in or near Watercourses and Lakes" (PB 2289) were published in September 1995 and advise on the precautions that should be taken especially where fresh water may be abstracted for the irrigation of crops, watering of livestock, fisheries or in fish farming. Situations defined for such treatment include drainage channels, streams, rivers, ponds, lakes, reservoirs, canals and dry ditches. These guidelines also cover the control of vegetation growing on the banks, or areas immediately adjacent to such water bodies, but not the control of vegetation growing on nearby cropped or amenity land. Copies can be obtained free from MAFF Publications, ADMAIL 6000, London SW1A 2XX (Telephone No. 0645 556000).

18. Products approved for use in or near water (other than for public hygiene or antifouling use) are listed in Annex H at the end of this chapter.

ANNEX A

FOOD AND ENVIRONMENT PROTECTION ACT 1985
PART III

SCHEDULES TO THE CONTROL OF PESTICIDES REGULATIONS (AS AMENDED) 1986 GIVEN BY MINISTERS

The advertisement, sale, supply, storage and use of pesticides is prohibited unless the conditions of consent set out in the appropriate Schedules to the Control Of Pesticides Regulations 1986 (as amended) are met.

Ministers gave their consent to the advertisment, sale, supply and storage and use of pesticides, subject to the conditions set out in the Schedules to COPR, on 31 January 1997 (these were published, on that date, in the London and Edinburgh Gazettes.) The schedules to COPR may be changed at any time following Parliamentary approval. The schedules to COPR are published in the "Pesticides Register" and notified to interested parties.

SCHEDULE 1 – CONDITIONS RELATING TO CONSENT TO THE ADVERTISMENT OF PESTICIDES

1.–(1) An advertisement of a pesticide shall relate only to such conditions as are permitted by the approval given in relation to that pesticide.

(2) No advertisement of a pesticide shall contain any claim for safety in relation to that pesticide which is not permitted by the approval given in relation to that pesticide to be on the label for the pesticide.

2. –(1) Any advertisement of a pesticide, other than a notice at the point of sale which is intended to draw attention solely to product name and price, shall include –

(a) a statement of each active ingredient of each pesticide mentioned in the advertisement, such statement being the name by which each active ingredient is identified in the approval given in relation to the pesticide in which it is contained;

(b) a general warning as follows:

"Always read the label. Use pesticides safely"; and

(c) where required by a condition of the approval given in relation to a pesticide mentioned in the advertisement, a statement of any special degree of risk to human beings, creatures, plants or the environment.

(2) Notwithstanding sub-paragraph (1)(a) above –

(a) any price list consisting only of an indication of product availability and price need not state the active ingredient of each pesticide;

(b) any advertisement of a range of pesticides need only state the active ingredients of those individual products which are identified by name.

(3) Any statement or warning given under this paragraph shall be –

(a) in the case of a printed or pictorial advertisement, clearly presented separately from any other text; and

(b) in the case of an advertisement which is broadcast or recorded or is stored or transmitted by electronic means, clearly spoken or shown separately.

3. In this Schedule "advertisement" means any printed, pictorial, broadcast or recorded advertisement and includes any advertisement which is stored or transmitted by electronic means

SCHEDULE 2 – CONDITIONS RELATING TO CONSENT TO THE SALE, SUPPLY AND STORAGE OF PESTICIDES

1. It shall be the duty of all employers to ensure that persons in their employment who may be required during the course of their employment to sell, supply or store pesticides are provided with such instruction, training and guidance as is necessary to enable those persons to comply with any requirements provided in and under these Regulations.

2.–(1) Any person who sells, supplies or stores a pesticide shall –

    (a) take all reasonable precautions, particularly with regard to storage and transport, to protect the health of human beings, creatures and plants, safeguard the environment and in particular avoid the pollution of water; and

    (b) be competent for the duties which that person is called upon to perform.

(2) In this paragraph "water" means –

    (a) any surface water;

    (b) any ground water.

3. No person shall sell, supply or otherwise market to the end-user an approved pesticide other than in the container which has been supplied for that purpose by the holder of the approval of that pesticide and labelled in a manner consistent with the approval.

4. No person shall store for the purpose of sale or supply a pesticide approved for agricultural use in a quantity in excess of, at any one time, 200 kg or 200 litres, or a similar mixed quantity, unless that person –

    (a) has obtained a certificate of competence recognised by the Ministers, or

    (b) stores that pesticide under the direct supervision of a person who holds such a certificate.

5. No person shall sell, supply or otherwise market to the end-user a pesticide approved for agricultural use unless that person –

    (a) has obtained a certificate of competence recognised by the Ministers, or

    (b) sells or supplies that pesticide under the direct supervision of a person who holds such a certificate.

6.–(1) In paragraphs 4 and 5 above –

"pesticide approved for agricultural use" means a pesticide (other than a pesticide with methyl bromide or chloropicrin as one of its active ingredients) approved for one or more of the following uses –

    (a) agriculture and horticulture (including amenity horticulture);

    (b) forestry;

    (c) in or near water other than for amateur, public hygiene or anti-fouling uses;

    (d) industrial herbicides, including weed-killers for use on land not intended for the production of any crop.

(2) In this paragraph "water" means any surface water.

SCHEDULE 3 – CONDITIONS RELATING TO CONSENT TO THE USE OF PESTICIDES

1. It shall be the duty of all employers to ensure that persons in their employment who may be required during the course of their employment to use pesticides are provided with such instruction, training and guidance as is necessary to enable those persons to comply with any requirements provided in and under these Regulations.

2.–(1) Any person who uses a pesticide shall take all reasonable precautions to protect the health of human beings, creatures and plants, safeguard the environment and in particular avoid the pollution of water.

(2) In this paragraph "water" means –

    (a) any surface water;

    (b) any ground water.

3. No person in the course of a business or employment shall use a pesticide, or give an instruction to others on the use of a pesticide, unless that person –

    (a) has received adequate instruction, training and guidance in the safe, efficient and humane use of pesticides, and

    (b) is competent for the duties which that person is called upon to perform.

4. Any person who uses a pesticide shall confine the application of that pesticide to the land, crop, structure, material or other area intended to be treated.

5.–(1) Subject to sub-paragraph (4) below, no person shall use a pesticide in conjunction with an adjuvant in any manner unless –

    (a) that adjuvant has been specified, upon application by any person (in this paragraph 5 referred to as "the applicant") to the Ministers, in a list of adjuvants published by the Ministers from time to time (in this paragraph 5 referred to as "the list"); and

    (b) the use of that pesticide with that adjuvant in that manner is in accordance with –

        (i) the conditions of the approval given in relation to that pesticide; and

        (ii) any requirements to which the use of that adjuvant with that pesticide is subject, as determined or amended under sub-paragraph (2)(a)(ii) or (iii) below.

(2) In the application of this paragraph –

    (a) the Ministers may, in relation to any adjuvant specified in the list, at any time –

        (i) determine data requirements (concerning human safety or environmental protection) to which the specification of that adjuvant in the list shall be subject;

        (ii) determine requirements to which the use of that adjuvant with approved pesticides shall be subject;

        (iii) for reasons of human safety or environmental protection, or with the consent of the applicant, amend any requirement which has been determined under sub-paragraph (ii) above;

    (b) the Ministers shall, in relation to any adjuvant specified in the list, also specify in that list any requirements which they have determined or amended under paragraph (a)(ii) or (iii) above.

(3) In the application of this paragraph –

    (a) the Ministers may, in relation to any adjuvant specified in the list, remove that adjuvant from the list –

(i) if it appears to them that the applicant has failed to comply with any data requirement which has been determined in relation to that adjuvant under sub-paragraph (2)(a)(i) above;

(ii) if it appears to them that any relevant literature relating to the adjuvant is not in accordance with any requirement to which the use of that adjuvant is subject, as determined or amended under sub-paragraph (2)(a)(ii) or (iii) above;

(iii) if it appears to them that –

(aa) any relevant literature relating to the adjuvant refers to a pesticide, and

(bb) the use of that adjuvant with that pesticide is not in accordance with the conditions of the approval given in relation to that pesticide;

(iv) for reasons of human safety or environmental protection;

(v) at the request of the applicant;

(b) the Ministers shall, upon a decision to remove an adjuvant from the list specify in the list –

(i) that decision, and

(ii) the date on which, and any conditions in accordance with which, the removal is to take effect;

(c) "relevant literature", in relation to any adjuvant, means –

(i) the labelling of the packaging in which the adjuvant is contained;

(ii) any leaflet accompanying that package;

(iii) any other literature produced by, or on behalf of, the applicant describing the adjuvant.

(4) This paragraph shall not apply where the use of an adjuvant with an approved pesticide is for the purpose of research or development and is carried out under the direct control of the person intending to place the adjuvant on the market.

(5) In this paragraph "adjuvant" means a substance other than water, without significant pesticidal properties, which enhances or is intended to enhance the effectiveness of a pesticide when it has been added to that pesticide.

6.–(1) No person shall combine or mix for use two or more pesticides which are anticholinesterase compounds unless such a mixture is expressly permitted by the conditions of the approval given in relation to at least one of those pesticides or by the labelling of the container in which at least one of those pesticides has been sold, supplied or otherwise marketed to that person.

(2) No person shall combine or mix for use two or more pesticides unless –

(a) all of the conditions of approval given in relation to each of those pesticides, and

(b) the labelling of the container in which each of those pesticides has been sold, supplied or otherwise marketed to that person,

can be complied with.

7.–(1)No person in the course of a commercial service shall use a pesticide approved for agricultural use unless that person –

(a) has obtained a certificate of competence recognised by the Ministers; or

(b) uses that pesticide under the direct and personal supervision of a person who holds such a certificate; or

(c) uses it in accordance with an approval, if any, for one or more of the following uses –

(i) home garden (amateur gardening);

(ii) animal husbandry;

(iii) food storage practice;

(iv) vertebrate control (including rodenticides and repellents);

(v) domestic use;

(vi) wood preservation;

(vii) as a surface biocide;

(viii) public hygiene or prevention of public nuisance;

(ix) other industrial biocides;

(x) as an anti-fouling product;

(xi) "other" (as may be defined by the Ministers).

(2) In this paragraph "commercial service" means the application of a pesticide by a person –

(a) to crops, land, produce, materials, buildings or the contents of buildings not in the ownership or occupation of that person or that person's employer;

(b) to seed other than seed intended solely for use by that person or that person's employer.

8. No person who was born later than 31 December 1964 shall use a pesticide approved for agricultural use unless that person –

(a) has obtained a certificate of competence recognised by the Ministers; or

(b) uses that pesticide under the direct and personal supervision of a person who holds such a certificate; or

(c) uses it in accordance with an approval, if any, for one of the uses specified in paragraph 7(1)(c) above.

9.–(1) In paragraphs 7 and 8 above "pesticide approved for agricultural use" means a pesticide (other than a pesticide with methyl bromide or chloropicrin as one of its active ingredients) approved for one or more of the following uses –

(a) agriculture and horticulture (including amenity horticulture);

(b) forestry;

(c) in or near water, other than for amateur, public hygiene or anti-fouling uses;

(d) industrial herbicides, including weed-killers for use on land not intended for the production of any crop.

(2) In this paragraph "water" means any surface water.

SCHEDULE 4 – CONDITIONS RELATING TO CONSENT TO THE USE OF PESTICIDES BY AERIAL APPLICATION

1. No person shall undertake an aerial application of a pesticide unless –

(a) an aerial application certificate granted under article 42(2) of the Air Navigation Order 1985 is held by that person, that person's employer or the main contractor undertaking the aerial application, and

(b) the pesticide to be used has been approved for the intended aerial application.

2.–(1) No person shall undertake an aerial application of a pesticide unless that person, or a person specifically designated in writing on that person's behalf, has –

(a) not less than 72 hours before the commencement of the aerial application consulted the relevant authority if any part of land which is a Local Nature Reserve, a Marine Nature Reserve, National Nature Reserve or Site of Special Scientific Interest lies within 1500 metres of any part of the land to which that pesticide is to be applied;

(b) not less than 72 hours before the commencement of the aerial application consulted the appropriate area office of the Environment Agency (if the area in which the intended aerial application is to take place is in England and Wales) or the appropriate area office of the Scottish Environment Protection Agency (if such area is in Scotland) if the land to which that pesticide is to be applied is adjacent to, or within 250 metres of, water;

(c) obtained the consent of such office if that pesticide is to be applied for the purpose of controlling aquatic weeds or weeds on the banks of watercourses or lakes;

(d) not less than 24 hours and (so far as is practicable) not more than 48 hours before the commencement of the aerial application, given notice of the intended aerial application to the Chief Environmental Health Officer for the district in which the intended aerial application is to take place;

(e) not less than 24 hours and (so far as is reasonably practicable) not more than 48 hours before the commencement of the aerial application given notice of the intended aerial application to the occupants or their agents of all property within 25 metres of the boundary of the land to which that pesticide is to be applied;

(f) not less than 24 hours and (so far as is practicable) not more than 48 hours before the commencement of the aerial application, given notice of the intended aerial application to the person in charge of any hospital, school or other institution any part of the curtilage of which lies within 150 metres of any flight path intended to be used for the aerial application; and

(g) not less than 48 hours before the commencement of the aerial application, given notice of the intended aerial application to the appropriate reporting point of the local beekeepers' spray warning scheme operating within the district in which the intended aerial application is to take place.

(2) A notice of an intended aerial application given under paragraph (e) or (f) of sub-paragraph (1) above shall be in writing and include details of–

(a) the name and address, and telephone number (if any), of the person intending to carry out the aerial application;

(b) the name of the pesticide to be applied and its active ingredient and approval registration number;

(c) the intended time and date of application; and

(d) an indication that the same details have been served on the Chief Environmental Health Officer for the district in which the intended aerial application is to take place.

3. No person shall undertake an aerial application of a pesticide unless –

(a) the wind velocity at the height of application at the place of intended aerial application does not exceed 10 knots, except where the approval given in relation to that pesticide permits aerial application when such wind velocity exceeds 10 knots;

(b) not less than 24 hours before the aerial application, that person has provided and put in place within 60 metres of the land to which that pesticide is to be applied signs, of adequate robustness and legibility, to warn pedestrians and drivers of vehicles of the time and place of the intended aerial application; and

(c) before the aerial application that person has provided ground markers in all circumstances where a ground marker will assist the pilot to comply with the provisions of paragraph 5 below.

4. Any person who undertakes the aerial application of a pesticide shall –

(a) keep and retain for not less than 3 years after each application records of –

   (i)   the nature, place and date of that application;

   (ii) the registration number of the aircraft used;

   (iii) the name and permanent address of the pilot of that aircraft;

   (iv) the name and quantity of the pesticide applied;

   (v) the dilution and volume of application of the pesticide applied;

   (vi) the type and specification of application system (which may include nozzle type and size);

   (vii) the method of application;

   (viii) the flight times of the aerial application;

   (ix) the speed and direction of the wind during the application; and

   (x) any unusual occurrences which affected the application;

(b) provide the Ministers with summaries of the records required by sub-paragraph (a) above, in any manner which they may require under section 16(11) of the 1985 Act, within 30 days after the end of the calendar month to which those records relate.

5. The pilot of an aircraft engaged in an aerial application shall –

(a) maintain the aircraft at a height of not less than 200 feet from ground level when flying over an occupied building or its curtilage;

(b) maintain the aircraft at a horizontal distance from any occupied building and its curtilage, children's playground, sports ground or building containing livestock of –

   (i)   not less than 30 metres, if the pilot has the written consent of the occupier; and

   (ii) not less than 60 metres, in any other case;

(c) maintain the aircraft at a height of not less than 250 feet from ground level over any motorway, or of not less than 100 feet from ground level over any other public highway, unless that motorway or public highway has been closed to traffic during the course of the application.

6. For the purposes of this Schedule –

"appropriate nature conservation agency" means English Nature, Scottish Natural Heritage and the Countryside Council for Wales;

"curtilage", in relation to any building, means the land attached to, and forming one enclosure with, that building;

"ground marker" includes a person who is instructed by a person intending to carry out an aerial application to be present on or near to the land to which the pesticide is to be applied so that that person is able to communicate with the pilot of the aircraft engaged in the aerial application for the purpose of ensuring the safe application of that pesticide;

"local beekeepers' spray warning scheme" means any scheme for the advance notification of the application of pesticides, organised by local beekeepers and notified to the Minister of Agriculture, Fisheries and Food, the Secretary of State for Scotland or the Secretary of State for Wales (being the Secretaries of State respectively concerned with agriculture in Scotland and Wales);

"Local Nature Reserve" means a nature reserve established by a local authority under section 21 of the National Parks and Access to the Countryside Act 1949 and "the relevant authority" in regard to such a reserve shall be the local authority which is providing or securing the provision of the reserve;

"Marine Nature Reserve" means an area designated as such by the Secretary of State under section 36 of the Wildlife and Countryside Act 1981 and the "relevant authority" in regard to such an area shall be the appropriate nature conservation agency;

"National Nature Reserve" means any land declared as such by the appropriate nature conservation agency under section 19 of the National Parks and Access to the Countryside Act 1949, or under section 35 of the Wildlife and Countryside Act 1981, and "the relevant authority" in regard to such land shall be the appropriate nature conservation agency;

"Site of Special Scientific Interest" means any area designated as such by the appropriate nature conservation agency under section 28 of the Wildlife and Countryside Act 1981, or in respect of which the Secretary of State has made an Order under section 29 of the Wildlife and Countryside Act 1981, and "the relevant authority" in regard to such an area shall be the appropriate nature conservation agency;

"water" means any surface water.

ANNEX B

PRODUCTS APPROVED FOR USE BY AERIAL APPLICATION

Products approved for use by aerial application, together with details of the crops on which they are approved for use are listed below. Only products approved for aerial application may be used for that purpose and regular returns have to be made to the Pesticides Usage Survey Group, MAFF, Central Science Laboratory, Sand Hutton Lane, Sand Hutton, York, YO4 1LW.

Note reference must be made to the label for conditions of use:

Active ingredient	Product (Reg No.)	Crops/Uses
Asulam	Asulox (06124)	Agricultural Grassland, Amenity Grassland, Forestry, Rough Upland Intended For Grazing
Benalaxyl + Mancozeb	Barclay Bezant (05914)	Potato (Early), Potato (Main)
	Clayton Benzeb (07081)	Potato (Early), Potato (Main)
	Galben M (07220)	Potato (Early), Potato (Main), Potato (Seed)
	Tairel (07767)	Potato (Early), Potato (Main), Potato (Seed)
Carbendazim	Ashlade Carbendazim Flowable (06213)	Oilseed Rape, Wheat (Winter)
	Delsene 50 Flo (09469)	Oilseed Rape, Wheat (Winter)
	HY-CARB (05933)	Wheat (Winter)
	MSS Mircarb (08788)	Oilseed Rape, Wheat (Winter)
	Quadrangle Hinge (08070)	Oilseed Rape, Wheat (Winter)
	Top Farm Carbendazim – 435 (05307)	Oilseed Rape, Wheat (Winter)
	Tripart Defensor FL (02752)	Oilseed Rape, Wheat (Winter)
Chlormequat	3C Chlormequat 460 (03916)	Barley (Winter), Oats, Wheat
	3C Chlormequat 600 (04079)	Barley (Winter), Oats, Wheat
	Agriguard Chlormequat 700 (09282)	Oats (Spring), Oats (Winter), Wheat (Spring), Wheat (Winter)
	Agriguard Chlormequat 700 (09782)	Oats (Spring), Oats (Winter), Wheat (Spring), Wheat (Winter)
	Allied Colloids Chlormequat 460 (07859)	Oats (Spring), Oats (Winter), Wheat (Spring), Wheat (Winter)
	Allied Colloids Chlormequat 460:320 (07861)	Oats (Spring), Oats (Winter), Wheat (Spring), Wheat (Winter)
	Allied Colloids Chlormequat 730 (07860)	Oats (Spring), Oats (Winter), Wheat (Spring), Wheat (Winter)
	Ashlade 460 CCC (06474)	Oats (Spring), Oats (Winter), Wheat (Spring), Wheat (Winter)

Active ingredient	Product (MAFF No.)	Crops/Uses
Chlormequat—*continued*	Ashlade 700 5C (07046)	Oats (Spring), Oats (Winter), Wheat (Spring), Wheat (Winter)
	Ashlade 700 CCC (06473)	Oats (Spring), Oats (Winter), Wheat (Spring), Wheat (Winter)
	Ashlade Brevis (08119)	Oats (Spring), Oats (Winter), Wheat (Spring), Wheat (Winter)
	Atlas 3C:645 Chlormequat (05710)	Oats, Wheat (Spring), Wheat (Winter)
	Atlas 3C:645 Chlormequat (07700)	Oats, Wheat (Spring), Wheat (Winter)
	Atlas 5C Chlormequat (03084)	Oats (Spring), Oats (Winter), Wheat (Spring), Wheat (Winter)
	Atlas 5C Chlormequat (07701)	Oats (Spring), Oats (Winter), Wheat (Spring), Wheat (Winter)
	Atlas Chlormequat 46 (07704)	Oats (Spring), Oats (Winter), Wheat (Spring), Wheat (Winter)
	Atlas Chlormequat 460:46 (06258)	Oats (Spring), Oats (Winter), Wheat (Spring), Wheat (Winter)
	Atlas Chlormequat 460:46 (07705)	Oats (Spring), Oats (Winter), Wheat (Spring), Wheat (Winter)
	Atlas Chlormequat 700 (03402)	Oats (Spring), Oats (Winter), Wheat (Spring), Wheat (Winter)
	Atlas Chlormequat 700 (07708)	Oats (Spring), Oats (Winter), Wheat (Spring), Wheat (Winter)
	Atlas Quintacel (07706)	Oats (Spring), Oats (Winter), Wheat (Winter)
	Atlas Terbine (06523)	Oats (Spring), Oats (Winter), Wheat (Spring), Wheat (Winter)
	Atlas Terbine (07709)	Oats (Spring), Oats (Winter), Wheat (Spring), Wheat (Winter)
	Atlas Tricol (07190)	Oats (Spring), Oats (Winter), Wheat (Spring), Wheat (Winter)
	Atlas Tricol (07707)	Oats (Spring), Oats (Winter), Wheat (Spring), Wheat (Winter)
	Atlas Tricol PCT (08015)	Barley (Winter), Wheat (Spring), Wheat (Winter)
	Barclay Holdup (06799)	Barley (Winter), Oats (Spring), Oats (Winter), Wheat (Spring), Wheat (Winter)
	Barclay Holdup 600 (08794)	Barley (Winter), Oats (Spring), Oats (Winter), Wheat (Spring), Wheat (Winter)
	Barclay Holdup 640 (08795)	Barley (Winter), Oats (Spring), Oats (Winter), Wheat (Spring), Wheat (Winter)
	Barclay Take 5 (08524)	Barley (Winter), Oats (Spring), Oats (Winter), Rye, Triticale, Wheat (Spring), Wheat (Winter)

Active ingredient	Product (MAFF No.)	Crops/Uses
Chlormequat—*continued*	BASF 3C Chlormequat 600 (04077)	Barley (Winter), Oats, Wheat
	BASF 3C Chlormequat 720 (06514)	Barley (Winter), Oats (Spring), Oats (Winter), Rye, Triticale, Wheat (Spring), Wheat (Winter)
	BASF 3C Chlormequat 750 (06878)	Barley (Winter), Oats (Spring), Oats (Winter), Rye, Triticale, Wheat (Spring), Wheat (Winter)
	Calypso (09505)	Oats (Spring), Oats (Winter), Wheat (Spring), Wheat (Winter)
	Ciba Chlormequat 460 (09525)	Oats (Spring), Oats (Winter), Wheat (Spring), Wheat (Winter)
	Ciba Chlormequat 5C 460:320 (09527)	Oats (Spring), Oats (Winter), Wheat (Spring), Wheat (Winter)
	Ciba Chlormequat 730 (09526)	Oats (Spring), Oats (Winter), Wheat (Spring), Wheat (Winter)
	Clayton Standup (08771)	Oats (Spring), Oats (Winter), Wheat (Spring), Wheat (Winter)
	Greencrop Carna (09403)	Barley (Winter), Oats (Spring), Oats (Winter), Wheat (Spring), Wheat (Winter)
	Greencrop Coolfin (09449)	Oats (Spring), Oats (Winter), Wheat (Spring), Wheat (Winter)
	Hyquat 70 (03364)	Barley (Winter), Oats (Spring), Oats (Winter), Wheat (Spring), Wheat (Winter)
	Intracrop Balance (08037)	Barley (Winter), Oats (Winter), Wheat (Spring), Wheat (Winter)
	Mandops Barleyquat B (06001)	Barley (Winter), Rye
	Mandops Bettaquat B (06004)	Barley (Winter)
	Mandops Chlormequat 700 (06002)	Oats, Wheat (Spring), Wheat (Winter)
	MSS Mircell (06939)	Oats (Spring), Oats (Winter), Wheat (Spring), Wheat (Winter)
	MSS Mirquat (08166)	Oats (Spring), Oats (Winter), Wheat (Spring), Wheat (Winter)
	New 5C Cycocel (01482)	Barley (Winter) (Autumn Application in), Barley (Winter) (Spring Application in), Oats (Spring), Oats (Winter), Rye, Triticale, Wheat (Spring), Wheat (Spring) (Autumn Drilled), Wheat (Winter)
	New 5C Cycocel (01483)	Oats, Rye, Triticale, Wheat (Spring), Wheat (Winter)
	Sigma PCT (08663)	Barley (Winter), Wheat (Spring), Wheat (Winter)

Active ingredient	Product (MAFF No.)	Crops/Uses
Chlormequat—*continued*	Stabilan 5C (08144)	Oats (Spring), Oats (Winter), Wheat (Spring), Wheat (Winter)
	Stabilan 750 (08004)	Barley (Winter), Oats (Winter), Wheat (Spring), Wheat (Winter)
	Stabilan 750 (09303)	Barley (Spring), Oats (Winter), Wheat (Spring), Wheat (Winter)
	Supaquat (09381)	Barley (Winter), Oats (Spring), Oats (Winter), Rye, Triticale, Wheat (Spring), Wheat (Winter)
	Top Farm Chlormequat 640 (05323)	Oats (Spring), Oats (Winter), Wheat (Spring), Wheat (Winter)
	Trio (08883)	Barley (Winter), Oats (Spring), Oats (Winter), Rye (Winter), Triticale, Wheat (Spring), Wheat (Winter)
	Tripart Brevis (03754)	Oats (Spring), Oats (Winter), Wheat (Spring), Wheat (Winter)
	Tripart Chlormequat 460 (03685)	Oats (Spring), Oats (Winter), Wheat (Spring), Wheat (Winter)
	Uplift (07527)	Barley (Winter), Oats (Spring), Oats (Winter), Wheat (Spring), Wheat (Winter)
	Whyte Chlormequat 700 (09641)	Oats (Spring), Oats (Winter), Wheat (Spring), Wheat (Winter)
2-Chloroethylphosphonic acid	Aventis Cerone (09748)	Barley (Winter)
	Barclay Coolmore (07917)	Barley (Winter)
	Cerone (06185)	Barley (Winter)
	Charger (08827)	Barley (Winter)
	EXP03149D (08828)	Barley (Winter)
	Stantion (06205)	Barley (Winter)
	Stefes Stance (06125)	Barley (Winter)
	Unistar Ethephon 480 (06282)	Barley (Winter)
Chlorothalonil	Agriguard Chlorothalonil (09390)	Potato
	Barclay Corrib (05886)	Broccoli, Brussels Sprout, Cabbage, Cauliflower, Celery, Field Bean, Kale, Oilseed Rape, Onion, Pea (Not Vining), Potato, Wheat (Winter)
	Barclay Corrib 500 (06392)	Barley (Spring), Barley (Winter), Broccoli, Brussels Sprout, Cabbage, Calabrese, Cauliflower, Celery, Field Bean, Onion, Potato, Wheat (Spring), Wheat (Winter)
	Barclay Corrib 500 (08981)	Potato
	Baton SC (07945)	Potato

Active ingredient	Product (MAFF No.)	Crops/Uses
Chlorothalonil—*continued*	BB Chlorothalonil (03320)	Blackcurrant, Potato, Wheat
	Bombardier (02675)	Potato
	Bombardier FL (07910)	Potato
	Bravo 500 (05637)	Potato
	Bravo 500 (05638)	Potato
	Bravo 500 (09059)	Potato
	Bravo 720 (09104)	Potato
	Bravo Star (08371)	Potato
	Bravo Star (09108)	Potato
	Chloronil (08273)	Potato
	Chloronil (09110)	Potato
	Clayton Turret (09400)	Potato
	Clortosip (06126)	Blackcurrant, Potato, Wheat
	Clortosip 500 (09320)	Potato
	Flute (08953)	Potato
	Greencrop Orchid (09566)	Potato
	ISK 375 (07455)	Potato
	ISK 375 (09103)	Potato
	Jupital (05554)	Potato
	Jupital (09109)	Potato
	Jupital DG (09181)	Potato
	Landgold Chlorothalonil FL (06335)	Broccoli, Brussels Sprout, Cabbage, Calabrese, Cauliflower, Field Bean, Potato
	Mainstay (05625)	Potato
	Repulse (07641)	Potato
	Sipcam UK Rover 500 (04165)	Blackcurrant, Potato
	Strada (08824)	Potato
	Top Farm Chlorothalonil 500 (05926)	Field Bean, Oilseed Rape (Winter), Potato, Wheat (Winter)
	Tripart Faber (04549)	Potato
	Tripart Faber (05505)	Potato, Wheat (Winter)
	Ultrafaber (05627)	Potato
Chlorothalonil + Cymoxanil	Ashlade Cyclops (04857)	Potato
	Cyclops (06650)	Potato
	Guardian (06676)	Potato
Chlorotoluron	Dicurane (08403)	Barley (Winter), Durum Wheat, Triticale, Wheat (Winter)
	Dicurane 500 SC (05836)	Barley (Winter), Triticale, Wheat (Winter)
	Dicurane 700 SC (04859)	Barley (Winter), Durum Wheat, Triticale, Wheat (Winter)
	Lentipur CL 500 (08743)	Barley (Winter), Durum Wheat, Triticale, Wheat (Winter)
	Luxan Chlorotoluron 500 Flowable (09165)	Barley (Winter), Durum Wheat, Triticale, Wheat (Winter)
	Talisman (03109)	Barley (Winter), Durum Wheat, Triticale, Wheat (Winter)

Active ingredient	Product (MAFF No.)	Crops/Uses
Copper oxychloride	Cuprokylt (00604)	Potato
Cymoxanil + Mancozeb	Clayton Krypton (06973)	Potato
	Fytospore (06517)	Potato
	Standon Cymoxanil Extra (06807)	Potato
	Stefes Blight Spray (05811)	Potato
Diflubenzuron	Dimilin Flo (07151)	Forestry
	Dimilin Flo (08769)	Forestry
	Dimilin Flo (08985)	Forestry
Dimethoate	Barclay Dimethosect (08538)	Barley (Winter), Oats (Winter), Pea, Potato (Ware), Rye (Winter), Sugar Beet, Triticale, Wheat (Winter)
	BASF Dimethoate 40 (00199)	Barley (Spring), Barley (Winter), Oats (Spring), Oats (Winter), Pea, Potato (Ware), Rye, Sugar Beet, Triticale, Wheat (Spring), Wheat (Winter)
	Danadim Dimethoate 40 (07351)	Barley (Winter), Oats (Winter), Pea, Potato (Ware), Rye (Winter), Sugar Beet, Triticale, Wheat (Winter)
	PA Dimethoate 40 (01527)	Cereal, Fodder Beet, Mangel, Pea, Potato (Chitting House), Potato (Seed Crop), Potato (Ware), Red Beet, Sugar Beet
	Rogor L40 (07611)	Barley (Spring), Barley (Winter), Oats (Spring), Oats (Winter), Pea, Potato (Ware), Rye, Sugar Beet, Triticale, Wheat (Spring), Wheat (Winter)
	Sector (08882)	Barley (Winter), Oats (Winter), Pea, Potato (Ware), Rye (Winter), Sugar Beet, Triticale, Wheat (Winter)
	Top Farm Dimethoate (05936)	Carrot, Fodder Beet, Mangel, Mangel (Steckling), Potato, Potato (Chitting House), Potato (Seed), Potato (Ware), Red Beet, Sugar Beet, Sugar Beet (Steckling)
Disulfoton	Disyston P-10 (00715)	Brussels Sprout, Cabbage, Carrot, Cauliflower, Sugar Beet
Fenitrothion	Dicofen (00693)	Barley (Spring), Barley (Winter), Pea, Wheat (Spring), Wheat (Winter)

Active ingredient	Product (MAFF No.)	Crops/Uses
Fenitrothion—*continued*	Dicofen (09598)	Barley (Spring), Barley (Winter), Pea, Wheat (Spring), Wheat (Winter)
	Unicrop Fenitrothion 50 (02267)	Barley (Spring), Barley (Winter), Pea, Wheat (Spring), Wheat (Winter)
Flamprop-M-ethyl	Stefes Flamprop (05789)	Barley (Spring), Barley (Winter), Durum Wheat, Rye, Triticale, Wheat (Spring), Wheat (Winter)
Flamprop-M-isopropyl	Commando (07005)	Barley (Spring), Barley (Winter), Durum Wheat, Rye, Triticale, Wheat (Spring), Wheat (Winter)
Iprodione	Landgold Iprodione 250 (06465)	Field Bean, Oilseed Rape
Mancozeb	Absezeb WDG (07797)	Potato
	Agrichem Mancozeb 80 (06354)	Potato
	Ashlade Mancozeb FL (06226)	Potato
	Barclay Manzeb 455 (07990)	Potato
	Barclay Manzeb 80 (05944)	Potato
	Dequiman MZ (06870)	Potato
	Dithane 945 (00719)	Potato
	Dithane 945 (04017)	Potato
	Dithane Dry Flowable (04251)	Potato
	Dithane Dry Flowable (04255)	Potato
	Dithane Dry Flowable (09754	Potato
	Dithane Superflo (06290)	Potato
	Headland Zebra Flo (07442)	Potato
	Headland Zebra WP (07441)	Potato
	Helm 75 WG Newtec (09757)	Potato
	Helm 75WG (08309)	Potato
	Kor DF (08979)	Potato
	Kor DF Newtec (09758)	Potato
	Kor Flo (08019)	Potato
	Landgold Mancozeb 80 W (06507)	Potato
	Luxan Mancozeb Flowable (06812)	Potato
	Manconex (09555)	Potato
	Mandate 80 WP (09080)	Potato
	Manex II (07637)	Potato
	Manzate 200 PI (07209)	Potato
	Manzate 200 PI (09480)	Potato
	Micene 80 (09112)	Potato
	Mortar Flo (09592)	Potato
	Opie 80 WP (08301)	Potato
	Penncozeb (07820)	Potato
	Penncozeb WDG (07833)	Potato

Active ingredient	Product (MAFF No.)	Crops/Uses
Mancozeb—*continued*	Penncozeb WDG (09690)	Potato
	Quell Flo (08317)	Potato
	Restraint DF (09499)	Potato
	Restraint DF Newtec (09755)	Potato
	Stefes Deny (08932)	Potato
	Stefes Mancozeb DF (08010)	Potato
	Stefes Mancozeb WP (07655)	Potato
	Stefes Restraint (08945)	Potato
	Tariff 75 WG Newtec (09756)	Potato
	Tariff 75WG (08308)	Potato
	Tridex (07922)	Potato
	Trimanzone (09278)	Potato
	Trimanzone (09584)	Potato
	Unicrop Mancozeb (05467)	Potato
	Unicrop Mancozeb 80 (07451)	Potato
Mancozeb + Metalaxyl	Fubol 75 WP (03462)	Potato
	Fubol 75 WP (08409)	Potato
	Osprey 58 WP (05717)	Potato (Early), Potato (Main)
	Osprey 58 WP (08428)	Potato (Early), Potato (Main)
	Ridomil MZ 75 (07640)	Potato
	Ridomil MZ 75 WP (08438)	Potato
Mancozeb + Oxadixyl	Recoil (04039)	Potato
	Recoil (08483)	Potato
Maneb	Agrichem Maneb 80 (05474)	Potato
	Ashlade Maneb Flowable (06477)	Potato
	Headland Spirit (04548)	Potato
	Maneb 80 (01276)	Potato
	Mazin (06061)	Potato
	RH Maneb 80 (01796)	Potato
	Stefes Maneb DF (06418)	Potato
	Trimangol 80 (06070)	Potato
	Trimangol 80 (06871)	Potato
	Trimangol WDG (06992)	Potato
	Unicrop Maneb 80 (06926)	Potato
	X-Spor SC (08077)	Potato
Metaldehyde	Aristo (09622)	Around Edible Crop (Outdoor), Around Edible Crop (Protected), Around Non-Edible Crop (Outdoor), Around Non-Edible Crop (Protected), Bare Soil
	Clartex (09213)	All Edible Crop, All Non-Edible Crop
	Doff Agricultural Slug Killer with Animal Repellent (06058)	Edible Crop (Around Outdoor), Edible Crop (Around Protected), Non-Edible Crop (Around Protected), Non-Edible Crop (Outdoor) (Around), Soil (Bare)

Active ingredient	Product (MAFF No.)	Crops/Uses
Metaldehyde—*continued*	Doff Horticultural Slug Killer Blue Mini Pellets (05688)	Edible Crop, Non-Edible Crop, Soil (Bare)
	Doff Horticultural Slug Killer Blue Mini Pellets (09666)	All Edible Crop, All Non-Edible Crop, Bare Soil
	Doff New Formula Metaldehyde Slug Killer Mini Pellets (09338)	Around Edible Crop (Outdoor), Around Edible Crop (Protected), Around Non-Edible Crop (Outdoor), Around Non-Edible Crop (Protected), Bare Soil
	Doff New Formula Metaldehyde Slug Killer Mini Pellets (09772)	Around Edible Crop (Outdoor), Around Edible Crop (Protected), Around Non-Edible Crop (Outdoor), Around Non-Edible Crop (Protected), Bare Soil
	EM 1617/01 (09344)	All Edible Crop, All Non-Edible Crop, Bare Soil
	ESP (09428)	All Edible Crop, All Non-Edible Crop
	FP 107 (06666)	Bare Soil, Edible Crop (Outdoor), Edible Plant (Around), Non-Edible Crop (Outdoor), Non-Edible Plant (Around)
	FP 107 (09060)	Around Edible Crop (Outdoor), Around Non-Edible Crop (Outdoor), Bare Soil
	Gastrotox 6G Slug Pellets (04066)	Edible Crop, Non-Edible Crop, Soil (Bare)
	Lynx (09137)	Around Edible Crop (Outdoor), Around Edible Crop (Protected), Around Non-Edible Crop (Outdoor)
	Lynx (09770)	Around Edible Crop (Outdoor), Around Edible Crop (Protected), Around Non-Edible Crop (Outdoor), Around Non-Edible Crop (Protected), Bare Soil
	Optimol (06688)	Bare Soil, Edible Plant (Around), Non-Edible Plant (Around)
	Optimol (09061)	Around Edible Crop (Outdoor), Around Non-Edible Crop (Outdoor), Bare Soil
	pbi Slug Pellets (09607)	All Edible Crop (Outdoor), All Non-Edible Crop (Outdoor), Bare Soil
	Quadrangle Mini Slug Pellets (01670)	Edible Crop, Non-Edible Crop

Active ingredient	Product (MAFF No.)	Crops/Uses
Metaldehyde—*continued*	Slug Pellets (01558)	All Edible Crop (Outdoor), All Non-Edible Crop (Outdoor), Bare Soil
	Super-flor 6% Metaldehyde Slug Killer Mini Pellets (05453)	Around Edible Crop (Outdoor), Around Edible Crop (Protected), Around Non-Edible Crop (Outdoor), Around Non-Edible Crop (Protected), Bare Soil
	Super-Flor 6% Metaldehyde Slug Killer Mini Pellets (09773)	Around Edible Crop (Outdoor), Around Edible Crop (Protected), Around Non-Edible Crop (Outdoor), Around Non-Edible Crop (Protected), Bare Soil
	Tripart Mini Slug Pellets (02207)	Agricultural Crop, Horticultural Crop
	Unicrop 6% Mini Slug Pellets (02275)	Around Edible Crop (Outdoor), Around Edible Crop (Protected), Around Non-Edible Crop (Outdoor), Around Non-Edible Crop (Protected), Bare Soil
	Unicrop 6% Mini Slug Pellets (09771)	Around Edible Crop (Outdoor), Around Edible Crop (Protected), Around Non-Edible Crop (Outdoor), Around Non-Edible Crop (Protected), Bare Soil
	Yeoman (09623)	Around Edible Crop (Outdoor), Around Edible Crop (Protected), Around Non-Edible Crop (Outdoor), Around Non-Edible Crop (Protected), Bare Soil
Methabenzthiazuron	Tribunil (02169)	Barley (Spring), Barley (Winter), Durum Wheat, Oats (Winter), Rye (Winter), Ryegrass (Perennial), Triticale, Wheat (Spring) (Autumn Sown), Wheat (Winter)
Methiocarb	Barclay Poacher (09031)	All Edible Crop (Outdoor), All Non-Edible Crop (Outdoor), Bare Soil
	Bayer UK 808 (09513)	Bare Soil, Cereal Seed (Admixture), Edible Crop (Outdoor), Non-Edible Crop (Outdoor), Ryegrass (Seed) (Admixture)

Active ingredient	Product (MAFF No.)	Crops/Uses
Methiocarb—*continued*	Bayer UK 809 (09514)	Bare Soil, Cereal Seed (Admixture), Edible Crop (Outdoor), Non-Edible Crop (Outdoor), Ryegrass (Seed) (Admixture)
	Bayer UK 892 (09540)	Bare Soil, Edible Crop (Outdoor), Non-Edible Crop (Outdoor)
	Bayer UK 935 (09541)	Bare Soil, Edible Crop (Outdoor), Non-Edible Crop (Outdoor)
	Club (07176)	Bare Soil, Edible Crop (Outdoor), Non-Edible Crop (Outdoor)
	Decoy (06535)	Bare Soil, Edible Crop (Outdoor), Non-Edible Crop (Outdoor)
	Decoy Plus (07615)	Bare Soil, Edible Crop (Outdoor), Non-Edible Crop (Outdoor)
	Decoy Wetex (09707)	Bare Soil, Cereal Seed (Admixture), Edible Crop (Outdoor), Non-Edible (Outdoor), Ryegrass (Seed) (Admixture)
	Draza (00765)	Bare Soil, Edible Crop (Outdoor), Non-Edible Crop (Outdoor)
	Draza 2 (04748)	Bare Soil, Edible Crop (Outdoor), Non-Edible Crop (Outdoor)
	Draza Plus (06553)	Bare Soil, Edible Crop (Outdoor), Non-Edible Crop (Outdoor)
	Draza Wetex (09704)	Bare Soil, Cereal Seed (Admixture), Edible Crop (Outdoor), Non-Edible Crop (Outdoor), Ryegrass (Seed) (Admixture)
	Elvitox (06738)	Bare Soil, Edible Crop (Outdoor), Non-Edible Crop (Outdoor)
	Epox (06737)	Bare Soil, Edible Crop (Outdoor), Non-Edible Crop (Outdoor)
	Exit (07632)	Bare Soil, Edible Crop (Outdoor), Non-Edible Crop

Active ingredient	Product (MAFF No.)	Crops/Uses
Methiocarb—*continued*	Karan (09637)	All Edible Crop (Outdoor), All Non-Edible Crop (Outdoor), Bare Soil, Cereal Seed (Admixture), Ryegrass (Seed) (Admixture)
	Lupus (09638)	All Edible Crop (Outdoor), All Non-Edible Crop (Outdoor), Bare Soil, Cereal Seed (Admixture), Ryegrass (Seed) (Admixture)
	Rescur (07942)	Bare Soil, Edible Crop (Outdoor), Non-Edible Crop (Outdoor)
	Rivet (09512)	Bare Soil, Cereal Seed (Admixture), Edible Crop (Outdoor), Non-Edible Crop (Outdoor), Ryegrass (Seed) (Admixture)
Monolinuron	Arresin (07303)	Dwarf French Bean, Potato
Phosalone	Zolone Liquid (06173)	Brassica (Seed Crop), Cereal, Oilseed Rape
Pirimicarb	Agriguard Pirimicarb (09620)	Barley, Durum Wheat, Oats, Rye, Triticale, Wheat
	Aphox (06633)	Barley, Durum Wheat, Oats, Rye, Triticale, Wheat
	Barclay Pirimisect (06929)	Barley (Spring), Barley (Winter), Durum Wheat, Oats (Spring), Oats (Winter), Rye, Triticale, Wheat (Spring), Wheat (Winter)
	Barclay Pirimisect (09057)	Barley (Spring), Barley (Winter), Oats (Spring), Oats (Winter), Wheat (Spring), Wheat (Winter)
	Clayton Pirimicarb 50 SG (06972)	Barley, Durum Wheat, Oats, Rye, Triticale, Wheat
	Clayton Pirimicarb 50SG (09221)	Barley, Durum Wheat, Oats, Rye, Triticale, Wheat
	Helocarb Granule 500 (08157)	Barley, Durum Wheat, Oats, Rye, Triticale, Wheat
	Phantom (04519)	Barley, Durum Wheat, Oats, Rye, Triticale, Wheat
	Pirimate (09568)	Barley, Oats, Wheat
	Portman Pirimicarb (06922)	Barley (Spring), Barley (Winter), Oats (Spring), Oats (Winter), Wheat (Spring), Wheat (Winter)

Active ingredient	Product (MAFF No.)	Crops/Uses
Pirimicarb—*continued*	Stefes Pirimicarb (05758)	Barley, Broad Bean, Broccoli (Including Calabrese), Brussels Sprout, Cabbage, Carrot, Cauliflower, Chinese Cabbage, Collard, Durum Wheat, Dwarf French Bean, Field Bean, Kale, Maize, Oats, Oilseed Rape, Pea, Potato, Runner Bean, Rye, Sugar Beet, Swede, Sweetcorn, Triticale, Turnip, Wheat
	Unistar Pirimicarb 500 (06975)	Barley, Durum Wheat, Oats, Rye, Triticale, Wheat
Propiconazole	Barclay Bolt (08341)	Barley (Spring), Barley (Winter), Oats (Spring), Oats (Winter), Rye, Wheat (Spring), Wheat (Winter)
	Clayton Propiconazole (06415)	Barley (Spring), Barley (Winter), Wheat (Spring), Wheat (Winter)
	Mantis (08423)	Barley (Spring), Barley (Winter), Oats (Spring), Oats (Winter), Rye, Wheat (Spring), Wheat (Winter)
	Mantis 250EC (06240)	Barley (Spring), Barley (Winter), Oats (Spring), Oats (Winter), Rye, Wheat (Spring), Wheat (Winter)
	Radar (06747)	Barley (Spring), Barley (Winter), Oats (Spring), Oats (Winter), Rye (Spring), Rye (Winter), Wheat (Spring), Wheat (Winter)
	Radar (09168)	Barley (Spring), Barley (Winter), Oats, Rye, Wheat (Spring), Wheat (Winter)
	Standon Propiconazole (07037)	Barley (Spring), Barley (Winter), Oats, Rye, Wheat (Spring), Wheat (Winter)
	Stefes Restore (06267)	Barley (Spring), Barley (Winter), Oats (Spring), Oats (Winter), Rye (Spring), Rye (Winter), Wheat (Spring), Wheat (Winter)
	Tilt (08456)	Barley (Spring), Barley (Winter), Oats (Spring), Oats (Winter), Rye, Wheat (Spring), Wheat (Winter)
	Tilt 250EC (02138)	Barley (Spring), Barley (Winter), Oats (Spring), Oats (Winter), Rye, Wheat (Spring), Wheat (Winter)

Sulphur	Stoller Flowable Sulphur (03760)	Barley (Spring), Barley (Winter), Sugar Beet, Wheat (Spring), Wheat (Winter)
	Thiovit (02125)	Sugar Beet
	Thiovit (05572)	Sugar Beet
	Thiovit (08493)	Sugar Beet
Terbutryn	Prebane (08432)	Barley (Winter), Wheat (Winter)
	Prebane SC (07634)	Barley (Winter), Wheat (Winter)
Thiometon	Ekatin (05281)	Barley, Barley (Spring), Barley (Winter), Durum Wheat, Oats, Oats (Spring), Oats (Winter), Rye, Triticale, Wheat, Wheat (Spring), Wheat (Winter)
	Ekatin (08474)	Barley (Spring), Barley (Winter), Durum Wheat, Oats (Spring), Oats (Winter), Rye, Triticale, Wheat (Spring), Wheat (Winter)
Tri-allate	Avadex BW Granular (00174)	Barley (Spring), Barley (Winter), Field Bean (Winter), Pea (Dried), Pea (Seed), Pea (Vining), Wheat (Winter)
	Avadex Excel 15G (07117)	Barley, Barley (Spring), Barley (Winter), Field Bean (Winter), Pea (Dried), Pea (Seed), Pea (Vining), Wheat (Winter)
Triadimefon	100-Plus (05112)	Barley (Spring), Barley (Winter), Oats (Spring), Oats (Winter), Rye (Spring), Rye (Winter), Wheat (Spring), Wheat (Winter)
	Bayleton (00221)	Barley (Spring), Barley (Winter), Oats (Spring), Oats (Winter), Rye, Rye (Spring), Rye (Winter), Wheat, Wheat (Spring), Wheat (Winter)
	Standon Triadimefon 25 (05673)	Barley (Spring), Barley (Winter), Oats (Spring), Oats (Winter), Rye (Spring), Rye (Winter), Wheat (Spring), Wheat (Winter)
Zineb	Unicrop Zineb (02279)	Potato
Zineb-ethylene thiuram disulphide adduct	Polyram DF (08234)	Potato

ANNEX C
THE LONG TERM ARRANGEMENTS FOR EXTENSION OF USE (2000)

PLEASE NOTE THAT THESE EXTENSIONS OF USE ARE AT ALL TIMES DONE AT THE USER'S CHOOSING, AND THE COMMERCIAL RISK IS ENTIRELY THEIRS.

SPECIFIC RESTRICTIONS FOR EXTENSION OF USE UNDER THESE ARRANGEMENTS

To ensure that the extension of use does not increase the risk to the operator, the consumer or the environment, **the following conditions MUST be followed** when applying pesticides under the terms of this scheme:

GENERAL RESTRICTIONS

1. These arrangements apply to label and specific off-label recommendations for use of ONLY products approved for use as Agricultural/Horticultural pesticides.

All safety precautions and statutory conditions relating to use (which are clearly identified in the statutory box on product labels) MUST be observed. If extrapolation from a specific off-label is to be used then in addition to all safety precautions and statutory conditions relating to use specified on the product's label, all conditions relating to use specified on the Notice of Approval for the specific off-label use MUST be observed.

Pesticides MUST only be used in the same situation (outdoor or protected) as that specified on the product label/specific off-label Notice of Approval for the use on which the extrapolation is to be based, specifically:

Pesticides must not be used on protected crops, i.e. crops grown in glasshouses, poly tunnels, cloches or polythene covers or in any other building, unless the product label specifically allows use under protection on the crop on which the extrapolation is to be based. Similarly, pesticides approved only for use in protected situations must not be applied outdoors.

PLEASE NOTE: Unless specifically restricted to outdoor crops only, pesticides approved for use on tomatoes, cucumbers, lettuce, chrysanthemum and mushrooms are assumed to be approved for use under protection. **For all other uses, if the label/specific off-label Notice of Approval does not specify a situation, then only extrapolation to an outdoor use is permitted.**

APPLICATION METHOD RESTRICTIONS

4. The method of application must be as stated on the pesticide label and in accordance with the relevant codes of practice and requirements under COSHH 1994 (Control of Substances Hazardous to Health).

5. When planning to use hand held equipment to apply a pesticide under these arrangements, users MUST ensure that hand held use is appropriate for the current on-label recommendations/ specific off-label Notice of Approval conditions. **Note:** unless otherwise stated spray applications to protected crops include hand held uses.

Where hand held use is not appropriate for the use on which the extrapolation is to be based, hand held application should NOT be made if the pesticide label/specific off-label Notice of Approval.

    (a) prohibits hand held use;

    (b) requires the use of personal protective clothing when using the pesticide diluted to the minimum volume rate recommended on the label/specific off-label Notice of Approval for the dose required;

(c) is classified with one of the following hazard warnings:

"Corrosive", "Very toxic", "Toxic", or "Risk of serious damage to eyes".

In other cases hand held application is permitted provided that:

(i) the concentration of the spray volume for the extension of use is no greater than the maximum concentration recommended on the pesticide label;

(ii) spray quality is at least as coarse as the British Crop Protection Council medium or coarse spray;

(iii) operators wear at least a coverall, gloves and rubber boots when applying pesticides below waist level. Use of a faceshield is also required for applications which are above waist height.

(iv) where there are label precautions with regards to buffer zone restrictions for vehicle mounted use, then users must observe a buffer zone distance of 1 m from the top of the bank of any static or flowing water body when applying by hand held equipment.

ENVIRONMENTAL RESTRICTIONS

6. When planning to apply a pesticide under these arrangements by broadcast air-assisted sprayer (any equipment which broadcasts spray droplets by means of fan assistance which carry outwards and upwards from the source of the spray), only pesticides with specific on-label/off-label recommendations for such use on the crop on which the extrapolation is to be based (e.g. on hops, bush, cane or top fruit) can be used. Any associated buffer zone restrictions must also be observed.

7. Pesticides classified as Harmful, Dangerous, Extremely Dangerous or High Risk to bees must not be used during flowering of any crop (i.e. from first flower to complete petal fall) unless otherwise permitted. Applications of such pesticides must also not be made when flowering weeds are present or where bees are actively foraging.

8. If there is an aquatic buffer zone restriction set for the on-label/off-label use, then where appropriate, users are also obliged to conduct a Local Environmental Risk Assessment for Pesticides (LERAP) for the extension of use.

9. All reasonable precautions MUST be taken to safeguard wildlife and the environment.

EXCLUSIONS

10. The following uses are NOT PERMITTED under these arrangements.

(a) Aerial applications

(b) Use in or near water (in or near water includes drainage channels, streams, rivers, ponds, lakes, reservoirs, canals, dry ditches, areas designated for water storage).

(c) Use in or near coastal waters.

(d) Use of rodenticides and other vertebrate control agents.

(e) Use on land not intended for cropping, land not intended to bear vegetation, amenity grassland, managed amenity turf and amenity vegetation (this includes areas such as paths, pavements, roads, ground around buildings, motorway verges, railway embankments, public parks, turf, sports fields, upland areas, moorland areas, nature reserves, etc.).

EXTENSIONS OF USE

I. NON-EDIBLE CROPS AND PLANTS

(a) Subject to the SPECIFIC RESTRICTIONS FOR EXTENSION OF USE set out above, pesticides approved for use on any growing crop may be used on commercial agricultural and horticultural holdings and in forest nurseries on the following crops and plants:

   (i) hardy ornamental nursery stock, ornamental plants, ornamental bulbs and flowers and ornamental crops grown for seed where neither the seed nor any part of the plant is to be consumed by humans or animals;

   (ii) forest nursery crops prior to final planting out.

(b) Subject to the SPECIFIC RESTRICTIONS FOR EXTENSION OF USE set out above, pesticides approved for use on any growing edible crop may be used on commercial agricultural and horticultural holdings on non-ornamental crops grown for seed where neither the seed nor any part of the plant is to be consumed by humans or animals. This extrapolation EXCLUDES use on potatoes, cereals, oilseeds, peas, beans and other pulses grown for seed. Seed treatments themselves are NOT included in this extension of use.

(c) Subject to the SPECIFIC RESTRICTIONS FOR EXTENSION OF USE set out above, pesticides approved for use on oilseed rape may be used on commercial agricultural and horticultural holdings on hemp grown for fibre. Seed treatments are NOT included in this extension of use.

(d) Subject to the SPECIFIC RESTRICTIONS FOR EXTENSION OF USE set out above, herbicides approved for use on cereals, grass and maize may be used on commercial agricultural and horticultural holdings on *Miscanthus spp* (Elephant grass). Applications must NOT be made after the crop has reached 1 metre in height. The crop or products of the crop must NOT be used for food or feed.

PLEASE NOTE:

For a – d above, all on-label/off-label conditions of use must be observed, including any harvest interval i.e. any interval between application and harvest/exposure to the public specified on the label for the use on which the extrapolation is to be based, must be observed.

Before making hand held applications see paragraph 5 of the SPECIFIC RESTRICTIONS FOR EXTENSION OF USE.

II. FARM FORESTRY AND ROTATIONAL COPPICING

Subject to the SPECIFIC RESTRICTIONS FOR EXTENSION OF USE set out above, herbicides approved for use on:

   (a) cereals may be used in the first five years of establishment in farm forestry (including short rotation coppicing) on land previously under arable cultivation or improved grassland (as defined by the Farm Woodland Scheme) and reclaimed brownfield sites;

   (b) cereals, oilseed rape, sugar beet, potatoes, peas and beans may be used in the first year of re-growth after cutting in coppices (short term, rotational, intensive wood production e.g. poplar or willow biofuel production) established on land previously under arable cultivation or improved grassland (as defined by the Farm Woodland Scheme) and reclaimed brownfield sites;

III NURSERY FRUIT CROPS

Subject to the SPECIFIC RESTRICTIONS FOR EXTENSION OF USE set out above, pesticides approved for use on any crop for human or animal consumption may be used on commercial agricultural and horticultural holdings on nursery fruit trees, nursery grape vines prior to final planting out, bushes, canes and non-fruiting strawberry plants provided any fruit harvested within 12 months of treatment is destroyed. Applications must NOT be made where there are fruit present.

If hand held or broadcast air assisted use is required see paragraphs 5 and 6 respectively of the SPECIFIC RESTRICTIONS FOR EXTENSION OF USE.

IV HOPS *(Humulus spp.)*

Subject to the SPECIFIC RESTRICTIONS FOR EXTENSION OF USE set out above, pesticides may be used on commercial agricultural and horticultural holdings on the following hop plants grown in the circumstance below:

(a) Mature stock or mother plants which are kept specifically for the supply of propagation material.

(b) Propagation of hop planting material- propagules prior to final planting out.

(c) "Nursery hops". First year plants not taken to harvest that year, in their final planting out position.

PLEASE NOTE:

For a – c above, treated hops must NOT be harvested for human or animal consumption (including idling) within 12 months of treatment.

If hand held or broadcast air assisted application is required, users must comply with paragraphs 5 and 6 respectively of the SPECIFIC RESTRICTIONS FOR EXTENSION OF USE.

V. CROPS USED PARTLY OR WHOLLY FOR HUMAN OR ANIMAL CONSUMPTION.

Subject to the SPECIFIC RESTRICTIONS FOR EXTENSION OF USE set out above, pesticides may be used on commercial agricultural or horticultural holdings on the crops listed in TABLE ONE and TWO below in the first column if they have been approved for use on the crop(s) listed opposite them in the second column.

HOWEVER, BEFORE USING ANY OF THE FOLLOWING EXTRAPOLATIONS (TABLES ONE AND TWO), THE USER MUST FIRST OBSERVE THE FOLLOWING:

(a) It is the responsibility of the user to ensure that the proposed use does not result in any statutory UK Maximum Residue Levels (MRLs) being exceeded. MRLs are set out in statutory instrument No. 1985 of 1994: >The Pesticides (Maximum Residue Levels in Crops, Food and Feeding Stuffs) Regulations 1994= (The Stationery Office, ISBN 0-11-044985-1) and any subsequent updates.

(b) These extrapolations DO NOT APPLY in the following situations:

(i) Where the MRL for the crop in column 1 is lower than the MRL for the crop in column 2.

(ii) Where the MRL for the crop in column 1 is set at the limit of determination.

(iii) Where no MRL is set for the crop in Column 2, but a MRL has been established for the crops in column 1.

In any of the above circumstances use on the crop in column 1 is NOT PERMITTED.

Column 1: Minor use	Column 2: Crops on which use is approved	Additional special conditions
A. ARABLE CROPS		
Poppy (grown for oilseed), Sesame	Sunflower	
Mustard, Linseed, Honesty, Evening primrose	Oilseed rape	
Borage (grown for oilseed) Canary flower e.g. *Echium vulgare/Echium plantaginium* (grown for oilseed)	Oilseed rape	Seed treatments are not permitted
Rye, Triticale	Barley	Treatments applied before first spikelet of inflorescence just visible
Rye, Triticale	Wheat	
Grass seed crop	Grass for grazing or fodder	
Grass seed crop	Wheat, barley, oats, rye, triticale	Treated crops must not be grazed or cut for fodder until 90 days after treatment. Seed treatments are not permitted Use of chlormequat-containing products is not permitted
Lupins	Combining peas or field beans	
B. FRUIT CROPS		
Almond, Chestnut, Walnut, Hazelnut	Apple or cherry or plum	For herbicides used on the orchard <u>floor</u> ONLY
Almond, Chestnut, Walnut, Hazelnut	Products approved for use on two of the following: almond, chestnut, hazelnut and walnut	
Quince, Crab apple	Apple or pear	
Nectarine, Apricot	Peach	
Blackberry, Dewberry Rubus species (e.g. tayberry, loganberry)	Raspberry	
Whitecurrant, Bilberry, Cranberry	Blackcurrant or redcurrant	
Redcurrant	Blackcurrant	
C. VEGETABLE CROPS		
Parsley root	Carrot or radish	
Fodder beet, Mangel	Sugar beet	

Column 1: Minor use	Column 2: Crops on which use is approved	Additional special conditions
Horseradish	Carrot or radish	
Parsnip	Carrot	
Salsify	Carrot or celeriac	
Swede	Turnip	
Turnip	Swede	
Garlic, Shallot	Bulb onion	
Aubergine	Tomato	
Squash, Pumpkin, Marrow, Watermelon	Melon	
Broccoli	Calabrese	
Calabrese	Broccoli	
Roscoff cauliflower	Cauliflower	
Collards	Kale	
Lamb's lettuce, frisee/frise, radicchio, cress, scarole	Lettuce	
Leaf herbs and edible flowers*	Lettuce or spinach or parsley or sage or mint or tarragon	
Beet leaves, Red chard, White chard, Yellow chard	Spinach	
Edible podded peas (e.g. mange-tout, sugar snap)	Edible podded beans	
Runner beans	Dwarf French beans	
Rhubarb, Cardoon	Celery	
Edible fungi other than mushroom (e.g. oyster mushroom)	Mushroom	

*This extension of use applies to the following leaf herbs and edible flowers: angelica, balm, basil, bay, borage, burnet (salad), caraway, camomile, chervil, chives, clary, coriander, dill, fennel, fenugreek, feverfew, hyssop, land cress, lovage, marjoram, marigold, mint, nasturtium, nettle, oregano, parsley, rocket, rosemary, rue, sage, savory, sorrel, tarragon, thyme, verbena (lemon), woodruff.

For applications in store on crops PARTLY OR WHOLLY FOR HUMAN OR ANIMAL CONSUMPTION, the following extensions of use apply:

TABLE TWO

Column 1: Minor Use	Column 2: Crops on which use is approved.
Rye, Barley, Oats, Buckwheat, Millet, Sorghum, Triticale	Wheat
Dried pea Dried bean	Dried bean Dried pea
Mustard, Sunflower, Honesty, Sesame, Linseed, Evening primrose, Poppy (grown for oilseed), Borage (grown for oilseed) Canary flower e.g. *Echium vulgare/Echium plantaginium* (grown for oilseed)	Oilseed rape

VI. CLARIFICATIONS:

Under these arrangement the following crops are considered to be synonymous or equivalent and as such, uses on crops in Column 1 can be read across to uses in Column 2.

Column 1:	Column 2: equivalent
Hazelnut	Cobnuts, Filberts
French bean	Navy bean
Vining pea	Picking pea, Shelling pea, Non-edible podded pea
Linseed	Linola, Flax
Wheat	Durum wheat

ANNEX D
COMMODITY SUBSTANCE APPROVALS
HEALTH AND SAFETY EXECUTIVE

Food and Environment Protection Act 1985

Schedule: COMMODITY SUBSTANCE: **4-CHLORO-*m*-CRESOL**

Date of issue: 18 February 1993

Date of expiry: 28 February 2001

This approval is subject to the following conditions:

1 *FIELD OF USE:*   ONLY AS A FUNGICIDE

2 *PEST AND USAGE AREA:* FOR THE CONTROL OF FUNGI ON INSECT SPECIMENS

3 *APPLICATION METHOD:*   4-CHLORO-*m*-CRESOL CRYSTALS IN A COLLECTING BOX

Operator protection:

(1) A written COSHH assessment must be made before using 4-chloro-*m*-cresol.

(2) Engineering control of operator exposure must be used where reasonably practicable in addition to the following items of personal protective equipment.

   Operators must wear suitable protective clothing, including protective gloves and eye protection and a dust mask, when handling or applying the material.

(3) However engineering controls may replace personal protective equipment if a COSHH assessment shows they provide an equal or higher standard of protection.

Other specific restrictions:

(1) Operators should be provided with adequate information about the hazards of the substance and the precautions necessary for safe use. Sources of information include the supplier's Safety Data Sheet.

(2) Unprotected persons and animals must be excluded from any areas where treatment is taking place, and such areas should be ventilated after treatment.

(3) Must be used only by operators who are suitably trained and competent to carry out this work.

HEALTH AND SAFETY EXECUTIVE

Food and Environment Protection Act 1985

Schedule: COMMODITY SUBSTANCE: **CAMPHOR**

Date of issue: 18 February 1993

Date of expiry: 28 February 2001

This approval is subject to the following conditions:

1 *FIELD OF USE*:   ONLY AS AN INSECT REPELLENT IN MUSEUMS AND
                               BUILDINGS OF CULTURAL, ARTISTIC AND HISTORICAL
                               INTEREST.

2 *PEST AND USAGE AREA*:   FOR THE CONTROL OF FLYING AND CRAWLING INSECTS

3 *APPLICATION METHOD*:   CRYSTALS OF CAMPHOR IN A SEALED SPECIMEN CASE

Operator protection:

(1) A written COSHH assessment must be made before using camphor. Operators should also observe the OES set out in HSE guidance note EH40/93 or subsequent issues.

(2) Engineering control of operator exposure must be used where reasonably practicable in addition to the following items of personal protective equipment.

Operators must wear suitable protective clothing, including protective gloves and eye protection, when handling or applying the material.

(3) However engineering controls may replace personal protective equipment if a COSHH assessment shows they provide an equal or higher standard of protection.

Other specific restrictions:

(1) Operators should be provided with adequate information about the hazards of the substance and the precautions necessary for safe use. Sources of information include the supplier's Safety Data Sheet.

(2) Unprotected persons and animals must be excluded from any areas where treatment is taking place, and such areas should be ventilated after treatment.

(3) Must be used only by operators who are suitably trained and competent to carry out this work.

Food and Environment Protection Act 1985
Control of Pesticides Regulations 1986 (SI 1986 No. 1510) : APPROVAL

In exercise of the powers conferred by Regulation 5 of the Control of Pesticides Regulations 1986 (SI 1986/1510) and of all other powers enabling them in that behalf, the Minister of Agriculture, Fisheries and Food and the Secretary of State hereby jointly give full approval for the use of:

Commodity substance: being 99.9% v/v **CARBON DIOXIDE** subject to the conditions set out below:

Date of issue: 8 October 1993

Use:

Field of use: **Only as a rodenticide**

Situations: Trapped rodents.

Operator protection:

(1) Engineering control of operator exposure must be used where reasonably practicable in addition to the following personal protective equipment:

Operators must wear self-contained breathing apparatus when CO_2 levels are greater than 0.5% v/v.

(2) However, engineering controls may replace personal protective equipment if a COSHH assessment shows they provide an equal or higher standard of protection.

Other specific restrictions:

(1) Unprotected persons and non-target animals must be excluded from the treatment enclosures and from the area surrounding the treatment enclosures unless CO_2 levels are below 0.5% v/v.

(2) This substance must only be used by operators who are suitably trained and competent to carry out this work.

ADVISORY NOTE

This approval allows the use of CO_2 to destroy trapped rodent pests.

Food and Environment Protection Act 1985
Control of Pesticides Regulations 1986 (SI 1986 No. 1510) : APPROVAL

In exercise of the powers conferred by Regulation 5 of the Control of Pesticides Regulations 1986 (SI 1986/1510) and of all other powers enabling them in that behalf, the Minister of Agriculture, Fisheries and Food and the Secretary of State hereby jointly give full approval for the use of:

Commodity substance: being 99.9% v/v **CARBON DIOXIDE** subject to the conditions set out below:

Date of issue: 8 October 1993

Use:

Field of use: **Only in vertebrate control**

Situations: Birds covered by general licences issued by the Agriculture and Environment Departments under Section 16(1) of the Wildlife and Countryside Act (1981) for the control of opportunistic bird species, where birds have been trapped or stupefied with alphachloralose/seconal.

Operator protection:

(1) Engineering control of operator exposure must be used where reasonably practicable in addition to the following personal protective equipment:

Operators must wear self-contained breathing apparatus when CO_2 levels are greater than 0.5% v/v.

(2) However, engineering controls may replace personal protective equipment if a COSHH assessment shows they provide an equal or higher standard of protection.

Other specific restrictions:

(1) Unprotected persons and non-target animals must be excluded from the treatment enclosures and from the area surrounding the treatment enclosures unless CO_2 levels are below 0.5% v/v.

(2) This substance must only be used by operators who are suitably trained and competent to carry out this work.

(3) Only to be used where a licence has been issued in accordance with Section 16(1) of the Wildlife and Countryside Act 1981 to permit the use of a substance otherwise prohibited under Section 5 of the Wildlife and Countryside Act 1981.

Food and Environment Protection Act 1985

Control of Pesticides Regulations 1986 (SI 1986 No. 1510) : APPROVAL

In exercise of the powers conferred by Regulation 5 of the Control of Pesticides Regulations 1986 (SI 1986/1510) and of all other powers enabling them in that behalf, the Minister of Agriculture, Fisheries and Food and the Secretary of State hereby jointly give full approval for the use of:

Commodity substance: being 99.9% v/v **CARBON DIOXIDE** subject to the conditions set out below:

Date of issue: 8 October 1993

Use:

Field of use: **Only as an insecticide, acaricide and rodenticide in food storage practice**

Situations: Raw and processed food commodities.

Operator protection:

(1) Engineering control of operator exposure must be used where reasonably practicable in addition to the following personal protective equipment:

Operators must wear self-contained breathing apparatus when CO_2 levels are greater than 0.5% v/v.

(2) However, engineering controls may replace personal protective equipment if a COSHH assessment shows they provide an equal or higher standard of protection.

Other specific restrictions:

(1) Unprotected persons and non-target animals must be excluded from the treatment enclosures and from the area surrounding the treatment enclosures unless CO_2 levels are below 0.5% v/v.

(2) This substance must only be used by operators who are suitably trained and competent to carry out this work.

ADVISORY NOTE

Ensure adequate ventilation of premises during all treatment and venting operations.

HEALTH AND SAFETY EXECUTIVE

Food and Environment Protection Act 1985

Schedule: COMMODITY SUBSTANCE: **ETHANOL**

Date of issue: 18 February 1993

Date of expiry: 28 February 2001

This approval is subject to the following conditions:

1 *FIELD OF USE:*	i) AS AN INSECTICIDE IN MUSEUMS AND BUILDINGS OF CULTURAL, ARTISTIC AND HISTORICAL INTEREST.
	ii) AS A PRESERVATIVE IN MUSEUMS, AND BUILDINGS OF CULTURAL ARTISTIC AND HISTORICAL INTEREST
2 *PEST AND USAGE AREA:*	FOR THE CONTROL OF FLYING AND CRAWLING INSECTS AND FUNGI
3 *APPLICATION METHOD:*	i) IMMERSION IN A TANK ENCLOSED IN A FUME CUPBOARD
	ii) STORAGE OF SPECIMENS IN MATERIAL.

Operator protection:

(1) A written COSHH assessment must be made before using ethanol. Operators should also observe the OES set out in HSE guidance note EH40/93 or subsequent issues.

(2) Engineering control of operator exposure must be used where reasonably practicable in addition to the following items of personal protective equipment.

Operators must wear suitable protective clothing, including protective gloves and eye protection, when handling or applying the material.

(3) However engineering controls may replace personal protective equipment if a COSHH assessment shows they provide an equal or higher standard of protection.

Other specific restrictions:

(1) Operators should be provided with adequate information about the hazards of the substance and the precautions necessary for safe use. Sources of information include the supplier's Safety Data Sheet.

(2) Unprotected persons and animals must be excluded from any areas where treatment is taking place, and such areas should be ventilated after treatment.

(3) Must be used only by operators who are suitably trained and competent to carry out this work.

HEALTH AND SAFETY EXECUTIVE

Food and Environment Protection Act 1985

Schedule: COMMODITY SUBSTANCE: **ETHYL ACETATE**

Date of issue: 18 February 1993

Date of expiry: 28 February 2001

This approval is subject to the following conditions;

1 *FIELD OF USE:*	ONLY AS AN INSECTICIDE IN MUSEUMS AND BUILDINGS OF CULTURAL, ARTISTIC AND HISTORICAL INTEREST.
2 *PEST AND USAGE AREA:*	FOR THE CONTROL OF FLYING AND CRAWLING INSECTS.
3 *APPLICATION METHOD:*	TREATMENT IN SEALED CONTAINERS.

Operation protection:

(1) A written COSHH assessment must be made before using ethyl acetate. Operators should also observe the OES set out in HSE guidance note EH40/93 or subsequent issues.

(2) Engineering control of operator exposure must be used where reasonably practicable in addition to the following items of personal protective equipment.

Operators must wear suitable protective clothing, including protective gloves and eye protection, when handling or applying the substance.

(3) However engineering controls may replace personal protective equipment if a COSHH assessment shows they provide an equal or higher standard of protection.

Other specific restrictions:

(1) Operators should be provided with adequate information about the hazards of the substance and the precautions necessary for safe use. Sources of information include the supplier's Safety Data Sheet.

(2) Unprotected persons and animals must be excluded from any areas where treatment is taking place, and such areas should be ventilated after treatment.

(3) Must be used only by operators who are suitably trained and competent to carry out this work.

Food and Environment Protection Act 1985
Control of Pesticides Regulations 1986 (SI 1986 NO 1510): APPROVAL

In exercise of the powers conferred by Regulation 5 of the Control of Pesticides Regulations 1986 (SI 1986/1510) and of all other powers enabling them in that behalf, the Minister of Agriculture, Fisheries and Food and the Secretary of State hereby jointly give full approval for the use of:

Commodity Substance: being **FORMALDEHYDE** (formalin 38–40% aqueous solution) subject to the conditions set out below:

Date of issue: 1 March 1991

Date of expiry: 28 February 2001 (see advisory note 3)

Use:

Field of Use: **Only as an agricultural/horticultural fungicide and sterilant**

Crops/Situations	Maximum individual dose	Other specific restrictions (1) Maximum concentration
Soil and compost sterilant, indoors and outdoors	As a drench: 0.5 litre formalin/m^2	1:4 parts water
Bulb dip	—	As a dip: 1:200 parts water
Mushroom houses	As a spray: 0.5 litre formalin/m^2	1:50 parts water
	As a fumigant: 400 ml formalin/$100m^3$ or 100 g of potassium permanganate added to 500 ml formalin/$100m^3$	
Greenhouse hygiene, boxes, pots etc.	As a fumigant: 400 ml formalin/$100m^3$ or 100 g of potassium permanganate added to 500 ml of formalin/$100m^3$	As a spray or dip: 1:50 parts water

Operator protection:

(1)   A written COSHH assessment must be made before using formaldehyde. Operators should observe the Maximum Exposure Limit set out in HSE guidance note EH40/93 or subsequent issues and COP 30 "The control of substances hazardous to health in fumigation operations".

(2)   Engineering control of operator exposure must be used where reasonably practicable in addition to the following personal protective equipment:

   (a)   Operators must wear suitable respiratory equipment and other suitable protective equipment when handling or applying the fumigant.

   (b)   Operators must wear suitable protective clothing and gloves when handling the concentrate.

(3) However, engineering controls may replace personal protective equipment if a COSHH assessment shows they provide an equal or higher standard of protection.

Other specific restrictions:

(1) Maximum concentration of formalin in water: see table.

(2) Operators must be supplied with a Section 6 (HSW) Safety Data Sheet before commencing work.

ADVISORY NOTES

1 Use as a disinfectant for equipment, greenhouse and public hygiene purposes are outside the Regulations.

2 Use of this substance for sterilising hatching eggs are outside the Regulations.

3 Data submitted by interested parties is being considered by PSD for approval of this commodity substance beyond the stated expiry date.

Food and Environment Protection Act 1985
Control of Pesticides Regulations 1986 (SI 1986 NO 1510): APPROVAL

In exercise of the powers conferred by Regulation 5 of the Control of Pesticides Regulations 1986 (SI 1986/1510) and of all other powers enabling them in that behalf, the Minister of Agriculture, Fisheries and Food and the Secretary of State hereby jointly give full approval for the use of:

Commodity Substance: being **FORMALDEHYDE** (paraformaldehyde) subject to the conditions set out below:

Date of issue: 1 March 1991

Date of expiry: 28 February 2001 (see advisory note)

Use:

Field of Use: **Only as an animal husbandry fungicide**

Situations: Animal houses

Maximum individual dose: As a fumigant: 5g paraformaldehyde/m^3 or 20 g potassium permanganate added to 40 ml of formalin/m^3.

Operator protection:

(1) A written COSHH assessment must be made before using formaldehyde. Operators should observe the Maximum Exposure Limit set out in the HSE guidance note EH40/93 or subsequent issues and COP 30 "The control of substances hazardous to health in fumigation operations".

(2) Engineering control of operator exposure must be used where reasonably practicable in addition to the following personal protective equipment:

Operators must wear suitable respiratory equipment and other suitable protective equipment when handling or applying the fumigant.

(3) However, engineering controls may replace personal protective equipment if a COSHH assessment shows they provide an equal or higher standard of protection.

Other specific restrictions:

(1) Operators must remove livestock, feed, exposed milk and water and collect eggs before application.

(2) Operators must be supplied with a Section 6 (HSW) safety data sheet before commencing work.

ADVISORY NOTE

Data submitted by interested parties is being considered by PSD for approval of this commodity substance beyond the stated expiry date.

HEALTH AND SAFETY EXECUTIVE

Food and Environment Protection Act 1985

Schedule: COMMODITY SUBSTANCE: **FORMALDEHYDE**

Date of issue: 18 February 1993

Date of expiry: 28 February 2001

This approval is subject to the following conditions:

1 *FIELD OF USE:*	i) AS AN INSECTICIDE IN MUSEUMS AND BUILDINGS OF CULTURAL, ARTISTIC AND HISTORICAL INTEREST ii) AS A PRESERVATIVE IN MUSEUMS AND BUILDINGS OF CULTURAL, ARTISTIC AND HISTORICAL INTEREST
2 *PEST AND USAGE AREA:*	FOR THE CONTROL OF FLYING AND CRAWLING INSECTS AND FUNGI
3 *APPLICATION METHOD:*	i) TREATMENT IN SEALED CONTAINERS ii) STORAGE IN 5-l0% AQUEOUS SOLUTION

Operator protection:

(1) A written COSHH assessment must be made before using formaldehyde. Operators should also observe the MEL set out in HSE guidance note EH40/93 or subsequent issues.

(2) Engineering control of operator exposure must be used where reasonably practicable in addition to the following items of personal protective equipment.

Operators must wear suitable protective clothing, including protective gloves, eye protection and suitable respiratory protective equipment, when handling or applying the substance.

(3) However engineering controls may replace personal protective equipment if a COSHH assessment shows they provide an equal or higher standard of protection.

Other specific restrictions:

(1) Operators should be provided with adequate information about the hazards of the substance and the precautions necessary for safe use. Sources of information include the supplier's Safety Data Sheet.

(2) Unprotected persons and animals must be excluded from any areas where treatment is taking place, and such areas should be ventilated after treatment.

(3) Must be used only by operators who are suitably trained and competent to carry out this work.

HEALTH AND SAFETY EXECUTIVE

Food and Environment Protection Act 1985

Schedule:　　　COMMODITY SUBSTANCE: **ISOPROPANOL**

Date of issue:　18 February 1993

Date of expiry:　28 February 2001

This approval is subject to the following conditions:

1 *FIELD OF USE:*　　　　ONLY AS A PRESERVATIVE IN MUSEUMS AND BUILDINGS
　　　　　　　　　　　　　OF CULTURAL, ARTISTIC AND HISTORICAL INTEREST.

2 *PEST AND USAGE AREA:*　FOR THE CONTROL OF FUNGI ON SPECIMENS

3 *APPLICATION METHOD:*　STORAGE OF SPECIMENS IN 50-60% AQUEOUS
　　　　　　　　　　　　　SOLUTION

Operator protection:

(1) A written COSHH assessment must be made before using isopropanol. Operators should also observe the OES set out in HSE guidance note EH40/93 or subsequent issues.

(2) Engineering control of operator exposure must be used where reasonably practicable in addition to the following items of personal protective equipment.

　　Operators must wear suitable protective clothing, including protective gloves and eye protection, when handling or applying the material.

(3) However engineering controls may replace personal protective equipment if a COSHH assessment shows they provide an equal or higher standard of protection.

Other specific restrictions:

(1) Operators should be provided with adequate information about the hazards of the substance and the precautions necessary for safe use. Sources of information include the supplier's Safety Data Sheet.

(2) Unprotected persons and animals must be excluded from any areas where treatment is taking place, and such areas should be ventilated after treatment.

(3) Must be used only by operators who are suitably trained and competent to carry out this work.

Food and Environment Protection Act 1985
Control of Pesticides Regulations 1986 (SI 1986 No. 1510) : APPROVAL

In exercise of the powers conferred by Regulation 5 of the Control of Pesticides Regulations 1986 (SI 1986/1510) and of all other powers enabling them in that behalf, the Minister of Agriculture, Fisheries and Food and the Secretary of State hereby jointly give full approval for the use of:

Commodity substance: being **METHYL BROMIDE** subject to the conditions set out below:

Date of issue: 13 March 1996

Date of expiry: 28 February 2001 (see advisory note 2)

Use:

Field of use: **Only as a fumigant in vertebrate control**

Situations: Aircraft, ship, stacks of commodity, enclosed structures (such as warehouses, factories, mills and other buildings, containers, chambers and other transport units).

Operator protection:

(1) A written COSHH assessment must be made before using methyl bromide.

(2) Engineering control of operator exposure must be used where reasonably practicable in addition to the following personal protective equipment:

Operators must wear suitable approved respiratory protective equipment and other suitable protective equipment when using the product.

Operators must *not* wear gloves or rubber boots when using the product.

(3) However, engineering controls may replace personal protective equipment if a COSHH assessment shows they provide an equal or higher standard of protection.

Other specific restrictions:

(1) Unprotected persons and animals must be kept out of aircraft holds or ship holds under fumigation or being aired following fumigation until any exposure levels are below the Occupational Exposure Standard.

(2) This product must only be used by professional operators of servicing companies, local authorities and Government departments who must be suitably trained and competent to carry out this work.

(3) Operators must refer to HSE guidance notes CS22, 'Fumigation with methyl bromide' and ACOP 30. 'The control of substances hazardous to health in fumigation operations' before using this product.

(4) Operators must be supplied with a Section 6 (HSW) safety data sheet before commencing work.

(5) The container must not be re-used for any purpose.

ADVISORY NOTES

1. Approved products containing methyl bromide are available for agricultural, horticultural, food storage and space spray uses.

2. Data submitted by interested parties is being considered by PSD for approval of this commodity substance beyond the stated expiry date.

3. This approval is intended to cover situations where other methods of control are unsuitable. Therefore due consideration should be given to other methods of vertebrate control before using methyl bromide.

4. This approval supersedes that issued for the use of methyl bromide as a fumigant in vertebrate control issued on 22 November 1994.

HEALTH AND SAFETY EXECUTIVE

Food and Environment Protection Act 1985

Schedule: COMMODITY SUBSTANCE: **METHYL BROMIDE**

Date of Issue: 1 March 1991

Date of expiry: 28 February 2001

This approval is subject to the following conditions:

1 *FIELD OF USE:*   ONLY AS A FUMIGANT IN PUBLIC HYGIENE

2 *PEST AND USAGE AREA:* FOR THE CONTROL OF FLYING AND CRAWLING INSECTS

3 *APPLICATION METHOD:*   USERS SHOULD FOLLOW THE GUIDANCE GIVEN IN HSE
GUIDANCE NOTE CS12 "FUMIGATION USING METHYL
BROMIDE (BROMOETHANE)" (ISSUED 1986 AND REVISED
IN 1991) AND COP 30 "THE CONTROL OF SUBSTANCES
HAZARDOUS TO HEALTH IN FUMIGATION OPERATIONS"

Operator protection:

(1) A written COSHH assessment must be made before using methyl bromide.

(2) Engineering control of operator exposure must be used where reasonably practicable in
addition to the following personal protective equipment:

Operators must wear suitable approved respiratory equipment and other suitable protective
equipment when handling or applying the fumigant.

(3) However, engineering controls may replace personal protective equipment if a COSHH
assessment shows they provide an equal or higher standard of protection.

Other specific restrictions:

(1) Must be used only by professional operators of servicing companies, local authorities and
Government departments who must be suitably trained and competent to carry out this work.

(2) Operators must be supplied with a Section 6 (HSW) safety data sheet before commencing
work.

(3) Unprotected persons and animals must be kept out of premises under fumigation or being
aired following fumigation until any exposure levels are below the Occupational Exposure
Standard.

ADVISORY NOTE

1 Approved products containing methyl bromide are available for agricultural, horticultural, food
storage, rodenticide and space spray uses.

2 This field of use covers pesticides used for the control of harmful organisms, chiefly insects,
detrimental to public health.

HEALTH AND SAFETY EXECUTIVE

Food and Environment Protection Act 1985

Schedule: COMMODITY SUBSTANCE: **LIQUID NITROGEN**

Date of Issue: 18 February 1993

Date of expiry: 28 February 2001

This approval is subject to the following conditions:

1 *FIELD OF USE:*   ONLY AS AN ACARICIDE IN DOMESTIC AND PUBLIC HYGIENE SITUATIONS.

2 *PEST AND USAGE AREA:* FOR THE CONTROL OF DUST MITES ON FURNISHINGS.

3 *APPLICATION METHOD:* SUITABLE LIQUID NITROGEN APPLICATOR.

Operator protection:

(1) An assessment under The Management of Health and Safety at Work Regulations 1992 must be made before using liquid nitrogen.

(2) Operators must wear suitable protective clothing, including protective gloves, eye protection a dust mask and an oxygen monitor, when handling or applying the material.

Other specific restrictions:

(1) Operators should be provided with adequate information about the hazards of the substance and the precautions necessary for safe use. Sources of information include the supplier's Safety Data Sheet.

(2) Safe working procedures must be specified where it is possible that an oxygen deficient atmosphere could develop. Guidance on such procedures is given in HSE Guidance Note GS5 "Entry into confined spaces".

(3) Unprotected persons and animals must be excluded from any areas where treatment is taking place, and such areas should be ventilated after treatment.

(4) Must be used only by operators who are suitably trained and competent to carry out this work.

Food and Environment Protection Act 1985
Control of Pesticides Regulations 1986 (SI 1986 No. 1510) : APPROVAL

In exercise of the powers conferred by Regulation 5 of the Control of Pesticides Regulations 1986 (SI 1986/1510) and of all other powers enabling them in that behalf, the Minister of Agriculture, Fisheries and Food and the Secretary of State, hereby jointly give full approval for the use of:

Commodity substance: being **PARAFFIN OIL** subject to the conditions set out below:

Date of issue: 25 April 1995

Use:

Field of use: **Only in vertebrate control**

Situations: Eggs of birds covered by licences issued by the Agriculture and Environment Departments under Section 16 (1) of the Wildlife and Countryside Act (1981)

Operator protection:

(1) Engineering control of operator exposure must be used where reasonably practicable in addition to the following personal protective equipment:

Operators must wear suitable protective gloves and faceshield when handling and applying the substance.

(2) However, engineering controls may replace personal protective equipment if a COSHH assessment shows they provide an equal or higher standard of protection.

Other specific restrictions:

Only to be used where a licence has been approved in accordance with Section 16 (1) of the Wildlife and Countryside Act 1981 to permit the use of a substance otherwise prohibited under Section 5 of the Wildlife and Countryside Act 1981.

ADVISORY NOTES

1. Egg treatment should be undertaken as soon as a clutch is complete.

2. Eggs should be treated once by complete immersion in liquid paraffin.

Food and Enviroment Protection Act 1985

Control of Pesticides Regulations 1986 (SI 1986 No. 1510): Commodity Substance Approval

In exercise of the powers conferred by Regulation 5 of the Control of Pesticides Regulations 1986 (SI 1986/1510) and of all other powers enabling in that behalf, the Minister of Agriculture, Fisheries and Food and the Secretary of State, hereby jointly give full approval for the use of:

Commodity substance: being **SODIUM CHLORIDE** subject to the conditions set out below:

Date of issue: 11 September 1996

Use:

Field of Use: **Only as an agricultural herbicide**

Crop: Sugar beet

ADVISORY NOTES:

1. Typical treatment consists of a single application of 1000 litres of saturated sodium chloride (salt) per hectare (or split applications) containing a non-ionic wetter (0.1% w/w) to the crop from after the emergence of three pairs of true leaves growth stage to the end of July for the control of volunteer potatoes and activity against other weeds.

2. Crop phytotoxicity may occur after treatments.

Food and Environment Protection Act 1985
Control of Pesticides Regulations 1986 (SI 1986 No. 1510): Approval

In exercise of the powers conferred by Regulation 5 of the Control of Pesticides Regulations 1986 (SI 1986/1510) and of all other powers enabling them in that behalf, the Minister of Agriculture, Fisheries and Food and the Secretary of State, hereby jointly give full approval for the use of:

Commodity substance: being **SODIUM HYPOCHLORITE** subject to the conditions set out below:

Date of issue: *5 December 1996*

Use:

Field of Use: **Only as an horticultural bactericide**

Crop: Mushroom

Maximum individual dose: See 'Other specific restrictions'

Latest time of application: One day before harvest

Operator protection: (1) Engineering control of operator exposure must be used where reasonably practicable in addition to the following personal protective equipment: Operators must wear suitable protective clothing (coveralls), suitable protective gloves and face protection (faceshield) when handling the concentrate.

(2) However, engineering controls may replace personal protective equipment if a COSHH assessment shows they provide an equal or higher standard of protection.

Environmental protection: Since this substance is harmful to fish or other aquatic life surface waters or ditches must not be contaminated with chemical or used container.

Other specific restrictions: (1) This substance must only be used by operators who are suitably trained and competent to carry out this work.

(2) The maximum concentration must not exceed 315 mg sodium hypochlorite per litre of water (equivalent to 150 mg available chlorine per litre of water).

ADVISORY NOTE:

Mixing and loading must only take place in a ventilated area.

Food and Environment Protection Act 1985
Control of Pesticides Regulations 1986 (SI 1986 No. 1510): APPROVAL

In exercise of the powers conferred by regulation 5 of the Control of Pesticides Regulations 1986 (SI 1986/1510) and of all other powers enabling them in that behalf, the Minister of Agriculture, Fisheries and Food and the Secretary of State, hereby jointly give full approval for the use of:

Commodity substance: being **STRYCHNINE HYDROCHLORIDE** subject to the conditions set out below:

Date of issue: 19 June 1997

Use:

Field of Use: **Only as a vertebrate control agent for the destruction of moles underground**

Crops/Situations: Commercial agricultural/horticultural land where public access is restricted; grassland associated with aircraft landing strips, horse paddocks, gallops and race courses; golf courses and other areas specifically approved by Agricultural Departments.

Sale and Supply:

Label: Substance to be supplied with a label in accordance in the Poisons Act 1972

Container: Substance to be supplied only in the original sealed packaging of the manufacturer and only in units of up to 2 g.

Other (a) Must only be supplied to the holders of an Authority to Purchase issued by the appropriate Agriculture Department (in England – the Ministry of Agriculture, Fisheries and Food; in Wales – the Welsh Office Agriculture Department; and in Scotland – the Department of Agriculture and Fisheries for Scotland).

(b) Quantities of more than 8 g must only be supplied to providers of a commercial service.

Storage: (a) Providers of a commercial service must not hold more strychnine than the amount specified by the authorising Agriculture Department.

(b) Substance must be stored in the original container under lock and key and only on the premises and under the control of the holder of an authority to purchase or a named individual who satisfies the appropriate Agriculture Department.

Advertisement: The substance must not be advertised except that individual pharmacists may provide details of availability and price to a person authorised to purchase the substance.

Operator protection:

(a) A written Control of Substances Hazardous to Health (COSHH) assessment must be made before using strychnine (see Advisory Note 1).

(b) Gloves must be worn when preparing and laying bait and when handling contaminated utensils.

Environmental protection:

(a) The substance must be prepared for applications with great care so that there is no contamination of the surface of the ground.

(b) Any prepared bait remaining at the end of the day must be buried (see Advisory Note 2).

Other specific restrictions:

(a) Authorities to purchase may only be issued to persons who satisfy the appropriate Agriculture Department that they are trained and competent in its use and can be entrusted with it (see Advisory Note 3).

(b) Access to strychnine must be restricted to those who hold the authority to purchase and to named individuals who satisfy the appropriate Agriculture Department.

(c) The substance must be used as and where directed by the appropriate Agriculture Department.

(d) Operators must be supplied with a CHIP Safety Data Sheet before commencing work (see Advisory Note 4).

(e) Providers of a commercial service must use suitable dedicated utensils capable of being washed clean. Such equipment must be cleaned after every treatment and washings disposed of in a mole run. The equipment must be stored securely between treatments.

(f) Non commercial users must use suitable disposal utensils which must be disposed of by burial on the land where the treatment takes place. They must not be retained for re-use.

(g) Providers of a commercial service must advise the local office of the appropriate Agriculture Department of the treatments applied in the previous year, report the quantity of the substance held in store and the arrangements for secure storage i) when applying for further quantities, or ii) within 12 months of last using it, whichever is the earlier.

ADVISORY NOTES:

1. Guidance on how to carry out a COSHH assessment may be obtained from HSE

2. Burial of contaminated materials should be in accordance with advice given in the Code of Practice for the Safe Use of Pesticides on Farms and Holdings.

3. When deciding whether a person is fit to be entrusted with the substance, account will be taken of character and expertise.

4. The Chemicals (Hazard Information and Packaging for Supply) Regulations 1994 as amended (CHIP) data sheet is obtainable from the supplier of strychnine eg the pharmacist.

Food and Enviroment Protection Act 1985
Control of Pesticides Regulations 1986 (SI 1986 No 1510): Approval

In exercise of the powers conferred by regulation 5 of the Control of Pesticides Regulations 1986 (SI 1986/1510) and of all other powers enabling them in that behalf, the Minister of Agriculture, Fisheries and Food and the Secretary of State, hereby jointly give full approval for the use of:

Commodity Substance: being 77% clean **SULPHURIC ACID** subject to the conditions set out below:

Date of issue: 23 November 1995

Date of expiry: 25 July 2003

Use:

Field of Use: **Only as an agricultural desiccant**

Crops	Maximum individual dose	Maximum number of treatments	Times of application
Potato (grown for canning)	800 litres substance/hectare	3 per crop (See 'Other specific restrictions')	1 May – 15 November
Potato (Other)	340 litres substance/hectare	3 per crop (See 'Other specific restrictions')	1 May – 15 November
Bulbs and Corms	280 litres substance/hectare	1 per year	1 May – 15 November
Peas	220 litres substance/hectare	1 per crop	1 May – 15 November

Operator protection:

(1) A written COSHH assessment must be made before using sulphuric acid. Operators should observe the Occupational Exposure Standard set out in HSE guidance note EH40/90 or subsequent issues.

(2) Engineering control of operator exposure must be used where reasonably practicable in addition to the following personal protective equipment:

Operators must wear suitable protective clothing as listed below:

(a) When filling sprayer; spraying carrying out adjustment to application equipment or cleaning equipment; re-entering the sprayed area within 24 hours.

Face shield of acid resistant type; acid resistant coveralls either single or combination garment (or for persons operating bulk installations, acid proof); gauntlet gloves either natural rubber or PVC material, rubber boots, acid proof apron and suitable respiratory protective equipment.

(b) When re-entering the sprayed area between 24 and 96 hours.

Acid resistant coveralls, gloves and boots.

(3) However, engineering controls may replace personal protective equipment if a COSHH assessment shows they provide an equal or higher standard of protection.

(4) Operators must have liquid suitable for eye irrigation immediately available at all times throughout the operation.

Other specific restrictions:

(1) Must only be used by operators holding a relevant recognised certificate of competence in the use of the equipment for the application of sulphuric acid.

(2) Must not be applied using pedestrian controlled applicators or hand held equipment.

(3) All equipment must be constructed of materials suitable for use with or exposure to sulphuric acid.

(4) Application must be confined to the land intended to be treated.

(5) Spray must not be deposited within one metre of public footpaths.

(6) At least 24 hours written notice of the intended operation and the possibility of a hazard must be given to occupants of any premises and to the owner, or his agent, of any livestock or crops within 25 metres of any boundary of the land intended to be treated.

(7) Before the spraying takes place, readable notices must be posted on adjacent roads and paths warning passers by and drivers of vehicles of the time and place of the intended application and possibility of hazard. Notices to be kept in place for 96 hours following treatment

(8) Unprotected persons must be kept out of treated areas for at least 96 hours following treatment.

(9) The maximum quantity to be applied to potatoes must not exceed 800 litres per hectare per crop.

(10) Do not apply to crops in which bees are actively foraging. Do not apply when flowering weeds are present.

(11) Only "sulphur burnt" sulphuric acid to be used.

HEALTH AND SAFETY EXECUTIVE

Food and Environment Protection Act 1985

Schedule: COMMODITY SUBSTANCE: **TETRACHLOROETHYLENE**

Date of issue: 18 February 1993

Date of expiry: 28 February 2001

This approval is subject to the following conditions:

1 *FIELD OF USE:* ONLY AS AN INSECTICIDE IN MUSEUMS AND BUILDINGS OF CULTURAL, ARTISTIC AND HISTORICAL INTEREST

2 *PEST AND USAGE AREA:* FOR THE CONTROL OF FLYING AND CRAWLING INSECTS ON TEXTILES.

3 *APPLICATION METHOD:* IMMERSION IN A TANK ENCLOSED IN A FUME CUPBOARD

Operator protection:

(1) A written COSHH assessment must be made before using tetrachloroethylene. Operators should also observe the OES set out in HSE guidance note EH40/93 or subsequent issues.

(2) Engineering controls of operator exposure must be used where reasonably practicable in addition to the following items of personal protective equipment.

Operators must wear suitable protective clothing, including protective gloves and eye protection, when handling and using the material.

(3) However engineering controls may replace personal protective equipment if a COSHH assessment shows they provide an equal or higher standard of protection.

Other specific restrictions

(1) Operators should be provided with adequate information about the hazards of the substance and the precautions necessary for safe use. Sources of information include the supplier's Safety Data Sheet.

(2) Unprotected persons and animals must be excluded from any areas where treatment is taking place, and such areas should be ventilated after treatment.

(3) Must be used only by operators who are suitably trained and competent to carry out this work.

HEALTH AND SAFETY EXECUTIVE
Food and Environment Protection Act 1985

Schedule: COMMODITY SUBSTANCE: **THYMOL**

Date of issue: 18 February 1993

Date of expiry: 28 February 2001

This approval is subject to the following conditions:

1 *FIELD OF USE:*   ONLY AS A FUNGICIDE IN MUSEUMS AND BUILDINGS OF
                               CULTURAL, ARTISTIC AND HISTORICAL INTEREST.

2 *PEST AND USAGE AREA:*   FOR THE CONTROL OF FUNGI.

3 *APPLICATION METHOD:*   TREATMENT IN SEALED TREATMENT CABINETS

Operator protection:

(1) A written COSHH assessment must be made before using thymol.

(2) Engineering control of operator exposure must be used where reasonably practicable in
addition to the following items of personal protective equipment.

Operators must wear suitable protective clothing, including protective gloves and eye
protection, when handling or applying the material.

(3) However engineering controls may replace personal protective equipment if a COSHH
assessment shows they provide an equal or higher standard of protection.

Other specific restrictions:

(1) Operators should be provided with adequate information about the hazards of the substance
and the precautions necessary for safe use. Sources of information include the supplier's
Safety Data Sheet.

(2) Unprotected persons and animals must be excluded from any areas where treatment is taking
place, and such areas should be ventilated after treatment.

(3) Must be used only by operators who are suitably trained and competent to carry out this work.

Food and Environment Protection Act 1985
Control of Pestices Regulations 1986 (SI 1986 NO 1510): APPROVAL

In exercise of the powers conferred by Regulation 5 of the Control of Pesticides Regulations 1986 (SI 1986/1510) and of all other powers enabling them in that behalf, the Minister of Agriculture, Fisheries and Food and the Secretary of State, hereby jointly give full approval for the use of:

Commodity Substance: being **UREA** subject to the conditions set out below:

Date of issue: 1 March 1991

Date of expiry: 28 February 2001 (see advisory note 4)

Use:

Field of Use: **Only as a home garden fungicide**

Crop/ Situation	*Maximum individual dose*	*Maximum number of treatments*	*Latest time of application*
Cut stumps of trees	1 litre of 37% w/v aqueous solution/m^2 of cut stump	One per cut stump	At felling
Apple and pear trees	0.07 litres of 7% w/v solution/m^2	One per tree per year	Post harvest, pre leaf-fall

Operator protection:

When used at work, the following must be observed:

(1) Engineering control of operator exposure must be used where reasonably practicable in addition to the following personal protective equipment:

Operators must wear suitable protective gloves when mixing urea solution.

(2) However, engineering controls may replace personal protective equipment if a COSHH assessment shows they provide an equal or higher standard of protection.

ADVISORY NOTES

1   Pesticides approved for amateur use may be used by professional operators without user certification.

2   For the treatment of cut stumps of trees the following dyes may be used at 0.04% w/v concentration when mixed with solution of urea:

(i) Kenacid Turquoise V5898 (CI Acid Blue 42045)

(ii) Denacid Turquoise AN 200 }
     Duasyn Acid Blue AE-20   } (CI Acid Blue 9 42090)

3   Operators should wear rubber or other chemical-proof gloves, a nuisance dust mask and a cotton/terylene overall when handling dye powder.

4   Data submitted by interested parties is being considered by PSD for approval of this commodity substance beyond the stated expiry date.

HEALTH AND SAFETY EXECUTIVE

Food and Environment Protection Act 1985

Schedule: COMMODITY SUBSTANCE: **WHITE SPIRIT**

Date of issue: 18 February 1993

Date of expiry: 28 February 2001

This approval is subject to the following conditions:

1 *FIELD OF USE:*	ONLY AS AN INSECTICIDE IN MUSEUMS AND BUILDINGS OF CULTURAL, ARTISTIC AND HISTORICAL INTEREST
2 *PEST AND USAGE AREA:*	FOR THE CONTROL OF FLYING AND CRAWLING INSECTS ON TEXTILES
3 *APPLICATION METHOD:*	IMMERSION IN A TANK ENCLOSED IN A FUME CUPBOARD

Operator protection:

(1) A written COSHH assessment must be made before using white spirit. Operators should also observe the OES set out in HSE guidance note EH40/93 or subsequent issues.

(2) Engineering controls of operator exposure must be used where reasonably practicable in addition to the following items of personal protective equipment.

Operators must wear suitable protective clothing, including protective gloves and eye protection, when handling and using the material.

(3) However engineering controls may replace personal protective equipment if a COSHH assessment shows they provide an equal or higher standard of protection.

Other specific restrictions:

(1) Operators must be provided with adequate information on hazards and precautions. Sources of information may include the supplier's Safety Data Sheet.

(2) Unprotected persons and animals must be excluded from any areas where treatment is taking place, and such areas should be ventilated after treatment.

(3) Must be used only by operators who are suitably trained and competent to carry out this work.

ANNEX E

BANNED AND SEVERELY RESTRICTED PESTICIDES IN THE UNITED KINGDOM

These lists have been partially reviewed for Pesticides 2000; a full review is planned for the 2001 edition.

ACTIVE SUBSTANCES BANNED IN THE EUROPEAN UNION UNDER COUNCIL DIRECTIVE 79/117/EEC*

Council Directive 79/117/EEC dated 21 December 1978 prohibits the placing on the market and the use of plant protection products containing certain active substances which, even if applied in an approved manner, could give rise to harmful effects on human health or the environment.

ACTIVE SUBSTANCE	EFFECTIVE DATE OF BAN
Mercury Compounds	
Mercuric oxide (mercury oxide)	1992
Mercurous chloride (calomel)	1992
Other inorganic mercury compounds	1981
Alkyl mercury compounds	1991
Alkoxyalkyl and aryl mercury compounds	1992
Persistent Organo-chlorine Compounds	
Aldrin	1991
Chlordane	1981
Dieldrin	1981
DDT	1986
Endrin	1991
HCH containing less than 99% of gamma isomer	1981
Heptachlor	1984
Camphechlor	1984
Hexachlorobenzene	1981
Other Compounds	
Ethylene oxide	1991
Nitrofen	1988
1,2 Dibromoethane (Ethylene dibromide)	1988
1,2 Dichloroethane (Ethylene dichloride)	1989
Dinoseb, its acetate and salts	1991
Binapacryl	1991
Captafol	1991
Dicofol containing less than 78% of ppl dicofol or more than 1g/kg DDT and DDT related compounds.	1991

(a) Maleic hydrazide and its salts, other than its choline   1991
potassium and sodium salts

(b) Choline, potassium and sodium salts of maleic hydrazide   1991
containing more than 1g/kg of free hydrazine expressed on
the basis of the acid equivalent.

Quintozene containing more than 1g/kg of HCB or more than   1991
10g/kg pentachlorobenzene

Note: Some of these active substances had already been banned in the UK prior to inclusion in Directive 79/117/EEC. Further details are available from PSD.

* Directive 79/117/EEC was published in the Official Journal of the European Communities L33 of 8 February 1979. It has been amended by Council Directives 83/131/EEC dated 14 March 1983, 85/298/EEC dated 22 May 1985, 86/214/EEC dated 26 May 1986, 86/355/EEC dated 21 July 1986, 87/181/EEC dated 9 March 1987, 87/477/EEC dated 9 September 1987, 89/365/EEC dated 30 May 1989, 90/335/EEC dated 7 June 1990. 90/533/EEC dated 15 October 1990, 91/188/EEC dated 19 March 1991. Copies of all these Directives are available from The Stationery Office, 51 Nine Elms Lane, London SW8 5DR.

ACTIVE SUBSTANCES BANNED IN THE UK

In addition to the substances listed above, the UK has also banned the following substances:

ACTIVE SUBSTANCE	EFFECTIVE DATE OF BAN	REASON FOR BAN
Antu (thiourea)	1966	Evidence of carcinogenicity
Azobenzene	1975	Evidence of carcinogenicity
Cadmium compounds	1965	Evidence of carcinogenicity
Calcium arsenate	1968	High acute toxicity; persistence in soil; evidence of carcinogenicity
Chlordecone	1977	Evidence of carcinogenicity
Cyhexatin	1988	Evidence of teratogenicity
Dinoterb	1988	Evidence of teratogenicity
DNOC	1989	Evidence of teratogenicity in related dinitro compounds
Methyl mercury	1971	Environmental hazard (accumulation in the food chain)
Phenylmercury salicylate	1972	Acute toxicity; accumulations in the Environment
Potassium arsenite	1961	Acute toxicity to wildlife and livestock
Selenium compounds eg sodium selenate	1962	Acute toxicity to humans and livestock
Sodium arsenite	1961	Acute toxicity to wildlife and livestock
1,1,2,2-tetrachloroethane	1969	Acute and chronic toxicity to humans

SUBSTANCES SEVERELY RESTRICED AS PESTICIDES IN THE UK

PESTICIDE	DATE LAST REVIEWED	RESTRICTION
2-Aminobutane	1990	Permitted only on seed potatoes
Atrazine	1992	Approval for use on non-crop land (*excluding home garden use) revoked. Approvals for aerial use revoked. Restrictions on number of applications to crops.
Bromoxynil	1995	Approval for home garden use and application via hand-held applicators revoked other than via hand-held lances and pedestrian controlled sprayers (use via knapsack sprayers is not permitted). Timing restriction on grassland, leeks and onions.
	1990	Timing restrictions on some crops.
Carbaryl	1995	Approval for non-agricultural, home garden and poultry house use revoked, with a one year period (to November 1996) allowed for products in the supply chain to be used up.
Inorganic fluorides e.g. sodium fluoride	1966	Not permitted for agricultural/horticultural food storage or amateur use.
Ioxynil	(as bromoxynil)	
Pentachlorophenol (PCP) Sodium Pentachlorophenoxide & Pentachlorophenyl Laurate	1995	Use as a wood preservative restricted to professional and industrial operators. Use as a surface biocide restricted to professional operators. Professional use permitted only in buildings of artistic, cultural and historic interest or in emergencies to treat against dry rot fungus or cubic rot fungus. Use by professionals must be notified prior to treatment, in accordance with agreed forms.
Simazine	(as atrazine)	
Tributyltin oxide (TBTO)	1987	Use in antifouling products restricted to boats over 25 metres in overall length.
	1990	Use as a wood preservative and surface biocide restricted to industrial process and in paste formulation applied by professional operators.
	1992	A maximum concentration of 1% only, within tributyltin copolymer antifouling products for professional (i.e. deep sea) use.

ANNEX F
ACTIVE INGREDIENTS SUBJECT TO THE POISONS LAW

Certain products in this book are subject to the provisions of the Poisons Act 1972, the Poisons List Order 1982, the Poisons Rules 1982 and amending Orders made to these statutory instruments (copies of all these are obtainable from The Stationery Office). These Rules include general and specific provisions for the storage and sale and supply of listed non-medicine poisons.

The Active Ingredients approved for use in the UK and included in this book are specified under Parts I and II of the Poisons List as follows:

Part I Poisons (sale restricted to registered retail pharmacists)

aluminium phosphide	methyl bromide
chloropicrin	sodium cyanide
fluoroacetamide	strychnine
magnesium phosphide	

Part II Poisons (sale restricted to registered retail pharmacists and listed sellers registered with local authority)

aldicarb	methomyl (f)
alphachloralose	mevinphos
ammonium bifluoride	nicotine (c)
azinphos-methyl	omethoate
carbofuran (a)	oxamyl (a)
chlorfenvinphos (a,b)	oxydemeton-methyl
demeton-S-methyl	paraquat (d)
demeton-S-methyl sulphone	phorate (a)
dichlorvos (a,e)	pirimiphos-ethyl (b)
disulfoton (a)	quinalphos
drazoxolon (b)	sodium fluoride
endosulfan	sulphuric acid
fentin acetate	thiofanox (a)
fentin hydroxide	thiometon
fonofos (a)	triazophos (a)
formaldehyde	vamidothion
mephosfolan	zinc phosphide

(a)   Granular formulations which do not contain more than 12% w/w of this, or a combination of similarly flagged poisons, are exempt.

(b)   Treatments on seeds are exempt

(c)   Formulations containing not more than 7.5% of nicotine are exempt

(d)   Pellets containing not more than 5% paraquat ion are exempt

(e)   Preparations in aerosol dispensers containing not more than 1% w/w ai are exempt. Materials impregnated with dichlorvos for slow release are exempt.

(f)   Solid substances containing not more than 1% w/w ai are exempt.

ANNEX G
PUBLISHED EVALUATIONS LIST

Title	Price	Document Number
Fenbuconazole	£25.00	(no.128)
Fenbutatin Oxide	£ 4.00	(no.26)
Fenoxaprop Ethyl	£10.50	(no.18)
Fenoxaprop-P-Ethyl	£ 8.00	(no.17)
Fenoxycarb	£25.00	(no.161)
Fenpiclonil	£18.50	(no.78)
Fenpropathrin	£ 9.50	(no.23)
Fenpropidin	£17.00	(no.67)
Fenpyroximate	£22.00	(no.130)
Fentin Acetate	£ 3.00	(no.27)
Fentin Hydroxide	£ 4.00	(no.25)
Flocoumafen	£ 3.00	(no.01)
Fluazifop-P-Butyl	£ 5.50	(no.10)
Fluazinam	£17.00	(no.100)
Fludioxonil	£20.00	(no.126)
Flufenoxuron	£11.00	(no.143)
Flufenoxuron 2 (Use as a Wood Preservative)	£14.50	(no.180)
Fluoroglycofen-Ethyl	£25.00	(no.50)
Flusilazole	£10.50	(no.11)
Flutriafol	£11.50	(no.158)
Fluquinconazole	£25.00	(no.184)
Fipronil: Use as a public Hygiene insecticide	£16.00	(no.187)
Fomesafen	£15.00	(no.118)

G

Title	Price	Document Number
Gamma-HCH (Lindane 1)	£ 4.00	(no.47)
Gamma-HCH (Lindane 2)	£18.50	(no.64)
Gamma-HCH (Lindane 3) (Agricultural Uses)	£25.00	(no.151)
Glufosinate-ammonium Grain Protectants in the UK	£21.50	(no.33)
Review of	£ 5.00	(no.59)
Guazatine	£ 9.50	(no.53)

H

Title	Price	Document Number
HOE 070542 Triazole Coformulant	£ 7.50	(no.19)
Hydramethylnon	£ 6.50	(no.102)
Hydroprene (1)	£ 5.50	(no.44)
Hydroprene (2)	£ 7.00	(no.145)

I

Title	Price	Document Number
Imazaquin	£17.00	(no.66)
Imazethapyr	£10.00	(no.149)
Imidacloprid	£23.00	(no.73)

Title	Price	Document Number
Ioxynil - Review of the Agricultural and Horticultural Uses	£11.00	(no.123)
IPBC (1)	£ 4.50	(no.13)
IPBC (2)	£ 8.50	(no. 115)
Iprodione	£ 4.00	(no.28)
Isoproturon	£23.00	(no.140)

K

Title	Price	Document Number
Kathon 886 (1)	£ 9.00	(no.90)
Kathon 886 (2)	£ 6.00	(no.176)
Kresoxim-Methyl	£25.00	(no.163)

L

Title	Price	Document Number
Lambda-Cyhalothrin	£ 3.00	(no.104)
Lindane - Reproductive Toxicity Effects in Dogs	£ 6.50	(no.164)
Lindane (Review of)	£13.50	(no.191)
Lindane 1 (Gamma-HCH)	£ 4.00	(no.47)
Lindane 2 (Gamma-HCH)	£18.50	(no.64)
Lindane 3 (Gamma-HCH) (Agricultural Uses)	£25.00	(no.151)
Linuron	£18.00	(no.132)

M

Title	Price	Document Number
Malathion	£19.50	(no.135)
MBC Fungicides - Benomyl and Carbendazim	£ 6.00	(no.170)
MBC Fungicide - Thiophanate Methyl	£ 4.50	(no. 181)
Mecoprop	£17.00	(no.95)
Mecoprop-P	£17.00	(no.96)
Metaldehyde	£19.50	(no.153)
Methiocarb (Review of)	£25.00	(no.179)
Methyl Bromide	£12.00	(no.63)
Metosulam	£22.00	(no.148)
Metsulfuron Methyl	£ 7.50	(no.38)
Metsulfuron-Methyl and Thifensulfuron-Methyl (Review of)	£ 5.50	(no.119)
Review of Environmental Persistence Monolinuron	£ 9.00	(no.133)

O

Title	Price	Document Number
Omethoate	£13.50	(no.83)
Oxine Copper	£ 3.50	(no.09)
Oxydemeton-Methyl	£11.00	(no.76)

Title	Price	Document Number
P		
Paclobutrazol	£11.00	(no.142)
Pentachlorophenol	£17.50	(no.114)
2-Phenyl Phenol	£ 8.00	(no.82)
Phlebiopsis Gigantea	£8.50	(no173)
Phorate	£14.00	(no.98)
Pirimicarb	£24.00	(no.134)
Pirimiphos-Methyl	£22.50	(no.167)
PP321 (Lambda-Cyhalothrin)	£ 5.00	(no.20)
Prallethrin	£ 6.50	(no.131)
Propamocarb Hydrochloride	£21.00	(no.62)
Propaquizafop	£23.00	(no.94)
Propiconazole	£ 8.00	(no.80)
Propyzamide	£ 3.50	(no.154)
Pyrimethanil	£18.00	(no.138)
Pyriproxyfen	£11.50	(no.147)
Q		
Quinmerac	£25.00	(no.177)
Quizalofop-Ethyl	£ 3.50	(no.02)
R		
RH 3866	£ 4.00	(no.06)
Rimsulfuron	£25.00	(no.146)
S		
SAN 619F (Cyproconazole)	£25.00	(no.45)
Simazine (1)	£ 9.50	(no.52)
Simazine (2)	£18.50	(no.72)
S-Methoprene (1)	£ 5.50	(no.85)
S-Methoprene (2)	£ 8.00	(no.166)
Sodium Cyanide	£11.00	(no.144)
Strychnine Hydrochloride	£ 6.00	(no.168)
Sulphuric Acid	£ 4.00	(no174)

Title	Price	Document Number
T		
Tau - Fluvalinate	£25.00	(no.162)
Tebuconazole (1)	£25.00	(no.65)
Tebuconazole (2)	£ 7.00	(no.88)
Tebufenpyrad	£25.00	(no.122)
Tecnazene (Review of)	£20.00	(no.127)
Teflubenzuron	£ 1.50	(no.40)
Tefluthrin	£13.00	(no.42)
Tetraconazole	£25.00	(no.185)
Thiabendazole	£10.00	(no.54)
Thifensulfuron-Methyl	£ 5.00	(no.39)
Thiodicarb	£25.00	(no.49)
Thiophanate-Methyl	£11.00	(no.56)
Tolclofos-Methyl	£16.00	(no.69)
Tolclofos-Methyl in the Products 'Rizolex'	£8.50	(no.178)
Tolylfluanid	£19.50	(no.136)
Tralkoxydim (PP604)	£25.00	(no.70)
Transfluthrin (1)	£10.00	(no.165)
Transfluthrin (2)	£ 7.50	(no.188)
Triasulfuron	£25.00	(no.48)
Triazamate	£25.00	(no.172)
Triazophos	£24.00	(no.84)
Triazoxide	£15.00	(no.97)
Tribenuron Methyl	£15.00	(no.61)
Tributyltin Naphthenate (1)	£ 3.00	(no.15)
Tributyltin Naphthenate (2)	£ 4.50	(no.110)
Tributyltin Oxide (1)	£ 7.00	(no.24)
Tributyltin Oxide (2)	£ 4.50	(no.74)
Tridemorph	£ 8.00	(no.190)
Triflusulfuron-Methyl	£25.00	(no.139)
Trinexapac Ethy	£18.00	(no.141)
Triorganotin Compounds (1)	£16.50	(no.111)
Triorganotin Compounds (2)	£ 8.50	(no.182)
V		
Vinclozolin (1)	£12.00	(no.34)
Vinclozolin(2)	£12.50	(no.113)

Evaluations are available on application in writing to the ACP Secretariat, Pesticides Safety Directorate, Room 308, Mallard House, Kings Pool, 3 Peasholme Green, York, YO1 7PX Tel. 01904 455705.

ANNEX H
PRODUCTS APPROVED FOR USE IN OR NEAR WATER
(Other than for Public Hygiene or Antifouling Use)

"Products approved for use in or near water as at 31 October 1999, are listed below. Before you use any product in or near water you should first consult the appropriate water regulatory body (Environment Agency/Scottish Environment Protection Agency). Always read the label before use."

Product name	Marketing company	Reg No.
Agricorn 2,4-D	Farmers Crop Chemicals Ltd	07349
Atlas 2,4-D	Atlas Crop Protection Ltd	07699
Azural	Monsanto Plc	09582
Barclay Gallup 360	Barclay Plant Protection	09127
Barclay Gallup Amenity	Barclay Chemicals (UK) Ltd	06753
Buggy SG	Sipcam UK Ltd	08573
Cardel Egret	Cardel	09703
Cardel Glyphosate	Cardel	09581
Casoron G	Miracle Professional	07926
Casoron G	Zeneca Crop Protection	08065
Casoron G	Uniroyal Chemical Ltd	09022
Casoron G	Nomix-Chipman Ltd	09023
Casoron G	Rigby Taylor Ltd	09326
Casoron GSR	Imperial Chemical Industries Plc	00451
Casoron GSR	Miracle Professional	07925
Clarosan	Novartis Crop Protection UK Ltd	08396
Clarosan	The Scotts Company (UK) Limited	09394
Clarosan 1FG	Ciba Agriculture	03859
Clayton Swath	Clayton Plant Protection (UK) Ltd	06715
CleanCrop Diquat	United Agri Products	09687
Clinic	Nufarm UK Ltd	08579
Clinic	Nufarm UK Ltd	09378
Danagri Glyphosate 360	Danagri ApS	06955
Dormone	Rhone-Poulenc Amenity	05412
Glyfos	Cheminova Agro (UK) Ltd	07109
Glyfos 480	Cheminova Agro (UK) Ltd	08014
Glyfos Proactive	Nomix-Chipman Ltd	07800
Glyper	Pan Britannica Industries Ltd	07968
Glyphogan	Makhteshim-Agan (UK) Ltd	05784
Glyphosate 360	Danagri ApS	08568
Glyphosate 360	Applyworld Limited	09151
Glyphosate 360	Danagri ApS	09233
Glyphosate 360A	Danagri ApS	09234
Glyphosate 360B	Danagri ApS	09235
Glyphosate Biactive	Monsanto Plc	08307
Helosate	Helm AG	06499
Krenite	Du Pont (UK) Ltd	01165
Levi	P.S.I. Phoenix Scientific Innovation UK Ltd	07845

Product name	Marketing company	Reg No.
Luxan Dichlobenil Granules	Luxan (UK) Ltd	09250
Marnoch Glyphosate	Marnoch Ventures Limited	09744
Midstream	Imperial Chemical Industries Plc	01348
Midstream	Zeneca Professional Products	06824
Midstream	Miracle Professional	07739
Midstream	The Scotts Company (UK) Limited	09267
MON 44068 Pro	Monsanto Plc	06815
MON 52276	Monsanto Plc	06949
Monty	Quadrangle Agrochemicals	07796
MSS 2,4-D Amine	Mirfield Sales Services Ltd	01391
MSS Glyfield	Mirfield Sales Services Ltd	08009
Reglone	Zeneca Crop Protection	06703
Reglone	Zeneca Crop Protection	09646
Regulox K	Rhone-Poulenc Amenity	05405
Rival	Monsanto Plc	09220
Roundup	Monsanto Plc	01828
Roundup A	Monsanto Plc	08375
Roundup Amenity	Monsanto Plc	08721
Roundup Biactive	Monsanto Plc	06941
Roundup Biactive Dry	Monsanto Plc	06942
Roundup Pro	Monsanto Plc	04146
Roundup Pro Biactive	Monsanto Plc	06954
Sierraron G	The Scotts Company (UK) Limited	09263
Sierraron G	The Scotts Company (UK) Limited	09675
Spasor	Rhone-Poulenc Amenity	07211
Spasor Biactive	Rhone-Poulenc Amenity	07651
Stetson	Monsanto Plc	06956
Typhoon 360	Feinchemie Schwebda GmbH	09322
Typhoon 360	Feinchemie (UK) Ltd	09792

PART B

PSD Registered Products

PROFESSIONAL PRODUCTS

Product Name	Marketing Company	Reg. No.	Expiry Date

1.1 Herbicides
including growth regulators, defoliants, rooting agents and desiccants

1 Amidosulfuron

C	Aventis Eagle	AgrEvo UK Ltd	09765
C	Barclay Cleave	Barclay Chemicals (UK) Ltd	09489
C	Druid	AgrEvo UK Ltd	08714
C	Eagle	AgrEvo UK Ltd	07318
C	Landgold Amidosulfuron	Landgold & Co Ltd	09021
C	Pursuit	AgrEvo UK Ltd	07333
C	Pursuit 50	AgrEvo UK Ltd	08716
C	Squire	AgrEvo UK Ltd	08715

2 Amidosulfuron + Metribuzin

C	Bayer UK 590	Bayer plc	08149	30/06/2001
C	Galis	Bayer plc	08039	30/06/2001

3 Amitrole

C	Loft	A H Marks & Co Ltd	06030	
C	MSS Aminotriazole Technical	Mirfield Sales Services Ltd	04645	31/12/2001
C	Weedazol-TL	A H Marks & Co Ltd	02349	
C	Weedazol-TL	Bayer plc	02979	

4 Amitrole + Bromacil + Diuron

C	BR Destral	Rhone-Poulenc Amenity	05184	31/12/2001

5 Amitrole + 2,4-D + Diuron

C	Trik	Mirfield Sales Services Ltd	07853

6 Amitrole + Simazine

C	Alpha Simazol	Makhteshim-Agan (UK) Ltd	04799

7 Ammonium sulphamate

C	Amcide	Battle Hayward & Bower Ltd	04246

8 Anthracene oils

C	Sterilite Hop Defoliant	Coventry Chemicals Ltd	05060

9 Asulam

C	Asulox	Rhone-Poulenc Agriculture	06124

C   These products are "approved for agricultural use". For further details refer to page vii.
A   These products are approved for use in or near water. For further details refer to page vii.

Product Name	Marketing Company	Reg. No.	Expiry Date

10 Atrazine

C Alpha Atrazine 50 SC	Makhteshim-Agan (UK) Ltd	04877	
C Alpha Atrazine 50 WP	Makhteshim-Agan (UK) Ltd	04793	
C Atlas Atrazine	Atlas Crop Protection Ltd	07702	
C Atlas Atrazine	Atlas Interlates Ltd	03097	30/06/2002
C Atrazine 90WG	Sipcam UK Ltd	09310	
C Atrazol	Sipcam UK Ltd	07598	
C Dapt Atrazine 50 SC	DAPT Agrochemicals Ltd	08031	
C Gesaprim	Novartis Crop Protection UK Ltd	08411	
C MSS Atrazine 50 FL	Mirfield Sales Services Ltd	01398	
C MSS Atrazine 80 WP	Mirfield Sales Services Ltd	04360	
C Unicrop Atrazine 50	Universal Crop Protection Ltd	02645	
C Unicrop Atrazine FL	Universal Crop Protection Ltd	08045	
C Unicrop Flowable Atrazine	Universal Crop Protection Ltd	05446	

11 Aziprotryne

C Brasoran	Novartis Crop Protection UK Ltd	08394	
C Brasoran 50 WP	Ciba Agriculture	00316	31/08/2000

12 Benazolin

C Aventis Galtak 50 SC	AgrEvo UK Ltd	09711	
C Galtak 50 SC	AgrEvo UK Ltd	07258	

13 Benazolin + Bromoxynil + Ioxynil

C Asset	AgrEvo UK Ltd	07243	

14 Benazolin + Clopyralid

C Benazalox	AgrEvo UK Ltd	07246	

15 Benazolin + 2,4-DB + MCPA

C Legumex Extra	AgrEvo UK Ltd	08676	
C Legumex Extra	AgrEvo UK Ltd	07901	31/05/2001
C Master Sward	Stefes Plant Protection Ltd	07883	
C Setter 33	AgrEvo UK Ltd	07282	
C Setter 33	Dow AgroSciences Ltd	05623	
C Stefes Clover Ley	Stefes Plant Protection Ltd	07842	
C Stefes Legumex Extra	Stefes Plant Protection Ltd	07841	

16 Bentazone

C Barclay Dingo	Barclay Chemicals (UK) Ltd	07642	30/08/2000
C Basagran	BASF plc	00188	
C Basagran SG	BASF plc	08360	
C IT Bentazone	IT Agro Ltd	08677	
C IT Bentazone 48	IT Agro Ltd	09283	

C These products are "approved for agricultural use". For further details refer to page vii.
A These products are approved for use in or near water. For further details refer to page vii.

Product Name	Marketing Company	Reg. No.	Expiry Date

16 Bentazone—continued

C	IT Bentazone 480	IT Agro Ltd	07458	29/02/2000
C	Osiris	P.S.I. Phoenix Scientific Innovation UK Ltd	07953	29/02/2000
C	Standon Bentazone	Standon Chemicals Ltd	09204	
C	Standon Bentazone	Standon Chemicals Ltd	06895	31/08/2000
C	Top Farm Bentazone	Top Farm Formulations Ltd	08026	29/02/2000

17 Bentazone + Cyanazine + 2,4-DB

C	Topshot	Cyanamid Agriculture Ltd	07178	28/02/2000

18 Bentazone + Dichlorprop-P

C	Quitt SL	BASF plc	08108	

19 Bentazone + MCPA + MCPB

C	Acumen	BASF plc	00028	
C	Headland Archer	Headland Agrochemicals Ltd	08814	

20 Bentazone + MCPB

C	Pulsar	BASF plc	04002	

21 Bentazone + Pendimethalin

C	Impuls	BASF plc	09720	

22 Bifenox + Chlorotoluron

C	Dicurane Duo 446 SC	Novartis Crop Protection UK Ltd	08404	
C	Dicurane Duo 446SC	Ciba Agriculture	05839	31/08/2000

23 Bifenox + Dicamba

C	Quickstep	Novartis Crop Protection UK Ltd	08557	31/10/2000
C	Quickstep	Sandoz Agro Ltd	07389	31/08/2000
C	RP 283	Rhone-Poulenc Agriculture	06178	31/10/2000

24 Bifenox + Isoproturon

C	Banco	Portman Agrochemicals Ltd	09027	
C	RP 4169	Rhone-Poulenc Agriculture	05801	

25 Bifenox + MCPA + Mecoprop-P

C	Sirocco	Rhone-Poulenc Amenity	09645	

26 Bromacil

C	Alpha Bromacil 80 WP	Makhteshim-Agan (UK) Ltd	04802	
C	Hyvar X	Du Pont (UK) Ltd	01105	

C These products are "approved for agricultural use". For further details refer to page vii.
A These products are approved for use in or near water. For further details refer to page vii.

Product Name	Marketing Company	Reg. No.	Expiry Date

27 Bromacil + Amitrole + Diuron
C   BR Destral   Rhone-Poulenc Amenity   05184   31/12/2001

28 Bromacil + Diuron
C   Borocil K   Rhone-Poulenc Amenity   05183

29 Bromacil + Picloram
C   Hydon   Chipman Ltd   01088

30 Bromoxynil
C	Alpha Bromolin 225 EC	Makhteshim-Agan (UK) Ltd	08255	
C	Alpha Bromotril P	Makhteshim-Agan (UK) Ltd	07099	
C	Barclay Mutiny	Barclay Chemicals (UK) Ltd	08933	
C	Barclay Mutiny	Barclay Chemicals (UK) Ltd	07773	29/02/2000
C	Flagon 400 EC	Makhteshim-Agan (UK) Ltd	08875	

31 Bromoxynil + Benazolin + Ioxynil
C   Asset   AgrEvo UK Ltd   07243

32 Bromoxynil + Clopyralid
C   Vindex   Dow AgroSciences Ltd   05470

33 Bromoxynil + Clopyralid + Fluroxypyr + Ioxynil
C   Crusader S   Dow AgroSciences Ltd   05174

34 Bromoxynil + Cyanazine
C   Greencrop Tassle   Greencrop Technology Ltd   09659

35 Bromoxynil + Dichlorprop + Ioxynil + MCPA
C   Atlas Minerva   Atlas Interlates Ltd   03046

36 Bromoxynil + Diflufenican + Ioxynil
C   Capture   Rhone-Poulenc Agriculture   06881

37 Bromoxynil + Ethofumesate + Ioxynil
| C | Leyclene | AgrEvo UK Ltd | 07263 |
| C | Stefes Leyclene | Stefes Plant Protection Ltd | 08173 |

38 Bromoxynil + Fluroxypyr
C   Sickle   Dow AgroSciences Ltd   05187

39 Bromoxynil + Fluroxypyr + Ioxynil
C   Advance   Dow AgroSciences Ltd   05173

C   These products are "approved for agricultural use". For further details refer to page vii.
A   These products are approved for use in or near water. For further details refer to page vii.

Product Name	Marketing Company	Reg. No.	Expiry Date

40 Bromoxynil + Ioxynil

C	Alpha Briotril	Makhteshim-Agan (UK) Ltd	04876	
C	Alpha Briotril Plus 19/19	Makhteshim-Agan (UK) Ltd	04740	
C	Deloxil	AgrEvo UK Ltd	07313	
C	Deloxil	Rhone-Poulenc Agriculture	07405	
C	Mextrol-Biox	Nufarm UK Ltd	09470	
C	Oxytril CM	Rhone-Poulenc Agriculture	08667	
C	Oxytril CM	Rhone-Poulenc Agriculture	06201	31/08/2000
C	Percept	MTM Agrochemicals Ltd	05481	
C	Status	Rhone-Poulenc Agriculture	08668	
C	Status	Rhone-Poulenc Agriculture	06906	31/08/2000
C	Stellox 380 EC	Ciba Agriculture	04169	31/08/2000
C	Stellox 380 EC	Novartis Crop Protection UK Ltd	08451	

41 Bromoxynil + Ioxynil + Isoproturon + Mecoprop

C	Terset	Rhone-Poulenc Agriculture	06171	30/04/2000

42 Bromoxynil + Ioxynil + Mecoprop

C	Swipe 560 EC	Ciba Agriculture	02057	30/04/2000

43 Bromoxynil + Ioxynil + Mecoprop-P

C	Swipe P	Ciba Agriculture	08150	31/12/2000
C	Swipe P	Novartis Crop Protection UK Ltd	08452	

44 Bromoxynil + Ioxynil + Triasulfuron

C	Teal	Ciba Agriculture	06117	31/08/2000
C	Teal	Novartis Crop Protection UK Ltd	08453	
C	Teal-M	Ciba Agriculture	08028	31/08/2000
C	Teal-M	Novartis Crop Protection UK Ltd	08454	

45 Bromoxynil + Ioxynil + Trifluralin

C	Masterspray	Pan Britannica Industries Ltd	02971	

46 Bromoxynil + Prosulfuron

C	Jester	Novartis Crop Protection UK Ltd	08681	

47 Bromoxynil + Terbuthylazine

C	Alpha Bromotril PT	Makhteshim-Agan (UK) Ltd	09435	

48 Carbendazim + Tecnazene

C	Hickstor 6 Plus MBC	Hickson & Welch Ltd	04176	
C	Hortag Tecnacarb	Avon Packers Ltd	02929	

C   These products are "approved for agricultural use". For further details refer to page vii.
A   These products are approved for use in or near water. For further details refer to page vii.

Product Name	Marketing Company	Reg. No.	Expiry Date

48 Carbendazim + Tecnazene—continued

C	New Arena Plus	Tripart Farm Chemicals	04598
C	New Hickstor 6 Plus MBC	Hickson & Welch Ltd	04599
C	Tripart Arena Plus	Tripart Farm Chemicals	05602

49 Carbetamide

C	Carbetamex	Rhone-Poulenc Agriculture	06186

50 Carbetamide + Diflufenican + Oxadiazon

C	Helmsman	Rhone-Poulenc Amenity	09696

51 Carfentrazone-ethyl

C	Platform	FMC Corporation (UK) Ltd	09661
C	Platform	FMC Europe NV	08628

52 Carfentrazone-ethyl + Flupyrsulfuron-methyl

C	Lexus Class	Du Pont (UK) Ltd	08636
C	Lexus Class WSB	Du Pont (UK) Ltd	08637

53 Carfentrazone-ethyl + Isoproturon

C	Affinity	FMC Europe NV	08639

54 Carfentrazone-ethyl + Mecoprop-P

C	Platform S	FMC Europe NV	08638

55 Carfentrazone-ethyl + Metsulfuron-methyl

C	Ally Express	Du Pont (UK) Ltd	08640
C	Ally Express WSB	Du Pont (UK) Ltd	08641
C	Standon Carfentrazone MM	Standon Chemicals Ltd	09159

56 Carfentrazone-ethyl + Thifensulfuron-methyl

C	Harmony Express	Du Pont (UK) Ltd	09467

57 Chlorbufam + Chloridazon

C	Alicep	BASF plc	00077	31/01/2000

58 Chloridazon

C	Barclay Champion	Barclay Chemicals (UK) Ltd	06903	29/02/2000
C	Barclay Claddagh	Barclay Chemicals (UK) Ltd	06330	29/02/2000
C	Better DF	Sipcam UK Ltd	06250	
C	Better Flowable	Sipcam UK Ltd	04924	
C	Burex 430 SC	Agricola Ltd	09494	
C	Gladiator	Tripart Farm Chemicals Ltd	00986	

C These products are "approved for agricultural use". For further details refer to page vii.
A These products are approved for use in or near water. For further details refer to page vii.

Product Name	Marketing Company	Reg. No.	Expiry Date

58 Chloridazon—continued

C	Gladiator DF	Tripart Farm Chemicals Ltd	06342	
C	Luxan Chloridazon	Luxan (UK) Ltd	06304	
C	Portman Weedmaster	Portman Agrochemicals Ltd	06018	
C	Pyramin DF	BASF plc	03438	
C	Pyramin FL	BASF plc	01661	
C	Questar	BASF plc	07955	
C	Sculptor	Sipcam UK Ltd	08836	
C	Starter Flowable	Truchem Ltd	03421	
C	Stefes Chloridazon	Stefes UK Ltd	07678	
C	Takron	BASF plc	06237	
C	Tripart Gladiator 2	Tripart Farm Chemicals Ltd	06618	
C	Weedmaster SC	Portman Agrochemicals Ltd	08793	

59 Chloridazon + Chlorbufam

C	Alicep	BASF plc	00077	31/01/2000

60 Chloridazon + Ethofumesate

C	Gremlin	Sipcam UK Ltd	09468	
C	Magnum	BASF plc	08635	
C	Magnum	BASF plc	01237	31/07/2000
C	Spectron	AgrEvo UK Ltd	07284	
C	Stefes Spectron	Stefes UK Ltd	09337	

61 Chloridazon + Lenacil

C	Advizor	Du Pont (UK) Ltd	06571	
C	Advizor S	Du Pont (UK) Ltd	09799	
C	Advizor S	Du Pont (UK) Ltd	06960	
C	Varmint	Pan Britannica Industries Ltd	01561	

62 Chloridazon + Propachlor

C	Ashlade CP	Ashlade Formulations Ltd	06481	

63 Chlormequat

C	3C Chlormequat 460	Pennine Chemical Services Ltd	03916	28/02/2001
C	3C Chlormequat 600	Pennine Chemical Services Ltd	04079	28/02/2001
C	Adjust	Mandops (UK) Ltd	05589	
C	Agriguard Chlormequat 700	Agriguard Ltd	09782	
C	Agriguard Chlormequat 700	Agriguard Ltd	09282	
C	Allied Colloids Chlormequat 460	Ciba Speciality Chemicals, Water Treatments Ltd	07859	30/06/2002
C	Allied Colloids Chlormequat 460:320	Ciba Speciality Chemicals, Water Treatments Ltd	07861	30/06/2002

C   These products are "approved for agricultural use". For further details refer to page vii.
A   These products are approved for use in or near water. For further details refer to page vii.

Product Name	Marketing Company	Reg. No.	Expiry Date

63 Chlormequat—continued

	Product Name	Marketing Company	Reg. No.	Expiry Date
C	Allied Colloids Chlormequat 730	Ciba Speciality Chemicals, Water Treatments Ltd	07860	30/06/2002
C	Alpha Chlormequat 460	Makhteshim-Agan (UK) Ltd	04804	
C	Alpha Pentagan	Makhteshim-Agan (UK) Ltd	04794	
C	Alpha Pentagan Extra	Makhteshim-Agan (UK) Ltd	04796	
C	Ashlade 460 CCC	Ashlade Formulations Ltd	06474	
C	Ashlade 5C	Ashlade Formulations Ltd	06227	
C	Ashlade 700 5C	Ashlade Formulations Ltd	07046	
C	Ashlade 700 CCC	Ashlade Formulations Ltd	06473	
C	Ashlade Brevis	Ashlade Formulations Ltd	08119	
C	Atlas 3C:645 Chlormequat	Atlas Crop Protection Ltd	07700	
C	Atlas 3C:645 Chlormequat	Atlas Interlates Ltd	05710	
C	Atlas 5C Chlormequat	Atlas Crop Protection Ltd	07701	
C	Atlas 5C Chlormequat	Atlas Interlates Ltd	03084	
C	Atlas Chlormequat 46	Atlas Crop Protection Ltd	07704	
C	Atlas Chlormequat 460:46	Atlas Crop Protection Ltd	07705	
C	Atlas Chlormequat 460:46	Atlas Interlates Ltd	06258	
C	Atlas Chlormequat 700	Atlas Crop Protection Ltd	07708	
C	Atlas Chlormequat 700	Atlas Interlates Ltd	03402	
C	Atlas Quintacel	Atlas Crop Protection Ltd	07706	
C	Atlas Terbine	Atlas Crop Protection Ltd	07709	
C	Atlas Terbine	Atlas Interlates Ltd	06523	
C	Atlas Tricol	Atlas Crop Protection Ltd	07707	
C	Atlas Tricol	Atlas Interlates Ltd	07190	
C	Atlas Tricol PCT	Atlas Crop Protection Ltd	08015	31/08/2000
C	Barclay Holdup	Barclay Chemicals (UK) Ltd	06799	
C	Barclay Holdup 600	Barclay Chemicals (UK) Ltd	08794	
C	Barclay Holdup 640	Barclay Chemicals (UK) Ltd	08795	
C	Barclay Take 5	Barclay Chemicals Manufacturing Ltd	08524	
C	Barleyquat B	Mandops (UK) Ltd	07051	
C	BASF 3C Chlormequat 600	BASF plc	04077	
C	BASF 3C Chlormequat 720	BASF plc	06514	
C	BASF 3C Chlormequat 750	BASF plc	06878	
C	Belcocel	UCB Chemicals	08248	
C	Bettaquat B	Mandops (UK) Ltd	07050	
C	Calypso	Griffin (Europe) Marketing NV (UK Branch)	09505	
C	Ciba Chlormequat 460	Ciba Speciality Chemicals, Water Treatments Ltd	09525	
C	Ciba Chlormequat 5C 460:320	Ciba Speciality Chemicals, Water Treatments Ltd	09527	
C	Ciba Chlormequat 730	Ciba Speciality Chemicals, Water Treatments Ltd	09526	

C   These products are "approved for agricultural use". For further details refer to page vii.
A   These products are approved for use in or near water. For further details refer to page vii.

Product Name	Marketing Company	Reg. No.	Expiry Date

63 Chlormequat—continued

	Product Name	Marketing Company	Reg. No.	Expiry Date
C	Clayton CCC 750	Clayton Plant Protection (UK) Ltd	07952	
C	Clayton Standup	Clayton Plant Protection Ltd	08771	
C	Cropsafe 5C Chlormequat	Hortichem Ltd	07897	
C	Fargro Chlormequat	Fargro Ltd	02600	
C	Greencrop Carna	Greencrop Technology Ltd	09403	
C	Greencrop Cong 750	Greencrop Technology Ltd	09383	
C	Greencrop Coolfin	Greencrop Technology Ltd	09449	
C	Hyquat 70	Agrichem (International) Ltd	03364	
C	Intracrop Balance	Intracrop	08037	
C	Intracrop MCCC	Intracrop	08506	
C	Landgold CC 720	Landgold & Co Ltd	08527	
C	Mandops Barleyquat B	Mandops (UK) Ltd	06001	
C	Mandops Bettaquat B	Mandops (UK) Ltd	06004	
C	Mandops Chlormequat 460	Mandops (UK) Ltd	06090	
C	Mandops Chlormequat 700	Mandops (UK) Ltd	06002	
C	Manipulator	Mandops (UK) Ltd	05871	
C	MSS Chlormequat 40	Mirfield Sales Services Ltd	01401	
C	MSS Chlormequat 460	Mirfield Sales Services Ltd	03935	
C	MSS Chlormequat 60	Mirfield Sales Services Ltd	03936	
C	MSS Chlormequat 70	Mirfield Sales Services Ltd	03937	
C	MSS Mircell	Mirfield Sales Services Ltd	06939	
C	MSS Mirquat	Mirfield Sales Services Ltd	08166	
C	New 5C Cycocel	BASF plc	01482	
C	New 5C Cycocel	Cyanamid Agriculture Ltd	01483	
C	PA Chlormequat 400	Portman Agrochemicals Ltd	01523	
C	PA Chlormequat 460	Portman Agrochemicals Ltd	02549	
C	Podquat	Mandops (UK) Ltd	03003	
C	Portman Chlormequat 700	Portman Agrochemicals Ltd	03465	
C	Portman Supaquat	Portman Agrochemicals Ltd	03466	
C	Quadrangle Chlormequat 700	Quadrangle Agrochemicals	03401	
C	Renown	Stefes UK Ltd	09058	
C	Sigma PCT	Atlas Crop Protection Ltd	08663	
C	Stabilan 460	Ashlade Formulations Ltd	08005	31/10/2001
C	Stabilan 5C	Ashlade Formulations Ltd	08144	
C	Stabilan 700	Ashlade Formulations Ltd	08006	31/10/2001
C	Stabilan 750	Ashlade Formulations Ltd	08004	31/10/2001
C	Stefes CCC	Stefes UK Ltd	05959	
C	Stefes CCC 640	Stefes UK Ltd	06993	
C	Stefes CCC 700	Stefes UK Ltd	07116	
C	Stefes CCC 720	Stefes UK Ltd	05834	
C	Stefes K2	Stefes Plant Protection Ltd	07054	
C	Supaquat	Portman Agrochemicals Ltd	09381	
C	Top Farm Chlormequat 640	Top Farm Formulations Ltd	05323	29/02/2000
C	Trio	Portman Agrochemicals Ltd	08883	31/12/2001

C   These products are "approved for agricultural use". For further details refer to page vii.
A   These products are approved for use in or near water. For further details refer to page vii.

Product Name	Marketing Company	Reg. No.	Expiry Date

63 Chlormequat—continued

C	Tripart 5C	Tripart Farm Chemicals Ltd	04726
C	Tripart Brevis	Tripart Farm Chemicals Ltd	03754
C	Tripart Brevis 2	Tripart Farm Chemicals Ltd	06612
C	Tripart Chlormequat 460	Tripart Farm Chemicals Ltd	03685
C	Uplift	United Phosphorus Ltd	07527
C	Whyte Chlormequat 700	Whyte Agrochemicals Ltd	09641

64 Chlormequat + 2-Chloroethylphosphonic acid

C	Barclay Banshee XL	Barclay Chemicals (UK) Ltd	09201
C	Greencrop Tycoon	Greencrop Technology Ltd	09571
C	Nomad	Rhone-Poulenc Agriculture	07888
C	Strate	Rhone-Poulenc Agriculture	07430
C	Sypex	BASF plc	04650
C	Terpal C	BASF plc	07062
C	Upgrade	Rhone-Poulenc Agriculture	06177

65 Chlormequat + 2-Chloroethylphosphonic acid + Imazaquin

C	Ice	Cyanamid Agriculture Ltd	08970
C	Satellite	Cyanamid Agriculture Ltd	08969

66 Chlormequat + 2-Chloroethylphosphonic acid + Mepiquat

C	Cyclade	BASF plc	08958

67 Chlormequat + Imazaquin

C	Meteor	Cyanamid Agriculture Ltd	06505
C	Standon Imazaquin 5C	Standon Chemicals Ltd	08813
C	Upright	Cyanamid Agriculture Ltd	08290

68 Chlormequat + Mepiquat

C	Cyter	BASF plc	09133
C	Stronghold	BASF plc	09134

69 2-Chloroethylphosphonic acid

C	Aventis Cerone	Rhone-Poulenc Agriculture	09748	
C	Barclay Coolmore	Barclay Chemicals (UK) Ltd	07917	
C	Cerone	A H Marks & Co Ltd	00462	20/04/2001
C	Cerone	Rhone-Poulenc Agriculture	06185	
C	Charger	Rhone-Poulenc Agriculture	08827	
C	Ethrel C	Hortichem Ltd	06995	
C	EXP03149D	Rhone-Poulenc Agriculture	08828	
C	Stantion	Rhone-Poulenc Agriculture	06205	31/01/2001
C	Stefes Stance	Stefes Plant Protection Ltd	06125	29/02/2000
C	Unistar Ethephon 480	Unistar Ltd	06282	20/04/2001

C   These products are "approved for agricultural use". For further details refer to page vii.
A   These products are approved for use in or near water. For further details refer to page vii.

Product Name	Marketing Company	Reg. No.	Expiry Date

70 2-Chloroethylphosphonic acid + Chlormequat

C	Barclay Banshee XL	Barclay Chemicals (UK) Ltd	09201
C	Greencrop Tycoon	Greencrop Technology Ltd	09571
C	Nomad	Rhone-Poulenc Agriculture	07888
C	Strate	Rhone-Poulenc Agriculture	07430
C	Sypex	BASF plc	04650
C	Terpal C	BASF plc	07062
C	Upgrade	Rhone-Poulenc Agriculture	06177

71 2-Chloroethylphosphonic acid + Chlormequat + Imazaquin

C	Ice	Cyanamid Agriculture Ltd	08970
C	Satellite	Cyanamid Agriculture Ltd	08969

72 2-Chloroethylphosphonic acid + Chlormequat + Mepiquat

C	Cyclade	BASF plc	08958

73 2-Chloroethylphosphonic acid + Mepiquat

C	Barclay Banshee	Barclay Chemicals (UK) Ltd	08175	
C	CleanCrop Fonic M	United Agri Products	09553	
C	Standon Mepiquat Plus	Standon Chemicals Ltd	09373	
C	Stefes Mepiquat	Stefes Plant Protection Ltd	06970	20/04/2001
C	Sypex M	BASF plc	06810	30/06/2001
C	Terpal	BASF plc	02103	
C	Terpal	Clayton Plant Protection (UK) Ltd	07626	
C	Terpitz	Me^2 Crop Protection Ltd	09634	

74 Chlorotoluron

C	Alpha Chlorotoluron 500	Makhteshim-Agan (UK) Ltd	04848	
C	Atol	Ashlade Formulations Ltd	07347	
C	Chlortoluron 500	Pan Britannica Industries Ltd	03686	
C	Clayton Chloron	Clayton Plant Protection (UK) Ltd	08148	
C	Dicurane	Novartis Crop Protection UK Ltd	08403	
C	Dicurane 500 SC	Ciba Agriculture	05836	31/08/2000
C	Dicurane 700 SC	Ciba Agriculture	04859	31/08/2000
C	Lentipur CL 500	Nufarm UK Ltd	08743	
C	Luxan Chlorotoluron 500 Flowable	Luxan (UK) Ltd	09165	
C	MSS Chlortoluron 500	Mirfield Sales Services Ltd	07871	
C	Portman Chlortoluron	Portman Agrochemicals Ltd	03068	
C	Stefes Toluron	Stefes Plant Protection Ltd	05779	
C	Talisman	Farmers Crop Chemicals Ltd	03109	
C	Tolugan 700	Makhteshim-Agan (UK) Ltd	08064	
C	Top Farm Toluron 500	Top Farm Formulations Ltd	05986	
C	Tripart Culmus	Tripart Farm Chemicals Ltd	06619	

C These products are "approved for agricultural use". For further details refer to page vii.
A These products are approved for use in or near water. For further details refer to page vii.

Product Name	Marketing Company	Reg. No.	Expiry Date

75 Chlorotoluron + Bifenox

C Dicurane Duo 446 SC	Novartis Crop Protection UK Ltd	08404	
C Dicurane Duo 446SC	Ciba Agriculture	05839	31/08/2000

76 Chlorotoluron + Isoproturon

C Tolugan Extra	Makhteshim-Agan (UK) Ltd	09393	

77 Chlorotoluron + Pendimethalin

C Totem	Cyanamid Agriculture Ltd	04670	

78 Chlorphonium

C Phosfleur 1.5	Perifleur Products Ltd	05750	
C Phosfleur 10% Liquid	Perifleur Products Ltd	01587	

79 Chlorpropham

C Atlas CIPC 40	Atlas Crop Protection Ltd	07710	
C Atlas CIPC 40	Atlas Interlates Ltd	03049	30/04/2000
C Atlas Herbon Pabrac	Atlas Crop Protection Ltd	07714	
C Atlas Herbon Pabrac	Atlas Interlates Ltd	03997	
BL500	Wheatley Chemical Co Ltd	00279	
C Croptex Pewter	Hortichem Ltd	02507	
Luxan Gro-Stop 300 EC	Luxan (UK) Ltd	08602	
Luxan Gro-Stop Basis	Luxan (UK) Ltd	08601	
Luxan Gro-Stop Fog	Luxan (UK) Ltd	09388	
Luxan Gro-Stop HN	Luxan (UK) Ltd	07689	
C MSS CIPC 40 EC	Mirfield Sales Services Ltd	01403	
MSS CIPC 50 LF	Mirfield Sales Services Ltd	03285	
MSS CIPC 50M	Mirfield Sales Services Ltd	01404	
MSS CIPC 5G	Mirfield Sales Services Ltd	01402	
C MTM CIPC 40	MTM Agrochemicals Ltd	05895	
Standon CIPC 300 HN	Standon Chemicals Ltd	09187	
C Triherbicide CIPC	Atochem Agri BV	06426	
C Triherbicide CIPC	Elf Atochem Agri SA	06874	
Warefog 25	Mirfield Sales Services Ltd	06776	

80 Chlorpropham + Fenuron

C Atlas Red	Atlas Crop Protection Ltd	07724	
C Atlas Red	Atlas Interlates Ltd	03091	
C Croptex Chrome	Hortichem Ltd	02415	

81 Chlorpropham + Linuron

C Profalon	AgrEvo UK Crop Protection Ltd	07331	31/01/2001
C Profalon	Hortichem Ltd	08900	

C These products are "approved for agricultural use". For further details refer to page vii.
A These products are approved for use in or near water. For further details refer to page vii.

Product Name	Marketing Company	Reg. No.	Expiry Date

82 Chlorpropham + Pentanochlor

C	Atlas Brown	Atlas Crop Protection Ltd	07703	
C	Atlas Brown	Atlas Interlates Ltd	03835	30/04/2000

83 Chlorthal-dimethyl

C	Dacthal W75	Hortichem Ltd	05500	
C	Dacthal W75	ISK Biosciences Ltd	05556	

84 Chlorthal-dimethyl + Propachlor

C	Decimate	ISK Biosciences Ltd	05626	

85 Cinidon-ethyl

C	Lotus	BASF plc	09231	

86 Clodinafop-propargyl

C	Landgold Clodinafop	Landgold & Co Ltd	09017	31/05/2002
C	Standon Clodinafop 240	Standon Chemicals Ltd	09343	31/05/2002
C	Topik	Ciba Agriculture	07763	28/02/2001
C	Topik	Novartis Crop Protection UK Ltd	08461	
C	Viscount	Ciba Agriculture	08127	28/02/2001
C	Viscount	Novartis Crop Protection UK Ltd	08462	

87 Clodinafop-propargyl + Diflufenican

C	Amazon	Novartis Crop Protection UK Ltd	08128	
C	Amazon TP	Ciba Agriculture	07681	30/09/2001
C	Amazon TP	Novartis Crop Protection UK Ltd	08384	30/06/2000
C	Invest	Rhone-Poulenc Agriculture	07682	30/06/2000
C	Lucifer	Rhone-Poulenc Agriculture	08129	

88 Clodinafop-propargyl + Diflufenican + Isoproturon

C	SL 520	Ciba Agriculture	07687	28/02/2001
C	SL 520	Novartis Crop Protection UK Ltd	08447	30/10/2001
C	Unite	Rhone-Poulenc Agriculture	07686	30/10/2001

89 Clodinafop-propargyl + Trifluralin

C	Hawk	Ciba Agriculture	08030	30/04/2000
C	Hawk	Novartis Crop Protection UK Ltd	08417	
C	Reserve	Ciba Agriculture	08169	28/02/2001
C	Reserve	Novartis Crop Protection UK Ltd	08435	

90 Clopyralid

C	Barclay Karaoke	Barclay Chemicals (UK) Ltd	08185	
C	Cliophar	Chimac-Agriphar SA	09430	
C	Dow Shield	Dow AgroSciences Ltd	05578	
C	Lontrel 100	Dow AgroSciences Ltd	05737	

C   These products are "approved for agricultural use". For further details refer to page vii.
A   These products are approved for use in or near water. For further details refer to page vii.

Product Name	Marketing Company	Reg. No.	Expiry Date

PSD HERBICIDES

91 Clopyralid + Benazolin
C Benazalox AgrEvo UK Ltd 07246

92 Clopyralid + Bromoxynil
C Vindex Dow AgroSciences Ltd 05470

93 Clopyralid + Bromoxynil + Fluroxypyr + Ioxynil
C Crusader S Dow AgroSciences Ltd 05174

94 Clopyralid + 2,4-D + MCPA
C Lonpar Dow AgroSciences Ltd 08686

95 Clopyralid + Dichlorprop + MCPA
C Lontrel Plus Dow AgroSciences Ltd 05269

96 Clopyralid + Diflufenican + MCPA
C Spearhead Rhone-Poulenc Amenity 07342

97 Clopyralid + Fluroxypyr + Ioxynil
C Hotspur Dow AgroSciences Ltd 05185

98 Clopyralid + Fluroxypyr + MCPA
C Greenor Rigby Taylor Ltd 07848
C Greenor Lo Dow AgroSciences Ltd 09585
C Greenor Lo Rigby Taylor Ltd 09586

99 Clopyralid + Fluroxypyr + Triclopyr
C Pastor Dow AgroSciences Ltd 07440

100 Clopyralid + Ioxynil
C Escort Dow AgroSciences Ltd 05466

101 Clopyralid + Propyzamide
C Matrikerb Pan Britannica Industries Ltd 01308
C Matrikerb pbi Agrochemicals Ltd 09604
C Matrikerb Rohm & Haas (UK) Ltd 02443

102 Clopyralid + Triclopyr
C Grazon 90 Dow AgroSciences Ltd 05456

103 Copper hydroxide
C Spin-Out Fargro Ltd 07610

C These products are "approved for agricultural use". For further details refer to page vii.
A These products are approved for use in or near water. For further details refer to page vii.

Product Name	Marketing Company	Reg. No.	Expiry Date

104 Cyanazine

C Barclay Canter	Barclay Chemicals (UK) Ltd	07530	
C Fortrol	Cyanamid Agriculture Ltd	07009	
C Reply	Cyanamid Agriculture Ltd	07084	
C Standon Cyanazine 50	Standon Chemicals Ltd	09230	

105 Cyanazine + Bentazone + 2,4-DB

C Topshot	Cyanamid Agriculture Ltd	07178	28/02/2000

106 Cyanazine + Bromoxynil

C Greencrop Tassle	Greencrop Technology Ltd	09659	

107 Cyanazine + Fluroxypyr

C Spitfire	DowElanco Ltd	05747	31/01/2000

108 Cyanazine + Mecoprop

C Cleaval	Cyanamid Agriculture Ltd	07142	30/04/2000
C FCC Topcorn Extra	Cyanamid Agriculture Ltd	07143	30/04/2000
C MTM Eminent	Cyanamid Agriculture Ltd	08053	30/04/2000
C MTM Eminent	MTM Agrochemicals Ltd	07144	30/04/2000

109 Cyanazine + Pendimethalin

C Activus	Cyanamid Agriculture Ltd	09174	
C Bullet	Cyanamid Agriculture Ltd	08049	

110 Cyanazine + Terbuthylazine

C Angle	Ciba Agriculture	08066	31/08/2000
C Angle	Novartis Crop Protection UK Ltd	08385	

111 Cycloxydim

C Landgold Cycloxydim	Landgold & Co Ltd	06269	
C Laser	BASF plc	05251	
C Standon Cycloxydim	Standon Chemicals Ltd	08830	
C Stratos	BASF plc	06891	

112 Cypermethrin

C MCC 25 EC	Chiltern Farm Chemicals Ltd	09115	

113 2,4-D

C Agrichem 2,4-D	Agrichem Ltd	04098	
C A Agricorn 2,4-D	Farmers Crop Chemicals Ltd	07349	
C Agricorn D	Farmers Crop Chemicals Ltd	00056	
C Agricorn D11	Nufarm UK Ltd	09415	

C These products are "approved for agricultural use". For further details refer to page vii.
A These products are approved for use in or near water. For further details refer to page vii.

Product Name	Marketing Company	Reg. No.	Expiry Date

113 2,4-D—continued

C A	Atlas 2,4-D	Atlas Crop Protection Ltd	07699	
C	Barclay Haybob	Barclay Chemicals (UK) Ltd	07597	29/02/2000
C	Barclay Haybob II	Barclay Plant Protection	08532	
C	Campbell's Dioweed 50	J D Campbell & Sons Ltd	00401	31/12/2000
C	Depitox	Nufarm UK Ltd	08000	
C	Dicotox Extra	Rhone-Poulenc Amenity	05330	
C	Dioweed 50	United Phosphorus Ltd	08050	
C A	Dormone	Rhone-Poulenc Amenity	05412	
C	For-ester	Synchemicals Ltd	00914	
C	GroWell 2,4-D Amine	GroWell Ltd	08750	
C	Headland Staff	Headland Agrochemicals Ltd	07189	
C	Herboxone	A H Marks & Co Ltd	09692	
C	Herboxone 60	A H Marks & Co Ltd	09693	
C	HY-D	Agrichem Ltd	06278	
C	Luxan 2,4-D	Luxan (UK) Ltd	09379	
C	Marks 2,4-D-A	A H Marks & Co Ltd	01282	
C A	MSS 2,4-D Amine	Mirfield Sales Services Ltd	01391	
C	MSS 2,4-D Ester	Mirfield Sales Services Ltd	01393	
C	Nomix 2,4-D Herbicide	Nomix-Chipman Ltd	06394	
C	Palormone D	Universal Crop Protection Ltd	01534	
C	Ritefeed 2,4-D Amine	Ritefeed Ltd	05309	
C	Silvapron D	BP Chemicals Ltd	01935	
C	Syford	Synchemicals Ltd	02062	

114 2,4-D + Amitrole + Diuron

C	Trik	Mirfield Sales Services Ltd	07853

115 2,4-D + Clopyralid + MCPA

C	Lonpar	Dow AgroSciences Ltd	08686

116 2,4-D + Dicamba

C	Lawn Builder Plus Weed Control	Scotts UK Professional	08499	
C	Longlife Plus	Miracle Professional	07867	31/07/2000
C	Longlife Plus	Zeneca Professional Products	06886	31/07/2000
C	New Estermone	Vitax Ltd	06336	

117 2,4-D + Dicamba + Ferrous sulphate

C	Longlife Renovator	Miracle Professional	07924	30/09/2000

118 2,4-D + Dicamba + Mecoprop

C	New Formulation Weed and Brushkiller	Vitax Ltd	07072	30/04/2000

C   These products are "approved for agricultural use". For further details refer to page vii.
A   These products are approved for use in or near water. For further details refer to page vii.

Product Name	Marketing Company	Reg. No.	Expiry Date

119 2,4-D + Dicamba + Triclopyr

C Broadshot	Cyanamid Agriculture Ltd	07141	30/09/2000
C Broadsword	United Phosphorus Ltd	09140	
C Nufarm Nu-Shot	Nufarm UK Ltd	09139	
C Terbel Triple	Chimac-Agriphar SA	09138	

120 2,4-D + Dichlorprop

C Marks Polytox-M	A H Marks & Co Ltd	01301	
C Polymone X	Universal Crop Protection Ltd	01613	

121 2,4-D + Dichlorprop + MCPA + Mecoprop

C Camppex	United Phosphorus Ltd	08266	31/10/2001

122 2,4-D + Dichlorprop + MCPA + Mecoprop-P

C UPL Camppex	United Phosphorus Ltd	09121	

123 2,4-D + Mecoprop

C BH CMPP/2,4-D	Rhone-Poulenc Amenity	05393	30/04/2000
C CDA Supertox 30	Rhone-Poulenc Amenity	04664	30/04/2000
C Mascot Selective Weedkiller	Rigby Taylor Ltd	03423	30/04/2000
C Nomix Turf Selective Herbicide	Nomix-Chipman Ltd	06777	31/03/2000
C Nomix Turf Selective LC Herbicide	Nomix-Chipman Ltd	05973	31/03/2000
C Selective Weedkiller	Yule Catto Consumer Chemicals Ltd	06579	30/04/2000
C Supertox 30	Rhone-Poulenc Amenity	05340	30/06/2001

124 2,4-D + Mecoprop-P

C Mascot Selective-P	Rigby Taylor Ltd	06105	
C Supertox 30	Rhone-Poulenc Amenity	09102	
C Sydex	Vitax Ltd	06412	

125 2,4-D + Picloram

C Atladox HI	Nomix-Chipman Ltd	05559	
C Tordon 101	Dow AgroSciences Ltd	05816	

126 Daminozide

C B-Nine	Hortichem Ltd	07844	
C B-Nine	Uniroyal Chemical Ltd	04468	
C Dazide	Fine Agrochemicals Ltd	02691	

127 2,4-DB

C DB Straight	United Phosphorus Ltd	07523	
C Marks 2,4-DB	A H Marks & Co Ltd	01283	

C These products are "approved for agricultural use". For further details refer to page vii.
A These products are approved for use in or near water. For further details refer to page vii.

Product Name	Marketing Company	Reg. No.	Expiry Date

128 2,4-DB + Benazolin + MCPA

C	Legumex Extra	AgrEvo UK Ltd	08676	
C	Legumex Extra	AgrEvo UK Ltd	07901	31/05/2001
C	Master Sward	Stefes Plant Protection Ltd	07883	
C	Setter 33	AgrEvo UK Ltd	07282	
C	Setter 33	Dow AgroSciences Ltd	05623	
C	Stefes Clover Ley	Stefes Plant Protection Ltd	07842	
C	Stefes Legumex Extra	Stefes Plant Protection Ltd	07841	

129 2,4-DB + Bentazone + Cyanazine

C	Topshot	Cyanamid Agriculture Ltd	07178	28/02/2000

130 2,4-DB + Linuron + MCPA

C	Alistell	Zeneca Crop Protection	06515	

131 2,4-DB + MCPA

C	Agrichem DB Plus	Agrichem Ltd	00044	
C	Marks 2,4-DB Extra	A H Marks & Co Ltd	01284	
C	MSS 2,4-DB + MCPA	Mirfield Sales Services Ltd	01392	
C	Redlegor	United Phosphorus Ltd	07519	

132 Desmedipham + Ethofumesate + Phenmedipham

C	Aventis Betanal Progress OF	AgrEvo UK Ltd	09722	
C	Betanal Congress	AgrEvo UK Ltd	07534	31/10/2000
C	Betanal Progress	AgrEvo UK Ltd	07533	31/05/2001
C	Betanal Progress OF	AgrEvo UK Ltd	07629	
C	Betanal Ultima	AgrEvo UK Ltd	07535	31/10/2000

133 Desmedipham + Phenmedipham

C	Betanal Compact	AgrEvo UK Ltd	07247	
C	Betanal Quorum	AgrEvo UK Ltd	07252	31/05/2001
C	Betanal Rostrum	AgrEvo UK Ltd	07253	31/05/2001

134 Desmetryn

C	Semeron	Novartis Crop Protection UK Ltd	08441	
C	Semeron 25 WP	Ciba Agriculture	01916	31/08/2000

135 Dicamba

C	Cadence	Barclay Chemicals (UK) Ltd	09578	
C	Cadence	Novartis Crop Protection UK Ltd	08796	
C	San 845H	Novartis Crop Protection UK Ltd	08797	

136 Dicamba + Bifenox

C	Quickstep	Novartis Crop Protection UK Ltd	08557	31/10/2000

C   These products are "approved for agricultural use". For further details refer to page vii.
A   These products are approved for use in or near water. For further details refer to page vii.

Product Name	Marketing Company	Reg. No.	Expiry Date

136 Dicamba + Bifenox—continued

C	Quickstep	Sandoz Agro Ltd	07389	31/08/2000
C	RP 283	Rhone-Poulenc Agriculture	06178	31/10/2000

137 Dicamba + 2,4-D

C	Lawn Builder Plus Weed Control	Scotts UK Professional	08499	
C	Longlife Plus	Miracle Professional	07867	31/07/2000
C	Longlife Plus	Zeneca Professional Products	06886	31/07/2000
C	New Estermone	Vitax Ltd	06336	

138 Dicamba + 2,4-D + Ferrous sulphate

C	Longlife Renovator	Miracle Professional	07924	30/09/2000

139 Dicamba + 2,4-D + Mecoprop

C	New Formulation Weed and Brushkiller	Vitax Ltd	07072	30/04/2000

140 Dicamba + 2,4-D + Triclopyr

C	Broadshot	Cyanamid Agriculture Ltd	07141	30/09/2000
C	Broadsword	United Phosphorus Ltd	09140	
C	Nufarm Nu-Shot	Nufarm UK Ltd	09139	
C	Terbel Triple	Chimac-Agriphar SA	09138	

141 Dicamba + Dichlorprop + Ferrous sulphate + MCPA

C	Longlife Renovator 2	Miracle Professional	08379	
C	Renovator 2	The Scotts Company (UK) Limited	09272	

142 Dicamba + Dichlorprop + MCPA

C	Cleanrun 2	The Scotts Company (UK) Limited	09271	
C	Intrepid	Miracle Professional	07819	
C	Intrepid	The Scotts Company (UK) Limited	09266	
C	Longlife Clearun 2	Miracle Professional	07851	

143 Dicamba + Maleic hydrazide + MCPA

C	Mazide Selective	Vitax Ltd	05753	

144 Dicamba + MCPA

C	Banvel M	Novartis Crop Protection UK Ltd	08469	
C	Banvel M	Sandoz Agro Ltd	05794	31/08/2000

145 Dicamba + MCPA + Mecoprop

C	Barclay Hat-Trick	Barclay Chemicals (UK) Ltd	07579	

C These products are "approved for agricultural use". For further details refer to page vii.
A These products are approved for use in or near water. For further details refer to page vii.

Product Name	Marketing Company	Reg. No.	Expiry Date

145 Dicamba + MCPA + Mecoprop—continued

C	Campbell's Field Marshall	United Phosphorus Ltd	08594	30/04/2000
C	Campbell's Field Marshall	J D Campbell & Sons Ltd	00406	30/06/2000
C	Campbell's Grassland Herbicide	MTM Agrochemicals Ltd	06157	30/04/2000
C	Fisons Tritox	Fisons Plc	03013	30/04/2000
C	Headland Relay	Headland Agrochemicals Ltd	07684	30/06/2000
C	Headland Relay	WBC Technology Ltd	03778	30/06/2000
C	Herrisol	Bayer plc	01048	30/04/2000
C	Hyprone	Agrichem (International) Ltd	01093	30/06/2001
C	Hysward	Agrichem (International) Ltd	01096	31/08/2001
C	Pasturol	Farmers Crop Chemicals Ltd	01545	30/06/2000
C	Premier Triple Selective	Amenity Land Services Limited	08529	31/05/2001
C	Stefes Banlene Plus	Stefes Plant Protection Ltd	07688	30/04/2000
C	Stefes Docklene	Stefes Plant Protection Ltd	07816	30/04/2000
C	Stefes Swelland	Stefes Plant Protection Ltd	08188	30/04/2000
C	Tritox	Levington Horticulture Ltd	07502	30/04/2000

146 Dicamba + MCPA + Mecoprop-P

C	ALS Premier Selective Plus	Amenity Land Services Limited	08940	
C	Banlene Super	AgrEvo UK Crop Protection Ltd	07168	
C	Field Marshal	MTM Agrochemicals Ltd	06077	31/03/2001
C	Field Marshal	United Phosphorus Ltd	08956	
C	Headland Relay P	Headland Agrochemicals Ltd	08580	
C	Headland Relay Turf	Headland Agrochemicals Ltd	08935	
C	Herrisol New	Bayer plc	07166	
C	Hyprone P	Agrichem (International) Ltd	09125	
C	Hysward P	Agrichem (International) Ltd	09052	
C	Mascot Super Selective-P	Rigby Taylor Ltd	06106	
C	MSS Mircam Plus	Mirfield Sales Services Ltd	01416	
C	MTM Grassland Herbicide	MTM Agrochemicals Ltd	06089	28/02/2001
C	Pasturol D	Farmers Crop Chemicals Ltd	07033	
C	Pasturol Plus	Farmers Crop Chemicals Ltd	08581	
C	Pasturol-P	Farmers Crop Chemicals Ltd	09099	
C	Premier Triple Selective P	Amenity Land Services Limited	09053	
C	Quadrangle Quadban	Quadrangle Agrochemicals	07090	
C	Stefes Banlene Super	Stefes Plant Protection Ltd	07661	
C	Stefes Docklene Super	Stefes Plant Protection Ltd	07696	
C	Stefes Swelland Super	Stefes Plant Protection Ltd	08189	
C	Tribute	Nomix-Chipman Ltd	06921	
C	Tribute Plus	Nomix-Chipman Ltd	09493	
C	Tritox	The Scotts Company (UK) Limited	07764	
C	UPL Grassland Herbicide	United Phosphorus Ltd	08934	

C   These products are "approved for agricultural use". For further details refer to page vii.
A   These products are approved for use in or near water. For further details refer to page vii.

Product Name	Marketing Company	Reg. No.	Expiry Date

147 Dicamba + Mecoprop

C	Banvel P	Nufarm UK Ltd	08346	30/04/2000
C	Condox	Zeneca Crop Protection	06519	31/05/2002
C	Hyban	Agrichem Ltd	01084	30/06/2001
C	Hygrass	Agrichem (International) Ltd	01090	30/06/2001
C	Pasturemaster D	Stefes Plant Protection Ltd	08481	28/02/2000
C	Pasturemaster D	Stefes Plant Protection Ltd	07954	30/04/2000

148 Dicamba + Mecoprop + Triclopyr

C	Fettel	Zeneca Crop Protection	06399	30/04/2000

149 Dicamba + Mecoprop-P

C	Condox	Zeneca Crop Protection	09429	
C	Di-Farmon R	Headland Agrochemicals Ltd	08472	
C	Di-Farmon R	Headland Agrochemicals Ltd	07482	30/04/2000
C	Foundation	Novartis Crop Protection UK Ltd	08475	
C	Foundation	Sandoz Agro Ltd	07465	31/08/2000
C	Hyban P	Agrichem (International) Ltd	09129	
C	Hygrass P	Agrichem (International) Ltd	09130	
C	MSS Mircam	Mirfield Sales Services Ltd	01415	
C	Optica Forte	A H Marks & Co Ltd	07432	

150 Dicamba + Paclobutrazol

C	Holdfast D	Imperial Chemical Industries Plc	05056	30/10/2001
C	Holdfast D	Miracle Professional	07864	31/03/2001

151 Dicamba + Triasulfuron

C	Banvel T	Novartis Crop Protection UK Ltd	08470	
C	Banvel T	Sandoz Agro Ltd	06497	31/08/2000
C	Framolene	Ciba Agriculture	06495	31/08/2000
C	Framolene	Novartis Crop Protection UK Ltd	08408	

152 Dichlobenil

C A	Casoron G	Miracle Professional	07926	
C A	Casoron G	Nomix-Chipman Ltd	09023	
C A	Casoron G	Rigby Taylor Ltd	09326	
C A	Casoron G	Uniroyal Chemical Ltd	09022	
C A	Casoron G	Zeneca Crop Protection	08065	
C	Casoron G4	ICI Agrochemicals	02406	31/05/2001
C	Casoron G4	Miracle Professional	07927	
C	Casoron G4	Uniroyal Chemical Ltd	09215	
C A	Casoron GSR	Imperial Chemical Industries Plc	00451	31/05/2001
C A	Casoron GSR	Miracle Professional	07925	
C A	Luxan Dichlobenil Granules	Luxan (UK) Ltd	09250	

C These products are "approved for agricultural use". For further details refer to page vii.
A These products are approved for use in or near water. For further details refer to page vii.

Product Name	Marketing Company	Reg. No.	Expiry Date

152 Dichlobenil—continued

C	Prefix D	Cyanamid Agriculture Ltd	07013	30/06/2001
C A	Sierraron G	The Scotts Company (UK) Limited	09675	
C A	Sierraron G	The Scotts Company (UK) Limited	09263	
C	Standon Dichlobenil 6G	Standon Chemicals Ltd	08874	

153 Dichlorophen

C	50-50 Liquid Mosskiller	Vitax Ltd	07191	
C	Enforcer	Miracle Professional	07866	30/06/2002
C	Enforcer	The Scotts Company (UK) Limited	09288	
C	Enforcer	Zeneca Professional Products	07079	31/05/2001
C	Mascot Mosskiller	Rigby Taylor Ltd	02439	
C	Mossicide	Pan Britannica Industries Ltd	08183	
C	Mossicide	pbi Agrochemicals Ltd	09606	
C	Nomix-Chipman Mosskiller	Nomix-Chipman Ltd	06271	
C	Panacide M	Coalite Chemicals	05611	
C	Panacide Technical Solution	Coalite Chemicals	05612	
C	Ritefeed Dichlorophen	Ritefeed Ltd	05265	
C	Super Mosstox	Rhone-Poulenc Amenity	05339	

154 Dichlorophen + Ferrous sulphate

C	Aitkens Lawn Sand Plus	RCC Agro	04542
C	SHL Lawn Sand Plus	Sinclair Horticulture & Leisure Ltd	04439

155 Dichlorprop

C	Marks Polytox-K	A H Marks & Co Ltd	01300
C	MSS 2,4-DP	Mirfield Sales Services Ltd	01394
C	Redipon	United Phosphorus Ltd	07524

156 Dichlorprop + Bromoxynil + Ioxynil + MCPA

C	Atlas Minerva	Atlas Interlates Ltd	03046

157 Dichlorprop + Clopyralid + MCPA

C	Lontrel Plus	Dow AgroSciences Ltd	05269

158 Dichlorprop + 2,4-D

C	Marks Polytox-M	A H Marks & Co Ltd	01301
C	Polymone X	Universal Crop Protection Ltd	01613

159 Dichlorprop + 2,4-D + MCPA + Mecoprop

C	Camppex	United Phosphorus Ltd	08266	31/10/2001

C These products are "approved for agricultural use". For further details refer to page vii.
A These products are approved for use in or near water. For further details refer to page vii.

Product Name	Marketing Company	Reg. No.	Expiry Date

160 Dichlorprop + 2,4-D + MCPA + Mecoprop-P

C	UPL Camppex	United Phosphorus Ltd	09121	

161 Dichlorprop + Dicamba + Ferrous sulphate + MCPA

C	Longlife Renovator 2	Miracle Professional	08379	
C	Renovator 2	The Scotts Company (UK) Limited	09272	

162 Dichlorprop + Dicamba + MCPA

C	Cleanrun 2	The Scotts Company (UK) Limited	09271	
C	Intrepid	Miracle Professional	07819	
C	Intrepid	The Scotts Company (UK) Limited	09266	
C	Longlife Clearun 2	Miracle Professional	07851	

163 Dichlorprop + Ferrous sulphate + MCPA

C	SHL Turf Feed and Weed and Mosskiller	William Sinclair Horticulture Ltd	04438	

164 Dichlorprop + MCPA

C	MSS 2,4-DP + MCPA	Mirfield Sales Services Ltd	01396	
C	Redipon Extra	United Phosphorus Ltd	07518	
C	Seritox 50	Rhone-Poulenc Agriculture	06170	
C	Seritox Turf	Rhone-Poulenc Environmental Products	06802	
C	SHL Turf Feed & Weed	William Sinclair Horticulture Ltd	04437	

165 Dichlorprop-P

C	Optica DP	A H Marks & Co Ltd	07818	

166 Dichlorprop-P + Bentazone

C	Quitt SL	BASF plc	08108	

167 Dichlorprop-P + MCPA + Mecoprop-P

C	Hymec Trio	Agrichem (International) Ltd	09764	
C	Optica Trio	A H Marks & Co Ltd	09747	

168 Diclofop-methyl

C	Hoegrass	AgrEvo UK Ltd	07323	31/07/2001
C	Landgold Diclofop	Landgold & Co Ltd	06359	29/02/2000

169 Diclofop-methyl + Fenoxaprop-P-ethyl

C	Corniche	AgrEvo UK Ltd	08947	
C	Tigress	AgrEvo UK Ltd	07337	31/12/2000
C	Tigress Ultra	AgrEvo UK Ltd	08946	

C These products are "approved for agricultural use". For further details refer to page vii.
A These products are approved for use in or near water. For further details refer to page vii.

Product Name	Marketing Company	Reg. No.	Expiry Date

170 Difenzoquat

C	Avenge 2	Cyanamid Agriculture Ltd	03241	

171 Diflufenican + Bromoxynil + Ioxynil

C	Capture	Rhone-Poulenc Agriculture	06881	

172 Diflufenican + Carbetamide + Oxadiazon

C	Helmsman	Rhone-Poulenc Amenity	09696	

173 Diflufenican + Clodinafop-propargyl

C	Amazon	Novartis Crop Protection UK Ltd	08128	
C	Amazon TP	Ciba Agriculture	07681	30/09/2001
C	Amazon TP	Novartis Crop Protection UK Ltd	08384	30/06/2000
C	Invest	Rhone-Poulenc Agriculture	07682	30/06/2000
C	Lucifer	Rhone-Poulenc Agriculture	08129	

174 Diflufenican + Clodinafop-propargyl + Isoproturon

C	SL 520	Ciba Agriculture	07687	28/02/2001
C	SL 520	Novartis Crop Protection UK Ltd	08447	30/10/2001
C	Unite	Rhone-Poulenc Agriculture	07686	30/10/2001

175 Diflufenican + Clopyralid + MCPA

C	Spearhead	Rhone-Poulenc Amenity	07342	

176 Diflufenican + Flurtamone

C	Bacara	Rhone-Poulenc Agriculture	08323	
C	Bonanza	Rhone-Poulenc Agriculture	08324	

177 Diflufenican + Flurtamone + Isoproturon

C	Buzzard	Rhone-Poulenc Agriculture	08322	
C	EXP 31165	Rhone-Poulenc Agriculture	08611	
C	Ingot	Rhone-Poulenc Agriculture	08610	
C	Roulette	Rhone-Poulenc Agriculture	08321	

178 Diflufenican + Isoproturon

C	Amulet	Rhone-Poulenc Agriculture	07529	31/07/2000
C	Barclay Slogan	Barclay Chemicals (UK) Ltd	09016	
C	Barracuda	Rhone-Poulenc Agriculture	08116	31/07/2000
C	Clayton Fenican 625	Clayton Plant Protection (UK) Ltd	09395	
C	Clayton Fenican-IPU	Clayton Plant Protection (UK) Ltd	06759	29/02/2000
C	Cougar	Rhone-Poulenc Ireland Ltd	06611	30/09/2001
C	DFF+IPU WDG	Rhone-Poulenc Agriculture	09700	
C	Gavel	Rhone-Poulenc Agriculture	08135	

C   These products are "approved for agricultural use". For further details refer to page vii.
A   These products are approved for use in or near water. For further details refer to page vii.

Product Name	Marketing Company	Reg. No.	Expiry Date

178 Diflufenican + Isoproturon—continued

C	Grenadier	Rhone-Poulenc Agriculture	08136	
C	Javelin	Rhone-Poulenc Agriculture	06192	
C	Javelin Gold	Rhone-Poulenc Agriculture	06200	
C	Landgold DFF 625	Landgold & Co Ltd	06274	
C	Landgold FF550	Landgold & Co Ltd	06053	29/02/2000
C	Oyster	Rhone-Poulenc Agriculture	07345	31/07/2000
C	Panther	Rhone-Poulenc Agriculture	06491	
C	Shire	Rhone-Poulenc Agriculture	08117	
C	Standon DFF-IPU	Standon Chemicals Ltd	08730	31/07/2001
C	Standon Diflufenican IPU	Standon Chemicals Ltd	09175	
C	Tolkan Turbo	Rhone-Poulenc Agriculture	06795	
C	Zodiac TX	Rhone-Poulenc Agriculture	06775	31/07/2000

179 Diflufenican + Terbuthylazine

C	Bolero	Ciba Agriculture	07436	31/07/2001
C	Bolero	Novartis Crop Protection UK Ltd	08392	

180 Diflufenican + Trifluralin

C	Ardent	Rhone-Poulenc Agriculture	06203	

181 Diquat

C	Barclay Desiquat	Barclay Chemicals (UK) Ltd	04969	29/02/2000
C	Barclay Desiquat	Barclay Chemicals (UK) Ltd	09063	31/07/2002
C	Clayton Diquat	Clayton Plant Protection (UK) Ltd	07650	31/07/2002
C	Clayton Quatrow	Clayton Plant Protection (UK) Ltd	07175	
C A	CleanCrop Diquat	United Agri Products	09687	
C	Diquat 200	Top Farm Formulations Ltd	05141	29/02/2000
C	Greencrop Boomerang	Greencrop Technology Ltd	09563	31/07/2002
C	Landgold Diquat	Landgold & Co Ltd	05974	30/09/2001
C	Landgold Diquat	Landgold & Co Ltd	09020	31/07/2002
C A	Levi	P.S.I. Phoenix Scientific Innovation UK Ltd	07845	31/07/2002
C A	Midstream	Imperial Chemical Industries Plc	01348	31/05/2001
C A	Midstream	Miracle Professional	07739	
C A	Midstream	The Scotts Company (UK) Limited	09267	
C A	Midstream	Zeneca Professional Products	06824	31/05/2001
C A	Reglone	Zeneca Crop Protection	09646	
C A	Reglone	Zeneca Crop Protection	06703	30/06/2002
C	Standon Diquat	Standon Chemicals Ltd	05587	31/07/2002
C	Stefes Diquat	Stefes Plant Protection Ltd	05493	29/02/2000
C	Top Farm Diquat-200	Top Farm Formulations Ltd	05552	29/02/2000

C These products are "approved for agricultural use". For further details refer to page vii.
A These products are approved for use in or near water. For further details refer to page vii.

Product Name		Marketing Company	Reg. No.	Expiry Date

182 Diquat + Paraquat

C	Parable	Zeneca Crop Protection	06692	31/07/2000
C	PDQ	Zeneca Crop Protection	06518	
C	Speedway 2	Miracle Professional	08164	31/10/2001
C	Speedway 2	The Scotts Company (UK) Limited	09273	

183 Diquat + Paraquat + Simazine

C	Pathclear S	Miracle Garden Care Ltd	08991	
C	Pathclear S	The Scotts Company (UK) Limited	09269	

184 Diuron

C	Atlas Diuron	Atlas Crop Protection Ltd	08214	
C	Chipko Diuron 80	Chipman Ltd	00497	
C	Chipman Diuron 80	Nomix-Chipman Ltd	08054	
C	Chipman Diuron Flowable	Nomix-Chipman Ltd	05701	
C	Diuron 50 FL	Staveley Chemicals Ltd	02814	
C	Diuron 80 WG	Rhone-Poulenc Amenity	08509	
C	Diuron 80 WP	Rhone-Poulenc Amenity	05199	
C	Diuron 80% WP	Staveley Chemicals Ltd	00730	
C	Epsilon	Rhone-Poulenc Amenity	08508	
C	Freeway	Rhone-Poulenc Amenity	06047	
C	Karmex	Du Pont (UK) Ltd	01128	28/02/2002
C	Karmex	Griffin (Europe) SA	09475	
C	MSS Diuron 500 FL	Mirfield Sales Services Ltd	08171	
C	MSS Diuron 50FL	Mirfield Sales Services Ltd	07160	
C	Rescind	Rhone-Poulenc Amenity	08036	
C	Sanuron	Dow AgroSciences Ltd	09236	
C	Sanuron	Sanachem International Ltd	08213	30/09/2001
C	Unicrop Flowable Diuron	Universal Crop Protection Ltd	02270	
C	UPL Diuron 80	United Phosphorus Ltd	07619	

185 Diuron + Amitrole + Bromacil

C	BR Destral	Rhone-Poulenc Amenity	05184	31/12/2001

186 Diuron + Amitrole + 2,4-D

C	Trik	Mirfield Sales Services Ltd	07853	

187 Diuron + Bromacil

C	Borocil K	Rhone-Poulenc Amenity	05183	

188 Diuron + Glyphosate

C	NX 3083	Nomix-Chipman Ltd	08712	
C	Total	Nomix-Chipman Ltd	07932	
C	Touche	Nomix-Chipman Ltd	07913	
C	Touche	Nomix-Chipman Ltd	06797	
C	Xanadu	Rhone-Poulenc Amenity	09228	

C These products are "approved for agricultural use". For further details refer to page vii.
A These products are approved for use in or near water. For further details refer to page vii.

Product Name	Marketing Company	Reg. No.	Expiry Date

189 Diuron + Paraquat

C	Dexuron	Nomix-Chipman Ltd	07169	

190 Ethofumesate

C	Agriguard Ethofumesate Flo	Tronsan Ltd	09478	
C	Atlas Thor	Atlas Crop Protection Ltd	07732	
C	Atlas Thor	Atlas Interlates Ltd	06966	
C	Barclay Keeper	Barclay Chemicals (UK) Ltd	08835	
C	Barclay Keeper 200	Barclay Chemicals (UK) Ltd	05266	
C	Barclay Keeper 500 FL	Barclay Chemicals (UK) Ltd	09438	
C	Ethosan	KVK Agro A/S	08555	30/06/2001
C	Ethosan	Pan Britannica Industries Ltd	07346	31/05/2000
C	Griffin Ethofumesate 200	Griffin (Bermuda) SA	08818	28/02/2002
C	Kubist	Griffin (Europe) Marketing NV (UK Branch)	09454	
C	Landgold Ethofumesate 200	Landgold & Co Ltd	08980	
C	Landgold Ethofumesate 200	Landgold & Co Ltd	06257	29/02/2000
C	MSS Primasan	KVK Agro A/S	08556	30/06/2001
C	MSS Primasan	Mirfield Sales Services Ltd	08048	31/05/2000
C	MSS Thor	Mirfield Sales Services Ltd	08817	
C	Nortron	AgrEvo UK Ltd	07266	
C	Nortron Flo	AgrEvo UK Ltd	08154	
C	Salute	United Phosphorus Ltd	07660	
C	Standon Ethofumesate 200	Standon Chemicals Ltd	09360	
C	Standon Ethofumesate 200	Standon Chemicals Ltd	08726	
C	Standon Ethofumesate 200	Standon Chemicals Ltd	05668	29/02/2000
C	Stefes Fumat	Stefes Plant Protection Ltd	05525	29/02/2000
C	Stefes Fumat 2	Stefes Plant Protection Ltd	07856	
C	Top Farm Ethofumesate 200	Top Farm Formulations Ltd	05987	29/02/2000

191 Ethofumesate + Bromoxynil + Ioxynil

C	Leyclene	AgrEvo UK Ltd	07263	
C	Stefes Leyclene	Stefes Plant Protection Ltd	08173	

192 Ethofumesate + Chloridazon

C	Gremlin	Sipcam UK Ltd	09468	
C	Magnum	BASF plc	08635	
C	Magnum	BASF plc	01237	31/07/2000
C	Spectron	AgrEvo UK Ltd	07284	
C	Stefes Spectron	Stefes UK Ltd	09337	

193 Ethofumesate + Desmedipham + Phenmedipham

C	Aventis Betanal Progress OF	AgrEvo UK Ltd	09722	
C	Betanal Congress	AgrEvo UK Ltd	07534	31/10/2000

C These products are "approved for agricultural use". For further details refer to page vii.
A These products are approved for use in or near water. For further details refer to page vii.

Product Name	Marketing Company	Reg. No.	Expiry Date

193 Ethofumesate + Desmedipham + Phenmedipham—continued

C	Betanal Progress	AgrEvo UK Ltd	07533	31/05/2001
C	Betanal Progress OF	AgrEvo UK Ltd	07629	
C	Betanal Ultima	AgrEvo UK Ltd	07535	31/10/2000

194 Ethofumesate + Metamitron

C	Stefes Duplex	Stefes UK Ltd	09252	
C	Stefes ST/STE	Stefes Plant Protection Ltd	09111	

195 Ethofumesate + Metamitron + Phenmedipham

C	Bayer UK 407 WG	Bayer UK Ltd	06313	28/02/2000
C	Bayer UK 540	Bayer plc	07399	
C	Betanal Trio OF	AgrEvo UK Crop Protection Ltd	08680	31/05/2001
C	Betanal Trio WG	AgrEvo UK Ltd	07537	
C	CX 171	AgrEvo UK Ltd	06803	28/02/2000
C	Goltix Triple	Bayer UK Ltd	06314	28/02/2000

196 Ethofumesate + Phenmedipham

C	Barclay Goalpost	Barclay Chemicals (UK) Ltd	09497	
C	Barclay Goalpost	Barclay Chemicals (UK) Ltd	09192	
C	Barclay Goalpost	Barclay Chemicals (UK) Ltd	08016	30/04/2000
C	Barclay Goldpost	Barclay Chemicals Manufacturing Ltd	08964	31/03/2002
C	Betanal Montage	AgrEvo UK Ltd	07250	28/02/2001
C	Betanal Quadrant	AgrEvo UK Crop Protection Ltd	07251	31/07/2000
C	Betanal Tandem	AgrEvo UK Ltd	07254	
C	Betosip Combi	Sipcam UK Ltd	08630	
C	Stefes Medimat	Stefes Plant Protection Ltd	07886	30/04/2001
C	Stefes Medimat 2	Stefes Plant Protection Ltd	07577	
C	Stefes Tandem	Stefes Plant Protection Ltd	08906	
C	Twin	Feinchemie (UK) Ltd	09784	
C	Twin	Feinchemie Schwebda GmbH	07374	31/10/2002

197 Fenoxaprop-ethyl

C	Cheetah R	AgrEvo UK Ltd	07308	31/07/2000
C	Clayton Fencer	Clayton Plant Protection (UK) Ltd	06525	29/02/2000
C	Landgold Fenoxaprop	Landgold & Co Ltd	06352	29/02/2000

198 Fenoxaprop-P-ethyl

C	Aventis Cheetah Super	AgrEvo UK Ltd	09730	
C	Cheetah Super	AgrEvo UK Ltd	08723	
C	Cheetah Super	AgrEvo UK Ltd	07309	30/09/2000
C	FPE 120	AgrEvo UK Ltd	08742	31/05/2001
C	FPE 55	AgrEvo UK Ltd	08725	31/05/2001
C	Gazelle	AgrEvo UK Ltd	08741	31/05/2001

C These products are "approved for agricultural use". For further details refer to page vii.
A These products are approved for use in or near water. For further details refer to page vii.

115

Product Name	Marketing Company	Reg. No.	Expiry Date

198 Fenoxaprop-P-ethyl—continued

C Triumph	AgrEvo UK Ltd	08740	
C Wildcat	AgrEvo UK Ltd	08724	31/05/2001
C Wildcat	AgrEvo UK Ltd	07340	30/09/2000

199 Fenoxaprop-P-ethyl + Diclofop-methyl

C Corniche	AgrEvo UK Ltd	08947	
C Tigress	AgrEvo UK Ltd	07337	31/12/2000
C Tigress Ultra	AgrEvo UK Ltd	08946	

200 Fenoxaprop-P-ethyl + Isoproturon

C Puma X	AgrEvo UK Ltd	08779	
C Puma X	AgrEvo UK Ltd	07332	31/12/2001
C Whip	AgrEvo UK Ltd	08780	31/05/2001
C Whip X	AgrEvo UK Ltd	07339	31/12/2001

201 Fenuron + Chlorpropham

C Atlas Red	Atlas Crop Protection Ltd	07724	
C Atlas Red	Atlas Interlates Ltd	03091	
C Croptex Chrome	Hortichem Ltd	02415	

202 Ferrous sulphate

C Aitkens Lawn Sand	RCC Agro	05253	
C Elliott's Lawn Sand	Thomas Elliott Ltd	04860	
C Elliott's Mosskiller	Thomas Elliott Ltd	04909	
C Fisons Greenmaster Autumn	Fisons Plc	03211	
C Fisons Greenmaster Mosskiller	Fisons Plc	00881	
C Greenmaster Autumn	The Scotts Company (UK) Limited	07508	
C Greenmaster Mosskiller	The Scotts Company (UK) Limited	07509	
C Maxicrop Mosskiller and Conditioner	Maxicrop International Ltd	04635	
C Moss Control Plus Lawn Fertilizer	Miracle Garden Care Ltd	07912	
C SHL Lawn Sand	Sinclair Horticulture & Leisure Ltd	05254	
C Taylor's Lawn Sand	Rigby Taylor Ltd	04451	
C Vitagrow Lawn Sand	Vitagrow (Fertilisers) Ltd	05097	
C Vitax Microgran 2	Vitax Ltd	04541	
C Vitax Turf Tonic	Vitax Ltd	04354	

203 Ferrous sulphate + 2,4-D + Dicamba

C Longlife Renovator	Miracle Professional	07924	30/09/2000

204 Ferrous sulphate + Dicamba + Dichlorprop + MCPA

C Longlife Renovator 2	Miracle Professional	08379	
C Renovator 2	The Scotts Company (UK) Limited	09272	

C   These products are "approved for agricultural use". For further details refer to page vii.
A   These products are approved for use in or near water. For further details refer to page vii.

Product Name	Marketing Company	Reg. No.	Expiry Date

205 Ferrous sulphate + Dichlorophen

C	Aitkens Lawn Sand Plus	RCC Agro	04542	
C	SHL Lawn Sand Plus	Sinclair Horticulture & Leisure Ltd	04439	

206 Ferrous sulphate + Dichlorprop + MCPA

C	SHL Turf Feed and Weed and Mosskiller	William Sinclair Horticulture Ltd	04438	

207 Flamprop-M-ethyl

C	Stefes Flamprop	Stefes Plant Protection Ltd	05789	29/02/2000

208 Flamprop-M-isopropyl

C	Commando	Cyanamid Agriculture Ltd	07005	

209 Fluazifop-P-butyl

C	Barclay Winner	Barclay Chemicals (UK) Ltd	06986	
C	Citadel	Zeneca Crop Protection	06762	30/09/2002
C	Corral	Zeneca Crop Protection	06647	30/09/2002
C	Fusilade 250 EW	Zeneca Crop Protection	06531	
C	Fusilade 5	Zeneca Crop Protection	06669	30/09/2002
C	Landgold Fluazifop-P	Landgold & Co Ltd	06020	29/02/2000
C	PP 007	Zeneca Crop Protection	06533	
C	Standon Fluazifop-P	Standon Chemicals Ltd	06060	29/02/2000
C	Stefes Slayer	Stefes Plant Protection Ltd	06277	29/02/2000
C	Wizzard	Zeneca Crop Protection	06521	

210 Fluoroglycofen-ethyl + Isoproturon

C	Compete Forte	AgrEvo UK Crop Protection Ltd	08708	
C	Competitor	AgrEvo UK Crop Protection Ltd	07310	
C	Effect	AgrEvo UK Ltd	07319	31/05/2001
C	Stefes Competitor	Stefes Plant Protection Ltd	08753	

211 Flupyrsulfuron-methyl

C	DPX-KE459 50DF	Du Pont (UK) Ltd	08539	
C	DPX-KE459 WSB	Du Pont (UK) Ltd	08540	
C	Lexus 50 DF	Du Pont (UK) Ltd	09026	

212 Flupyrsulfuron-methyl + Metsulfuron-methyl

C	Lexus XPE	Du Pont (UK) Ltd	08541	
C	Lexus XPE WSB	Du Pont (UK) Ltd	08542	

213 Flupyrsulfuron-methyl + Carfentrazone-ethyl

C	Lexus Class	Du Pont (UK) Ltd	08636	
C	Lexus Class WSB	Du Pont (UK) Ltd	08637	

C These products are "approved for agricultural use". For further details refer to page vii.
A These products are approved for use in or near water. For further details refer to page vii.

Product Name	Marketing Company	Reg. No.	Expiry Date

214 Flupyrsulfuron-methyl + Metsulfuron-methyl

C	Standon Flupyrsulfuron MM	Standon Chemicals Ltd	09098

215 Flupyrsulfuron-methyl + Thifensulfuron-methyl

C	Lexus Millenium	Du Pont (UK) Ltd	09206
C	Lexus Millenium WSB	Du Pont (UK) Ltd	09207

216 Fluroxypyr

C	Agriguard Fluroxypyr	Agriguard Ltd	09298	
C	Barclay Hurler	Barclay Chemicals (UK) Ltd	08791	
C	Barclay Hurler	Barclay Chemicals (UK) Ltd	06952	29/02/2000
C	Clayton Fluroxypyr	Clayton Plant Protection (UK) Ltd	06356	
C	DAS 143	Dow AgroSciences Ltd	08943	
C	Gala	Dow AgroSciences Ltd	09793	
C	Greencrop Reaper	Greencrop Technology Ltd	09359	
C	Landgold Fluroxypyr	Landgold & Co Ltd	06080	29/02/2000
C	Standon Fluroxypyr	Standon Chemicals Ltd	08923	
C	Standon Fluroxypyr	Standon Chemicals Ltd	07055	29/02/2000
C	Starane 2	Dow AgroSciences Ltd	09405	
C	Starane 2	Dow AgroSciences Ltd	05496	
C	Stefes Fluroxypyr	Stefes Plant Protection Ltd	05977	29/02/2000
C	Tomahawk	Makhteshim-Agan (UK) Ltd	09249	
C	Tomahawk 2000	Makhteshim-Agan (UK) Ltd	09307	

217 Fluroxypyr + Bromoxynil

C	Sickle	Dow AgroSciences Ltd	05187

218 Fluroxypyr + Bromoxynil + Clopyralid + Ioxynil

C	Crusader S	Dow AgroSciences Ltd	05174

219 Fluroxypyr + Bromoxynil + Ioxynil

C	Advance	Dow AgroSciences Ltd	05173

220 Fluroxypyr + Clopyralid + Ioxynil

C	Hotspur	Dow AgroSciences Ltd	05185

221 Fluroxypyr + Clopyralid + MCPA

C	Greenor	Rigby Taylor Ltd	07848
C	Greenor Lo	Dow AgroSciences Ltd	09585
C	Greenor Lo	Rigby Taylor Ltd	09586

C   These products are "approved for agricultural use". For further details refer to page vii.
A   These products are approved for use in or near water. For further details refer to page vii.

Product Name	Marketing Company	Reg. No.	Expiry Date

222 Fluroxypyr + Clopyralid + Triclopyr

C	Pastor	Dow AgroSciences Ltd	07440	

223 Fluroxypyr + Cyanazine

C	Spitfire	DowElanco Ltd	05747	31/01/2000

224 Fluroxypyr + Ioxynil

C	Stexal	Dow AgroSciences Ltd	05188	31/07/2000

225 Fluroxypyr + Mecoprop-P

C	Bastion T	Rigby Taylor Ltd	06011	

226 Fluroxypyr + Metosulam

C	EF 1166	Dow AgroSciences Ltd	07966	

227 Fluroxypyr + Thifensulfuron-methyl + Tribenuron-methyl

C	DP 353	Du Pont (UK) Ltd	09626	
C	Starane Super	Dow AgroSciences Ltd	09625	

228 Fluroxypyr + Triclopyr

C	Doxstar	Dow AgroSciences Ltd	06050	
C	Evade	Dow AgroSciences Ltd	08071	

229 Flurtamone + Diflufenican

C	Bacara	Rhone-Poulenc Agriculture	08323	
C	Bonanza	Rhone-Poulenc Agriculture	08324	

230 Flurtamone + Diflufenican + Isoproturon

C	Buzzard	Rhone-Poulenc Agriculture	08322	
C	EXP 31165	Rhone-Poulenc Agriculture	08611	
C	Ingot	Rhone-Poulenc Agriculture	08610	
C	Roulette	Rhone-Poulenc Agriculture	08321	

231 Fomesafen

C	Flex	Novartis Crop Protection UK Ltd	08885	
C	Flex	Zeneca Crop Protection	07457	30/04/2000

232 Fomesafen + Terbutryn

C	Reflex T	Novartis Crop Protection UK Ltd	08884	
C	Reflex T	Zeneca Crop Protection	07423	30/04/2000
C	Reflex T	Zeneca Crop Protection	08035	31/05/2000

C   These products are "approved for agricultural use". For further details refer to page vii.
A   These products are approved for use in or near water. For further details refer to page vii.

Product Name	Marketing Company	Reg. No.	Expiry Date

233 Fosamine-ammonium
C A Krenite	Du Pont (UK) Ltd	01165	

234 Gibberellins
C Berelex	Whyte Chemicals Ltd	08903	
C Berelex	Zeneca Crop Protection	06637	31/01/2001
C Novagib	Fine Agrochemicals Ltd	08954	
C Regulex	Zeneca Crop Protection	07997	

235 Glufosinate-ammonium
C Aventis Challenge	AgrEvo UK Ltd	09721	
C Challenge	AgrEvo UK Ltd	07306	
C Challenge 2	AgrEvo UK Ltd	07307	
C Challenge 60	AgrEvo UK Ltd	08236	
C Dash	Nomix-Chipman Ltd	05177	
C Harvest	AgrEvo UK Ltd	07321	
C Headland Sword	Headland Agrochemicals Ltd	07676	
C Nomix Touchweed	Nomix-Chipman Ltd	09596	

236 Glyphosate
C Agriguard Glyphosate 360	Agriguard Ltd	09184	
C Amega Pro TMF	Nufarm UK Ltd	09630	
C Apache	Zeneca Crop Protection	06748	
C Azural	Monsanto Plc	09582	
C Barbarian	Barclay Chemicals (UK) Ltd	07980	
C Barbarian	Barclay Chemicals (UK) Ltd	07625	
C Barclay Cleanup	Barclay Plant Protection	09179	
C Barclay Dart	Barclay Chemicals (UK) Ltd	05129	
C Barclay Gallup	Barclay Chemicals (UK) Ltd	05161	
C A Barclay Gallup 360	Barclay Plant Protection	09127	
C A Barclay Gallup Amenity	Barclay Chemicals (UK) Ltd	06753	
C Barclay Garryowen	Barclay Chemicals (UK) Ltd	08599	
C Barclay Trustee	Barclay Plant Protection	09177	
C A Buggy SG	Sipcam UK Ltd	08573	
C A Cardel Egret	Cardel	09703	
C Cardel Glyphosate	Cardel	09581	
C CDA Spasor	Rhone-Poulenc Amenity	06414	30/11/2000
C CDA Vanquish	Rhone-Poulenc Amenity	08577	
C Clarion	Zeneca Crop Protection	08272	
C Clayton Glyphosate	Clayton Plant Protection (UK) Ltd	06608	
C Clayton Glyphosate - 360	Clayton Plant Protection (UK) Ltd	05316	29/02/2000
C Clayton Rhizeup	Clayton Plant Protection Ltd	08920	
C A Clayton Swath	Clayton Plant Protection (UK) Ltd	06715	
C Clean-Up-360	Top Farm Formulations Ltd	05076	

C These products are "approved for agricultural use". For further details refer to page vii.
A These products are approved for use in or near water. For further details refer to page vii.

Product Name	Marketing Company	Reg. No.	Expiry Date

236 Glyphosate—continued

C A	Clinic	Nufarm UK Ltd	09378	
C A	Clinic	Nufarm UK Ltd	08579	
C	Clinic Pro TMF	Nufarm UK Ltd	09443	30/06/2002
C A	Danagri Glyphosate 360	Danagri ApS	06955	31/05/2000
C	Economix	Nomix-Chipman Ltd	08008	
C	Glyfonex 480	Danagri ApS	07464	
C A	Glyfos	Cheminova Agro (UK) Ltd	07109	
C A	Glyfos 480	Cheminova Agro (UK) Ltd	08014	
C A	Glyfos Proactive	Nomix-Chipman Ltd	07800	
C	Glyfosate-360	Top Farm Formulations Ltd	05319	
C A	Glyper	Pan Britannica Industries Ltd	07968	
C A	Glyphogan	Makhteshim-Agan (UK) Ltd	05784	
C	Glyphosate 120B	Monsanto Plc	08292	
C A	Glyphosate 360	Applyworld Limited	09151	
C A	Glyphosate 360	Danagri ApS	09233	
C A	Glyphosate 360	Danagri ApS	08568	31/07/2001
C A	Glyphosate 360A	Danagri ApS	09234	
C A	Glyphosate 360B	Danagri ApS	09235	
C A	Glyphosate Biactive	Monsanto Plc	08307	
C	Gorgon	Portman Agrochemicals Ltd	09182	
C	Greencrop Gypsy	Greencrop Technology Ltd	09432	
C A	Helosate	Helm AG	06499	
C	Hilite	Nomix-Chipman Ltd	06261	
C	Hilite 120	Nomix-Chipman Ltd	08600	30/11/2000
C	IT Glyphosate	IT Agro Ltd	07212	
C	Landgold Glyphosate 360	Landgold & Co Ltd	05929	
C A	Marnoch Glyphosate	Marnoch Ventures Limited	09744	
C	Mogul	Monsanto Plc	07076	
C	MON 240	Monsanto Plc	04538	
C A	MON 44068 Pro	Monsanto Plc	06815	
C A	MON 52276	Monsanto Plc	06949	
C A	Monty	Quadrangle Agrochemicals	07796	
C A	MSS Glyfield	Mirfield Sales Services Ltd	08009	
C	Muster	Zeneca Crop Protection	06685	
C	Muster LA	Zeneca Professional Products	05762	
C	Nomix G	Nomix-Chipman Ltd	08781	
C	Nomix Nova	Nomix-Chipman Ltd	08647	
C	Nomix Revenge	Nomix-Chipman Ltd	09483	
C	Nomix Supernova	Nomix-Chipman Ltd	09473	
C	Poise	Universal Crop Protection Ltd	08276	
C	Portman Glider	Portman Agrochemicals Ltd	04695	
C	Portman Glyphosate	Portman Agrochemicals Ltd	05891	
C	Portman Glyphosate 360	Portman Agrochemicals Ltd	04699	
C	Portman Glyphosate 480	Portman Agrochemicals Ltd	07194	31/08/2001

C   These products are "approved for agricultural use". For further details refer to page vii.
A   These products are approved for use in or near water. For further details refer to page vii.

Product Name	Marketing Company	Reg. No.	Expiry Date

236 Glyphosate—continued

	Product Name	Marketing Company	Reg. No.	Expiry Date
C A	Rival	Monsanto Plc	09220	
C A	Roundup	Monsanto Plc	01828	
C	Roundup 2000	Monsanto Plc	08069	
C A	Roundup A	Monsanto Plc	08375	31/08/2001
C	Roundup Accuply	Accuply Ltd	09544	
C A	Roundup Amenity	Monsanto Plc	08721	
C A	Roundup Biactive	Monsanto Plc	06941	
C A	Roundup Biactive Dry	Monsanto Plc	06942	
C	Roundup CF	Monsanto Plc	09547	
C	Roundup Four 80	Monsanto Plc	03176	
C	Roundup GT	Monsanto Plc	08068	
C	Roundup MX	Monsanto Plc	09631	
C A	Roundup Pro	Monsanto Plc	04146	
C A	Roundup Pro Biactive	Monsanto Plc	06954	
C	Roundup Rapide	Monsanto Plc	08067	
C	Smart 360	Mastra Europe Limited	09476	
C A	Spasor	Rhone-Poulenc Amenity	07211	
C A	Spasor Biactive	Rhone-Poulenc Amenity	07651	
C	Stacato	Sipcam UK Ltd	05892	
C	Stampede	Zeneca Crop Protection	06327	
C	Standon Glyphosate 360	Standon Chemicals Ltd	05582	
C	Stefes Complete	Stefes Plant Protection Ltd	06084	29/02/2000
C	Stefes Glyphosate	Stefes Plant Protection Ltd	05819	
C	Stefes Kickdown	Stefes Plant Protection Ltd	06329	
C	Stefes Kickdown 2	Stefes Plant Protection Ltd	06548	
C A	Stetson	Monsanto Plc	06956	31/03/2000
C	Sting	Monsanto Plc	02789	
C	Sting CT	Monsanto Plc	04754	
C	Sting Eco	Monsanto Plc	08291	
C	Stirrup	Chipman Ltd	04174	30/06/2000
C	Stirrup	Nomix-Chipman Ltd	06132	
C	Stride	Zeneca Crop Protection	08282	
C	Touchdown	Zeneca Crop Protection	06326	
C	Touchdown LA	Miracle Professional	07747	30/09/2001
C	Touchdown LA	The Scotts Company (UK) Limited	09270	
C	Touchdown LA	Zeneca Professional Products	06444	31/08/2000
C A	Typhoon 360	Feinchemie (UK) Ltd	09792	
C A	Typhoon 360	Feinchemie Schwebda GmbH	09322	31/10/2002
C	Unistar Glyfosate 360	Unistar Ltd	05928	
C	Unistar Glyphosate 360	Unistar Ltd	06332	

237 Glyphosate + Diuron

	Product Name	Marketing Company	Reg. No.	Expiry Date
C	NX 3083	Nomix-Chipman Ltd	08712	
C	Total	Nomix-Chipman Ltd	07932	

C   These products are "approved for agricultural use". For further details refer to page vii.
A   These products are approved for use in or near water. For further details refer to page vii.

Product Name	Marketing Company	Reg. No.	Expiry Date

237 Glyphosate + Diuron—continued

C	Touche	Nomix-Chipman Ltd	07913	
C	Touche	Nomix-Chipman Ltd	06797	
C	Xanadu	Rhone-Poulenc Amenity	09228	

238 Glyphosate + Oxadiazon

| C | Zapper | Rhone-Poulenc Amenity | 08605 | |

239 Imazamethabenz-methyl

| C | Assert | Cyanamid Agriculture Ltd | 08656 | |
| C | Dagger | Cyanamid Agriculture Ltd | 03737 | |

240 Imazapyr

C	Arsenal	Cyanamid Agriculture Ltd	04064	
C	Arsenal	Nomix-Chipman Ltd	05537	
C	Arsenal 50	Cyanamid Agriculture Ltd	04070	
C	Arsenal 50	Nomix-Chipman Ltd	05567	

241 Imazaquin + Chlormequat

C	Meteor	Cyanamid Agriculture Ltd	06505	
C	Standon Imazaquin 5C	Standon Chemicals Ltd	08813	
C	Upright	Cyanamid Agriculture Ltd	08290	

242 Imazaquin + 2-Chloroethylphosphonic acid + Chlormequat

| C | Ice | Cyanamid Agriculture Ltd | 08970 | |
| C | Satellite | Cyanamid Agriculture Ltd | 08969 | |

243 Imazethapyr

	Im-Pede	Cyanamid Agriculture Ltd	07602	25/07/2003
C	Im-Pede LC	Cyanamid Agriculture Ltd	07956	25/07/2003
C	Imp 3	Cyanamid Agriculture Ltd	07957	25/07/2003

244 Indol-3-ylacetic acid

C	Rhizopon A Powder	Fargro Ltd	07131	
C	Rhizopon A Powder	Rhizopon UK Limited	09087	
C	Rhizopon A Tablets	Fargro Ltd	07132	
C	Rhizopon A Tablets	Rhizopon UK Limited	09088	

245 Indol-3-ylbutyric-4 acid

C	Chryzoplus Grey 0.8%	Fargro Ltd	07984	
C	Chryzoplus Grey 0.8%	Rhizopon BV	09094	
C	Chryzopon Rose 0.1%	Fargro Ltd	07982	
C	Chryzopon Rose 0.1%	Rhizopon UK Limited	09092	

C   These products are "approved for agricultural use". For further details refer to page vii.
A   These products are approved for use in or near water. For further details refer to page vii.

Product Name	Marketing Company	Reg. No.	Expiry Date

245 Indol-3-ylbutyric-4 acid—continued

C	Chryzosan White 0.6%	Fargro Ltd	07983
C	Chryzosan White 0.6%	Rhizopon UK Limited	09093
C	Chryzotek Beige	Fargro Ltd	07125
C	Chryzotek Beige	Rhizopon UK Limited	09081
C	Chryzotop Green	Fargro Ltd	07129
C	Chryzotop Green	Rhizopon UK Limited	09085
C	Rhizopon AA Powder (0.5%)	Fargro Ltd	07126
C	Rhizopon AA Powder (1 %)	Fargro Ltd	07127
C	Rhizopon AA Powder (1 %)	Rhizopon UK Limited	09084
C	Rhizopon AA Powder 2%	Fargro Ltd	07128
C	Rhizopon AA Powder 2%	Rhizopon UK Limited	09083
C	Rhizopon AA Tablets	Fargro Ltd	07130
C	Rhizopon AA Tablets	Rhizopon UK Limited	09086
C	Seradix 1	Rhone-Poulenc Agriculture	06191
C	Seradix 2	Rhone-Poulenc Agriculture	06193
C	Seradix 3	Rhone-Poulenc Agriculture	06194

246 4-Indol-3-ylbutyric acid + 1-Naphthylacetic acid

C	Synergol	Hortichem Ltd	07386

247 Ioxynil

C	Actrilawn 10	Rhone-Poulenc Amenity	05247
C	Totril	Rhone-Poulenc Agriculture	06116

248 Ioxynil + Benazolin + Bromoxynil

C	Asset	AgrEvo UK Ltd	07243

249 Ioxynil + Bromoxynil

C	Alpha Briotril	Makhteshim-Agan (UK) Ltd	04876	
C	Alpha Briotril Plus 19/19	Makhteshim-Agan (UK) Ltd	04740	
C	Deloxil	AgrEvo UK Ltd	07313	
C	Deloxil	Rhone-Poulenc Agriculture	07405	
C	Mextrol-Biox	Nufarm UK Ltd	09470	
C	Oxytril CM	Rhone-Poulenc Agriculture	08667	
C	Oxytril CM	Rhone-Poulenc Agriculture	06201	31/08/2000
C	Percept	MTM Agrochemicals Ltd	05481	
C	Status	Rhone-Poulenc Agriculture	08668	
C	Status	Rhone-Poulenc Agriculture	06906	31/08/2000
C	Stellox 380 EC	Ciba Agriculture	04169	31/08/2000
C	Stellox 380 EC	Novartis Crop Protection UK Ltd	08451	

250 Ioxynil + Bromoxynil + Clopyralid + Fluroxypyr

C	Crusader S	Dow AgroSciences Ltd	05174

C   These products are "approved for agricultural use". For further details refer to page vii.
A   These products are approved for use in or near water. For further details refer to page vii.

Product Name	Marketing Company	Reg. No.	Expiry Date

251 Ioxynil + Bromoxynil + Dichlorprop + MCPA

C   Atlas Minerva   Atlas Interlates Ltd   03046

252 Ioxynil + Bromoxynil + Diflufenican

C   Capture   Rhone-Poulenc Agriculture   06881

253 Ioxynil + Bromoxynil + Ethofumesate

C   Leyclene   AgrEvo UK Ltd   07263
C   Stefes Leyclene   Stefes Plant Protection Ltd   08173

254 Ioxynil + Bromoxynil + Fluroxypyr

C   Advance   Dow AgroSciences Ltd   05173

255 Ioxynil + Bromoxynil + Isoproturon + Mecoprop

C   Terset   Rhone-Poulenc Agriculture   06171   30/04/2000

256 Ioxynil + Bromoxynil + Mecoprop

C   Swipe 560 EC   Ciba Agriculture   02057   30/04/2000

257 Ioxynil + Bromoxynil + Mecoprop-P

C   Swipe P   Ciba Agriculture   08150   31/12/2000
C   Swipe P   Novartis Crop Protection UK Ltd 08452

258 Ioxynil + Bromoxynil + Triasulfuron

C   Teal   Ciba Agriculture   06117   31/08/2000
C   Teal   Novartis Crop Protection UK Ltd 08453
C   Teal-M   Ciba Agriculture   08028   31/08/2000
C   Teal-M   Novartis Crop Protection UK Ltd 08454

259 Ioxynil + Bromoxynil + Trifluralin

C   Masterspray   Pan Britannica Industries Ltd 02971

260 Ioxynil + Clopyralid

C   Escort   Dow AgroSciences Ltd   05466

261 Ioxynil + Clopyralid + Fluroxypyr

C   Hotspur   Dow AgroSciences Ltd   05185

262 Ioxynil + Fluroxypyr

C   Stexal   Dow AgroSciences Ltd   05188   31/07/2000

C   These products are "approved for agricultural use". For further details refer to page vii.
A   These products are approved for use in or near water. For further details refer to page vii.

Product Name	Marketing Company	Reg. No.	Expiry Date

263 Isoproturon

	Product Name	Marketing Company	Reg. No.	Expiry Date
C	Agriguard Isoproturon	Tronsan Ltd	09769	
C	Aligran WDG	Portman Agrochemicals Ltd	08931	
C	Alpha Isoproturon 500	Makhteshim-Agan (UK) Ltd	05882	
C	Alpha Isoproturon 650	Makhteshim-Agan (UK) Ltd	07034	
C	Arelon	AgrEvo UK Crop Protection Ltd	06716	31/05/2001
C	Arelon 500	AgrEvo UK Ltd	08100	
C	Atlas Fieldguard	Atlas Crop Protection Ltd	08582	
C	Atlas Protall	Atlas Interlates Ltd	06340	
C	Atum	Rhone-Poulenc Agriculture	07379	
C	Atum WDG	Rhone-Poulenc Agriculture	07778	
C	Auger	Rhone-Poulenc Agriculture	06581	
C	Auger WDG	Rhone-Poulenc Agriculture	07777	31/03/2001
C	Barclay Guideline	Barclay Chemicals (UK) Ltd	06743	29/02/2000
C	Barclay Guideline 500	Barclay Chemicals (UK) Ltd	09257	
C	Bison	Gharda Chemicals Ltd	08699	
C	Clayton IPU	Clayton Plant Protection (UK) Ltd	07661	31/08/2001
C	Clayton Siptu 50 FL	Clayton Plant Protection (UK) Ltd	08751	
C	Clayton Siptu 500	Clayton Plant Protection (UK) Ltd	08735	
C	DAPT Isoproturon 500 FL	DAPT Agrochemicals Ltd	08092	
C	Hytane 500 SC	Ciba Agriculture	05848	31/08/2000
C	Hytane 500 SC	Novartis Crop Protection UK Ltd	08419	
C	I.P.U Sun	Phytorus UK Ltd	08798	
C	IPU 500	Rhone-Poulenc Agriculture	07380	
C	IPU 500 WDG	Rhone-Poulenc Agriculture	07780	31/03/2001
C	ISO 500	Quality Assured Marketing Ltd	08775	31/08/2001
C	Isoguard	Chiltern Farm Chemicals Ltd	08497	
C	Isoguard SVR	Cyanamid Agriculture Ltd	08696	
C	Isoproturon 500	AgrEvo UK Ltd	06718	
C	Isoproturon 553	AgrEvo UK Crop Protection Ltd	06719	31/05/2001
C	Isotop SC	Portman Agrochemicals Ltd	07663	
C	Ki-Hara IPU 500	Ki-Hara Chemicals Ltd	06102	29/02/2000
C	Landgold Isoproturon	Landgold & Co Ltd	06012	
C	Landgold Isoproturon FC	Landgold & Co Ltd	06021	29/02/2000
C	Landgold Isoproturon FL	Landgold & Co Ltd	06034	29/02/2000
C	Landgold Isoproturon SC	Landgold & Co Ltd	05989	29/02/2000
C	Luxan Isoproturon 500 Flowable	Luxan (UK) Ltd	07437	
C	MSS Iprofile	Mirfield Sales Services Ltd	06341	
C	Mysen	Portman Agrochemicals Ltd	07695	
C	Mysen WDG	Portman Agrochemicals Ltd	09141	
C	Portman Isotop	Portman Agrochemicals Ltd	03434	
C	Quintil 500	Phytorus UK Ltd	06800	
C	RP 500	Rhone-Poulenc Agriculture	07378	
C	RP 500 WDG	Rhone-Poulenc Agriculture	07779	31/03/2001

C These products are "approved for agricultural use". For further details refer to page vii.
A These products are approved for use in or near water. For further details refer to page vii.

Product Name	Marketing Company	Reg. No.	Expiry Date

263 Isoproturon—continued

C	Sabre	AgrEvo UK Crop Protection Ltd	06717	31/05/2001
C	Sabre 500	AgrEvo UK Ltd	08101	31/05/2001
C	Standon IPU	Standon Chemicals Ltd	08671	
C	Stefes IPU	Stefes Plant Protection Ltd	05776	
C	Stefes IPU 500	Stefes UK Ltd	08102	
C	Tolkan	Rhone-Poulenc Agriculture	07365	
C	Tolkan Liquid	Rhone-Poulenc Agriculture	06172	
C	Tolkan WDG	Rhone-Poulenc Agriculture	07776	31/03/2001
C	Top Farm IPU 500	Top Farm Formulations Ltd	05978	29/02/2000
C	Tripart Pugil	Tripart Farm Chemicals Ltd	06153	28/02/2000
C	Tripart Pugil 5	Tripart Farm Chemicals Ltd	08606	

264 Isoproturon + Bifenox

C	Banco	Portman Agrochemicals Ltd	09027	
C	RP 4169	Rhone-Poulenc Agriculture	05801	

265 Isoproturon + Bromoxynil + Ioxynil + Mecoprop

C	Terset	Rhone-Poulenc Agriculture	06171	30/04/2000

266 Isoproturon + Carfentrazone-ethyl

C	Affinity	FMC Europe NV	08639	

267 Isoproturon + Chlorotoluron

C	Tolugan Extra	Makhteshim-Agan (UK) Ltd	09393	

268 Isoproturon + Clodinafop-propargyl + Diflufenican

C	SL 520	Ciba Agriculture	07687	28/02/2001
C	SL 520	Novartis Crop Protection UK Ltd	08447	30/10/2001
C	Unite	Rhone-Poulenc Agriculture	07686	30/10/2001

269 Isoproturon + Diflufenican

C	Amulet	Rhone-Poulenc Agriculture	07529	31/07/2000
C	Barclay Slogan	Barclay Chemicals (UK) Ltd	09016	
C	Barracuda	Rhone-Poulenc Agriculture	08116	31/07/2000
C	Clayton Fenican 625	Clayton Plant Protection (UK) Ltd	09395	
C	Clayton Fenican-IPU	Clayton Plant Protection (UK) Ltd	06759	29/02/2000
C	Cougar	Rhone-Poulenc Ireland Ltd	06611	30/09/2001
C	DFF+IPU WDG	Rhone-Poulenc Agriculture	09700	
C	Gavel	Rhone-Poulenc Agriculture	08135	
C	Grenadier	Rhone-Poulenc Agriculture	08136	
C	Javelin	Rhone-Poulenc Agriculture	06192	
C	Javelin Gold	Rhone-Poulenc Agriculture	06200	
C	Landgold DFF 625	Landgold & Co Ltd	06274	

C These products are "approved for agricultural use". For further details refer to page vii.
A These products are approved for use in or near water. For further details refer to page vii.

Product Name	Marketing Company	Reg. No.	Expiry Date

269 Isoproturon + Diflufenican—continued

C Landgold FF550	Landgold & Co Ltd	06053	29/02/2000
C Oyster	Rhone-Poulenc Agriculture	07345	31/07/2000
C Panther	Rhone-Poulenc Agriculture	06491	
C Shire	Rhone-Poulenc Agriculture	08117	
C Standon DFF-IPU	Standon Chemicals Ltd	08730	31/07/2001
C Standon Diflufenican IPU	Standon Chemicals Ltd	09175	
C Tolkan Turbo	Rhone-Poulenc Agriculture	06795	
C Zodiac TX	Rhone-Poulenc Agriculture	06775	31/07/2000

270 Isoproturon + Diflufenican + Flurtamone

C Buzzard	Rhone-Poulenc Agriculture	08322	
C EXP 31165	Rhone-Poulenc Agriculture	08611	
C Ingot	Rhone-Poulenc Agriculture	08610	
C Roulette	Rhone-Poulenc Agriculture	08321	

271 Isoproturon + Fenoxaprop-P-ethyl

C Puma X	AgrEvo UK Ltd	08779	
C Puma X	AgrEvo UK Ltd	07332	31/12/2001
C Whip	AgrEvo UK Ltd	08780	31/05/2001
C Whip X	AgrEvo UK Ltd	07339	31/12/2001

272 Isoproturon + Fluoroglycofen-ethyl

C Compete Forte	AgrEvo UK Crop Protection Ltd	08708	
C Competitor	AgrEvo UK Crop Protection Ltd	07310	
C Effect	AgrEvo UK Ltd	07319	31/05/2001
C Stefes Competitor	Stefes Plant Protection Ltd	08753	

273 Isoproturon + Isoxaben

C Fanfare 469 SC	Ciba Agriculture	05841	31/08/2000
C IPSO	Dow AgroSciences Ltd	05736	

274 Isoproturon + Pendimethalin

C Artillery	Cyanamid Agriculture Ltd	07768	
C Artillery SC	Cyanamid Agriculture Ltd	08504	
C Clayton Pendalin-IPU 37	Clayton Plant Protection (UK) Ltd	09154	
C Clayton Pendalin-IPU 47	Clayton Plant Protection (UK) Ltd	09170	
C Encore	Cyanamid Agriculture Ltd	04737	
C Encore SC	Cyanamid Agriculture Ltd	08502	
C Jolt	Cyanamid Agriculture Ltd	05488	
C Jolt SC	Cyanamid Agriculture Ltd	08503	
C Orient	Cyanamid Agriculture Ltd	07798	
C Orient SC	Cyanamid Agriculture Ltd	08501	
C Trump	Cyanamid Agriculture Ltd	03687	
C Trump SC	Cyanamid Agriculture Ltd	08500	

C These products are "approved for agricultural use". For further details refer to page vii.
A These products are approved for use in or near water. For further details refer to page vii.

Product Name	Marketing Company	Reg. No.	Expiry Date

275 Isoproturon + Simazine

C	Alpha Protugan Plus	Makhteshim-Agan (UK) Ltd	08799	
C	Harlequin 500 SC	Ciba Agriculture	05847	31/08/2000
C	Harlequin 500 SC	Makhteshim-Agan (UK) Ltd	09779	
C	Harlequin 500 SC	Novartis Crop Protection UK Ltd	08416	30/09/2002

276 Isoproturon + Trifluralin

| C | Autumn Kite | AgrEvo UK Ltd | 07119 | |
| C | Stefes Union | Stefes Plant Protection Ltd | 08146 | |

277 Isoxaben

C	Flexidor	Dow AgroSciences Ltd	05121	
C	Flexidor 125	Dow AgroSciences Ltd	05104	
C	Gallery 125	Rigby Taylor Ltd	06889	
C	Knot Out	Vitax Ltd	05163	

278 Isoxaben + Isoproturon

| C | Fanfare 469 SC | Ciba Agriculture | 05841 | 31/08/2000 |
| C | IPSO | Dow AgroSciences Ltd | 05736 | |

279 Isoxaben + Methabenzthiazuron

| C | Glytex | Bayer plc | 04230 | |

280 Isoxaben + Terbuthylazine

| C | Skirmish | Ciba Agriculture | 08079 | 31/08/2000 |
| C | Skirmish | Novartis Crop Protection UK Ltd | 08444 | |

281 Isoxaben + Trifluralin

| C | Axit GR | Dow AgroSciences Ltd | 08892 | |
| C | Premiere Granules | Dow AgroSciences Ltd | 07987 | |

282 Lenacil

C	Agricola Lenacil FL	Agricola Ltd	09481	
C	Clayton Lenacil 80W	Clayton Plant Protection (UK) Ltd	09488	
C	Clayton Lenacil 80W	Clayton Plant Protection (UK) Ltd	07074	29/02/2000
C	Stefes Lenacil	Stefes Plant Protection Ltd	07103	29/02/2000
C	Venzar Flowable	Du Pont (UK) Ltd	06907	
C	Vizor	Du Pont (UK) Ltd	06572	

283 Lenacil + Chloridazon

C	Advizor	Du Pont (UK) Ltd	06571	
C	Advizor S	Du Pont (UK) Ltd	09799	
C	Advizor S	Du Pont (UK) Ltd	06960	
C	Varmint	Pan Britannica Industries Ltd	01561	

C   These products are "approved for agricultural use". For further details refer to page vii.
A   These products are approved for use in or near water. For further details refer to page vii.

Product Name	Marketing Company	Reg. No.	Expiry Date

284 Lenacil + Linuron

C Lanslide	Pan Britannica Industries Ltd	01184	

285 Lenacil + Phenmedipham

C DUK-880	Du Pont (UK) Ltd	04121	
C Stefes 880	Stefes Plant Protection Ltd	08914	
C Stefes Excel	Stefes Plant Protection Ltd	08913	

286 Linuron

C Afalon	AgrEvo UK Crop Protection Ltd	07299	
C Afalon	AgrEvo UK Ltd	08186	
C Alpha Linuron 50 SC	Makhteshim-Agan (UK) Ltd	06967	
C Alpha Linuron 50 WP	Makhteshim-Agan (UK) Ltd	04870	
C Ashlade Linuron FL	Ashlade Formulations Ltd	06221	
C DAPT Linuron 50 SC	DAPT Agrochemicals Ltd	08058	
C Linurex 50SC	Makhteshim-Agan (UK) Ltd	07950	
C Linuron Flowable	Pan Britannica Industries Ltd	02965	
C Linuron Flowable	pbi Agrochemicals Ltd	09602	
C Liquid Linuron	Pan Britannica Industries Ltd	01556	
C Masterspray	pbi Agrochemicals Ltd	09603	
C MSS Linuron 50	Mirfield Sales Services Ltd	07862	31/01/2000
C MSS Linuron 500	Mirfield Sales Services Ltd	08893	
C Stefes Linuron	Stefes Plant Protection Ltd	08187	
C Stefes Linuron	Stefes Plant Protection Ltd	07902	
C UPL Linuron 45% Flowable	United Phosphorus Ltd	07435	

287 Linuron + Chlorpropham

C Profalon	AgrEvo UK Crop Protection Ltd	07331	31/01/2001
C Profalon	Hortichem Ltd	08900	

288 Linuron + 2,4-DB + MCPA

C Alistell	Zeneca Crop Protection	06515	

289 Linuron + Lenacil

C Lanslide	Pan Britannica Industries Ltd	01184	

290 Linuron + Trifluralin

C Chandor	Dow AgroSciences Ltd	05631	
C Ipicombi TL	I Pi Ci	04608	31/01/2000
C Linnet	Pan Britannica Industries Ltd	01555	
C Linnet	pbi Agrochemicals Ltd	09601	
C Neminfest	Montedison UK Ltd	02546	
C Neminfest	Sipcam UK Ltd	07219	
C Trifluron	United Phosphorus Ltd	07854	28/02/2000
C Triplen Combi	Sipcam UK Ltd	05939	30/06/2000

C These products are "approved for agricultural use". For further details refer to page vii.
A These products are approved for use in or near water. For further details refer to page vii.

Product Name	Marketing Company	Reg. No.	Expiry Date

291 Maleic hydrazide

C	Bos MH 180	Uniroyal Chemical Ltd	06502	
C	Fazor	DowElanco Ltd	05558	
C	Fazor	Uniroyal Chemical Ltd	05461	
C	Mazide 25	Synchemicals Ltd	02067	
C A	Regulox K	Rhone-Poulenc Amenity	05405	
C	Royal MH 180	Mirfield Sales Services Ltd	07840	
C	Royal MH180	Uniroyal Chemical Ltd	07043	
C	Source	Chiltern Farm Chemicals Ltd	07427	31/08/2000
C	Source II	Chiltern Farm Chemicals Ltd	08314	

292 Maleic hydrazide + Dicamba + MCPA

C	Mazide Selective	Vitax Ltd	05753

293 MCPA

C	Agrichem MCPA 50	Agrichem Ltd	04097
C	Agricorn 500	Farmers Crop Chemicals Ltd	00055
C	Agricorn 500 II	Farmers Crop Chemicals Ltd	09155
C	Agritox 50	Nufarm UK Ltd	07400
C	Agritox D	Nufarm UK Ltd	07061
C	Agroxone	A H Marks & Co Ltd	08536
C	AGROXONE 50	Mirfield Sales Services Ltd	08345
C	Agroxone 75	A H Marks & Co Ltd	09208
C	Atlas MCPA	Atlas Crop Protection Ltd	07717
C	Atlas MCPA	Atlas Interlates Ltd	03055
C	Barclay Meadowman	Barclay Chemicals (UK) Ltd	07639
C	Barclay Meadowman II	Barclay Plant Protection	08525
C	BASF MCPA Amine 50	BASF plc	00209
C	Campbell's MCPA 50	J D Campbell & Sons Ltd	00381
C	Cirsium	Mirfield Sales Services Ltd	08729
C	Empal	Universal Crop Protection Ltd	00795
C	FCC Agricorn 50M	Farmers Crop Chemicals Ltd	09032
C	Headland Spear	Headland Agrochemicals Ltd	07115
C	HY-MCPA	Agrichem Ltd	06293
C	Luxan MCPA 500	Luxan (UK) Ltd	07470
C	Marks MCPA 50A	A H Marks & Co Ltd	02710
C	Marks MCPA P30	A H Marks & Co Ltd	01292
C	Marks MCPA S25	A H Marks & Co Ltd	01294
C	Marks MCPA SP	A H Marks & Co Ltd	01293
C	MCPA 25%	Nufarm UK Ltd	07998
C	MCPA 500	Nufarm UK Ltd	08655
C	MSS MCPA 50	Mirfield Sales Services Ltd	01412
C	Nufarm MCPA DMA 480	Nufarm UK Ltd	09686
C	Nufarm MCPA DMA 500	Nufarm UK Ltd	09685
C	Quadrangle MCPA 50	Quadrangle Agrochemicals	06935

C   These products are "approved for agricultural use". For further details refer to page vii.
A   These products are approved for use in or near water. For further details refer to page vii.

Product Name	Marketing Company	Reg. No.	Expiry Date

293 MCPA—continued

	Product Name	Marketing Company	Reg. No.	Expiry Date
C	Sanaphen M	Dow AgroSciences Ltd	09238	
C	Sanaphen M	Sanachem International Ltd	08728	30/09/2001
C	Stefes Phenoxylene 50	Stefes Plant Protection Ltd	07612	
C	Top Farm MCPA 500	Top Farm Formulations Ltd	05930	29/02/2000
C	Tripart MCPA 50	Tripart Farm Chemicals Ltd	02206	
C	Vitax MCPA Ester	Vitax Ltd	07353	31/05/2001

294 MCPA + Bentazone + MCPB

C	Acumen	BASF plc	00028
C	Headland Archer	Headland Agrochemicals Ltd	08814

295 MCPA + Bifenox + Mecoprop-P

C	Sirocco	Rhone-Poulenc Amenity	09645

296 MCPA + Bromoxynil + Dichlorprop + Ioxynil

C	Atlas Minerva	Atlas Interlates Ltd	03046

297 MCPA + Clopyralid + Dichlorprop

C	Lontrel Plus	Dow AgroSciences Ltd	05269

298 MCPA + Clopyralid + Diflufenican

C	Spearhead	Rhone-Poulenc Amenity	07342

299 MCPA + Clopyralid + Fluroxypyr

C	Greenor	Rigby Taylor Ltd	07848
C	Greenor Lo	Dow AgroSciences Ltd	09585
C	Greenor Lo	Rigby Taylor Ltd	09586

300 MCPA + 2,4-D + Clopyralid

C	Lonpar	Dow AgroSciences Ltd	08686

301 MCPA + 2,4-D + Dichlorprop + Mecoprop

C	Camppex	United Phosphorus Ltd	08266	31/10/2001

302 MCPA + 2,4-D + Dichlorprop + Mecoprop-P

C	UPL Camppex	United Phosphorus Ltd	09121

303 MCPA + 2,4-DB

C	Agrichem DB Plus	Agrichem Ltd	00044
C	Marks 2,4-DB Extra	A H Marks & Co Ltd	01284
C	MSS 2,4-DB + MCPA	Mirfield Sales Services Ltd	01392
C	Redlegor	United Phosphorus Ltd	07519

C   These products are "approved for agricultural use". For further details refer to page vii.

A   These products are approved for use in or near water. For further details refer to page vii.

Product Name	Marketing Company	Reg. No.	Expiry Date

304 MCPA + 2,4-DB + Benazolin

C Legumex Extra	AgrEvo UK Ltd	08676	
C Legumex Extra	AgrEvo UK Ltd	07901	31/05/2001
C Master Sward	Stefes Plant Protection Ltd	07883	
C Setter 33	AgrEvo UK Ltd	07282	
C Setter 33	Dow AgroSciences Ltd	05623	
C Stefes Clover Ley	Stefes Plant Protection Ltd	07842	
C Stefes Legumex Extra	Stefes Plant Protection Ltd	07841	

305 MCPA + 2,4-DB + Linuron

C Alistell	Zeneca Crop Protection	06515	

306 MCPA + Dicamba

C Banvel M	Novartis Crop Protection UK Ltd	08469	
C Banvel M	Sandoz Agro Ltd	05794	31/08/2000

307 MCPA + Dicamba + Dichlorprop

C Cleanrun 2	The Scotts Company (UK) Limited	09271	
C Intrepid	Miracle Professional	07819	
C Intrepid	The Scotts Company (UK) Limited	09266	
C Longlife Clearun 2	Miracle Professional	07851	

308 MCPA + Dicamba + Dichlorprop + Ferrous sulphate

C Longlife Renovator 2	Miracle Professional	08379	
C Renovator 2	The Scotts Company (UK) Limited	09272	

309 MCPA + Dicamba + Maleic hydrazide

C Mazide Selective	Vitax Ltd	05753	

310 MCPA + Dicamba + Mecoprop

C Barclay Hat-Trick	Barclay Chemicals (UK) Ltd	07579	
C Campbell's Field Marshal	United Phosphorus Ltd	08594	30/04/2000
C Campbell's Field Marshall	J D Campbell & Sons Ltd	00406	30/06/2000
C Campbell's Grassland Herbicide	MTM Agrochemicals Ltd	06157	30/04/2000
C Fisons Tritox	Fisons Plc	03013	30/04/2000
C Headland Relay	Headland Agrochemicals Ltd	07684	30/06/2000
C Headland Relay	WBC Technology Ltd	03778	30/06/2000
C Herrisol	Bayer plc	01048	30/04/2000
C Hyprone	Agrichem (International) Ltd	01093	30/06/2001
C Hysward	Agrichem (International) Ltd	01096	31/08/2001
C Pasturol	Farmers Crop Chemicals Ltd	01545	30/06/2000
C Premier Triple Selective	Amenity Land Services Limited	08529	31/05/2001
C Stefes Banlene Plus	Stefes Plant Protection Ltd	07688	30/04/2000

C These products are "approved for agricultural use". For further details refer to page vii.
A These products are approved for use in or near water. For further details refer to page vii.

Product Name	Marketing Company	Reg. No.	Expiry Date

310 MCPA + Dicamba + Mecoprop—continued

C	Stefes Docklene	Stefes Plant Protection Ltd	07816	30/04/2000
C	Stefes Swelland	Stefes Plant Protection Ltd	08188	30/04/2000
C	Tritox	Levington Horticulture Ltd	07502	30/04/2000

311 MCPA + Dicamba + Mecoprop-P

C	ALS Premier Selective Plus	Amenity Land Services Limited	08940	
C	Banlene Super	AgrEvo UK Crop Protection Ltd	07168	
C	Field Marshal	MTM Agrochemicals Ltd	06077	31/03/2001
C	Field Marshal	United Phosphorus Ltd	08956	
C	Headland Relay P	Headland Agrochemicals Ltd	08580	
C	Headland Relay Turf	Headland Agrochemicals Ltd	08935	
C	Herrisol New	Bayer plc	07166	
C	Hyprone P	Agrichem (International) Ltd	09125	
C	Hysward P	Agrichem (International) Ltd	09052	
C	Mascot Super Selective-P	Rigby Taylor Ltd	06106	
C	MSS Mircam Plus	Mirfield Sales Services Ltd	01416	
C	MTM Grassland Herbicide	MTM Agrochemicals Ltd	06089	28/02/2001
C	Pasturol D	Farmers Crop Chemicals Ltd	07033	
C	Pasturol Plus	Farmers Crop Chemicals Ltd	08581	
C	Pasturol-P	Farmers Crop Chemicals Ltd	09099	
C	Premier Triple Selective P	Amenity Land Services Limited	09053	
C	Quadrangle Quadban	Quadrangle Agrochemicals	07090	
C	Stefes Banlene Super	Stefes Plant Protection Ltd	07691	
C	Stefes Docklene Super	Stefes Plant Protection Ltd	07696	
C	Stefes Swelland Super	Stefes Plant Protection Ltd	08189	
C	Tribute	Nomix-Chipman Ltd	06921	
C	Tribute Plus	Nomix-Chipman Ltd	09493	
C	Tritox	The Scotts Company (UK) Limited	07764	
C	UPL Grassland Herbicide	United Phosphorus Ltd	08934	

312 MCPA + Dichlorprop

C	MSS 2,4-DP + MCPA	Mirfield Sales Services Ltd	01396	
C	Redipon Extra	United Phosphorus Ltd	07518	
C	Seritox 50	Rhone-Poulenc Agriculture	06170	
C	Seritox Turf	Rhone-Poulenc Environmental Products	06802	
C	SHL Turf Feed & Weed	William Sinclair Horticulture Ltd	04437	

313 MCPA + Dichlorprop + Ferrous sulphate

C	SHL Turf Feed and Weed and Mosskiller	William Sinclair Horticulture Ltd	04438	

314 MCPA + Dichlorprop-P + Mecoprop-P

C	Hymec Trio	Agrichem (International) Ltd	09764	
C	Optica Trio	A H Marks & Co Ltd	09747	

C These products are "approved for agricultural use". For further details refer to page vii.
A These products are approved for use in or near water. For further details refer to page vii.

Product Name	Marketing Company	Reg. No.	Expiry Date

315 MCPA + MCPB

C	Bellmac Plus	United Phosphorus Ltd	07521
C	MSS MCPB + MCPA	Mirfield Sales Services Ltd	01413
C	Trifolex-Tra	Cyanamid Agriculture Ltd	07147
C	Tropotox Plus	Rhone-Poulenc Agriculture	06156

316 MCPA + Mecoprop

C	Fisons Greenmaster Extra	Fisons Plc	03130	30/04/2000
C	Greenmaster Extra	Levington Horticulture Ltd	07491	30/04/2000

317 MCPA + Mecoprop-P

C	Greenmaster Extra	The Scotts Company (UK) Limited	07594

318 MCPB

C	Bellmac Straight	United Phosphorus Ltd	07522
C	Marks MCPB	A H Marks & Co Ltd	03497
C	Tropotox	Rhone-Poulenc Agriculture	06179

319 MCPB + Bentazone

C	Pulsar	BASF plc	04002

320 MCPB + Bentazone + MCPA

C	Acumen	BASF plc	00028
C	Headland Archer	Headland Agrochemicals Ltd	08814

321 MCPB + MCPA

C	Bellmac Plus	United Phosphorus Ltd	07521
C	MSS MCPB + MCPA	Mirfield Sales Services Ltd	01413
C	Trifolex-Tra	Cyanamid Agriculture Ltd	07147
C	Tropotox Plus	Rhone-Poulenc Agriculture	06156

322 Mecoprop

C	Amos CMPP	Luxan (UK) Ltd	06556	29/02/2000
C	Atlas CMPP	Atlas Interlates Ltd	03050	30/04/2000
C	Barclay Mecrop	Barclay Chemicals (UK) Ltd	07578	
C	Barclay Mecrop II	Barclay Plant Protection	08562	30/04/2000
C	BASF CMPP Amine 60	BASF plc	07837	30/04/2000
C	BH CMPP Extra	Rhone-Poulenc Amenity	05392	30/04/2000
C	Campbell's CMPP	J D Campbell & Sons Ltd	02918	31/03/2001
C	Clenecorn	Farmers Crop Chemicals Ltd	00542	31/03/2001
C	Clenecorn II	Farmers Crop Chemicals Ltd	08880	30/04/2000
C	CMPP 60%	Nufarm UK Ltd	07999	31/12/2000
C	Compitox Extra	Nufarm UK Ltd	07398	30/04/2000
C	Headland Charge	Headland Agrochemicals Ltd	04495	30/04/2000

C These products are "approved for agricultural use". For further details refer to page vii.
A These products are approved for use in or near water. For further details refer to page vii.

Product Name	Marketing Company	Reg. No.	Expiry Date

322 Mecoprop—continued

C	Hymec	Agrichem Ltd	01091	30/06/2001
C	Landgold CMPP F	Landgold & Co Ltd	06276	29/02/2000
C	Luxan CMPP	Luxan (UK) Ltd	06037	30/04/2000
C	Luxan CMPP 600	Luxan (UK) Ltd	05909	29/02/2000
C	Marks Mecoprop K	A H Marks & Co Ltd	01297	30/04/2000
C	Marks Mecoprop-E	A H Marks & Co Ltd	01296	30/04/2000
C	Mascot Clover Killer	Rigby Taylor Ltd	02438	30/04/2000
C	MSS CMPP	Mirfield Sales Services Ltd	01405	30/04/2000
C	Top Farm CMPP	Top Farm Formulations Ltd	05468	29/02/2000
C	Unistar CMPP	Unistar Ltd	06353	29/02/2000

323 Mecoprop + Bromoxynil + Ioxynil

C	Swipe 560 EC	Ciba Agriculture	02057	30/04/2000

324 Mecoprop + Bromoxynil + Ioxynil + Isoproturon

C	Terset	Rhone-Poulenc Agriculture	06171	30/04/2000

325 Mecoprop + Cyanazine

C	Cleaval	Cyanamid Agriculture Ltd	07142	30/04/2000
C	FCC Topcorn Extra	Cyanamid Agriculture Ltd	07143	30/04/2000
C	MTM Eminent	Cyanamid Agriculture Ltd	08053	30/04/2000
C	MTM Eminent	MTM Agrochemicals Ltd	07144	30/04/2000

326 Mecoprop + 2,4-D

C	BH CMPP/2,4-D	Rhone-Poulenc Amenity	05393	30/04/2000
C	CDA Supertox 30	Rhone-Poulenc Amenity	04664	30/04/2000
C	Mascot Selective Weedkiller	Rigby Taylor Ltd	03423	30/04/2000
C	Nomix Turf Selective Herbicide	Nomix-Chipman Ltd	06777	31/03/2000
C	Nomix Turf Selective LC Herbicide	Nomix-Chipman Ltd	05973	31/03/2000
C	Selective Weedkiller	Yule Catto Consumer Chemicals Ltd	06579	30/04/2000
C	Supertox 30	Rhone-Poulenc Amenity	05340	30/06/2001

327 Mecoprop + 2,4-D + Dicamba

C	New Formulation Weed and Brushkiller	Vitax Ltd	07072	30/04/2000

328 Mecoprop + 2,4-D + Dichlorprop + MCPA

C	Camppex	United Phosphorus Ltd	08266	31/10/2001

C These products are "approved for agricultural use". For further details refer to page vii.
A These products are approved for use in or near water. For further details refer to page vii.

Product Name	Marketing Company	Reg. No.	Expiry Date

329 Mecoprop + Dicamba

C	Banvel P	Nufarm UK Ltd	08346	30/04/2000
C	Condox	Zeneca Crop Protection	06519	31/05/2002
C	Hyban	Agrichem Ltd	01084	30/06/2001
C	Hygrass	Agrichem (International) Ltd	01090	30/06/2001
C	Pasturemaster D	Stefes Plant Protection Ltd	08481	28/02/2000
C	Pasturemaster D	Stefes Plant Protection Ltd	07954	30/04/2000

330 Mecoprop + Dicamba + MCPA

C	Barclay Hat-Trick	Barclay Chemicals (UK) Ltd	07579	
C	Campbell's Field Marshal	United Phosphorus Ltd	08594	30/04/2000
C	Campbell's Field Marshall	J D Campbell & Sons Ltd	00406	30/06/2000
C	Campbell's Grassland Herbicide	MTM Agrochemicals Ltd	06157	30/04/2000
C	Fisons Tritox	Fisons Plc	03013	30/04/2000
C	Headland Relay	Headland Agrochemicals Ltd	07684	30/06/2000
C	Headland Relay	WBC Technology Ltd	03778	30/06/2000
C	Herrisol	Bayer plc	01048	30/04/2000
C	Hyprone	Agrichem (International) Ltd	01093	30/06/2001
C	Hysward	Agrichem (International) Ltd	01096	31/08/2001
C	Pasturol	Farmers Crop Chemicals Ltd	01545	30/06/2000
C	Premier Triple Selective	Amenity Land Services Limited	08529	31/05/2001
C	Stefes Banlene Plus	Stefes Plant Protection Ltd	07688	30/04/2000
C	Stefes Docklene	Stefes Plant Protection Ltd	07816	30/04/2000
C	Stefes Swelland	Stefes Plant Protection Ltd	08188	30/04/2000
C	Tritox	Levington Horticulture Ltd	07502	30/04/2000

331 Mecoprop + Dicamba + Triclopyr

C	Fettel	Zeneca Crop Protection	06399	30/04/2000

332 Mecoprop + MCPA

C	Fisons Greenmaster Extra	Fisons Plc	03130	30/04/2000
C	Greenmaster Extra	Levington Horticulture Ltd	07491	30/04/2000

333 Mecoprop-P

C	Astix	Rhone-Poulenc Agriculture	06174	
C	Astix K	Rhone-Poulenc Agriculture	06904	
C	Barclay Melody	Barclay Chemicals (UK) Ltd	08790	31/03/2002
C	BAS 03729H	Mirfield Sales Services Ltd	05912	
C	Clenecorn Super	Farmers Crop Chemicals Ltd	09382	31/03/2002
C	Clovotox	Rhone-Poulenc Amenity	05354	
C	Compitox Extra P	Nufarm UK Ltd	09558	
C	Compitox Plus	Nufarm UK Ltd	07930	
C	Compitox Plus	Nufarm UK Ltd	08997	31/12/2001

C   These products are "approved for agricultural use". For further details refer to page vii.
A   These products are approved for use in or near water. For further details refer to page vii.

Product Name	Marketing Company	Reg. No.	Expiry Date

333 Mecoprop-P—continued

C Duplosan	BASF plc	05889	28/02/2002
C Duplosan 500	BASF plc	07889	
C Duplosan KV	BASF plc	09431	
C Duplosan KV 500	BASF plc	08027	
C Duplosan New System CMPP	BASF plc	04481	
C Landgold Mecoprop-P	Landgold & Co Ltd	06052	
C Landgold Mecoprop-P 600	Landgold & Co Ltd	09014	
C Landgold Mecoprop-P 600	Landgold & Co Ltd	06461	31/05/2000
C Mascot Cloverkiller-P	Rigby Taylor Ltd	06099	30/11/2000
C MSS Mirprop	Mirfield Sales Services Ltd	05911	
C MSS Optica	Mirfield Sales Services Ltd	04973	
C Optica	A H Marks & Co Ltd	05814	
C Optica	Bayer UK Ltd	04922	
C Optica 50	A H Marks & Co Ltd	08283	
C Standon Mecoprop-P	Standon Chemicals Ltd	05651	29/02/2000
C Stefes Mecoprop-P	Stefes Plant Protection Ltd	05780	29/02/2000
C Stefes Mecoprop-P2	Stefes Plant Protection Ltd	06239	31/05/2000
C Unistar CMPP	Unistar Ltd	06306	29/02/2000

334 Mecoprop-P + Bifenox + MCPA

C Sirocco	Rhone-Poulenc Amenity	09645	

335 Mecoprop-P + Bromoxynil + Ioxynil

C Swipe P	Ciba Agriculture	08150	31/12/2000
C Swipe P	Novartis Crop Protection UK Ltd	08452	

336 Mecoprop-P + Carfentrazone-ethyl

C Platform S	FMC Europe NV	08638	

337 Mecoprop-P + 2,4-D

C Mascot Selective-P	Rigby Taylor Ltd	06105	
C Supertox 30	Rhone-Poulenc Amenity	09102	
C Sydex	Vitax Ltd	06412	

338 Mecoprop-P + 2,4-D + Dichlorprop + MCPA

C UPL Camppex	United Phosphorus Ltd	09121	

339 Mecoprop-P + Dicamba

C Condox	Zeneca Crop Protection	09429	
C Di-Farmon R	Headland Agrochemicals Ltd	08472	
C Di-Farmon R	Headland Agrochemicals Ltd	07482	30/04/2000
C Foundation	Novartis Crop Protection UK Ltd	08475	
C Foundation	Sandoz Agro Ltd	07465	31/08/2000

C These products are "approved for agricultural use". For further details refer to page vii.
A These products are approved for use in or near water. For further details refer to page vii.

Product Name	Marketing Company	Reg. No.	Expiry Date

339 Mecoprop-P + Dicamba—continued

C	Hyban P	Agrichem (International) Ltd	09129
C	Hygrass P	Agrichem (International) Ltd	09130
C	MSS Mircam	Mirfield Sales Services Ltd	01415
C	Optica Forte	A H Marks & Co Ltd	07432

340 Mecoprop-P + Dicamba + MCPA

C	ALS Premier Selective Plus	Amenity Land Services Limited	08940	
C	Banlene Super	AgrEvo UK Crop Protection Ltd	07168	
C	Field Marshal	MTM Agrochemicals Ltd	06077	31/03/2001
C	Field Marshal	United Phosphorus Ltd	08956	
C	Headland Relay P	Headland Agrochemicals Ltd	08580	
C	Headland Relay Turf	Headland Agrochemicals Ltd	08935	
C	Herrisol New	Bayer plc	07166	
C	Hyprone P	Agrichem (International) Ltd	09125	
C	Hysward P	Agrichem (International) Ltd	09052	
C	Mascot Super Selective-P	Rigby Taylor Ltd	06106	
C	MSS Mircam Plus	Mirfield Sales Services Ltd	01416	
C	MTM Grassland Herbicide	MTM Agrochemicals Ltd	06089	28/02/2001
C	Pasturol D	Farmers Crop Chemicals Ltd	07033	
C	Pasturol Plus	Farmers Crop Chemicals Ltd	08581	
C	Pasturol-P	Farmers Crop Chemicals Ltd	09099	
C	Premier Triple Selective P	Amenity Land Services Limited	09053	
C	Quadrangle Quadban	Quadrangle Agrochemicals	07090	
C	Stefes Banlene Super	Stefes Plant Protection Ltd	07691	
C	Stefes Docklene Super	Stefes Plant Protection Ltd	07696	
C	Stefes Swelland Super	Stefes Plant Protection Ltd	08189	
C	Tribute	Nomix-Chipman Ltd	06921	
C	Tribute Plus	Nomix-Chipman Ltd	09493	
C	Tritox	The Scotts Company (UK) Limited	07764	
C	UPL Grassland Herbicide	United Phosphorus Ltd	08934	

341 Mecoprop-P + Dichlorprop-P + MCPA

C	Hymec Trio	Agrichem (International) Ltd	09764
C	Optica Trio	A H Marks & Co Ltd	09747

342 Mecoprop-P + Fluroxypyr

C	Bastion T	Rigby Taylor Ltd	06011

343 Mecoprop-P + MCPA

C	Greenmaster Extra	The Scotts Company (UK) Limited	07594

344 Mecoprop-P + Metribuzin

C	Bayer UK 683	Bayer plc	08113
C	Centra	Bayer plc	08112

C   These products are "approved for agricultural use". For further details refer to page vii.
A   These products are approved for use in or near water. For further details refer to page vii.

Product Name	Marketing Company	Reg. No.	Expiry Date

344 Mecoprop-P + Metribuzin—continued

C	Optica Plus	A H Marks & Co Ltd	09325	

345 Mecoprop-P + Triasulfuron

C	Raven	Ciba Agriculture	06119	31/08/2000
C	Raven	Novartis Crop Protection UK Ltd	08434	

346 Mefluidide

C	Check Turf II	Certified Laboratories Ltd	04463	
C	Embark	Gordon International Corp	04810	
C	Gro-Tard II	Chemsearch	04462	

347 Mepiquat + Chlormequat

C	Cyter	BASF plc	09133	
C	Stronghold	BASF plc	09134	

348 Mepiquat + 2-Chloroethylphosphonic acid

C	Barclay Banshee	Barclay Chemicals (UK) Ltd	08175	
C	CleanCrop Fonic M	United Agri Products	09553	
C	Standon Mepiquat Plus	Standon Chemicals Ltd	09373	
C	Stefes Mepiquat	Stefes Plant Protection Ltd	06970	20/04/2001
C	Sypex M	BASF plc	06810	30/06/2001
C	Terpal	BASF plc	02103	
C	Terpal	Clayton Plant Protection (UK) Ltd	07626	
C	Terpitz	Me^2 Crop Protection Ltd	09634	

349 Mepiquat + 2-Chloroethylphosphonic acid + Chlormequat

C	Cyclade	BASF plc	08958	

350 Metamitron

C	Barclay Seismic	Barclay Chemicals (UK) Ltd	09323	
C	Barclay Seismic XL	Barclay Chemicals (UK) Ltd	09328	
C	Goltix 90	Bayer plc	08654	
C	Goltix Flowable	Bayer plc	08986	
C	Goltix WG	Bayer plc	02430	
C	Landgold Metamitron	Landgold & Co Ltd	06287	
C	Lektan	Bayer plc	06111	
C	Marquise	Makhteshim-Agan (UK) Ltd	08738	
C	MM70	Gharda Chemicals Ltd	09490	
C	Quartz	Quadrangle Agrochemicals	05167	
C	Skater	Feinchemie (UK) Ltd	09790	
C	Skater	Feinchemie Schwebda GmbH	08623	31/10/2002
C	Standon Metamitron	Standon Chemicals Ltd	07885	
C	Stefes 7G	Stefes Plant Protection Ltd	06350	

C These products are "approved for agricultural use". For further details refer to page vii.
A These products are approved for use in or near water. For further details refer to page vii.

Product Name	Marketing Company	Reg. No.	Expiry Date

PSD HERBICIDES

350 Metamitron—continued

C	Stefes Metamitron	Stefes Plant Protection Ltd	05821
C	Tripart Accendo	Tripart Farm Chemicals Ltd	06110
C	Valiant	Stefes UK Ltd	09251
C	Volcan	Sipcam UK Ltd	09295

351 Metamitron + Ethofumesate

C	Stefes Duplex	Stefes UK Ltd	09252
C	Stefes ST/STE	Stefes Plant Protection Ltd	09111

352 Metamitron + Ethofumesate + Phenmedipham

C	Bayer UK 407 WG	Bayer UK Ltd	06313	28/02/2000
C	Bayer UK 540	Bayer plc	07399	
C	Betanal Trio OF	AgrEvo UK Crop Protection Ltd	08680	31/05/2001
C	Betanal Trio WG	AgrEvo UK Ltd	07537	
C	CX 171	AgrEvo UK Ltd	06803	28/02/2000
C	Goltix Triple	Bayer UK Ltd	06314	28/02/2000

353 Metazachlor

C	Barclay Metaza	Barclay Chemicals (UK) Ltd	09169	
C	Barclay Metaza	Barclay Chemicals (UK) Ltd	07354	31/05/2000
C	Butisan S	BASF plc	00357	
C	Clayton Metazachlor	Clayton Plant Protection (UK) Ltd	09688	
C	Clayton Metazachlor	Clayton Plant Protection (UK) Ltd	07679	31/05/2000
C	CleanCrop MTZ 500	United Agri Products	09222	
C	Comet	BASF plc	06817	
C	Landgold Metazachlor 50	Landgold & Co Ltd	09726	
C	Landgold Metazachlor 50	Landgold & Co Ltd	08062	31/05/2000
C	Marnoch Metazachlor	Marnoch Ventures Limited	09761	
C	Standon Metazachlor 50	Standon Chemicals Ltd	05581	

354 Metazachlor + Quinmerac

C	Katamaran	BASF plc	09049
C	Standon Metazachlor-Q	Standon Chemicals Ltd	09676

355 Methabenzthiazuron

C	Clayton Benson	Clayton Plant Protection (UK) Ltd	08687
C	Tribunil	Bayer plc	02169
C	Tribunil WG	Bayer plc	03260

356 Methabenzthiazuron + Isoxaben

C	Glytex	Bayer plc	04230

C These products are "approved for agricultural use". For further details refer to page vii.
A These products are approved for use in or near water. For further details refer to page vii.

Product Name	Marketing Company	Reg. No.	Expiry Date

357 Metosulam

C	EF 1077	Dow AgroSciences Ltd	07965	

358 Metosulam + Fluroxypyr

C	EF 1166	Dow AgroSciences Ltd	07966	

359 Metoxuron

C	Deftor	Novartis Crop Protection UK Ltd	08471	
C	Deftor	Sandoz Agro Ltd	05545	31/08/2000
C	Dosaflo	Novartis Crop Protection UK Ltd	09351	
C	Dosaflo	Novartis Crop Protection UK Ltd	08473	28/02/2002
C	Dosaflo	Sandoz Agro Ltd	00754	31/08/2000
C	Endspray II	Pan Britannica Industries Ltd	05079	

360 Metribuzin

C	Bayer UK093	Bayer plc	07903	
C	Citation	IT Agro Ltd	08685	30/11/2001
C	Citation 70	United Phosphorus Ltd	09370	
C	CleanCrop Metribuzin	United Agri Products	09621	
C	Inter-Metribuzin WG	Agriguard Ltd	09336	
C	Lexone 70 DF	Du Pont (UK) Ltd	04991	
C	Python	Feinchemie (UK) Ltd	09791	
C	Python	Headland Agrochemicals Ltd	09243	31/10/2002
C	Sencorex WG	Bayer plc	03755	

361 Metribuzin + Amidosulfuron

C	Bayer UK 590	Bayer plc	08149	30/06/2001
C	Galis	Bayer plc	08039	30/06/2001

362 Metribuzin + Mecoprop-P

C	Bayer UK 683	Bayer plc	08113	
C	Centra	Bayer plc	08112	
C	Optica Plus	A H Marks & Co Ltd	09325	

363 Metsulfuron-methyl

C	Agriguard Metsulfuron	Tronsan Ltd	09569	
C	Ally	Du Pont (UK) Ltd	02977	
C	Ally WSB	Du Pont (UK) Ltd	06588	
C	Associate	Nufarm UK Ltd	08652	
C	Barclay Flumen	Barclay Chemicals (UK) Ltd	08752	
C	Clayton Metsulfuron	Clayton Plant Protection (UK) Ltd	06734	
C	Gem 690	Griffin (Europe) HQ NV	09436	
C	Jubilee	Du Pont (UK) Ltd	06082	
C	Jubilee 20DF	Du Pont (UK) Ltd	06136	

C   These products are "approved for agricultural use". For further details refer to page vii.
A   These products are approved for use in or near water. For further details refer to page vii.

Product Name	Marketing Company	Reg. No.	Expiry Date

363 Metsulfuron-methyl—continued

C	Landgold Metsulfuron	Landgold & Co Ltd	06280	
C	Lorate 20DF	Du Pont (UK) Ltd	06135	
C	Simba 20 DF	Griffin (Europe) Marketing NV (UK Branch)	09437	
C	Standon Metsulfuron	Standon Chemicals Ltd	05670	

364 Metsulfuron-methyl + Carfentrazone-ethyl

C	Ally Express	Du Pont (UK) Ltd	08640	
C	Ally Express WSB	Du Pont (UK) Ltd	08641	
C	Standon Carfentrazone MM	Standon Chemicals Ltd	09159	

365 Metsulfuron-methyl + Flupyrsulfuron-methyl

C	Lexus XPE	Du Pont (UK) Ltd	08541	
C	Lexus XPE WSB	Du Pont (UK) Ltd	08542	
C	Standon Flupyrsulfuron MM	Standon Chemicals Ltd	09098	

366 Metsulfuron-methyl + Thifensulfuron-methyl

C	DP 928	Du Pont (UK) Ltd	09632	
C	Harmony M	Du Pont (UK) Ltd	03990	
C	Landgold TM 75	Landgold & Co Ltd	08376	31/10/2000

367 Monolinuron

| C | Arresin | AgrEvo UK Ltd | 07303 | |

368 Monolinuron + Paraquat

| C | Gramonol Five | Zeneca Crop Protection | 06673 | 31/10/2000 |

369 1-Naphthylacetic acid

C	Rhizopon B Powder (0.1%)	Fargro Ltd	07133	
C	Rhizopon B Powder (0.1%)	Rhizopon UK Limited	09089	
C	Rhizopon B Powder (0.2 %)	Rhizopon UK Limited	09090	
C	Rhizopon B Powder (0.2%)	Fargro Ltd	07134	
C	Rhizopon B Tablets	Fargro Ltd	07135	
C	Rhizopon B Tablets	Rhizopon UK Limited	09091	
C	Tipoff	Universal Crop Protection Ltd	05878	

370 1-Naphthylacetic acid + 4-Indol-3-ylbutyric acid

| C | Synergol | Hortichem Ltd | 07386 | |

371 2-Naphthyloxyacetic Acid

| C | Betapal Concentrate | Synchemicals Ltd | 00234 | |

C These products are "approved for agricultural use". For further details refer to page vii.
A These products are approved for use in or near water. For further details refer to page vii.

Product Name	Marketing Company	Reg. No.	Expiry Date

372 Napropamide

C	Agriguard Napropamide	Agriguard Ltd	09223	30/04/2002
C	Banweed	United Phosphorus Ltd	09376	
C	Banweed	Zeneca Crop Protection	06672	31/12/2001
C	Devrinol	Rhone-Poulenc Agriculture	06195	31/12/2001
C	Devrinol	United Phosphorus Ltd	09375	
C	Devrinol	United Phosphorus Ltd	09374	
C	Devrinol	Zeneca Crop Protection	06653	31/12/2001

373 Oxadiazon

C	Ronstar 2G	Rhone-Poulenc Agriculture	06492	
C	Ronstar Liquid	Hortichem Ltd	08974	
C	Ronstar Liquid	Rhone-Poulenc Agriculture	06766	30/09/2001

374 Oxadiazon + Carbetamide + Diflufenican

C	Helmsman	Rhone-Poulenc Amenity	09696	

375 Oxadiazon + Glyphosate

C	Zapper	Rhone-Poulenc Amenity	08605	

376 Paclobutrazol

C	Bonzi	Zeneca Crop Protection	06640	
C	Cultar	Zeneca Crop Protection	06649	

377 Paclobutrazol + Dicamba

C	Holdfast D	Imperial Chemical Industries Plc	05056	30/10/2001
C	Holdfast D	Miracle Professional	07864	31/03/2001

378 Paraquat

C	Barclay Total	Barclay Chemicals (UK) Ltd	08822	
C	Barclay Total	Barclay Chemicals (UK) Ltd	05260	29/02/2000
C	Dextrone X	Chipman Ltd	00687	
C	Gramoxone 100	Zeneca Crop Protection	06674	
C	Landgold Paraquat	Landgold & Co Ltd	06025	29/02/2000
C	Scythe LC	Cyanamid Agriculture Ltd	05877	30/06/2001
C	Speedway	Miracle Professional	07743	30/10/2001
C	Speedway Liquid	Imperial Chemical Industries Plc	04365	31/05/2001
C	Speedway Liquid	Miracle Professional	07744	
C	Speedway Liquid	Zeneca Professional Products	06825	31/05/2001
C	Standon Paraquat	Standon Chemicals Ltd	05621	29/02/2000
C	Stefes Paraquat	Stefes Plant Protection Ltd	05134	29/02/2000
C	Top Farm Paraquat	Top Farm Formulations Ltd	05075	29/02/2000
C	Top Farm Paraquat 200	Top Farm Formulations Ltd	05519	29/02/2000

C   These products are "approved for agricultural use". For further details refer to page vii.
A   These products are approved for use in or near water. For further details refer to page vii.

Product Name	Marketing Company	Reg. No.	Expiry Date

379 Paraquat + Diquat

C	Parable	Zeneca Crop Protection	06692	31/07/2000
C	PDQ	Zeneca Crop Protection	06518	
C	Speedway 2	Miracle Professional	08164	31/10/2001
C	Speedway 2	The Scotts Company (UK) Limited	09273	

380 Paraquat + Diquat + Simazine

C	Pathclear S	Miracle Garden Care Ltd	08991	
C	Pathclear S	The Scotts Company (UK) Limited	09269	

381 Paraquat + Diuron

C	Dexuron	Nomix-Chipman Ltd	07169	

382 Paraquat + Monolinuron

C	Gramonol Five	Zeneca Crop Protection	06673	31/10/2000

383 Pendimethalin

C	Aspire	Cyanamid Agriculture Ltd	05919	31/01/2000
C	Barclay Tremor	Barclay Chemicals (UK) Ltd	09188	
C	Claymore	Cyanamid Agriculture Ltd	09232	
C	Clayton Pendalin	Clayton Plant Protection (UK) Ltd	09708	
C	Ipimethalin	pbi Agrochemicals Ltd	09701	
C	Plinth	pbi Agrochemicals Ltd	09702	
C	Sovereign	Ciba Agriculture	08152	31/08/2000
C	Sovereign	Cyanamid Agriculture Ltd	08151	
C	Sovereign	Novartis Crop Protection UK Ltd	08533	
C	Stomp 400 SC	Cyanamid Agriculture Ltd	04183	

384 Pendimethalin + Bentazone

C	Impuls	BASF plc	09720	

385 Pendimethalin + Chlorotoluron

C	Totem	Cyanamid Agriculture Ltd	04670	

386 Pendimethalin + Cyanazine

C	Activus	Cyanamid Agriculture Ltd	09174	
C	Bullet	Cyanamid Agriculture Ltd	08049	

387 Pendimethalin + Isoproturon

C	Artillery	Cyanamid Agriculture Ltd	07768	
C	Artillery SC	Cyanamid Agriculture Ltd	08504	
C	Clayton Pendalin-IPU 37	Clayton Plant Protection (UK) Ltd	09154	
C	Clayton Pendalin-IPU 47	Clayton Plant Protection (UK) Ltd	09170	

C   These products are "approved for agricultural use". For further details refer to page vii.
A   These products are approved for use in or near water. For further details refer to page vii.

Product Name	Marketing Company	Reg. No.	Expiry Date

387 Pendimethalin + Isoproturon—continued

C	Encore	Cyanamid Agriculture Ltd	04737	
C	Encore SC	Cyanamid Agriculture Ltd	08502	
C	Jolt	Cyanamid Agriculture Ltd	05488	
C	Jolt SC	Cyanamid Agriculture Ltd	08503	
C	Orient	Cyanamid Agriculture Ltd	07798	
C	Orient SC	Cyanamid Agriculture Ltd	08501	
C	Trump	Cyanamid Agriculture Ltd	03687	
C	Trump SC	Cyanamid Agriculture Ltd	08500	

388 Pendimethalin + Prometryn

C	Monarch	Cyanamid Agriculture Ltd	05160	31/05/2000

389 Pendimethalin + Simazine

C	Deuce	Cyanamid Agriculture Ltd	06746	
C	Merit	Cyanamid Agriculture Ltd	04976	

390 Pentanochlor

C	Atlas Solan 40	Atlas Crop Protection Ltd	07726	
C	Croptex Bronze	Hortichem Ltd	04087	

391 Pentanochlor + Chlorpropham

C	Atlas Brown	Atlas Crop Protection Ltd	07703	
C	Atlas Brown	Atlas Interlates Ltd	03835	30/04/2000

392 Phenmedipham

C	Atlas Protrum K	Atlas Crop Protection Ltd	07723	
C	Atlas Protrum K	Atlas Interlates Ltd	03089	
C	Barclay Punter XL	Barclay Chemicals (UK) Ltd	08047	
C	Beetup	United Phosphorus Ltd	07520	
C	Betanal E	AgrEvo UK Ltd	07248	
C	Betanal Flo	AgrEvo UK Ltd	08898	
C	Betosip	Sipcam UK Ltd	06787	
C	Betosip 114	Sipcam UK Ltd	05910	
C	Cirrus	Quadrangle Ltd	06367	31/05/2001
C	Cleancrop Phenmedipham	United Agri Products	09495	
C	Griffin Phenmedipham 114	Griffin (Bermuda) SA	08343	28/02/2002
C	Herbasan	KVK Agro A/S	08553	
C	Herbasan	Pan Britannica Industries Ltd	07161	31/05/2000
C	Hickson Phenmedipham	Hickson & Welch Ltd	02825	
C	Kemifam E	AgrEvo UK Ltd	08978	
C	Kemifam E	Kemira Cropcare Ltd	06104	
C	Kemifam E	Pan Britannica Industries Ltd	07150	31/05/2000
C	Kemifam Flow	AgrEvo UK Ltd	07991	31/05/2001
C	Luxan Phenmedipham	Luxan (UK) Ltd	06933	

C   These products are "approved for agricultural use". For further details refer to page vii.
A   These products are approved for use in or near water. For further details refer to page vii.

Product Name	Marketing Company	Reg. No.	Expiry Date

392 Phenmedipham—continued

C	Mandolin	Griffin (Europe) Marketing NV (UK Branch)	09456	
C	MSS Betaren Flow	KVK Agro A/S	08554	
C	MSS Betaren Flow	Mirfield Sales Services Ltd	08022	31/05/2000
C	MSS Protrum G	Mirfield Sales Services Ltd	08342	
C	Rubenal Flow	AgrEvo UK Ltd	09227	
C	Rubenal Flow	Pan Britannica Industries Ltd	07647	30/09/2001
C	Stefes Forte	Stefes Plant Protection Ltd	06427	
C	Stefes Forte 2	Stefes Plant Protection Ltd	08204	
C	Stefes Medipham	Stefes Plant Protection Ltd	07152	31/10/2001
C	Stefes Medipham 2	Stefes Plant Protection Ltd	08203	
C	Tripart Beta	Tripart Farm Chemicals Ltd	03111	
C	Tripart Beta 2	Tripart Farm Chemicals Ltd	06510	31/10/2001
C	Tripart Beta S	Tripart Farm Chemicals Ltd	08202	30/06/2001

393 Phenmedipham + Desmedipham

C	Betanal Compact	AgrEvo UK Ltd	07247	
C	Betanal Quorum	AgrEvo UK Ltd	07252	31/05/2001
C	Betanal Rostrum	AgrEvo UK Ltd	07253	31/05/2001

394 Phenmedipham + Desmedipham + Ethofumesate

C	Aventis Betanal Progress OF	AgrEvo UK Ltd	09722	
C	Betanal Congress	AgrEvo UK Ltd	07534	31/10/2000
C	Betanal Progress	AgrEvo UK Ltd	07533	31/05/2001
C	Betanal Progress OF	AgrEvo UK Ltd	07629	
C	Betanal Ultima	AgrEvo UK Ltd	07535	31/10/2000

395 Phenmedipham + Ethofumesate

C	Barclay Goalpost	Barclay Chemicals (UK) Ltd	09497	
C	Barclay Goalpost	Barclay Chemicals (UK) Ltd	09192	
C	Barclay Goalpost	Barclay Chemicals (UK) Ltd	08016	30/04/2000
C	Barclay Goldpost	Barclay Chemicals Manufacturing Ltd	08964	31/03/2002
C	Betanal Montage	AgrEvo UK Ltd	07250	28/02/2001
C	Betanal Quadrant	AgrEvo UK Crop Protection Ltd	07251	31/07/2000
C	Betanal Tandem	AgrEvo UK Ltd	07254	
C	Betosip Combi	Sipcam UK Ltd	08630	
C	Stefes Medimat	Stefes Plant Protection Ltd	07886	30/04/2001
C	Stefes Medimat 2	Stefes Plant Protection Ltd	07577	
C	Stefes Tandem	Stefes Plant Protection Ltd	08906	
C	Twin	Feinchemie (UK) Ltd	09784	
C	Twin	Feinchemie Schwebda GmbH	07374	31/10/2002

C   These products are "approved for agricultural use". For further details refer to page vii.
A   These products are approved for use in or near water. For further details refer to page vii.

Product Name	Marketing Company	Reg. No.	Expiry Date

396 Phenmedipham + Ethofumesate + Metamitron

C	Bayer UK 407 WG	Bayer UK Ltd	06313	28/02/2000
C	Bayer UK 540	Bayer plc	07399	
C	Betanal Trio OF	AgrEvo UK Crop Protection Ltd	08680	31/05/2001
C	Betanal Trio WG	AgrEvo UK Ltd	07537	
C	CX 171	AgrEvo UK Ltd	06803	28/02/2000
C	Goltix Triple	Bayer UK Ltd	06314	28/02/2000

397 Phenmedipham + Lenacil

C	DUK-880	Du Pont (UK) Ltd	04121
C	Stefes 880	Stefes Plant Protection Ltd	08914
C	Stefes Excel	Stefes Plant Protection Ltd	08913

398 Picloram

C	Tordon 22K	Dow AgroSciences Ltd	05083
C	Tordon 22K	Nomix-Chipman Ltd	05790

399 Picloram + Bromacil

C	Hydon	Chipman Ltd	01088

400 Picloram + 2,4-D

C	Atladox HI	Nomix-Chipman Ltd	05559
C	Tordon 101	Dow AgroSciences Ltd	05816

401 Prometryn

C	Alpha Prometryne 50 WP	Makhteshim-Agan (UK) Ltd	04871	
C	Alpha Prometryne 80 WP	Makhteshim-Agan (UK) Ltd	04795	
C	Atlas Prometryn 50 WP	Atlas Interlates Ltd	03502	
C	Gesagard	Novartis Crop Protection UK Ltd	08410	
C	Gesagard 50 WP	Ciba Agriculture	00981	30/06/2000

402 Prometryn + Pendimethalin

C	Monarch	Cyanamid Agriculture Ltd	05160	31/05/2000

403 Prometryn + Terbutryn

C	P-Weed	Pan Britannica Industries Ltd	08574	
C	P-Weed	pbi Agrochemicals Ltd	09608	
C	Peaweed	Pan Britannica Industries Ltd	03248	30/06/2000

404 Propachlor

C	Alpha Propachlor 50 SC	Makhteshim-Agan (UK) Ltd	04873
C	Alpha Propachlor 65 WP	Makhteshim-Agan (UK) Ltd	04807
C	Atlas Orange	Atlas Interlates Ltd	03096

C These products are "approved for agricultural use". For further details refer to page vii.
A These products are approved for use in or near water. For further details refer to page vii.

Product Name	Marketing Company	Reg. No.	Expiry Date

404 Propachlor—continued

C	Atlas Propachlor	Atlas Interlates Ltd	06462	
C	Portman Brasson	Portman Agrochemicals Ltd	08158	
C	Portman Propachlor 50 FL	Portman Agrochemicals Ltd	06892	
C	Portman Propachlor SC	Portman Agrochemicals Ltd	08159	
C	Ramrod 20 Granular	Hortichem Ltd	05806	
C	Ramrod 20 Granular	Monsanto Plc	01687	
C	Ramrod Flowable	Monsanto Plc	01688	
C	Sentinel 2	Tripart Farm Chemicals Ltd	05140	
C	Tripart Sentinel	Tripart Farm Chemicals Ltd	03250	

405 Propachlor + Chloridazon

C	Ashlade CP	Ashlade Formulations Ltd	06481

406 Propachlor + Chlorthal-dimethyl

C	Decimate	ISK Biosciences Ltd	05626

407 Propaquizafop

C	Barclay Rebel	Barclay Chemicals (UK) Ltd	08608	
C	Barclay Rebel II	Barclay Chemicals (UK) Ltd	08897	
C	CleanCrop PropaQ	United Agri Products	09297	
C	Falcon	Cyanamid Agriculture Ltd	08288	
C	Falcon	Cyanamid of GB Ltd	07025	31/01/2000
C	Falcon	Novartis Crop Protection UK Ltd	09384	
C	Landgold PQF 100	Landgold & Co Ltd	08976	
C	Shogun	Ciba Agriculture	08250	31/12/2000
C	Shogun	Cyanamid Agriculture Ltd	08251	
C	Shogun	Novartis Crop Protection UK Ltd	08443	
C	Shogun 100EC	Ciba Agriculture	07026	31/01/2000
C	Standon Propaquizafop	Standon Chemicals Ltd	09120	

408 Propyzamide

C	Barclay Piza 400 FL	Barclay Chemicals (UK) Ltd	09633	
C	Barclay Piza 500	Barclay Chemicals (UK) Ltd	05283	30/09/2000
C	Bulwark Flo	Headland Agrochemicals Ltd	09135	
C	Clayton Propel	Clayton Plant Protection (UK) Ltd	09783	
C	Clayton Propel	Clayton Plant Protection (UK) Ltd	06073	30/09/2000
C	Headland Judo	Headland Agrochemicals Ltd	08339	
C	Headland Judo	Headland Agrochemicals Ltd	07387	
C	Interfarm Base	Interfarm (UK) Limited	09733	
C	Interfarm Quaver	Interfarm (UK) Limited	09728	
C	IT Propyzamide	IT Agro Ltd	08271	30/09/2000
C	Kerb 50 W	Pan Britannica Industries Ltd	01133	
C	Kerb 50 W	Rohm & Haas (UK) Ltd	02986	

C   These products are "approved for agricultural use". For further details refer to page vii.
A   These products are approved for use in or near water. For further details refer to page vii.

Product Name	Marketing Company	Reg. No.	Expiry Date

408 Propyzamide—continued

C	Kerb 80 EDF	Pan Britannica Industries Ltd	09256	
C	Kerb Flo	Pan Britannica Industries Ltd	04521	
C	Kerb Flo	Pan Britannica Industries Ltd	02759	
C	Kerb Granules	Pan Britannica Industries Ltd	01135	30/09/2000
C	Kerb Granules	pbi Agrochemicals Ltd	08917	
C	Kerb Granules	Rohm & Haas (UK) Ltd	01136	
C	Kerb Pro Flo	Pan Britannica Industries Ltd	08679	
C	Kerb Pro Granules	Pan Britannica Industries Ltd	08698	
C	Kerb Pro Granules	pbi Agrochemicals Ltd	09600	
C	Landgold Propyzamide 50	Landgold & Co Ltd	05916	30/09/2000
C	Menace 80 EDF	Headland Agrochemicals Ltd	09255	
C	Mithras 80 EDF	Headland Agrochemicals Ltd	09219	
C	Precis	Pan Britannica Industries Ltd	08678	
C	Precis Flo	Pan Britannica Industries Ltd	08530	
C	Rapier	MTM Agrochemicals Ltd	05314	
C	Redeem Flo	Headland Agrochemicals Ltd	08563	
C	Standon Propyzamide 50	Standon Chemicals Ltd	09054	
C	Standon Propyzamide 50	Standon Chemicals Ltd	05569	30/09/2000
C	Stefes Pride	Stefes Plant Protection Ltd	08899	
C	Stefes Pride	Stefes Plant Protection Ltd	05616	30/09/2000
C	Stefes Pride Flo	Stefes Plant Protection Ltd	07812	
C	Top Farm Propyzamide 500	Top Farm Formulations Ltd	05484	

409 Propyzamide + Clopyralid

C	Matrikerb	Pan Britannica Industries Ltd	01308
C	Matrikerb	pbi Agrochemicals Ltd	09604
C	Matrikerb	Rohm & Haas (UK) Ltd	02443

410 Prosulfuron

C	SL 600	Novartis Crop Protection UK Ltd	08633

411 Prosulfuron + Bromoxynil

C	Jester	Novartis Crop Protection UK Ltd	08681

412 Pyridate

C	Barclay Pirate	Barclay Chemicals (UK) Ltd	07104	29/02/2000
C	Lentagran EC	Novartis Crop Protection UK Ltd	08987	
C	Lentagran WP	Novartis Crop Protection UK Ltd	08478	
C	Lentagran WP	Sandoz Agro Ltd	07556	31/07/2001

413 Quinmerac + Metazachlor

C	Katamaran	BASF plc	09049
C	Standon Metazachlor-Q	Standon Chemicals Ltd	09676

C These products are "approved for agricultural use". For further details refer to page vii.
A These products are approved for use in or near water. For further details refer to page vii.

Product Name	Marketing Company	Reg. No.	Expiry Date

414 Quizalofop-ethyl

C Everest	AgrEvo UK Ltd	07341	30/04/2001
C Mission	AgrEvo UK Ltd	07264	30/04/2001
C Pilot	AgrEvo UK Ltd	07268	31/07/2001
C Stefes Biggles	Stefes Plant Protection Ltd	07083	29/02/2000

415 Quizalofop-P-ethyl

C Alias	AgrEvo UK Ltd	08044	31/05/2001
C CoPilot	AgrEvo UK Ltd	08042	
C Pilot D	AgrEvo UK Ltd	08041	
C Sceptre	AgrEvo UK Ltd	08043	

416 Rimsulfuron

C Landgold Rimsulfuron	Landgold & Co Ltd	08959	
C Tarot	Du Pont (UK) Ltd	07909	
C Titus	Du Pont (UK) Ltd	07908	

417 Sethoxydim

C Checkmate	Hortichem Ltd	09122	
C Checkmate	Rhone-Poulenc Agriculture	06129	30/06/2001

418 Simazine

C Alpha Siamzine 50 WP	Makhteshim-Agan (UK) Ltd	04879	
C Alpha Simazine 50 SC	Makhteshim-Agan (UK) Ltd	04801	
C Alpha Simazine 80 WP	Makhteshim-Agan (UK) Ltd	04800	
C Atlas Simazine	Atlas Crop Protection Ltd	07725	
C Atlas Simazine	Atlas Interlates Ltd	05610	
C Gesatop	Novartis Crop Protection UK Ltd	08412	
C Gesatop 500 SC	Ciba Agriculture	05846	31/08/2000
C MSS Simazine 50 FL	Mirfield Sales Services Ltd	01418	
C Sipcam Simazine Flowable	Sipcam UK Ltd	07622	
C Unicrop Flowable Simazine	Universal Crop Protection Ltd	05447	
C Unicrop Simazine 50	Universal Crop Protection Ltd	02646	
C Unicrop Simazine FL	Universal Crop Protection Ltd	08032	

419 Simazine + Amitrole

C Alpha Simazol	Makhteshim-Agan (UK) Ltd	04799	

420 Simazine + Diquat + Paraquat

C Pathclear S	Miracle Garden Care Ltd	08991	
C Pathclear S	The Scotts Company (UK) Limited	09269	

421 Simazine + Isoproturon

C Alpha Protugan Plus	Makhteshim-Agan (UK) Ltd	08799	

C These products are "approved for agricultural use". For further details refer to page vii.
A These products are approved for use in or near water. For further details refer to page vii.

Product Name	Marketing Company	Reg. No.	Expiry Date

421 Simazine + Isoproturon—continued

C Harlequin 500 SC	Ciba Agriculture	05847	31/08/2000
C Harlequin 500 SC	Makhteshim-Agan (UK) Ltd	09779	
C Harlequin 500 SC	Novartis Crop Protection UK Ltd	08416	30/09/2002

422 Simazine + Pendimethalin

C Deuce	Cyanamid Agriculture Ltd	06746	
C Merit	Cyanamid Agriculture Ltd	04976	

423 Simazine + Trietazine

C Remtal SC	AgrEvo UK Ltd	07270	

424 Sodium chlorate

C Ace-Sodium Chlorate (Fire Suppressed) Weedkiller	Ace Chemicals Ltd	06413	
C Atlacide Soluble Powder Weedkiller	Chipman Ltd	00125	
C Centex	Chemsearch	00456	31/07/2000
C Cooke's Professional Sodium Chlorate Weedkiller with Fire Depressant	Cooke's Chemicals (Sales) Ltd	06796	
C Cooke's Weedclear	Cooke's Chemicals (Sales) Ltd	06512	
C Deosan Chlorate Weedkiller	Deosan Ltd	08275	31/03/2000
C Deosan Chlorate Weedkiller	DiverseyLever Ltd	08521	
C Doff Sodium Chlorate Weedkiller	Doff Portland Ltd	06049	
C Gem Sodium Chlorate Weedkiller	Joseph Metcalf Ltd	04276	
C Morgan's Sodium Chlorate Weedkiller	David Morgan (Nottingham) Ltd	08927	
C Sodium Chlorate	Marlow Chemical Co Ltd	06294	
C Strathclyde Sodium Chlorate Weedclear	Strathclyde Chemical Co Ltd	07420	
C TWK-Total Weedkiller	Yule Catto Consumer Chemicals Ltd	06393	

425 Sodium chloroacetate

C Atlas Somon	Atlas Crop Protection Ltd	07727	
C Atlas Somon	Atlas Interlates Ltd	03045	

426 Sodium monochloroacetate

C Croptex Steel	Hortichem Ltd	02418	

C These products are "approved for agricultural use". For further details refer to page vii.
A These products are approved for use in or near water. For further details refer to page vii.

Product Name	Marketing Company	Reg. No.	Expiry Date

427 Sodium silver thiosulphate

C Argylene	Fargro Ltd	03386	

428 Tebutam

C Comodor 600	Agrichem (International) Ltd	08398	
C Comodor 600	Agrichem (International) Ltd	06808	31/08/2000

429 Tecnazene

C Atlas Tecgran 100	Atlas Crop Protection Ltd	07730	
C Atlas Tecgran 100	Atlas Interlates Ltd	05574	31/01/2001
C Atlas Tecnazene 6% dust	Atlas Crop Protection Ltd	07731	
C Atlas Tecnazene 6% Dust	Atlas Interlates Ltd	06351	
C Bygran F	Wheatley Chemical Co Ltd	00365	
C Fusarex 10G	Hickson & Welch Ltd	09727	
C Hickstor 10	Hickson & Welch Ltd	03121	
Hickstor 3	Hickson & Welch Ltd	03105	
C Hickstor 5	Hickson & Welch Ltd	03180	
Hickstor 6	Hickson & Welch Ltd	03106	
C Hystore 10	Agrichem (International) Ltd	03581	
Hytec	Agrichem Ltd	01099	
C Hytec 6	Agrichem Ltd	03580	
C New Hickstor 6	Hickson & Welch Ltd	04221	
C Tripart Arena 10G	Tripart Farm Chemicals Ltd	05603	
Tripart Arena 3	Tripart Farm Chemicals Ltd	05605	
C Tripart Arena 5G	Tripart Farm Chemicals Ltd	05604	
C Tripart New Arena 6	Tripart Farm Chemicals Ltd	05813	

430 Tecnazene + Carbendazim

C Hickstor 6 Plus MBC	Hickson & Welch Ltd	04176	
C Hortag Tecnacarb	Avon Packers Ltd	02929	
C New Arena Plus	Tripart Farm Chemicals	04598	
C New Hickstor 6 Plus MBC	Hickson & Welch Ltd	04599	
C Tripart Arena Plus	Tripart Farm Chemicals	05602	

431 Tecnazene + Thiabendazole

C Hytec Super	Agrichem (International) Ltd	01100	
New Arena TBZ 6	Tripart Farm Chemicals Ltd	05606	

432 Terbacil

C Sinbar	Du Pont (UK) Ltd	01956	

433 Terbuthylazine + Bromoxynil

C Alpha Bromotril PT	Makhteshim-Agan (UK) Ltd	09435	

C These products are "approved for agricultural use". For further details refer to page vii.
A These products are approved for use in or near water. For further details refer to page vii.

Product Name	Marketing Company	Reg. No.	Expiry Date

434 Terbuthylazine + Cyanazine

C	Angle	Ciba Agriculture	08066	31/08/2000
C	Angle	Novartis Crop Protection UK Ltd	08385	

435 Terbuthylazine + Diflufenican

C	Bolero	Ciba Agriculture	07436	31/07/2001
C	Bolero	Novartis Crop Protection UK Ltd	08392	

436 Terbuthylazine + Isoxaben

C	Skirmish	Ciba Agriculture	08079	31/08/2000
C	Skirmish	Novartis Crop Protection UK Ltd	08444	

437 Terbuthylazine + Terbutryn

C	Batallion	Makhteshim-Agan (UK) Ltd	08305	
C	Opogard	Novartis Crop Protection UK Ltd	08427	
C	Opogard 500SC	Ciba Agriculture	05850	31/08/2000

438 Terbutryn

C	Alpha Terbutryne 50 SC	Makhteshim-Agan (UK) Ltd	04809	
C	Alpha Terbutryne 50 WP	Makhteshim-Agan (UK) Ltd	04875	
C A	Clarosan	Novartis Crop Protection UK Ltd	08396	31/08/2001
C A	Clarosan	The Scotts Company (UK) Limited	09394	
C A	Clarosan 1FG	Ciba Agriculture	03859	31/08/2000
C	Prebane	Novartis Crop Protection UK Ltd	08432	
C	Prebane SC	Ciba Agriculture	07634	31/08/2000

439 Terbutryn + Fomesafen

C	Reflex T	Novartis Crop Protection UK Ltd	08884	
C	Reflex T	Zeneca Crop Protection	07423	30/04/2000
C	Reflex T	Zeneca Crop Protection	08035	31/05/2000

440 Terbutryn + Prometryn

C	P-Weed	Pan Britannica Industries Ltd	08574	
C	P-Weed	pbi Agrochemicals Ltd	09608	
C	Peaweed	Pan Britannica Industries Ltd	03248	30/06/2000

441 Terbutryn + Terbuthylazine

C	Batallion	Makhteshim-Agan (UK) Ltd	08305	
C	Opogard	Novartis Crop Protection UK Ltd	08427	
C	Opogard 500SC	Ciba Agriculture	05850	31/08/2000

442 Terbutryn + Trietazine

C	Senate	AgrEvo UK Ltd	07279	

C   These products are "approved for agricultural use". For further details refer to page vii.
A   These products are approved for use in or near water. For further details refer to page vii.

Product Name	Marketing Company	Reg. No.	Expiry Date

443 Terbutryn + Trifluralin

C	Alpha Terbalin 35 SC	Makhteshim-Agan (UK) Ltd	04792	
C	Alpha Terbalin 350 CS	Makhteshim-Agan (UK) Ltd	07118	31/03/2000
C	Ashlade Summit	Ashlade Formulations Ltd	06214	

444 Thiabendazole + Tecnazene

| C | Hytec Super | Agrichem (International) Ltd | 01100 | |
| | New Arena TBZ 6 | Tripart Farm Chemicals Ltd | 05606 | |

445 Thifensulfuron-methyl

C	DUK 118	Du Pont (UK) Ltd	04596	
C	Landgold Thifensulfuron	Landgold & Co Ltd	08523	
C	Prospect	Du Pont (UK) Ltd	06541	31/12/2001
C	Prospect	Nufarm UK Ltd	09389	

446 Thifensulfuron-methyl + Carfentrazone-ethyl

| C | Harmony Express | Du Pont (UK) Ltd | 09467 | |

447 Thifensulfuron-methyl + Flupyrsulfuron-methyl

| C | Lexus Millenium | Du Pont (UK) Ltd | 09206 | |
| C | Lexus Millenium WSB | Du Pont (UK) Ltd | 09207 | |

448 Thifensulfuron-methyl + Fluroxypyr + Tribenuron-methyl

| C | DP 353 | Du Pont (UK) Ltd | 09626 | |
| C | Starane Super | Dow AgroSciences Ltd | 09625 | |

449 Thifensulfuron-methyl + Metsulfuron-methyl

C	DP 928	Du Pont (UK) Ltd	09632	
C	Harmony M	Du Pont (UK) Ltd	03990	
C	Landgold TM 75	Landgold & Co Ltd	08376	31/10/2000

450 Thifensulfuron-methyl + Tribenuron-methyl

C	Calibre	Du Pont (UK) Ltd	07795	
C	DUK 110	Du Pont (UK) Ltd	09189	
C	DUK 110	Du Pont (UK) Ltd	06266	31/08/2001

451 Tralkoxydim

C	Grasp	Zeneca Crop Protection	06675	
C	Landgold Tralkoxydim	Landgold & Co Ltd	08604	
C	Standon Tralkoxydim	Standon Chemicals Ltd	09579	
C	Standon Tralkoxydim	Standon Chemicals Ltd	08326	30/04/2000

C   These products are "approved for agricultural use". For further details refer to page vii.
A   These products are approved for use in or near water. For further details refer to page vii.

Product Name	Marketing Company	Reg. No.	Expiry Date

452 Tri-allate

C	Avadex BW	Monsanto Plc	00173	31/07/2001
C	Avadex BW 480	Monsanto Plc	04742	31/07/2001
C	Avadex BW Granular	Monsanto Plc	00174	
C	Avadex Excel 15G	Monsanto Plc	07117	
C	Landgold Triallate 480	Landgold & Co Ltd	08505	31/07/2001

453 Triasulfuron

C	Lo-Gran	Novartis Crop Protection UK Ltd	08421	
C	Lo-Gran 20WG	Ciba Agriculture	05993	31/08/2000

454 Triasulfuron + Bromoxynil + Ioxynil

C	Teal	Ciba Agriculture	06117	31/08/2000
C	Teal	Novartis Crop Protection UK Ltd	08453	
C	Teal-M	Ciba Agriculture	08028	31/08/2000
C	Teal-M	Novartis Crop Protection UK Ltd	08454	

455 Triasulfuron + Dicamba

C	Banvel T	Novartis Crop Protection UK Ltd	08470	
C	Banvel T	Sandoz Agro Ltd	06497	31/08/2000
C	Framolene	Ciba Agriculture	06495	31/08/2000
C	Framolene	Novartis Crop Protection UK Ltd	08408	

456 Triasulfuron + Mecoprop-P

C	Raven	Ciba Agriculture	06119	31/08/2000
C	Raven	Novartis Crop Protection UK Ltd	08434	

457 Tribenuron-methyl

C	Quantum	Du Pont (UK) Ltd	06270	
C	Quantum 75 DF	Du Pont (UK) Ltd	09340	

458 Tribenuron-methyl + Fluroxypyr + Thifensulfuron-methyl

C	DP 353	Du Pont (UK) Ltd	09626	
C	Starane Super	Dow AgroSciences Ltd	09625	

459 Tribenuron-methyl + Thifensulfuron-methyl

C	Calibre	Du Pont (UK) Ltd	07795	
C	DUK 110	Du Pont (UK) Ltd	09189	
C	DUK 110	Du Pont (UK) Ltd	06266	31/08/2001

460 Triclopyr

C	Chipman Garlon 4	Nomix-Chipman Ltd	06016	
C	Convoy	Nufarm UK Ltd	09185	
C	Garlon 2	Dow AgroSciences Ltd	05682	

C These products are "approved for agricultural use". For further details refer to page vii.
A These products are approved for use in or near water. For further details refer to page vii.

Product Name	Marketing Company	Reg. No.	Expiry Date

460 Triclopyr—continued

C Garlon 2	Zeneca Crop Protection	06616	
C Garlon 4	Dow AgroSciences Ltd	05090	
C Timbrel	Dow AgroSciences Ltd	05815	
C Tribel 480	United Phosphorus Ltd	08977	31/10/2001
C Triptic 48EC	United Phosphorus Ltd	09294	

461 Triclopyr + Clopyralid

C Grazon 90	Dow AgroSciences Ltd	05456	

462 Triclopyr + Clopyralid + Fluroxypyr

C Pastor	Dow AgroSciences Ltd	07440	

463 Triclopyr + 2,4-D + Dicamba

C Broadshot	Cyanamid Agriculture Ltd	07141	30/09/2000
C Broadsword	United Phosphorus Ltd	09140	
C Nufarm Nu-Shot	Nufarm UK Ltd	09139	
C Terbel Triple	Chimac-Agriphar SA	09138	

464 Triclopyr + Dicamba + Mecoprop

C Fettel	Zeneca Crop Protection	06399	30/04/2000

465 Triclopyr + Fluroxypyr

C Doxstar	Dow AgroSciences Ltd	06050	
C Evade	Dow AgroSciences Ltd	08071	

466 Trietazine + Simazine

C Remtal SC	AgrEvo UK Ltd	07270	

467 Trietazine + Terbutryn

C Senate	AgrEvo UK Ltd	07279	

468 Trifluralin

C Alpha Trifluralin 48 EC	Makhteshim-Agan (UK) Ltd	07406	
C Ashlade Trifluralin	Ashlade Formulations Ltd	08303	
C Ashlade Trimaran	Ashlade Formulations Ltd	06228	
C Atlas Trifluralin	Atlas Crop Protection Ltd	08498	
C Atlas Trifluralin	Atlas Interlates Ltd	03051	30/04/2000
C DAPT Trifluralin 48 EC	DAPT Agrochemicals Ltd	07906	
C Digermin	Montedison UK Ltd	00701	
C Digermin	Sipcam UK Ltd	07221	
C FCC Trigard	Farmers Crop Chemicals Ltd	09030	
C Ipifluor	I Pi Ci	04692	

C   These products are "approved for agricultural use". For further details refer to page vii.
A   These products are approved for use in or near water. For further details refer to page vii.

Product Name	Marketing Company	Reg. No.	Expiry Date

468 Trifluralin—continued

C	MSS Trifluralin 48 EC	Mirfield Sales Services Ltd	07753	
C	MTM Trifluralin	MTM Agrochemicals Ltd	05313	30/04/2001
C	Nufarm Triflur	Nufarm UK Ltd	08311	
C	Portman Trifluralin	Portman Agrochemicals Ltd	05751	
C	Treflan	Dow AgroSciences Ltd	05817	
C	Triflur	Nufarm UK Ltd	09670	
C	Triflurex 48EC	Makhteshim-Agan (UK) Ltd	07947	
C	Trifsan	Dow AgroSciences Ltd	09237	
C	Trifsan	Sanachem International Ltd	08233	30/09/2001
C	Trigard	Farmers Crop Chemicals Ltd	02178	31/08/2002
C	Trigard	pbi Agrochemicals Ltd	09699	
C	Trilogy	United Phosphorus Ltd	08996	
C	Tripart Trifluralin 48 EC	Tripart Farm Chemicals Ltd	02215	
C	Triplen	Sipcam UK Ltd	05897	
C	Tristar	Pan Britannica Industries Ltd	02219	
C	Tristar	pbi Agrochemicals Ltd	09612	
C	Whyte Trifluralin	Whyte Agrochemicals Ltd	09286	

469 Trifluralin + Bromoxynil + Ioxynil

C	Masterspray	Pan Britannica Industries Ltd	02971

470 Trifluralin + Clodinafop-propargyl

C	Hawk	Ciba Agriculture	08030	30/04/2000
C	Hawk	Novartis Crop Protection UK Ltd	08417	
C	Reserve	Ciba Agriculture	08169	28/02/2001
C	Reserve	Novartis Crop Protection UK Ltd	08435	

471 Trifluralin + Diflufenican

C	Ardent	Rhone-Poulenc Agriculture	06203

472 Trifluralin + Isoproturon

C	Autumn Kite	AgrEvo UK Ltd	07119
C	Stefes Union	Stefes Plant Protection Ltd	08146

473 Trifluralin + Isoxaben

C	Axit GR	Dow AgroSciences Ltd	08892
C	Premiere Granules	Dow AgroSciences Ltd	07987

474 Trifluralin + Linuron

C	Chandor	Dow AgroSciences Ltd	05631	
C	Ipicombi TL	I Pi Ci	04608	31/01/2000
C	Linnet	Pan Britannica Industries Ltd	01555	
C	Linnet	pbi Agrochemicals Ltd	09601	
C	Neminfest	Montedison UK Ltd	02546	

C These products are "approved for agricultural use". For further details refer to page vii.
A These products are approved for use in or near water. For further details refer to page vii.

Product Name	Marketing Company	Reg. No.	Expiry Date

474 Trifluralin + Linuron—continued

C	Neminfest	Sipcam UK Ltd	07219	
C	Trifluron	United Phosphorus Ltd	07854	28/02/2000
C	Triplen Combi	Sipcam UK Ltd	05939	30/06/2000

475 Trifluralin + Terbutryn

C	Alpha Terbalin 35 SC	Makhteshim-Agan (UK) Ltd	04792	
C	Alpha Terbalin 350 CS	Makhteshim-Agan (UK) Ltd	07118	31/03/2000
C	Ashlade Summit	Ashlade Formulations Ltd	06214	

476 Triflusulfuron-methyl

C	Debut	Du Pont (UK) Ltd	07804	
C	Debut WSB	Du Pont (UK) Ltd	07809	
C	DUK 440	Du Pont (UK) Ltd	07811	
C	DUK 550	Du Pont (UK) Ltd	07810	
C	Landgold TFS 50	Landgold & Co Ltd	08941	
C	Standon Triflusulfuron	Standon Chemicals Ltd	09487	

477 Trinexapac-ethyl

C	Moddus	Ciba Agriculture	07830	30/11/2000
C	Moddus	Novartis Crop Protection UK Ltd	08801	
C	Moddus	Novartis Crop Protection UK Ltd	08426	30/11/2000
C	Moddus	Novartis Crop Protection UK Ltd	09035	31/03/2001

C   These products are "approved for agricultural use". For further details refer to page vii.
A   These products are approved for use in or near water. For further details refer to page vii.

Product Name	Marketing Company	Reg. No.	Expiry Date

1.2 Fungicides
including bactericides

478 2-Aminobutane

C	Hortichem 2-Aminobutane	Hortichem Ltd	06147

479 Azaconazole + Imazalil

C	Nectec Paste	Hortichem Ltd	08510

480 Azoxystrobin

C	Amistar	Zeneca Crop Protection	08517
C	Barclay ZX	Barclay Chemicals (UK) Ltd	09570
C	Clayton Stobik	Clayton Plant Protection (UK) Ltd	09440
C	Gemstone	Zeneca Crop Protection	08519
C	Landgold Strobilurin 250	Landgold & Co Ltd	09595
C	Me^2 Azoxystrobin	Me^2 Crop Protection Ltd	09654
C	Olympus	Zeneca Crop Protection	08520
C	Priori	Zeneca Crop Protection	08516
C	Standon Azoxystrobin	Standon Chemicals Ltd	09515

481 Azoxystrobin + Fenpropimorph

C	Amistar Pro	Zeneca Crop Protection	08871
C	Aspect	Zeneca Crop Protection	08872
C	Barclay ZA	Barclay Chemicals (UK) Ltd	09462

482 Azoxystrobin + Flutriafol

C	Amigo	Zeneca Crop Protection	08834
C	Amistar Gem	Zeneca Crop Protection	08833

483 Benalaxyl + Mancozeb

C	Barclay Bezant	Barclay Chemicals (UK) Ltd	05914	29/02/2000
C	Clayton Benzeb	Clayton Plant Protection (UK) Ltd	07081	29/02/2000
C	Galben M	Sipcam UK Ltd	07220	
C	Tairel	Sipcam UK Ltd	07767	

484 Benodanil

C	Calirus	BASF plc	00368

485 Benomyl

C	Benlate Fungicide	Du Pont (UK) Ltd	00229

C   These products are "approved for agricultural use". For further details refer to page vii.
A   These products are approved for use in or near water. For further details refer to page vii.

Product Name	Marketing Company	Reg. No.	Expiry Date

486 Bitertanol + Fuberidazole

C	Sibutol	Zeneca Crop Protection	07487	31/12/2000
C	Sibutol	Bayer plc	07238	
C	Sibutol CF	Bayer plc	08174	
C	Sibutol LS	Bayer plc	08109	
C	Sibutol New Formula	Zeneca Crop Protection	08526	31/12/2000
C	Sibutol New Formula	Bayer plc	08270	

487 Bitertanol + Fuberidazole + Imidacloprid + Triadimenol

C	Cereline Secur	Bayer plc	09511

488 Bitertanol + Fuberidazole + Triadimenol

C	Cereline	Bayer plc	07239

489 Bromuconazole

C	Granit	Rhone-Poulenc Agriculture	08268

490 Bromuconazole + Fenpropimorph

C	Granit M	Rhone-Poulenc Agriculture	08269

491 Bupirimate

C	Nimrod	Zeneca Crop Protection	06686

492 Bupirimate + Triforine

C	Nimrod T	The Scotts Company (UK) Limited	09268	
C	Nimrod T	Miracle Professional	07865	
C	Nimrod T	Zeneca Professional Products	06859	31/01/2001
C	Nimrod T	ICI Agrochemicals	01499	31/05/2001

493 Captan

C	Alpha Captan 50 WP	Makhteshim-Agan (UK) Ltd	04797	
C	Alpha Captan 80 WDG	Makhteshim-Agan (UK) Ltd	07096	
C	Alpha Captan 83 WP	Makhteshim-Agan (UK) Ltd	04806	
C	PP Captan 80 WG	Tomen (UK) Plc	08971	
C	PP Captan 80 WG	Zeneca Crop Protection	06696	31/03/2001
C	PP Captan 83	Tomen (UK) Plc	08768	

494 Captan + Penconazole

C	Topas C 50 WP	Novartis Crop Protection UK Ltd	08459	
C	Topas C 50 WP	Ciba Agriculture	03232	31/08/2000

C   These products are "approved for agricultural use". For further details refer to page vii.
A   These products are approved for use in or near water. For further details refer to page vii.

Product Name	Marketing Company	Reg. No.	Expiry Date

495 Carbendazim

	Product Name	Marketing Company	Reg. No.	Expiry Date
C	Ashlade Carbendazim Flowable	Ashlade Formulations Ltd	06213	
C	Barclay Shelter	Barclay Chemicals (UK) Ltd	09000	
C	BASF Turf Systemic Fungicide	BASF plc	05774	
C	Bavistin	BASF plc	00217	
C	Bavistin DF	BASF plc	03848	
C	Bavistin FL	BASF plc	00218	
C	Carbate Flowable	pbi Agrochemicals Ltd	08957	
C	Carbate Flowable	Pan Britannica Industries Ltd	03341	31/03/2001
C	Clayton Chizm	Clayton Plant Protection (UK) Ltd	09124	
C	Delsene 50 Flo	Griffin (Europe) Marketing NV (UK Branch)	09469	
C	Derosal Liquid	AgrEvo UK Ltd	07315	
C	Derosal WDG	AgrEvo UK Ltd	07316	
C	Focal Liquid	AgrEvo UK Crop Protection Ltd	08763	31/05/2001
C	Focal WDG	AgrEvo UK Crop Protection Ltd	08762	31/05/2001
C	Greencrop Mooncoin	Greencrop Technology Ltd	09392	
C	Headland Addstem	Headland Agrochemicals Ltd	06755	
C	Headland Addstem DF	Headland Agrochemicals Ltd	08904	
C	Headland Regain	Headland Agrochemicals Ltd	08675	
C	HY-CARB	Agrichem Ltd	05933	
C	Mascot Systemic	Levington Horticulture Ltd	09132	
C	Mascot Systemic	Rigby Taylor Ltd	08776	
C	Mascot Systemic	Rigby Taylor Ltd	07654	30/11/2001
C	MSS Mircarb	Mirfield Sales Services Ltd	08788	
C	pbi Turf Fungicide	Pan Britannica Industries Ltd	08194	30/11/2001
C	pbi Turf Systemic Fungicide	pbi Agrochemicals Ltd	09349	
C	Quadrangle Hinge	Quadrangle Agrochemicals	08070	
C	Stefes C-Flo	Stefes Plant Protection Ltd	07052	29/02/2000
C	Stefes C-Flo 2	Stefes Plant Protection Ltd	08059	
C	Stefes Carbendazim Flo	Stefes Plant Protection Ltd	05677	
C	Stefes Derosal Liquid	Stefes Plant Protection Ltd	07649	
C	Stefes Derosal WDG	Stefes Plant Protection Ltd	07658	
C	Supercarb	pbi Agrochemicals Ltd	09610	
C	Supercarb	Pan Britannica Industries Ltd	01560	
C	Top Farm Carbendazim - 435	Top Farm Formulations Ltd	05307	
C	Tripart Defensor FL	Tripart Farm Chemicals Ltd	02752	
C	Tripart Defensor Liq	Tripart Farm Chemicals Ltd	07857	30/06/2001
C	Tripart Defensor WDG	Tripart Farm Chemicals Ltd	07855	30/06/2001
C	Turfclear	The Scotts Company (UK) Limited	07506	
C	Turfclear WDG	The Scotts Company (UK) Limited	07490	
C	Twincarb	Vitax Ltd	08777	
C	UPL Carbendazim 500 FL	United Phosphorus Ltd	07472	30/09/2001

C   These products are "approved for agricultural use". For further details refer to page vii.
A   These products are approved for use in or near water. For further details refer to page vii.

Product Name	Marketing Company	Reg. No.	Expiry Date

496 Carbendazim + Chlorothalonil

C	Bravocarb	Zeneca Crop Protection	09105
C	Bravocarb	ISK Biosciences Ltd	05119
C	Greenshield	The Scotts Company (UK) Limited	07988

497 Carbendazim + Chlorothalonil + Maneb

C	Ashlade Mancarb Plus	Ashlade Formulations Ltd	08160
C	Tripart Victor	Tripart Farm Chemicals Ltd	08161

498 Carbendazim + Cymoxanil + Oxadixyl + Thiram

C	Apron Elite	Novartis Crop Protection UK Ltd	08770

499 Carbendazim + Cyproconazole

C	Alto Combi	Novartis Crop Protection UK Ltd	08465	30/04/2001
C	Alto Combi	Sandoz Agro Ltd	05066	31/08/2000

500 Carbendazim + Flusilazole

C	Contrast	Du Pont (UK) Ltd	06150
C	Landgold Flusilazole MBC	Landgold & Co Ltd	08528
C	Punch C	Du Pont (UK) Ltd	06801
C	Punch CA	Du Pont (UK) Ltd	09671
C	Standon Flusilazole Plus	Standon Chemicals Ltd	07403

501 Carbendazim + Flutriafol

C	Early Impact	Zeneca Crop Protection	06659
C	Pacer	Zeneca Crop Protection	06690
C	Palette	Zeneca Crop Protection	06691

502 Carbendazim + Iprodione

C	Calidan	Rhone-Poulenc Agriculture	06536
C	Vitesse	Rhone-Poulenc Amenity	06537

503 Carbendazim + Mancozeb

C	Headland Maple	Headland Agrochemicals Ltd	09044	
C	Kombat WDG	AgrEvo UK Ltd	07201	31/10/2000
C	Kombat WDG	Rohm & Haas (UK) Ltd	05509	
C	Stefes Kombat WDG	Stefes Plant Protection Ltd	07985	30/04/2000

504 Carbendazim + Maneb

C	Ashlade Mancarb FL	Ashlade Formulations Ltd	07977	
C	Headland Dual	Headland Agrochemicals Ltd	03782	
C	MC Flowable	United Phosphorus Ltd	08198	
C	MC Flowable	MTM Agrochemicals Ltd	06295	31/05/2000

C These products are "approved for agricultural use". For further details refer to page vii.
A These products are approved for use in or near water. For further details refer to page vii.

Product Name	Marketing Company	Reg. No.	Expiry Date

504 Carbendazim + Maneb—continued

C	Multi-W FL	pbi Agrochemicals Ltd	09447	
C	Multi-W FL	Pan Britannica Industries Ltd	04131	28/02/2002
C	New Squadron	Quadrangle Agrochemicals Ltd	06756	31/08/2000
C	Protector	Procam Agriculture Ltd	09448	
C	Protector	Procam Agriculture Ltd	07075	28/02/2002
C	Tripart 147	Tripart Farm Chemicals Ltd	07978	
C	Tripart Legion	Tripart Farm Chemicals Ltd	06113	31/08/2000

505 Carbendazim + Maneb + Sulphur

C	Bolda FL	Atlas Crop Protection Ltd	07653	
C	Legion S	Ashlade Formulations Ltd	08121	30/06/2002

506 Carbendazim + Maneb + Tridemorph

C	Cosmic FL	BASF plc	03473	

507 Carbendazim + Metalaxyl

C	Ridomil MBC 60 WP	Novartis Crop Protection UK Ltd	08437	
C	Ridomil MBC 60 WP	Ciba Agriculture	01804	31/08/2000

508 Carbendazim + Prochloraz

C	Novak	AgrEvo UK Ltd	08020	
C	Sportak Alpha	AgrEvo UK Ltd	07222	31/05/2001
C	Sportak Alpha HF	AgrEvo UK Ltd	07225	

509 Carbendazim + Propiconazole

C	Hispor 45 WP	Novartis Crop Protection UK Ltd	08418	
C	Hispor 45WP	Ciba Agriculture	01050	31/08/2000
C	Sparkle 45 WP	Novartis Crop Protection UK Ltd	08450	
C	Sparkle 45 WP	Ciba Agriculture	04968	31/10/2000

510 Carbendazim + Tebuconazole

C	Bayer UK413	Bayer plc	08277	

511 Carbendazim + Tecnazene

C	Hickstor 6 Plus MBC	Hickson & Welch Ltd	04176	
C	Hortag Tecnacarb	Avon Packers Ltd	02929	
C	New Arena Plus	Tripart Farm Chemicals	04598	
C	New Hickstor 6 Plus MBC	Hickson & Welch Ltd	04599	
C	Tripart Arena Plus	Tripart Farm Chemicals	05602	

512 Carbendazim + Vinclozolin

C	Konker	BASF plc	03988	

C   These products are "approved for agricultural use". For further details refer to page vii.
A   These products are approved for use in or near water. For further details refer to page vii.

Product Name	Marketing Company	Reg. No.	Expiry Date

513 Carboxin + Imazalil + Thiabendazole

C	Vitaflo Extra	Uniroyal Chemical Ltd	07048	

514 Carboxin + Prochloraz

C	Prelude Universal LS	AgrEvo UK Crop Protection Ltd	08650	31/05/2001
C	Provax	Uniroyal Chemical Company	08651	31/05/2001

515 Carboxin + Thiabendazole

C	Vitaflo	Uniroyal Chemical Ltd	06379	31/05/2000

516 Carboxin + Thiram

C	Anchor	Uniroyal Chemical Ltd	08684	
C	Anchor	Uniroyal Chemical Ltd	07949	30/09/2000

517 Chlormequat

C	Stabilan 460	Nufarm UK Ltd	09304	
C	Stabilan 640	Nufarm UK Ltd	09401	
C	Stabilan 670	Nufarm UK Ltd	09402	
C	Stabilan 700	Nufarm UK Ltd	09302	
C	Stabilan 750	Nufarm UK Ltd	09303	

518 Chlorothalonil

C	Agriguard Chlorothalonil	Tronsan Ltd	09390	
C	Atlas Cropguard	Mirfield Sales Services Ltd	09123	
C	Barclay Corrib	Barclay Chemicals (UK) Ltd	05886	29/02/2000
C	Barclay Corrib 500	Barclay Chemicals (UK) Ltd	08981	
C	Barclay Corrib 500	Barclay Chemicals (UK) Ltd	06392	29/02/2000
C	Baton SC	Bayer plc	07945	
C	Baton WG	Bayer plc	07944	
C	BB Chlorothalonil	Brown Butlin Group	03320	
C	Bombardier	Universal Crop Protection Ltd	02675	
C	Bombardier FL	Universal Crop Protection Ltd	07910	
C	Bravo 500	Zeneca Crop Protection	09059	
C	Bravo 500	ISK Biosciences Ltd	05638	31/05/2001
C	Bravo 500	BASF plc	05637	
C	Bravo 720	Zeneca Crop Protection	09104	
C	Bravo 720	ISK Biosciences Ltd	05544	30/06/2002
C	Bravo Star	Zeneca Crop Protection	09108	31/10/2000
C	Bravo Star	ISK Biosciences Ltd	08371	
C	Chloronil	Zeneca Crop Protection	09110	31/10/2000
C	Chloronil	ISK Biosciences Ltd	08273	
C	Clayton Turret	Clayton Plant Protection (UK) Ltd	09400	
C	Clayton Turret	Clayton Plant Protection (UK) Ltd	07662	31/03/2000
C	Clortosip	Sipcam UK Ltd	06126	
C	Clortosip 500	Sipcam UK Ltd	09320	

C   These products are "approved for agricultural use". For further details refer to page vii.
A   These products are approved for use in or near water. For further details refer to page vii.

Product Name	Marketing Company	Reg. No.	Expiry Date

518 Chlorothalonil—continued

	Product Name	Marketing Company	Reg. No.	Expiry Date
C	Contact 75	Zeneca Crop Protection	09107	31/10/2000
C	Contact 75	ISK Biosciences Ltd	05563	
C	Daconil Turf	The Scotts Company (UK) Limited	09265	
C	Daconil Turf	Miracle Professional	07929	28/02/2002
C	Daconil Turf	Imperial Chemical Industries Plc	03658	31/10/2000
C	Duomo	Sipcam UK Ltd	06152	29/02/2000
C	Flute	Sipcam UK Ltd	08953	
C	Fusonil Turf	Rigby Taylor Ltd	09695	
C	Greencrop Orchid	Greencrop Technology Ltd	09566	
C	ISK 375	Zeneca Crop Protection	09103	
C	ISK 375	ISK Biosciences Ltd	07455	30/06/2002
C	Jupital	Zeneca Crop Protection	09109	
C	Jupital	ISK Biosciences Ltd	05554	30/06/2002
C	Jupital DG	Zeneca Crop Protection	09181	
C	Landgold Chlorothalonil 50	Landgold & Co Ltd	06265	29/02/2000
C	Landgold Chlorothalonil FL	Landgold & Co Ltd	06335	
C	Mainstay	Quadrangle Agrochemicals	05625	
C	Marnoch Chlorothalonil	Marnoch Ventures Limited	09763	
C	Miros DF	Sipcam UK Ltd	04966	
C	MSS Chlorothalonil	Mirfield Sales Services Ltd	08366	30/06/2001
C	Mycoguard	Chiltern Farm Chemicals Ltd	08115	
C	Repulse	Hortichem Ltd	07641	
C	Repulse	Zeneca Crop Protection	06705	
C	Rover DF	Sipcam UK Ltd	06151	
C	Sipcam Echo 75	Sipcam UK Ltd	08302	
C	Sipcam UK Rover 500	Sipcam UK Ltd	04165	
C	Standon Chlorothalonil 50	Standon Chemicals Ltd	05922	29/02/2000
C	Standon Chlorothalonil 500	Standon Chemicals Ltd	08597	
C	Strada	Sipcam UK Ltd	08824	25/07/2003
C	Top Farm Chlorothalonil 500	Top Farm Formulations Ltd	05926	29/02/2000
C	Tripart Faber	Tripart Farm Chemicals Ltd	05505	31/10/2000
C	Tripart Faber	Tripart Farm Chemicals Ltd	04549	
C	Ultrafaber	Tripart Farm Chemicals Ltd	05627	
C	Visclor 500 SC	Sipcam UK Ltd	09404	
C	Visclor 75 DF	Vischim SRL	09361	

519 Chlorothalonil + Carbendazim

	Product Name	Marketing Company	Reg. No.	Expiry Date
C	Bravocarb	Zeneca Crop Protection	09105	
C	Bravocarb	ISK Biosciences Ltd	05119	
C	Greenshield	The Scotts Company (UK) Limited	07988	

520 Chlorothalonil + Carbendazim + Maneb

	Product Name	Marketing Company	Reg. No.	Expiry Date
C	Ashlade Mancarb Plus	Ashlade Formulations Ltd	08160	
C	Tripart Victor	Tripart Farm Chemicals Ltd	08161	

C   These products are "approved for agricultural use". For further details refer to page vii.
A   These products are approved for use in or near water. For further details refer to page vii.

Product Name	Marketing Company	Reg. No.	Expiry Date

521 Chlorothalonil + Cymoxanil

C	Ashlade Cyclops	ICI Agrochemicals	04857	30/06/2001
C	Cyclops	Zeneca Crop Protection	06650	31/10/2000
C	DUK 44	Du Pont (UK) Ltd	07475	
C	Guardian	Zeneca Crop Protection	06676	31/10/2000

522 Chlorothalonil + Cyproconazole

C	Alto Elite	Novartis Crop Protection UK Ltd	08467	
C	Alto Elite	Sandoz Agro Ltd	05069	31/08/2000
C	CleanCrop Cyprothal	United Agri Products	09580	
C	Octolan	Novartis Crop Protection UK Ltd	08480	
C	Octolan	Sandoz Agro Ltd	06256	31/08/2000
C	SAN 703	Novartis Crop Protection UK Ltd	08487	
C	SAN 703	Sandoz Agro Ltd	06255	31/08/2000

523 Chlorothalonil + Fenpropimorph

C	BAS 438	BASF plc	03451
C	Corbel CL	BASF plc	04196

524 Chlorothalonil + Fluquinconazole

C	Trident	AgrEvo UK Ltd	09369
C	Vista CT	AgrEvo UK Ltd	09368

525 Chlorothalonil + Flutriafol

C	Halo	Zeneca Crop Protection	06520	
C	Impact Excel	Zeneca Crop Protection	06680	
C	PP 375	Zeneca Crop Protection	07898	31/10/2000
C	Prospa	Zeneca Crop Protection	07899	31/10/2000

526 Chlorothalonil + Mancozeb

C	Adagio	pbi Agrochemicals Ltd	09597
C	Adagio	Pan Britannica Industries Ltd	07832
C	Adagio C	pbi Agrochemicals Ltd	09309
C	Dreadnought Flo	pbi Agrochemicals Ltd	09599
C	Dreadnought Flo	Rohm & Haas (UK) Ltd	07600
C	Dreadnought Flo	Pan Britannica Industries Ltd	07599
C	Sipcam Flo	Sipcam UK Ltd	07601

527 Chlorothalonil + Metalaxyl

C	Folio	Novartis Crop Protection UK Ltd	08547	
C	Folio 575 SC	Novartis Crop Protection UK Ltd	08406	30/11/2000
C	Folio 575 SC	Ciba Agriculture	05843	31/08/2000

C These products are "approved for agricultural use". For further details refer to page vii.
A These products are approved for use in or near water. For further details refer to page vii.

Product Name	Marketing Company	Reg. No.	Expiry Date

528 Chlorothalonil + Propamocarb hydrochloride

C	Aventis Merlin	AgrEvo UK Ltd	09719	
C	Merlin	AgrEvo UK Ltd	07943	
C	Tattoo C	AgrEvo UK Ltd	07623	

529 Chlorothalonil + Propiconazole

C	Sambarin 312.5 SC	Novartis Crop Protection UK Ltd	08439	
C	Sambarin 312.5 SC	Ciba Agriculture	05809	31/08/2000

530 Chlorothalonil + Tetraconazole

C	TC Elite	Monsanto Plc	09412	
C	Voodoo	Sipcam UK Ltd	09414	

531 Chlorothalonil + Vinclozolin

C	Curalan CL	BASF plc	07174	

532 Copper Ammonium Carbonate

C	Croptex Fungex	Hortichem Ltd	02888	

533 Copper complex - bordeaux

C	Wetcol 3 Copper Fungicide	Ford Smith & Co Ltd	02360	

534 Copper oxychloride

C	Cuprokylt	Universal Crop Protection Ltd	00604	
C	Cuprokylt FL	Universal Crop Protection Ltd	08299	
C	Cuprokylt L	Universal Crop Protection Ltd	02769	31/01/2000
C	Cuprosana H	Universal Crop Protection Ltd	00605	
C	Headland Inorganic Liquid Copper	Headland Agrochemicals Ltd	07799	

535 Copper oxychloride + Maneb + Sulphur

C	Ashlade SMC Flowable	Ashlade Formulations Ltd	06494	

536 Copper oxychloride + Metalaxyl

C	Ridomil Plus	Novartis Crop Protection UK Ltd	08353	
C	Ridomil Plus 50 WP	Ciba Agriculture	01803	31/03/2001

537 Cymoxanil + Carbendazim + Oxadixyl + Thiram

C	Apron Elite	Novartis Crop Protection UK Ltd	08770	

538 Cymoxanil + Chlorothalonil

C	Ashlade Cyclops	ICI Agrochemicals	04857	30/06/2001
C	Cyclops	Zeneca Crop Protection	06650	31/10/2000
C	DUK 44	Du Pont (UK) Ltd	07475	
C	Guardian	Zeneca Crop Protection	06676	31/10/2000

C These products are "approved for agricultural use". For further details refer to page vii.

A These products are approved for use in or near water. For further details refer to page vii.

Product Name	Marketing Company	Reg. No.	Expiry Date

PSD FUNGICIDES

539 Cymoxanil + Mancozeb

C	Ashlade Solace	Ashlade Formulations Ltd	08087	
C	Besiege	Du Pont (UK) Ltd	08086	
C	Besiege WSB	Du Pont (UK) Ltd	08075	
C	Clayton Krypton	Clayton Plant Protection (UK) Ltd	09398	
C	Clayton Krypton	Clayton Plant Protection (UK) Ltd	06973	29/02/2000
C	Curzate M68	Du Pont (UK) Ltd	08072	
C	Curzate M68 WSB	Du Pont (UK) Ltd	08073	
C	Fytospore	Zeneca Crop Protection	06517	28/02/2000
C	Fytospore 68	Du Pont (UK) Ltd	08649	
C	Me^2 Cymoxeb	Me^2 Crop Protection Ltd	09486	
C	Rhythm	Interfarm (UK) Limited	09636	
C	Standon Cymoxanil Extra	Standon Chemicals Ltd	09442	
C	Standon Cymoxanil Extra	Standon Chemicals Ltd	06807	29/02/2000
C	Stefes Blight Spray	Stefes Plant Protection Ltd	05811	29/02/2000
C	Systol M	Quadrangle Agrochemicals	08085	
C	Systol M	Quadrangle Agrochemicals	03480	
C	Systol M WSB	Quadrangle Agrochemicals	08074	

540 Cymoxanil + Mancozeb + Oxadixyl

C	Ripost	Novartis Crop Protection UK Ltd	08484	31/08/2000
C	Ripost	Sandoz Agro Ltd	04890	31/08/2000
C	Ripost Pepite	Bayer Ltd	08895	
C	Ripost Pepite	Novartis Crop Protection UK Ltd	08485	
C	Ripost Pepite	Sandoz Agro Ltd	06485	31/03/2001
C	Trustan	Du Pont (UK) Ltd	05022	31/12/2000
C	Trustan WDG	Du Pont (UK) Ltd	05050	

541 Cyproconazole

C	Agriguard Cyproconazole	AgriGuard	09406	
C	Alto 100 SL	Novartis Crop Protection UK Ltd	08350	30/04/2001
C	Alto 100 SL	Sandoz Agro Ltd	05065	29/02/2000
C	Alto 240 EC	Novartis Crop Protection UK Ltd	08354	
C	Alto 240 EC	Sandoz Agro Ltd	08243	31/03/2000
C	Aplan	Novartis Crop Protection UK Ltd	08351	31/03/2001
C	Aplan	Sandoz Agro Ltd	06121	29/02/2000
C	Aplan 240 EC	Novartis Crop Protection UK Ltd	08355	
C	Aplan 240 EC	Sandoz Agro Ltd	08244	31/03/2000
C	Barclay Shandon	Barclay Chemicals (UK) Ltd	06464	30/04/2002
C	Clayton Cyprocon	Clayton Plant Protection (UK) Ltd	07668	30/04/2002
C	Greencrop Gentian	Greencrop Technology Ltd	09380	30/04/2002
C	Landgold Cyproconazole 100	Landgold & Co Ltd	06463	
C	SAN 619	Sandoz Agro Ltd	06120	31/08/2000
C	SAN 619F 240 EC	Novartis Crop Protection UK Ltd	08357	
C	Standon Cyproconazole	Standon Chemicals Ltd	07751	30/04/2002

C These products are "approved for agricultural use". For further details refer to page vii.
A These products are approved for use in or near water. For further details refer to page vii.

Product Name	Marketing Company	Reg. No.	Expiry Date

541 Cyproconazole—continued

C Star Cyproconazole	Star Agrochem Ltd	09163	30/04/2002

542 Cyproconazole + Carbendazim

C Alto Combi	Novartis Crop Protection UK Ltd	08465	30/04/2001
C Alto Combi	Sandoz Agro Ltd	05066	31/08/2000

543 Cyproconazole + Chlorothalonil

C Alto Elite	Novartis Crop Protection UK Ltd	08467	
C Alto Elite	Sandoz Agro Ltd	05069	31/08/2000
C CleanCrop Cyprothal	United Agri Products	09580	
C Octolan	Novartis Crop Protection UK Ltd	08480	
C Octolan	Sandoz Agro Ltd	06256	31/08/2000
C SAN 703	Novartis Crop Protection UK Ltd	08487	
C SAN 703	Sandoz Agro Ltd	06255	31/08/2000

544 Cyproconazole + Cyprodinil

C Radius	Novartis Crop Protection UK Ltd	09387	

545 Cyproconazole + Mancozeb

C Alto Eco	Novartis Crop Protection UK Ltd	08466	
C Alto Eco	Sandoz Agro Ltd	05068	31/08/2000

546 Cyproconazole + Prochloraz

C Profile	AgrEvo UK Ltd	08134	
C SAN 710	Novartis Crop Protection UK Ltd	08488	31/05/2001
C SAN 710	Sandoz Agro Ltd	07918	31/08/2000
C SAN 710 HF	Novartis Crop Protection UK Ltd	08489	
C SAN 710 HF	Sandoz Agro Ltd	07919	31/08/2000
C Sportak Delta 460	AgrEvo UK Crop Protection Ltd	07224	31/05/2001
C Sportak Delta 460 HF	AgrEvo UK Ltd	07431	
C Tiptor	Novartis Crop Protection UK Ltd	08495	
C Tiptor	AgrEvo UK Ltd	07295	31/05/2001
C Tiptor	Sandoz Agro Ltd	05105	31/08/2000

547 Cyproconazole + Propiconazole

C Menara	Novartis Crop Protection UK Ltd	09321	

548 Cyproconazole + Quinoxyfen

C Divora	Novartis Crop Protection UK Ltd	08960	
C DOE 1762	Dow AgroSciences Ltd	08962	
C NOV 1390	Novartis Crop Protection UK Ltd	08961	

C These products are "approved for agricultural use". For further details refer to page vii.
A These products are approved for use in or near water. For further details refer to page vii.

Product Name	Marketing Company	Reg. No.	Expiry Date

549 Cyproconazole + Tridemorph

C	Alto Major	Novartis Crop Protection UK Ltd	08468	30/09/2001
C	Alto Major	Sandoz Agro Ltd	06979	30/09/2001
C	Moot	Novartis Crop Protection UK Ltd	08479	30/09/2001
C	Moot	Sandoz Agro Ltd	06990	30/09/2001
C	SAN 735	Novartis Crop Protection UK Ltd	08490	30/09/2001
C	San 735	Sandoz Agro Ltd	06991	30/09/2001

550 Cyprodinil

C	Barclay Amtrak	Barclay Chemicals (UK) Ltd	09562	
C	Chieftain	Novartis Crop Protection UK Ltd	09301	
C	CleanCrop Cyprodinil	United Agri Products	09668	
C	Helmet	Novartis Crop Protection UK Ltd	08765	
C	NOV 219	Novartis Crop Protection UK Ltd	08766	
C	Skua	Novartis Crop Protection UK Ltd	08767	
C	Standon Cyprodinil	Standon Chemicals Ltd	09345	
C	Unix	Novartis Crop Protection UK Ltd	08764	

551 Cyprodinil + Cyproconazole

C	Radius	Novartis Crop Protection UK Ltd	09387	

552 Dichlofluanid

C	Elvaron	Bayer plc	00789	31/12/2000
C	Elvaron WG	Bayer plc	04855	

553 Dichlorophen

C	Halophen RE 49	McMillan Technical Services Ltd	04636	
C	Super Mosstox	Rhone-Poulenc Amenity	05339	

554 Dicloran

C	Fumite Dicloran Smoke	Octavius Hunt Ltd	09291	
C	Fumite Dicloran Smoke	Octavius Hunt Ltd	00930	31/10/2001

555 Difenoconazole

C	Plover	Novartis Crop Protection UK Ltd	08429	
C	Plover	Ciba Agriculture	07232	31/08/2000

556 Difenzoquat

C	Match	Cyanamid Agriculture Ltd	07186	
C	Match SL	Cyanamid Agriculture Ltd	07185	31/05/2000

557 Dimethomorph + Mancozeb

C	Invader	Cyanamid Agriculture Ltd	06989	

C   These products are "approved for agricultural use". For further details refer to page vii.
A   These products are approved for use in or near water. For further details refer to page vii.

Product Name	Marketing Company	Reg. No.	Expiry Date

557 Dimethomorph + Mancozeb—continued

C	Invader WDG	Cyanamid Agriculture Ltd	09556	
C	Saracen	Cyanamid Agriculture Ltd	07872	
C	Saracen WDG	Cyanamid Agriculture Ltd	09557	

558 Dinocap

C	Karathane Liquid	Landseer Limited	09262	
C	Karathane Liquid	Rohm & Haas (UK) Ltd	07198	30/09/2001

559 Dithianon

C	Barclay Cluster	Barclay Chemicals (UK) Ltd	08792	
C	Barclay Cluster	Barclay Chemicals (UK) Ltd	07467	29/02/2000
C	Dithianon Flowable	Cyanamid Agriculture Ltd	07007	

560 Dithianon + Penconazole

C	Topas D275 SC	Novartis Crop Protection UK Ltd	08460	
C	Topas D275 SC	Ciba Agriculture	05855	31/08/2000

561 Dodemorph

C	F238	BASF plc	00206	

562 Dodine

C	Barclay Dodex	Barclay Chemicals (UK) Ltd	09055	
C	Barclay Dodex	Barclay Chemicals (UK) Ltd	05655	29/02/2000
C	Radspor FL	Truchem Ltd	01685	

563 Epoxiconazole

C	Agriguard Epoxiconazole	Tronsan Ltd	09407	
C	Epic	BASF plc	08320	
C	Epic	BASF plc	07237	30/11/2000
C	Opus	BASF plc	08319	
C	Opus	BASF plc	07236	30/11/2000
C	Standon Epoxiconazole	Standon Chemicals Ltd	09517	

564 Epoxiconazole + Fenpropimorph

C	Barclay Riverdance	Barclay Chemicals (UK) Ltd	09658	
C	Eclipse	BASF plc	07361	
C	Greencrop Galore	Greencrop Technology Ltd	09561	
C	Landgold Epoxiconazole FM	Landgold & Co Ltd	08806	
C	Opus Team	BASF plc	07362	
C	Standon Epoxifen	Standon Chemicals Ltd	08972	

C These products are "approved for agricultural use". For further details refer to page vii.
A These products are approved for use in or near water. For further details refer to page vii.

Product Name	Marketing Company	Reg. No.	Expiry Date

565 Epoxiconazole + Fenpropimorph + Kresoxim-methyl

C	CleanCrop Kresoxazole Plus	United Agri Products	09742	
C	Mantra	BASF plc	08886	
C	Standon Kresoxim Super	Standon Chemicals Ltd	09794	

566 Epoxiconazole + Kresoxim-methyl

C	Barclay Avalon	Barclay Chemicals (UK) Ltd	09466	
C	Clayton Gantry	Clayton Plant Protection (UK) Ltd	09482	
C	CleanCrop Kresoxazole	United Agri Products	09698	
C	Landmark	BASF plc	08889	
C	Me^2 KME	Me^2 Crop Protection Ltd	09594	
C	Standon Kresoxim-Epoxiconazole	Standon Chemicals Ltd	09281	

568 Epoxiconazole + Tridemorph

C	BAS 47807F	BASF plc	08147	31/12/2000
C	Opus Plus	BASF plc	07363	

569 Ethirimol + Flutriafol + Thiabendazole

C	Ferrax	Zeneca Crop Protection	06662	
C	Ferrax	Bayer plc	05284	

570 Ethirimol + Fuberidazole + Triadimenol

C	Bay UK 292	Bayer plc	04335	

571 Etridiazole

C	AAterra WP	Zeneca Crop Protection	06625	
C	Standon Etridiazole 35	Standon Chemicals Ltd	08778	

572 Fenarimol

C	Rimidin	Dow AgroSciences Ltd	07938	
C	Rimidin	Rigby Taylor Ltd	05907	
C	Rubigan	Dow AgroSciences Ltd	05489	

573 Fenbuconazole

C	Indar 5EC	Rohm & Haas (UK) Ltd	07581	
C	Indar 5EW	T P Whelehan Son & Co Ltd	09644	
C	Indar 5EW	Landseer Limited	09518	
C	Indar 5EW	Rohm & Haas (UK) Ltd	08011	31/03/2002
C	Indar 5EW	AgrEvo UK Crop Protection Ltd	07580	
C	Kruga 5EC	Headland Agrochemicals Ltd	08756	
C	Reward 5EC	Headland Agrochemicals Ltd	08757	
C	Scarab 5EW	Headland Agrochemicals Ltd	08012	31/03/2002
C	Surpass 5EC	Headland Agrochemicals Ltd	08758	

C   These products are "approved for agricultural use". For further details refer to page vii.
A   These products are approved for use in or near water. For further details refer to page vii.

Product Name	Marketing Company	Reg. No.	Expiry Date

574 Fenbuconazole + Fenpropimorph

C Cherokee 318.5 EC	Novartis Crop Protection UK Ltd	08395	31/03/2000
C Cherokee 318.5 EC	Ciba Agriculture	07582	31/03/2000
C Indar Must	Rohm & Haas (UK) Ltd	08718	31/03/2000
C Indar Must	Rohm & Haas (UK) Ltd	07583	31/03/2000
C Myriad	Rhone-Poulenc Agriculture Ltd	08719	31/03/2000
C Myriad	Rhone-Poulenc Agriculture	07584	31/03/2000

575 Fenbuconazole + Prochloraz

C Mirage Extra	Stefes UK Ltd	07958	
C Stefes Fortune	Stefes Plant Protection Ltd	08537	31/01/2001
C Stefes Inception	Stefes Plant Protection Ltd	08829	

576 Fenbuconazole + Propiconazole

C Graphic	Novartis Crop Protection UK Ltd	08415	
C Graphic	Ciba Agriculture	07585	31/03/2000
C Indar CG	Novartis Crop Protection UK Ltd	08578	
C Indar CG	Ciba Agriculture	07586	31/03/2000

577 Fenbuconazole + Tridemorph

C Unison	Pan Britannica Industries Ltd	08318	30/04/2002

578 Fenhexamid

C Lattice	Bayer plc	09198	
C Teldor	Bayer plc	08955	

579 Fenpiclonil

C Gambit	Novartis Crop Protection UK Ltd	08535	
C Gambit	Ciba Agriculture	08182	31/08/2000

580 Fenpropidin

C Landgold Fenpropidin 750	Landgold & Co Ltd	08973	
C Mallard	Novartis Crop Protection UK Ltd	08662	
C Mallard	Novartis Crop Protection UK Ltd	08422	31/08/2000
C Mallard	Ciba Agriculture	07934	31/08/2000
C Patrol	Novartis Crop Protection UK Ltd	08661	
C Patrol	Zeneca Crop Protection	06693	31/08/2000
C Tern	Novartis Crop Protection UK Ltd	08660	
C Tern	Novartis Crop Protection UK Ltd	08455	31/08/2000
C Tern	Ciba Agriculture	07933	31/08/2000

581 Fenpropidin + Fenpropimorph

C Agrys	Novartis Crop Protection UK Ltd	08382	
C Agrys	Ciba Agriculture	08083	31/08/2000

C These products are "approved for agricultural use". For further details refer to page vii.
A These products are approved for use in or near water. For further details refer to page vii.

Product Name	Marketing Company	Reg. No.	Expiry Date

581 Fenpropidin + Fenpropimorph—continued

C Boscor	Novartis Crop Protection UK Ltd	08682	
C Boscor	Novartis Crop Protection UK Ltd	08393	31/08/2000
C Boscor	Ciba Agriculture	07416	31/08/2000

582 Fenpropidin + Prochloraz

C SL 552A	Novartis Crop Protection UK Ltd	08673	
C SL 552A	Novartis Crop Protection UK Ltd	08448	31/08/2000
C SL 552A	Ciba Agriculture	07137	31/08/2000
C Sponsor	AgrEvo UK Ltd	08674	
C Sponsor	AgrEvo UK Crop Protection Ltd	07136	31/08/2000

583 Fenpropidin + Propiconazole

C Prophet	Novartis Crop Protection UK Ltd	08433	
C Prophet 500EC	Ciba Agriculture	07447	31/08/2000
C Sheen	Novartis Crop Protection UK Ltd	08442	
C Sheen	Ciba Agriculture	07828	31/08/2000
C Zulu	Novartis Crop Protection UK Ltd	08464	
C Zulu	Ciba Agriculture	07829	31/08/2000

584 Fenpropidin + Propiconazole + Tebuconazole

C Bayer UK 593	Bayer plc	08285	
C Gladio	Novartis Crop Protection UK Ltd	08413	
C Gladio	Ciba Agriculture	08284	31/08/2000

585 Fenpropidin + Tebuconazole

C Monicle	Bayer plc	07375	
C SL 556 500 EC	Ciba Agriculture	07376	31/08/2000
C SL-556 500 EC	Novartis Crop Protection UK Ltd	08449	

586 Fenpropimorph

C Aura	Novartis Crop Protection UK Ltd	08388	
C Aura 750 EC	Ciba Agriculture	05705	31/08/2000
C BAS 421F	BASF plc	06127	
C CleanCrop Fenpropimorph	United Agri Products	09445	
C Corbel	BASF plc	00578	
C Keetak	BASF plc	06950	
C Landgold Fenpropimorph 750	Landgold & Co Ltd	06319	29/02/2000
C Mistral	Novartis Crop Protection UK Ltd	08425	
C Mistral	Ciba Agriculture	06943	31/08/2000
C Standon Fenpropimorph 750	Standon Chemicals Ltd	08965	
C Standon Fenpropimorph 750	Standon Chemicals Ltd	05654	29/02/2000
C Widgeon	Novartis Crop Protection UK Ltd	08463	
C Widgeon 750EC	Ciba Agriculture	06101	31/08/2000

C These products are "approved for agricultural use". For further details refer to page vii.
A These products are approved for use in or near water. For further details refer to page vii.

Product Name	Marketing Company	Reg. No.	Expiry Date

587 Fenpropimorph + Azoxystrobin

C	Amistar Pro	Zeneca Crop Protection	08871
C	Aspect	Zeneca Crop Protection	08872
C	Barclay ZA	Barclay Chemicals (UK) Ltd	09462

588 Fenpropimorph + Bromuconazole

C	Granit M	Rhone-Poulenc Agriculture	08269

589 Fenpropimorph + Chlorothalonil

C	BAS 438	BASF plc	03451
C	Corbel CL	BASF plc	04196

590 Fenpropimorph + Epoxiconazole

C	Barclay Riverdance	Barclay Chemicals (UK) Ltd	09658
C	Eclipse	BASF plc	07361
C	Greencrop Galore	Greencrop Technology Ltd	09561
C	Landgold Epoxiconazole FM	Landgold & Co Ltd	08806
C	Opus Team	BASF plc	07362
C	Standon Epoxifen	Standon Chemicals Ltd	08972

591 Fenpropimorph + Epoxiconazole + Kresoxim-methyl

C	Cleancrop Kresoxazole Plus	United Agri Products	09742
C	Mantra	BASF plc	08886
C	Standon Kresoxim Super	Standon Chemicals Ltd	09794

592 Fenpropimorph + Fenbuconazole

C	Cherokee 318.5 EC	Novartis Crop Protection UK Ltd	08395	31/03/2000
C	Cherokee 318.5 EC	Ciba Agriculture	07582	31/03/2000
C	Indar Must	Rohm & Haas (UK) Ltd	08718	31/03/2000
C	Indar Must	Rohm & Haas (UK) Ltd	07583	31/03/2000
C	Myriad	Rhone-Poulenc Agriculture Ltd	08719	31/03/2000
C	Myriad	Rhone-Poulenc Agriculture	07584	31/03/2000

593 Fenpropimorph + Fenpropidin

C	Agrys	Novartis Crop Protection UK Ltd	08382	
C	Agrys	Ciba Agriculture	08083	31/08/2000
C	Boscor	Novartis Crop Protection UK Ltd	08682	
C	Boscor	Novartis Crop Protection UK Ltd	08393	31/08/2000
C	Boscor	Ciba Agriculture	07416	31/08/2000

594 Fenpropimorph + Flusilazole

C	BAS 48500F	BASF plc	06784
C	Colstar	Du Pont (UK) Ltd	06783
C	DUK 7876	Du Pont (UK) Ltd	09588

C These products are "approved for agricultural use". For further details refer to page vii.
A These products are approved for use in or near water. For further details refer to page vii.

Product Name	Marketing Company	Reg. No.	Expiry Date

595 Fenpropimorph + Flusilazole + Tridemorph

C	Bingo	BASF plc	06920	
C	DUK 51	Du Pont (UK) Ltd	06764	30/09/2001
C	Justice	Du Pont (UK) Ltd	07963	30/09/2001

596 Fenpropimorph + Kresoxim-methyl

C	Ensign	BASF plc	08362	
C	Greencrop Monsoon	Greencrop Technology Ltd	09573	
C	Standon Kresoxim FM	Standon Chemicals Ltd	08922	

597 Fenpropimorph + Prochloraz

C	Jester	Ciba Agriculture	06586	31/08/2000
C	Mirage Super 600EC	Makhteshim-Agan (UK) Ltd	08531	
C	SL571A	Novartis Crop Protection UK Ltd	08420	
C	Sprint	AgrEvo UK Ltd	07291	31/05/2001
C	Sprint HF	AgrEvo UK Ltd	07292	

598 Fenpropimorph + Propiconazole

C	Belvedere	Makhteshim-Agan (UK) Ltd	08084	
C	Decade	Novartis Crop Protection UK Ltd	08402	
C	Decade 500 EC	Ciba Agriculture	05757	31/08/2000
C	Glint	Novartis Crop Protection UK Ltd	08414	
C	Glint 500 EC	Ciba Agriculture	04126	31/08/2000
C	Mantle	Novartis Crop Protection UK Ltd	08424	
C	Mantle 425 EC	Ciba Agriculture	05715	30/04/2000

599 Fenpropimorph + Quinoxyfen

C	EF-1288	Dow AgroSciences Ltd	08241	
C	Orka	Dow AgroSciences Ltd	08879	

600 Fenpropimorph + Tebuconazole

C	Bayer UK506	Bayer plc	07540	31/03/2001
C	BUK 84000F	BASF plc	07541	31/03/2001

601 Fenpropimorph + Tridemorph

C	BAS 46402F	BASF plc	03313	
C	Gemini	BASF plc	05684	

602 Fentin acetate + Maneb

C	Brestan 60 SP	AgrEvo UK Ltd	07305	

603 Fentin hydroxide

C	Ashlade Flotin	Ashlade Formulations Ltd	06223	30/09/2000

C   These products are "approved for agricultural use". For further details refer to page vii.
A   These products are approved for use in or near water. For further details refer to page vii.

177

Product Name	Marketing Company	Reg. No.	Expiry Date

603 Fentin hydroxide—continued

C	Ashlade Flotin 2	Ashlade Formulations Ltd	06224	
C	Barclay Fentin Flow	Barclay Chemicals (UK) Ltd	07914	
C	Barclay Fentin Flow 532	Barclay Chemicals (UK) Ltd	09434	
C	Du-Ter 50	AgrEvo UK Ltd	07317	31/03/2000
C	Farmatin 560	AgrEvo UK Ltd	07320	
C	Greencrop Rosette	Greencrop Technology Ltd	09648	
C	Keytin	Chiltern Farm Chemicals Ltd	08894	
C	MSS Flotin 480	Mirfield Sales Services Ltd	07616	
C	Stefes Blytin	Stefes UK Ltd	07907	
C	Super-Tin 4L	Whyte Agrochemicals Ltd	09028	
C	Super-Tin 4L	Chiltern Farm Chemicals Ltd	02995	
C	Super-Tin 80WP	Chiltern Farm Chemicals Ltd	07606	31/05/2001
C	Super-Tin 80WP	Pan Britannica Industries Ltd	07605	
C	Supertin 4L	Griffin (Europe) Marketing NV (UK Branch)	09559	
C	Supertin 80 WP	Griffin (Europe) Marketing NV (UK Branch)	09560	

604 Fentin hydroxide + Glufosinate-ammonium

C	Safran	AgrEvo UK Crop Protection Ltd	08559	31/05/2001

605 Fluazinam

C	Barclay Cobbler	Barclay Chemicals (UK) Ltd	08349	
C	FD 4058	Zeneca Crop Protection	07093	30/11/2000
C	Landgold Fluazinam	Landgold & Co Ltd	08060	
C	Legacy	ISK Biosciences Ltd	07401	
C	Salvo	Zeneca Crop Protection	07092	
C	Shirlan	Zeneca Crop Protection	07091	
C	Shirlan Programme	Zeneca Crop Protection	08761	
C	Standon Fluazinam 500	Standon Chemicals Ltd	08670	
C	Top Farm Fluazinam	Top Farm Formulations Ltd	07683	

606 Fludioxonil

C	Beret Gold	Novartis Crop Protection UK Ltd	08390	
C	Beret Gold	Ciba Agriculture	07531	31/12/2000
C	Celest	Novartis Crop Protection UK Ltd	08617	

607 Fluquinconazole

C	Aventis Flamenco	AgrEvo UK Ltd	09729	
C	Diablo	AgrEvo UK Ltd	09386	
C	Flamenco	AgrEvo UK Ltd	09385	

C These products are "approved for agricultural use". For further details refer to page vii.
A These products are approved for use in or near water. For further details refer to page vii.

Product Name	Marketing Company	Reg. No.	Expiry Date

608 Fluquinconazole + Chlorothalonil

C	Trident	AgrEvo UK Ltd	09369
C	Vista CT	AgrEvo UK Ltd	09368

609 Fluquinconazole + Prochloraz

C	Aventis Foil	AgrEvo UK Ltd	09709
C	Baron	AgrEvo UK Ltd	09364
C	Flamenco Plus	AgrEvo UK Ltd	09362
C	Foil	AgrEvo UK Ltd	09363

610 Flusilazole

C	Benocap	Dow AgroSciences Ltd	07646	31/08/2000
C	DP 241	Du Pont (UK) Ltd	08253	
C	DUK 747	Du Pont (UK) Ltd	08239	
C	Genie	Du Pont (UK) Ltd	08238	
C	Landgold Flusilazole 400	Landgold & Co Ltd	05908	29/02/2000
C	Lyric	Du Pont (UK) Ltd	08252	
C	Sanction	Du Pont (UK) Ltd	08237	

611 Flusilazole + Carbendazim

C	Contrast	Du Pont (UK) Ltd	06150
C	Landgold Flusilazole MBC	Landgold & Co Ltd	08528
C	Punch C	Du Pont (UK) Ltd	06801
C	Punch CA	Du Pont (UK) Ltd	09671
C	Standon Flusilazole Plus	Standon Chemicals Ltd	07403

612 Flusilazole + Fenpropimorph

C	BAS 48500F	BASF plc	06784
C	Colstar	Du Pont (UK) Ltd	06783
C	DUK 7876	Du Pont (UK) Ltd	09588

613 Flusilazole + Fenpropimorph + Tridemorph

C	Bingo	BASF plc	06920	
C	DUK 51	Du Pont (UK) Ltd	06764	30/09/2001
C	Justice	Du Pont (UK) Ltd	07963	30/09/2001

614 Flusilazole + Tridemorph

C	Fusion	Du Pont (UK) Ltd	04908	30/09/2001
C	Meld	BASF plc	04914	
C	Option	Du Pont (UK) Ltd	07951	30/09/2001

615 Flutriafol

C	Barclay Rascal	Barclay Chemicals (UK) Ltd	08921	
C	Clayton Flutriafol	Clayton Plant Protection (UK) Ltd	05972	29/02/2000

C   These products are "approved for agricultural use". For further details refer to page vii.
A   These products are approved for use in or near water. For further details refer to page vii.

Product Name	Marketing Company	Reg. No.	Expiry Date

615 Flutriafol—continued

C	Impact	Zeneca Crop Protection	06679	
C	Landgold Flutriafol	Landgold & Co Ltd	06244	29/02/2000
C	Pointer	Zeneca Crop Protection	06695	
C	PP 450	Zeneca Crop Protection	06700	

616 Flutriafol + Azoxystrobin

C	Amigo	Zeneca Crop Protection	08834
C	Amistar Gem	Zeneca Crop Protection	08833

617 Flutriafol + Carbendazim

C	Early Impact	Zeneca Crop Protection	06659
C	Pacer	Zeneca Crop Protection	06690
C	Palette	Zeneca Crop Protection	06691

618 Flutriafol + Chlorothalonil

C	Halo	Zeneca Crop Protection	06520	
C	Impact Excel	Zeneca Crop Protection	06680	
C	PP 375	Zeneca Crop Protection	07898	31/10/2000
C	Prospa	Zeneca Crop Protection	07899	31/10/2000

619 Flutriafol + Ethirimol + Thiabendazole

C	Ferrax	Zeneca Crop Protection	06662
C	Ferrax	Bayer plc	05284

620 Fosetyl-aluminium

C	Aliette	Rhone-Poulenc Agriculture	05648
C	Aliette	Hortichem Ltd	02484
C	Aliette 80 WG	Hortichem Ltd	09156

621 Fuberidazole + Bitertanol

C	Sibutol	Zeneca Crop Protection	07487	31/12/2000
C	Sibutol	Bayer plc	07238	
C	Sibutol CF	Bayer plc	08174	
C	Sibutol LS	Bayer plc	08109	
C	Sibutol New Formula	Zeneca Crop Protection	08526	31/12/2000
C	Sibutol New Formula	Bayer plc	08270	

622 Fuberidazole + Bitertanol + Imidacloprid + Triadimenol

C	Cereline Secur	Bayer plc	09511

623 Fuberidazole + Bitertanol + Triadimenol

C	Cereline	Bayer plc	07239

C   These products are "approved for agricultural use". For further details refer to page vii.
A   These products are approved for use in or near water. For further details refer to page vii.

Product Name	Marketing Company	Reg. No.	Expiry Date

624 Fuberidazole + Ethirimol + Triadimenol

C   Bay UK 292   Bayer plc   04335

625 Fuberidazole + Imazalil + Triadimenol

C   Baytan IM   Bayer plc   00226

626 Fuberidazole + Imidacloprid + Triadimenol

C   Baytan Secur   Bayer plc   09510

627 Fuberidazole + Triadimenol

C	Baytan	Zeneca Crop Protection	06636	31/12/2000
C	Baytan	Bayer plc	00225	
C	Baytan CF	Bayer plc	08193	
C	Baytan Flowable	Zeneca Ltd	06635	31/12/2000
C	Baytan Flowable	Bayer plc	02593	

628 Furalaxyl

C	Fongarid	Novartis Crop Protection UK Ltd	08407	
C	Fongarid 25 WP	Ciba Agriculture	03595	31/08/2000

629 Glufosinate-ammonium + Fentin hydroxide

C   Safran   AgrEvo UK Crop Protection Ltd   08559   31/05/2001

630 Guazatine

C	Panoctine	Rhone-Poulenc Agriculture	06207
C	Ravine	Rhone-Poulenc Agriculture	07193

631 Guazatine + Imazalil

C	Panoctine Plus	Rhone-Poulenc Agriculture	06208
C	Ravine Plus	Rhone-Poulenc Agriculture	07343

632 Hymexazol

C   Tachigaren 70 WP   Sumitomo Corporation (UK) Plc   02649

633 Imazalil

C	Fungaflor	Brinkman UK Ltd	05968
C	Fungaflor	Hortichem Ltd	05967
C	Fungaflor Smoke	Brinkman UK Ltd	06009
C	Fungaflor Smoke	Hortichem Ltd	05969
C	Fungazil 100 SL	Rhone-Poulenc Agriculture	06202
C	Sphinx	Rhone-Poulenc Agriculture	07607
C	Stryper	Uniroyal Chemical Ltd	08310

C   These products are "approved for agricultural use". For further details refer to page vii.

A   These products are approved for use in or near water. For further details refer to page vii.

Product Name	Marketing Company	Reg. No.	Expiry Date

634 Imazalil + Azaconazole

C Nectec Paste	Hortichem Ltd	08510	

635 Imazalil + Carboxin + Thiabendazole

C Vitaflo Extra	Uniroyal Chemical Ltd	07048	

636 Imazalil + Fuberidazole + Triadimenol

C Baytan IM	Bayer plc	00226	

637 Imazalil + Guazatine

C Panoctine Plus	Rhone-Poulenc Agriculture	06208	
C Ravine Plus	Rhone-Poulenc Agriculture	07343	

638 Imazalil + Pencycuron

C Monceren IM	Bayer plc	06259	
C Monceren IM Flowable	Bayer plc	06731	

639 Imazalil + Thiabendazole

C Extratect Flowable	Seedcote Systems Ltd	09157	
C Extratect Flowable	Novartis Crop Protection UK Ltd	08704	
C Extratect Flowable	MSD Agvet	05507	31/08/2000

640 Imidacloprid + Bitertanol + Fuberidazole + Triadimenol

C Cereline Secur	Bayer plc	09511	

641 Imidacloprid + Fuberidazole + Triadimenol

C Baytan Secur	Bayer plc	09510	

642 Imidacloprid + Tebuconazole + Triazoxide

C Raxil Secur	Bayer plc	08966	

643 Iprodione

C Aventis Rovral Flo	Rhone-Poulenc Agriculture	09767	
C CDA Rovral	Rhone-Poulenc Amenity	04679	
C IT Iprodione	IT Agro Ltd	08267	
C Landgold Iprodione 250	Landgold & Co Ltd	06465	
C Rovral Flo	Rhone-Poulenc Agriculture	06328	
C Rovral Green	Rhone-Poulenc Amenity	05702	
C Rovral Liquid FS	Rhone-Poulenc Agriculture	09776	
C Rovral Liquid FS	Rhone-Poulenc Agriculture	06366	
C Rovral WP	Rhone-Poulenc Agriculture	06091	
C Turbair Rovral	Pan Britannica Industries Ltd	02248	

C These products are "approved for agricultural use". For further details refer to page vii.
A These products are approved for use in or near water. For further details refer to page vii.

PSD
FUNGICIDES

644 Iprodione + Carbendazim

C	Calidan	Rhone-Poulenc Agriculture	06536
C	Vitesse	Rhone-Poulenc Amenity	06537

645 Iprodione + Thiophanate-methyl

C	Aventis Compass	Rhone-Poulenc Agriculture	09760
C	Compass	Rhone-Poulenc Agriculture	06190
C	Snooker	Rhone-Poulenc Agriculture	07940

646 Kresoxim-methyl

C	Landgold Strobilurin KF	Landgold & Co Ltd	09196
C	Stroby WG	BASF plc	08653

647 Kresoxim-methyl + Epoxiconazole

C	Barclay Avalon	Barclay Chemicals (UK) Ltd	09466
C	Clayton Gantry	Clayton Plant Protection (UK) Ltd	09482
C	CleanCrop Kresoxazole	United Agri Products	09698
C	Landmark	BASF plc	08889
C	Me^2 KME	Me^2 Crop Protection Ltd	09594
C	Standon Kresoxim-Epoxiconazole	Standon Chemicals Ltd	09281

648 Kresoxim-methyl + Epoxiconazole + Fenpropimorph

C	CleanCrop Kresoxazole Plus	United Agri Products	09742
C	Mantra	BASF plc	08886
C	Standon Kresoxim Super	Standon Chemicals Ltd	09794

650 Kresoxim-methyl + Fenpropimorph

C	Ensign	BASF plc	08362
C	Greencrop Monsoon	Greencrop Technology Ltd	09573
C	Standon Kresoxim FM	Standon Chemicals Ltd	08922

651 Mancozeb

C	Absezeb WDG	Mark Simpson Agrochemicals	07797	29/02/2000
C	Agrichem Mancozeb 80	Agrichem (International) Ltd	06354	
C	Ashlade Mancozeb FL	Ashlade Formulations Ltd	06226	
C	Barclay Manzeb 455	Barclay Chemicals (UK) Ltd	07990	
C	Barclay Manzeb 80	Barclay Chemicals (UK) Ltd	05944	29/02/2000
C	Barclay Manzeb Flow	Barclay Chemicals (UK) Ltd	05872	
C	Dequiman MZ	Elf Atochem Agri SA	06870	
C	Dithane 945	Pan Britannica Industries Ltd	04017	
C	Dithane 945	Pan Britannica Industries Ltd	00719	

C These products are "approved for agricultural use". For further details refer to page vii.
A These products are approved for use in or near water. For further details refer to page vii.

183

Product Name	Marketing Company	Reg. No.	Expiry Date

651 Mancozeb—continued

	Product Name	Marketing Company	Reg. No.	Expiry Date
C	Dithane Dry Flowable	pbi Agrochemicals Ltd	09754	
C	Dithane Dry Flowable	Pan Britannica Industries Ltd	04255	
C	Dithane Dry Flowable	Pan Britannica Industries Ltd	04251	30/09/2002
C	Dithane Superflo	Pan Britannica Industries Ltd	06593	
C	Dithane Superflo	Pan Britannica Industries Ltd	06290	31/05/2002
C	Headland Zebra Flo	Headland Agrochemicals Ltd	07442	
C	Headland Zebra WP	Headland Agrochemicals Ltd	07441	
C	Helm 75 WG Newtec	pbi Agrochemicals Ltd	09757	
C	Helm 75WG	Pan Britannica Industries Ltd	08309	30/09/2002
C	Karamate Dry Flo	Landseer Limited	09259	
C	Karamate Dry Flo	Rohm & Haas (UK) Ltd	04250	30/09/2001
C	Karamate Dry Flo Newtec	Landseer Limited	09759	
C	Karamate N	Rohm & Haas (UK) Ltd	01125	
C	Kor DF	Headland Agrochemicals Ltd	08979	30/09/2002
C	Kor DF Newtec	Headland Agrochemicals Ltd	09758	
C	Kor Flo	Headland Agrochemicals Ltd	08019	
C	Landgold Mancozeb 80 W	Landgold & Co Ltd	06507	
C	Luxan Mancozeb Flowable	Luxan (UK) Ltd	06812	
C	Manconex	Griffin (Europe) Marketing NV (UK Branch)	09555	
C	Mandate 75 WDG	Portman Agrochemicals Ltd	09051	
C	Mandate 80 WP	Portman Agrochemicals Ltd	09080	
C	Manex II	Agrichem (International) Ltd	07637	
C	Manzate 200 DF	Du Pont (UK) Ltd	06010	
C	Manzate 200 PI	Griffin (Europe) SA	09480	
C	Manzate 200 PI	Du Pont (UK) Ltd	07209	28/02/2002
C	Micene 80	Sipcam UK Ltd	09112	
C	Micene 80	Sipcam UK Ltd	08560	
C	Mortar Flo	Griffin (Europe) Marketing NV (UK Branch)	09592	
C	Nemispor	Sipcam UK Ltd	07348	
C	Opie 80 WP	Pan Britannica Industries Ltd	08301	
C	Penncozeb	Elf Atochem Agri SA	07820	
C	Penncozeb WDG	Whyte Agrochemicals Ltd	09690	
C	Penncozeb WDG	Elf Atochem Agri SA	07833	31/07/2002
C	Portman Mandate 80	Portman Agrochemicals Ltd	06320	29/02/2000
C	Quell Flo	Headland Agrochemicals Ltd	08317	
C	Restraint DF	Griffin (Europe) Marketing NV (UK Branch)	09499	30/09/2002
C	Restraint DF Newtec	Griffin (Europe) Marketing NV (UK Branch)	09755	
C	Stefes Deny	Stefes Plant Protection Ltd	08932	
C	Stefes Mancozeb DF	Stefes Plant Protection Ltd	08010	29/02/2000
C	Stefes Mancozeb WP	Stefes Plant Protection Ltd	07655	29/02/2000

C   These products are "approved for agricultural use". For further details refer to page vii.
A   These products are approved for use in or near water. For further details refer to page vii.

Product Name	Marketing Company	Reg. No.	Expiry Date

651 Mancozeb—continued

C	Stefes Restraint	Stefes Plant Protection Ltd	08945	31/03/2001
C	Tariff 75 WG Newtec	pbi Agrochemicals Ltd	09756	
C	Tariff 75WG	Pan Britannica Industries Ltd	08308	30/09/2002
C	Tridex	Elf Atochem Agri SA	07922	31/12/2001
C	Trimanzone	Intracrop	09584	
C	Trimanzone	Brian Lewis Agriculture Ltd	09278	31/05/2002
C	Unicrop Mancozeb	Universal Crop Protection Ltd	05467	
C	Unicrop Mancozeb 80	Universal Crop Protection Ltd	07451	

652 Mancozeb + Benalaxyl

C	Barclay Bezant	Barclay Chemicals (UK) Ltd	05914	29/02/2000
C	Clayton Benzeb	Clayton Plant Protection (UK) Ltd	07081	29/02/2000
C	Galben M	Sipcam UK Ltd	07220	
C	Tairel	Sipcam UK Ltd	07767	

653 Mancozeb + Carbendazim

C	Headland Maple	Headland Agrochemicals Ltd	09044	
C	Kombat WDG	AgrEvo UK Ltd	07201	31/10/2000
C	Kombat WDG	Rohm & Haas (UK) Ltd	05509	
C	Stefes Kombat WDG	Stefes Plant Protection Ltd	07985	30/04/2000

654 Mancozeb + Chlorothalonil

C	Adagio	pbi Agrochemicals Ltd	09597	
C	Adagio	Pan Britannica Industries Ltd	07832	
C	Adagio C	pbi Agrochemicals Ltd	09309	
C	Dreadnought Flo	pbi Agrochemicals Ltd	09599	
C	Dreadnought Flo	Rohm & Haas (UK) Ltd	07600	
C	Dreadnought Flo	Pan Britannica Industries Ltd	07599	
C	Sipcam Flo	Sipcam UK Ltd	07601	

655 Mancozeb + Cymoxanil

C	Ashlade Solace	Ashlade Formulations Ltd	08087	
C	Besiege	Du Pont (UK) Ltd	08086	
C	Besiege WSB	Du Pont (UK) Ltd	08075	
C	Clayton Krypton	Clayton Plant Protection (UK) Ltd	09398	
C	Clayton Krypton	Clayton Plant Protection (UK) Ltd	06973	29/02/2000
C	Curzate M68	Du Pont (UK) Ltd	08072	
C	Curzate M68 WSB	Du Pont (UK) Ltd	08073	
C	Fytospore	Zeneca Crop Protection	06517	28/02/2000
C	Fytospore 68	Du Pont (UK) Ltd	08649	
C	Me^2 Cymoxeb	Me^2 Crop Protection Ltd	09486	
C	Rhythm	Interfarm (UK) Limited	09636	
C	Standon Cymoxanil Extra	Standon Chemicals Ltd	09442	
C	Standon Cymoxanil Extra	Standon Chemicals Ltd	06807	29/02/2000

C   These products are "approved for agricultural use". For further details refer to page vii.
A   These products are approved for use in or near water. For further details refer to page vii.

Product Name	Marketing Company	Reg. No.	Expiry Date

655 Mancozeb + Cymoxanil—continued

C	Stefes Blight Spray	Stefes Plant Protection Ltd	05811	29/02/2000
C	Systol M	Quadrangle Agrochemicals	08085	
C	Systol M	Quadrangle Agrochemicals	03480	
C	Systol M WSB	Quadrangle Agrochemicals	08074	

656 Mancozeb + Cymoxanil + Oxadixyl

C	Ripost	Novartis Crop Protection UK Ltd	08484	31/08/2000
C	Ripost	Sandoz Agro Ltd	04890	31/08/2000
C	Ripost Pepite	Bayer Ltd	08895	
C	Ripost Pepite	Novartis Crop Protection UK Ltd	08485	
C	Ripost Pepite	Sandoz Agro Ltd	06485	31/03/2001
C	Trustan	Du Pont (UK) Ltd	05022	31/12/2000
C	Trustan WDG	Du Pont (UK) Ltd	05050	

657 Mancozeb + Cyproconazole

C	Alto Eco	Novartis Crop Protection UK Ltd	08466	
C	Alto Eco	Sandoz Agro Ltd	05068	31/08/2000

658 Mancozeb + Dimethomorph

C	Invader	Cyanamid Agriculture Ltd	06989	
C	Invader WDG	Cyanamid Agriculture Ltd	09556	
C	Saracen	Cyanamid Agriculture Ltd	07872	
C	Saracen WDG	Cyanamid Agriculture Ltd	09557	

659 Mancozeb + Metalaxyl

C	Fubol 58 WP	Novartis Crop Protection UK Ltd	08534	
C	Fubol 58 WP	Ciba Agriculture	00927	31/08/2000
C	Fubol 75 WP	Novartis Crop Protection UK Ltd	08409	
C	Fubol 75 WP	Ciba Agriculture	03462	31/08/2000
C	Osprey 58 WP	Novartis Crop Protection UK Ltd	08428	
C	Osprey 58 WP	Ciba Agriculture	05717	31/08/2000
C	Ridomil MZ 75	Ciba Agriculture	07640	31/08/2000
C	Ridomil MZ 75 WP	Novartis Crop Protection UK Ltd	08438	

660 Mancozeb + Metalaxyl-M

C	Fubol Gold	Novartis Crop Protection UK Ltd	08812	

661 Mancozeb + Ofurace

C	Patafol	Mirfield Sales Services Ltd	07397	

662 Mancozeb + Oxadixyl

C	Recoil	Novartis Crop Protection UK Ltd	08483	
C	Recoil	Sandoz Agro Ltd	04039	31/08/2000

C These products are "approved for agricultural use". For further details refer to page vii.
A These products are approved for use in or near water. For further details refer to page vii.

Product Name	Marketing Company	Reg. No.	Expiry Date

662 Mancozeb + Oxadixyl—continued

C	Sandofan Pepite	Sandoz Agro Ltd	08061	31/08/2000

663 Mancozeb + Propamocarb hydrochloride

C	Tattoo	AgrEvo UK Ltd	07293	

664 Maneb

C	Agrichem Maneb 80	Agrichem (International) Ltd	05474	
C	Amos Maneb 80	Luxan (UK) Ltd	06560	29/02/2000
C	Ashlade Maneb Flowable	Ashlade Formulations Ltd	06477	
C	Headland Spirit	Headland Agrochemicals Ltd	04548	
C	Luxan Maneb 80	Luxan (UK) Ltd	06570	
C	Luxan Maneb 800 WP	Luxan (UK) Ltd	06561	29/02/2000
C	Maneb 80	Rohm & Haas (UK) Ltd	01276	
C	Manex	Griffin (Europe) Marketing NV (UK Branch)	09554	
C	Manex	Agrichem (International) Ltd	07935	
C	Manex	Chiltern Farm Chemicals Ltd	05731	
C	Mazin	Universal Crop Protection Ltd	06061	31/05/2000
C	RH Maneb 80	Rohm & Haas (UK) Ltd	01796	
C	Stefes Maneb DF	Stefes Plant Protection Ltd	06418	29/02/2000
C	Trimangol 80	Elf Atochem Agri SA	06871	
C	Trimangol 80	Atochem Agri BV	06070	
C	Trimangol WDG	Elf Atochem Agri SA	06992	
C	Unicrop Flowable Maneb	Universal Crop Protection Ltd	05546	29/02/2000
C	Unicrop Maneb 80	Universal Crop Protection Ltd	06926	
C	Unicrop Maneb FL	Universal Crop Protection Ltd	08025	
C	X-Spor SC	United Phosphorus Ltd	08077	

665 Maneb + Carbendazim

C	Ashlade Mancarb FL	Ashlade Formulations Ltd	07977	
C	Headland Dual	Headland Agrochemicals Ltd	03782	
C	MC Flowable	United Phosphorus Ltd	08198	
C	MC Flowable	MTM Agrochemicals Ltd	06295	31/05/2000
C	Multi-W FL	pbi Agrochemicals Ltd	09447	
C	Multi-W FL	Pan Britannica Industries Ltd	04131	28/02/2002
C	New Squadron	Quadrangle Agrochemicals Ltd	06756	31/08/2000
C	Protector	Procam Agriculture Ltd	09448	
C	Protector	Procam Agriculture Ltd	07075	28/02/2002
C	Tripart 147	Tripart Farm Chemicals Ltd	07978	
C	Tripart Legion	Tripart Farm Chemicals Ltd	06113	31/08/2000

666 Maneb + Carbendazim + Chlorothalonil

C	Ashlade Mancarb Plus	Ashlade Formulations Ltd	08160	

C These products are "approved for agricultural use". For further details refer to page vii.
A These products are approved for use in or near water. For further details refer to page vii.

Product Name	Marketing Company	Reg. No.	Expiry Date

666 Maneb + Carbendazim + Chlorothalonil—continued
| C | Tripart Victor | Tripart Farm Chemicals Ltd | 08161 | |

667 Maneb + Carbendazim + Sulphur
| C | Bolda FL | Atlas Crop Protection Ltd | 07653 | |
| C | Legion S | Ashlade Formulations Ltd | 08121 | 30/06/2002 |

668 Maneb + Carbendazim + Tridemorph
| C | Cosmic FL | BASF plc | 03473 | |

669 Maneb + Copper oxychloride + Sulphur
| C | Ashlade SMC Flowable | Ashlade Formulations Ltd | 06494 | |

670 Maneb + Fentin acetate
| C | Brestan 60 SP | AgrEvo UK Ltd | 07305 | |

671 Metalaxyl
C	Polycote Universal	Seedcote Systems Ltd	08431	
C	Polycote Universal	Seedcote Systems Ltd	05942	31/03/2000
C	Ridomil 5 FG	Novartis Crop Protection UK Ltd	08436	31/03/2000
C	Ridomil 5 FG	Ciba Agriculture	07407	31/03/2000

672 Metalaxyl + Carbendazim
| C | Ridomil MBC 60 WP | Novartis Crop Protection UK Ltd | 08437 | |
| C | Ridomil MBC 60 WP | Ciba Agriculture | 01804 | 31/08/2000 |

673 Metalaxyl + Chlorothalonil
C	Folio	Novartis Crop Protection UK Ltd	08547	
C	Folio 575 SC	Novartis Crop Protection UK Ltd	08406	30/11/2000
C	Folio 575 SC	Ciba Agriculture	05843	31/08/2000

674 Metalaxyl + Copper oxychloride
| C | Ridomil Plus | Novartis Crop Protection UK Ltd | 08353 | |
| C | Ridomil Plus 50 WP | Ciba Agriculture | 01803 | 31/03/2001 |

675 Metalaxyl + Mancozeb
C	Fubol 58 WP	Novartis Crop Protection UK Ltd	08534	
C	Fubol 58 WP	Ciba Agriculture	00927	31/08/2000
C	Fubol 75 WP	Novartis Crop Protection UK Ltd	08409	
C	Fubol 75 WP	Ciba Agriculture	03462	31/08/2000
C	Osprey 58 WP	Novartis Crop Protection UK Ltd	08428	
C	Osprey 58 WP	Ciba Agriculture	05717	31/08/2000
C	Ridomil MZ 75	Ciba Agriculture	07640	31/08/2000

C These products are "approved for agricultural use". For further details refer to page vii.
A These products are approved for use in or near water. For further details refer to page vii.

Product Name	Marketing Company	Reg. No.	Expiry Date

675 Metalaxyl + Mancozeb—continued

C Ridomil MZ 75 WP	Novartis Crop Protection UK Ltd	08438	

676 Metalaxyl + Thiabendazole

C Apron T69 WS	Novartis Crop Protection UK Ltd	08387	
C Apron T69 WS	Ciba Agriculture	06725	31/08/2000
C Polycote Select	Germains UK Ltd	09718	
C Polycote Select	Seedcote Systems Ltd	08430	
C Polycote Select	Seedcote Systems Ltd	06727	31/08/2000

677 Metalaxyl + Thiabendazole + Thiram

C Apron Combi FS	Novartis Crop Protection UK Ltd	08386	
C Apron Combi FS	Ciba Agriculture	07203	30/11/2000

678 Metalaxyl + Thiram

C Favour 600 SC	Novartis Crop Protection UK Ltd	08405	
C Favour 600 SC	Ciba Agriculture	05842	31/08/2000

679 Metalaxyl-M

C SL 567A	Novartis Crop Protection UK Ltd	08811	

680 Metalaxyl-M + Mancozeb

C Fubol Gold	Novartis Crop Protection UK Ltd	08812	

681 Myclobutanil

C Systhane 20 EW	T P Whelehan Son & Co Ltd	09397	
C Systhane 20 EW	Landseer Limited	09396	
C Systhane 6 Flo	AgrEvo UK Ltd	07334	
C Systhane 6 Flo	T P Whelehan Son & Co Ltd	06551	
C Systhane 6W	pbi Agrochemicals Ltd	09611	
C Systhane 6W	Pan Britannica Industries Ltd	04571	
C Systhane 6W	Pan Britannica Industries Ltd	04570	
C Systhane ST	Uniroyal Chemical Ltd	08034	30/10/2001

682 Nuarimol

C Triminol	Dow AgroSciences Ltd	05818	

683 Octhilinone

C Pancil T	Landseer Limited	09261	
C Pancil T	Rohm & Haas (UK) Ltd	01540	30/09/2001

684 Ofurace + Mancozeb

C Patafol	Mirfield Sales Services Ltd	07397	

C These products are "approved for agricultural use". For further details refer to page vii.
A These products are approved for use in or near water. For further details refer to page vii.

Product Name	Marketing Company	Reg. No.	Expiry Date

685 Oxadixyl + Carbendazim + Cymoxanil + Thiram

C	Apron Elite	Novartis Crop Protection UK Ltd	08770	

686 Oxadixyl + Cymoxanil + Mancozeb

C	Ripost	Novartis Crop Protection UK Ltd	08484	31/08/2000
C	Ripost	Sandoz Agro Ltd	04890	31/08/2000
C	Ripost Pepite	Bayer Ltd	08895	
C	Ripost Pepite	Novartis Crop Protection UK Ltd	08485	
C	Ripost Pepite	Sandoz Agro Ltd	06485	31/03/2001
C	Trustan	Du Pont (UK) Ltd	05022	31/12/2000
C	Trustan WDG	Du Pont (UK) Ltd	05050	

687 Oxadixyl + Mancozeb

C	Recoil	Novartis Crop Protection UK Ltd	08483	
C	Recoil	Sandoz Agro Ltd	04039	31/08/2000
C	Sandofan Pepite	Sandoz Agro Ltd	08061	31/08/2000

688 Oxadixyl + Thiabendazole + Thiram

C	Leap	Uniroyal Chemical Ltd	08477	
C	Leap	Uniroyal Chemical Ltd	07884	
C	Leap	Uniroyal Chemical Ltd	07822	31/08/2000

689 Oxycarboxin

C	Plantvax 20	Uniroyal Chemical Ltd	01600	31/05/2001
C	Plantvax 75	Fargro Ltd	01601	
C	Ringmaster	Rhone-Poulenc Amenity	05334	30/09/2001

690 Parahydroxy-phenylsalicylamide

	Fumispore	Laboratoire de Chimie et Biologie	08300	

691 Penconazole

C	Topas	Novartis Crop Protection UK Ltd	09717	
C	Topas	Novartis Crop Protection UK Ltd	08458	
C	Topas 100 EC	Ciba Agriculture	03231	31/05/2000

692 Penconazole + Captan

C	Topas C 50 WP	Novartis Crop Protection UK Ltd	08459	
C	Topas C 50 WP	Ciba Agriculture	03232	31/08/2000

693 Penconazole + Dithianon

C	Topas D275 SC	Novartis Crop Protection UK Ltd	08460	
C	Topas D275 SC	Ciba Agriculture	05855	31/08/2000

C   These products are "approved for agricultural use". For further details refer to page vii.
A   These products are approved for use in or near water. For further details refer to page vii.

Product Name	Marketing Company	Reg. No.	Expiry Date

694 Pencycuron

C Monceren DS	Bayer plc	04160	
C Monceren FS	Bayer plc	04907	
C Standon Pencycuron DP	Standon Chemicals Ltd	08774	

695 Pencycuron + Imazalil

C Monceren IM	Bayer plc	06259	
C Monceren IM Flowable	Bayer plc	06731	

696 Permethrin + Thiram

C Combinex	Pan Britannica Industries Ltd	00562	

697 Prochloraz

C Admiral	AgrEvo UK Ltd	08162	31/05/2001
C Barclay Eyetak	Barclay Chemicals (UK) Ltd	06813	30/06/2001
C Barclay Eyetak 40	Barclay Chemicals (UK) Ltd	07843	
C Barclay Eyetak 450	Barclay Chemicals (UK) Ltd	09484	
C Fisons Octave	Fisons Plc	03416	
C Levington Octave	Levington Horticulture Ltd	07505	
C Mirage 40 EC	Makhteshim-Agan (UK) Ltd	06770	
C Octave	AgrEvo UK Ltd	07267	
C Prelude 20 LF	AgrEvo UK Ltd	07269	
C Prelude 20 LF	Agrichem (International) Ltd	04371	
C Scotts Octave	The Scotts Company (UK) Limited	09275	
C Sporgon 50 WP	Sylvan Spawn Ltd	08802	
C Sporgon 50 WP	Darmycel	03829	31/12/2000
C Sportak	AgrEvo UK Ltd	07285	30/09/2000
C Sportak 45	AgrEvo UK Ltd	07286	30/09/2000
C Sportak 45 EW	AgrEvo UK Ltd	07996	
C Sportak 45 HF	AgrEvo UK Ltd	07287	
C Sportak Focus EW	AgrEvo UK Ltd	08002	30/09/2001
C Sportak Focus HF	AgrEvo UK Ltd	07288	30/09/2000
C Sportak HF	AgrEvo UK Ltd	07289	30/09/2000
C Sportak Sierra EW	AgrEvo UK Ltd	08003	31/05/2001
C Sportak Sierra HF	AgrEvo UK Ltd	07290	31/05/2001
C Stefes Poraz	Stefes UK Ltd	07528	

698 Prochloraz + Carbendazim

C Novak	AgrEvo UK Ltd	08020	
C Sportak Alpha	AgrEvo UK Ltd	07222	31/05/2001
C Sportak Alpha HF	AgrEvo UK Ltd	07225	

C   These products are "approved for agricultural use". For further details refer to page vii.
A   These products are approved for use in or near water. For further details refer to page vii.

Product Name	Marketing Company	Reg. No.	Expiry Date

699 Prochloraz + Carboxin

C	Prelude Universal LS	AgrEvo UK Crop Protection Ltd	08650	31/05/2001
C	Provax	Uniroyal Chemical Company	08651	31/05/2001

700 Prochloraz + Cyproconazole

C	Profile	AgrEvo UK Ltd	08134	
C	SAN 710	Novartis Crop Protection UK Ltd	08488	31/05/2001
C	SAN 710	Sandoz Agro Ltd	07918	31/08/2000
C	SAN 710 HF	Novartis Crop Protection UK Ltd	08489	
C	SAN 710 HF	Sandoz Agro Ltd	07919	31/08/2000
C	Sportak Delta 460	AgrEvo UK Crop Protection Ltd	07224	31/05/2001
C	Sportak Delta 460 HF	AgrEvo UK Ltd	07431	
C	Tiptor	Novartis Crop Protection UK Ltd	08495	
C	Tiptor	AgrEvo UK Ltd	07295	31/05/2001
C	Tiptor	Sandoz Agro Ltd	05105	31/08/2000

701 Prochloraz + Fenbuconazole

C	Mirage Extra	Stefes UK Ltd	07958	
C	Stefes Fortune	Stefes Plant Protection Ltd	08537	31/01/2001
C	Stefes Inception	Stefes Plant Protection Ltd	08829	

702 Prochloraz + Fenpropidin

C	SL 552A	Novartis Crop Protection UK Ltd	08673	
C	SL 552A	Novartis Crop Protection UK Ltd	08448	31/08/2000
C	SL 552A	Ciba Agriculture	07137	31/08/2000
C	Sponsor	AgrEvo UK Ltd	08674	
C	Sponsor	AgrEvo UK Crop Protection Ltd	07136	31/08/2000

703 Prochloraz + Fenpropimorph

C	Jester	Ciba Agriculture	06586	31/08/2000
C	Mirage Super 600EC	Makhteshim-Agan (UK) Ltd	08531	
C	SL571A	Novartis Crop Protection UK Ltd	08420	
C	Sprint	AgrEvo UK Ltd	07291	31/05/2001
C	Sprint HF	AgrEvo UK Ltd	07292	

704 Prochloraz + Fluquinconazole

C	Aventis Foil	AgrEvo UK Ltd	09709	
C	Baron	AgrEvo UK Ltd	09364	
C	Flamenco Plus	AgrEvo UK Ltd	09362	
C	Foil	AgrEvo UK Ltd	09363	

705 Prochloraz + Propiconazole

C	Bumper P	Makhteshim-Agan (UK) Ltd	08548	
C	Greencrop Twinstar	Greencrop Technology Ltd	09516	

C These products are "approved for agricultural use". For further details refer to page vii.
A These products are approved for use in or near water. For further details refer to page vii.

Product Name	Marketing Company	Reg. No.	Expiry Date

706 Prochloraz + Tebuconazole

C	Agate	Bayer plc	08826	

707 Prochloraz + Tolclofos-methyl

C	Rizolex Nova	AgrEvo UK Ltd	07434	31/01/2000

708 Propamocarb hydrochloride

C	Filex	The Scotts Company (UK) Limited	07631	
C	Previcur N	AgrEvo UK Ltd	08575	
C	Proplant	Chimac-Agriphar SA	08572	

709 Propamocarb hydrochloride + Chlorothalonil

C	Aventis Merlin	AgrEvo UK Ltd	09719	
C	Merlin	AgrEvo UK Ltd	07943	
C	Tattoo C	AgrEvo UK Ltd	07623	

710 Propamocarb hydrochloride + Mancozeb

C	Tattoo	AgrEvo UK Ltd	07293	

711 Propiconazole

C	Barclay Bolt	Barclay Chemicals (UK) Ltd	08341	
C	Bumper 250 EC	Makhteshim-Agan (UK) Ltd	09039	
C	Clayton Propiconazole	Clayton Plant Protection (UK) Ltd	06415	29/02/2000
C	Controller Fungicide	Novartis Crop Protection UK Ltd	08399	
C	Controller Fungicide	Ciba Agriculture	08170	31/07/2001
C	Landgold Propiconazole	Landgold & Co Ltd	06291	
C	Mantis	Novartis Crop Protection UK Ltd	08423	
C	Mantis 250EC	Ciba Agriculture	06240	31/07/2001
C	Radar	Zeneca Crop Protection	09168	
C	Radar	Zeneca Crop Protection	06747	31/08/2000
C	Standon Propiconazole	Standon Chemicals Ltd	07037	
C	Stefes Restore	Stefes Plant Protection Ltd	06267	29/02/2000
C	Tilt	Novartis Crop Protection UK Ltd	08456	
C	Tilt 250EC	Ciba Agriculture	02138	31/08/2000
C	Top Farm Propiconazole 250	Top Farm Formulations Ltd	05938	29/02/2000

712 Propiconazole + Carbendazim

C	Hispor 45 WP	Novartis Crop Protection UK Ltd	08418	
C	Hispor 45WP	Ciba Agriculture	01050	31/08/2000
C	Sparkle 45 WP	Novartis Crop Protection UK Ltd	08450	
C	Sparkle 45 WP	Ciba Agriculture	04968	31/10/2000

713 Propiconazole + Chlorothalonil

C	Sambarin 312.5 SC	Novartis Crop Protection UK Ltd	08439	

C   These products are "approved for agricultural use". For further details refer to page vii.
A   These products are approved for use in or near water. For further details refer to page vii.

Product Name	Marketing Company	Reg. No.	Expiry Date

713 Propiconazole + Chlorothalonil—continued

C	Sambarin 312.5 SC	Ciba Agriculture	05809	31/08/2000

714 Propiconazole + Cyproconazole

C	Menara	Novartis Crop Protection UK Ltd	09321	

715 Propiconazole + Fenbuconazole

C	Graphic	Novartis Crop Protection UK Ltd	08415	
C	Graphic	Ciba Agriculture	07585	31/03/2000
C	Indar CG	Novartis Crop Protection UK Ltd	08578	
C	Indar CG	Ciba Agriculture	07586	31/03/2000

716 Propiconazole + Fenpropidin

C	Prophet	Novartis Crop Protection UK Ltd	08433	
C	Prophet 500EC	Ciba Agriculture	07447	31/08/2000
C	Sheen	Novartis Crop Protection UK Ltd	08442	
C	Sheen	Ciba Agriculture	07828	31/08/2000
C	Zulu	Novartis Crop Protection UK Ltd	08464	
C	Zulu	Ciba Agriculture	07829	31/08/2000

717 Propiconazole + Fenpropidin + Tebuconazole

C	Bayer UK 593	Bayer plc	08285	
C	Gladio	Novartis Crop Protection UK Ltd	08413	
C	Gladio	Ciba Agriculture	08284	31/08/2000

718 Propiconazole + Fenpropimorph

C	Belvedere	Makhteshim-Agan (UK) Ltd	08084	
C	Decade	Novartis Crop Protection UK Ltd	08402	
C	Decade 500 EC	Ciba Agriculture	05757	31/08/2000
C	Glint	Novartis Crop Protection UK Ltd	08414	
C	Glint 500 EC	Ciba Agriculture	04126	31/08/2000
C	Mantle	Novartis Crop Protection UK Ltd	08424	
C	Mantle 425 EC	Ciba Agriculture	05715	30/04/2000

719 Propiconazole + Prochloraz

C	Bumper P	Makhteshim-Agan (UK) Ltd	08548	
C	Greencrop Twinstar	Greencrop Technology Ltd	09516	

720 Propiconazole + Tebuconazole

C	Cogito	Novartis Crop Protection UK Ltd	08397	
C	Cogito	Ciba Agriculture	07384	31/12/2000
C	Endeavour	Bayer plc	07385	

C These products are "approved for agricultural use". For further details refer to page vii.
A These products are approved for use in or near water. For further details refer to page vii.

Product Name	Marketing Company	Reg. No.	Expiry Date

721 Propiconazole + Tridemorph

C	Joust	Universal Crop Protection Ltd	08122	
C	Tilt Turbo 475 EC	Novartis Crop Protection UK Ltd	08457	30/09/2001
C	Tilt Turbo 475 EC	Ciba Agriculture	03476	30/09/2001

722 Propineb

C	Antracol	Bayer plc	00104	

723 Propoxur

C	Fumite Propoxur Smoke	Octavius Hunt Ltd	09290	

724 Pyrazophos

C	Afugan	AgrEvo UK Ltd	07301	20/04/2001

725 Pyrifenox

C	Dorado	Zeneca Crop Protection	06657	
C	SL 471 200 EC	Novartis Crop Protection UK Ltd	08446	31/07/2001
C	SL 471 200 EC	Ciba Agriculture	06035	31/08/2000

726 Pyrimethanil

C	Scala	AgrEvo UK Ltd	07806	

727 Quinomethionate

C	Morestan	Bayer plc	01376	

728 Quinoxyfen

C	Apres	Dow AgroSciences Ltd	08881
C	Barclay Foran	Barclay Chemicals (UK) Ltd	09479
C	DOE 81673	DowElanco Ltd	08280
C	EF-1186	Dow AgroSciences Ltd	08278
C	Erysto	Dow AgroSciences Ltd	08697
C	Fortress	Dow AgroSciences Ltd	08279
C	Standon Quinoxyfen 500	Standon Chemicals Ltd	08924

729 Quinoxyfen + Cyproconazole

C	Divora	Novartis Crop Protection UK Ltd	08960
C	DOE 1762	Dow AgroSciences Ltd	08962
C	NOV 1390	Novartis Crop Protection UK Ltd	08961

730 Quinoxyfen + Fenpropimorph

C	EF-1288	Dow AgroSciences Ltd	08241
C	Orka	Dow AgroSciences Ltd	08879

C   These products are "approved for agricultural use". For further details refer to page vii.
A   These products are approved for use in or near water. For further details refer to page vii.

Product Name	Marketing Company	Reg. No.	Expiry Date

731 Quintozene

C	Quintozene Wettable Powder	Rhone-Poulenc Amenity	05404
C	Terraclor 20D	Uniroyal Chemical Ltd	06578
C	Terraclor Flo	Hortichem Ltd	08666
C	Terraclor Flo	Uniroyal Chemical Ltd	08665

732 Spiroxamine

C	Accrue	Bayer plc	08335
C	Bayer UK 477	Bayer plc	08330
C	Neon	Bayer plc	08337
C	Standon Spiroxamin 500	Standon Chemicals Ltd	08916
C	Talvin	Bayer plc	09253
C	Torch	Bayer plc	08336
C	Zenon	Bayer plc	09193

733 Spiroxamine + Tebuconazole

C	Array	Bayer plc	09352
C	Bayer UK 552	Bayer plc	08331
C	Beam	Bayer plc	08332
C	Bronze	Bayer plc	08333
C	Draco	Bayer plc	09353
C	Sage	Bayer plc	08334

734 Sulphur

C	Ashlade Sulphur FL	Ashlade Formulations Ltd	06478	
C	Atlas Sulphur 80 FL	Atlas Crop Protection Ltd	07729	31/07/2000
C	Atlas Sulphur 80 FL	Atlas Agrochemicals Ltd	03802	31/07/2000
C	Headland Sulphur	Headland Agrochemicals Ltd	03714	
C	Headland Venus	Headland Agrochemicals Ltd	09572	
C	Kumulus DF	BASF plc	04707	
C	Kumulus S	BASF plc	01170	
C	Luxan Micro-Sulphur	Luxan (UK) Ltd	06565	
C	Microsul Flowable Sulphur	Stoller Chemical Ltd	03907	
C	Microthiol Special	Elf Atochem Agri SA	06268	
C	MSS Sulphur 80	Mirfield Sales Services Ltd	05752	
C	Solfa	Atlas Interlates Ltd	06959	
C	Stoller Flowable Sulphur	Stoller Chemical Ltd	03760	
C	Sulphur Flowable	United Phosphorus Ltd	07526	
C	Thiovit	Novartis Crop Protection UK Ltd	08493	
C	Thiovit	Sandoz Agro Ltd	05572	31/08/2000
C	Thiovit	Pan Britannica Industries Ltd	02125	31/01/2000
C	Tripart Imber	Tripart Farm Chemicals Ltd	04050	

C These products are "approved for agricultural use". For further details refer to page vii.
A These products are approved for use in or near water. For further details refer to page vii.

Product Name	Marketing Company	Reg. No.	Expiry Date

PSD
FUNGICIDES

735 Sulphur + Carbendazim + Maneb

C	Bolda FL	Atlas Crop Protection Ltd	07653	
C	Legion S	Ashlade Formulations Ltd	08121	30/06/2002

736 Sulphur + Copper oxychloride + Maneb

C	Ashlade SMC Flowable	Ashlade Formulations Ltd	06494	

737 Tar acids

C	Bray's Emulsion	Fargro Ltd	08316	

738 Tar oils

C	Sterilite Tar Oil Winter Wash 60% Stock Emulsion	Coventry Chemicals Ltd	05061	
C	Sterilite Tar Oil Winter Wash 80% Miscible Quality	Coventry Chemicals Ltd	05062	

739 Tebuconazole

C	Barclay Busker	Barclay Chemicals (UK) Ltd	08994	
C	Bayer UK226	Bayer plc	09504	
C	Clayton Tebucon	Clayton Plant Protection (UK) Ltd	08707	
C	Clayton Tebucon	Clayton Plant Protection (UK) Ltd	07045	
C	Folicur	Bayer plc	08691	
C	Folicur	Bayer plc	06386	31/08/2000
C	Gainer	Bayer plc	08692	
C	Gainer	Bayer plc	07643	31/08/2000
C	Halt	Bayer plc	08693	
C	Halt	Bayer plc	07443	31/08/2000
C	Landgold Tebuconazole	Landgold & Co Ltd	08063	31/08/2000
C	Me^2 Tebuconazole	Me^2 Crop Protection Ltd	09751	
C	Raxil	Bayer plc	06460	
C	Standon Tebuconazole	Standon Chemicals Ltd	09056	
C	Standon Tebuconazole	Standon Chemicals Ltd	07408	31/08/2000
C	UK 200	Bayer plc	08246	

740 Tebuconazole + Carbendazim

C	Bayer UK413	Bayer plc	08277	

741 Tebuconazole + Fenpropidin

C	Monicle	Bayer plc	07375	
C	SL 556 500 EC	Ciba Agriculture	07376	31/08/2000
C	SL-556 500 EC	Novartis Crop Protection UK Ltd	08449	

C   These products are "approved for agricultural use". For further details refer to page vii.
A   These products are approved for use in or near water. For further details refer to page vii.

Product Name	Marketing Company	Reg. No.	Expiry Date

742 Tebuconazole + Fenpropidin + Propiconazole

C	Bayer UK 593	Bayer plc	08285	
C	Gladio	Novartis Crop Protection UK Ltd	08413	
C	Gladio	Ciba Agriculture	08284	31/08/2000

743 Tebuconazole + Fenpropimorph

C	Bayer UK506	Bayer plc	07540	31/03/2001
C	BUK 84000F	BASF plc	07541	31/03/2001

744 Tebuconazole + Imidacloprid + Triazoxide

C	Raxil Secur	Bayer plc	08966	

745 Tebuconazole + Prochloraz

C	Agate	Bayer plc	08826	

746 Tebuconazole + Propiconazole

C	Cogito	Novartis Crop Protection UK Ltd	08397	
C	Cogito	Ciba Agriculture	07384	31/12/2000
C	Endeavour	Bayer plc	07385	

747 Tebuconazole + Spiroxamine

C	Array	Bayer plc	09352	
C	Bayer UK 552	Bayer plc	08331	
C	Beam	Bayer plc	08332	
C	Bronze	Bayer plc	08333	
C	Draco	Bayer plc	09353	
C	Sage	Bayer plc	08334	

748 Tebuconazole + Triadimenol

C	Garnet	Bayer plc	06391	
C	Ruby	Bayer plc	06389	
C	Silvacur	Bayer plc	06387	
C	Silvacur 300	Bayer plc	08056	
C	Veto F	Bayer plc	08057	

749 Tebuconazole + Triazoxide

C	Raxil S	Zeneca Crop Protection	07484	31/12/2000
C	Raxil S	Bayer plc	06974	
C	Raxil S CF	Bayer plc	08192	

750 Tebuconazole + Tridemorph

C	Allicur	Bayer plc	06468	30/09/2001
C	BAS 91580F	BASF plc	06469	

C   These products are "approved for agricultural use". For further details refer to page vii.
A   These products are approved for use in or near water. For further details refer to page vii.

Product Name	Marketing Company	Reg. No.	Expiry Date

751 Tecnazene

	Product Name	Marketing Company	Reg. No.	Expiry Date
C	Atlas Tecgran 100	Atlas Crop Protection Ltd	07730	
C	Atlas Tecgran 100	Atlas Interlates Ltd	05574	31/01/2001
C	Atlas Tecnazene 6% dust	Atlas Crop Protection Ltd	07731	
C	Atlas Tecnazene 6% Dust	Atlas Interlates Ltd	06351	
C	Bygran F	Wheatley Chemical Co Ltd	00365	
	Bygran S	Wheatley Chemical Co Ltd	00366	
	Fusarex 10% Granules	Zeneca Crop Protection	06668	18/02/2001
C	Fusarex 10G	Hickson & Welch Ltd	09727	
	Fusarex 6% Dust	Zeneca Crop Protection	06667	18/02/2001
C	Hickstor 10	Hickson & Welch Ltd	03121	
	Hickstor 3	Hickson & Welch Ltd	03105	
C	Hickstor 5	Hickson & Welch Ltd	03180	
	Hickstor 6	Hickson & Welch Ltd	03106	
C	Hortag Tecnazene 10% Granules	Hortag Chemicals Ltd	03966	
	Hortag Tecnazene Double Dust	Hortag Chemicals Ltd	01072	
	Hortag Tecnazene Potato Dust	Hortag Chemicals Ltd	01074	
	Hortag Tecnazene Potato Granules	Hortag Chemicals Ltd	01075	
C	Hystore 10	Agrichem (International) Ltd	03581	
C	Hystore 10G	Agrichem (International) Ltd	06899	
	Hytec	Agrichem Ltd	01099	
C	Hytec 6	Agrichem Ltd	03580	
C	Nebulin	Wheatley Chemical Co Ltd	01469	31/03/2001
C	New Hickstor 6	Hickson & Welch Ltd	04221	
	New Hystore	Agrichem Ltd	01485	
C	Tripart Arena 10G	Tripart Farm Chemicals Ltd	05603	
	Tripart Arena 3	Tripart Farm Chemicals Ltd	05605	
C	Tripart Arena 5G	Tripart Farm Chemicals Ltd	05604	
C	Tripart New Arena 6	Tripart Farm Chemicals Ltd	05813	

752 Tecnazene + Carbendazim

	Product Name	Marketing Company	Reg. No.	Expiry Date
C	Hickstor 6 Plus MBC	Hickson & Welch Ltd	04176	
C	Hortag Tecnacarb	Avon Packers Ltd	02929	
C	New Arena Plus	Tripart Farm Chemicals	04598	
C	New Hickstor 6 Plus MBC	Hickson & Welch Ltd	04599	
C	Tripart Arena Plus	Tripart Farm Chemicals	05602	

753 Tecnazene + Thiabendazole

	Product Name	Marketing Company	Reg. No.	Expiry Date
C	Hytec Super	Agrichem (International) Ltd	01100	
	New Arena TBZ 6	Tripart Farm Chemicals Ltd	05606	
C	Storite SS	Novartis Crop Protection UK Ltd	08702	30/09/2000
C	Storite SS	MSD Agvet	02034	31/08/2000

C   These products are "approved for agricultural use". For further details refer to page vii.
A   These products are approved for use in or near water. For further details refer to page vii.

Product Name	Marketing Company	Reg. No.	Expiry Date

754 Tetraconazole

C	Digit	Monsanto Plc	09409	
C	Eminent	Monsanto Plc	09410	
C	Juggler	Sipcam UK Ltd	09391	
C	MON 10EC	Monsanto Plc	09347	
C	Omen	Sipcam UK Ltd	09413	
C	Tetra 5634	Monsanto Plc	09411	

755 Tetraconazole + Chlorothalonil

C	TC Elite	Monsanto Plc	09412	
C	Voodoo	Sipcam UK Ltd	09414	

756 Thiabendazole

C	Hykeep	Agrichem (International) Ltd	06744	
C	Hymush	Agrichem (International) Ltd	07937	
C	Storite Clear Liquid	Novartis Crop Protection UK Ltd	09503	
C	Storite Clear Liquid	Seedcote Systems Ltd	08982	
C	Storite Clear Liquid	Novartis Crop Protection UK Ltd	08700	31/03/2001
C	Storite Clear Liquid	MSD Agvet	02032	31/03/2001
C	Storite Excel	Novartis Crop Protection UK Ltd	09542	
C	Storite Excel	Seedcote Systems Ltd	08993	
C	Storite Flowable	Novartis Crop Protection UK Ltd	08703	
C	Storite Flowable	MSD Agvet	02033	31/08/2000
C	Tecto Flowable Turf Fungicide	Vitax Ltd	06273	
C	Tecto Superflowable Turf Fungicide	Vitax Ltd	09258	

757 Thiabendazole + Carboxin

C	Vitaflo	Uniroyal Chemical Ltd	06379	31/05/2000

758 Thiabendazole + Carboxin + Imazalil

C	Vitaflo Extra	Uniroyal Chemical Ltd	07048	

759 Thiabendazole + Ethirimol + Flutriafol

C	Ferrax	Zeneca Crop Protection	06662	
C	Ferrax	Bayer plc	05284	

760 Thiabendazole + Imazalil

C	Extratect Flowable	Seedcote Systems Ltd	09157	
C	Extratect Flowable	Novartis Crop Protection UK Ltd	08704	
C	Extratect Flowable	MSD Agvet	05507	31/08/2000

761 Thiabendazole + Metalaxyl

C	Apron T69 WS	Novartis Crop Protection UK Ltd	08387	
C	Apron T69 WS	Ciba Agriculture	06725	31/08/2000

C   These products are "approved for agricultural use". For further details refer to page vii.
A   These products are approved for use in or near water. For further details refer to page vii.

	Product Name	Marketing Company	Reg. No.	Expiry Date

761 Thiabendazole + Metalaxyl—continued

C	Polycote Select	Germains UK Ltd	09718	
C	Polycote Select	Seedcote Systems Ltd	08430	
C	Polycote Select	Seedcote Systems Ltd	06727	31/08/2000

762 Thiabendazole + Metalaxyl + Thiram

C	Apron Combi FS	Novartis Crop Protection UK Ltd	08386	
C	Apron Combi FS	Ciba Agriculture	07203	30/11/2000

763 Thiabendazole + Oxadixyl + Thiram

C	Leap	Uniroyal Chemical Ltd	08477	
C	Leap	Uniroyal Chemical Ltd	07884	
C	Leap	Uniroyal Chemical Ltd	07822	31/08/2000

764 Thiabendazole + Tecnazene

C	Hytec Super	Agrichem (International) Ltd	01100	
	New Arena TBZ 6	Tripart Farm Chemicals Ltd	05606	
C	Storite SS	Novartis Crop Protection UK Ltd	08702	30/09/2000
C	Storite SS	MSD Agvet	02034	31/08/2000

765 Thiabendazole + Thiram

C	HY-TL	Agrichem (International) Ltd	06246	
C	HY-VIC	Agrichem (International) Ltd	06247	

766 Thiophanate-methyl

C	Cercobin Liquid	Rhone-Poulenc Agriculture	06188	
C	Mildothane Liquid	Rhone-Poulenc Agriculture	06211	
C	Mildothane Turf Liquid	Rhone-Poulenc Amenity	05331	

767 Thiophanate-methyl + Iprodione

C	Aventis Compass	Rhone-Poulenc Agriculture	09760	
C	Compass	Rhone-Poulenc Agriculture	06190	
C	Snooker	Rhone-Poulenc Agriculture	07940	

768 Thiram

C	Agrichem Flowable Thiram	Agrichem (International) Ltd	06245	
C	FS Thiram 15% Dust	Ford Smith & Co Ltd	07428	31/10/2001
C	Hy-Flo	Agrichem (International) Ltd	04637	
C	Robinson's Thiram 60	Agrichem (International) Ltd	04638	
C	Thiraflo	Uniroyal Chemical Ltd	09496	
C	Unicrop Thianosan DG	Universal Crop Protection Ltd	05454	

769 Thiram + Carbendazim + Cymoxanil + Oxadixyl

C	Apron Elite	Novartis Crop Protection UK Ltd	08770	

C   These products are "approved for agricultural use". For further details refer to page vii.
A   These products are approved for use in or near water. For further details refer to page vii.

Product Name	Marketing Company	Reg. No.	Expiry Date

770 Thiram + Carboxin

C	Anchor	Uniroyal Chemical Ltd	08684	
C	Anchor	Uniroyal Chemical Ltd	07949	30/09/2000

771 Thiram + Metalaxyl

C	Favour 600 SC	Novartis Crop Protection UK Ltd	08405	
C	Favour 600 SC	Ciba Agriculture	05842	31/08/2000

772 Thiram + Metalaxyl + Thiabendazole

C	Apron Combi FS	Novartis Crop Protection UK Ltd	08386	
C	Apron Combi FS	Ciba Agriculture	07203	30/11/2000

773 Thiram + Oxadixyl + Thiabendazole

C	Leap	Uniroyal Chemical Ltd	08477	
C	Leap	Uniroyal Chemical Ltd	07884	
C	Leap	Uniroyal Chemical Ltd	07822	31/08/2000

774 Thiram + Permethrin

C	Combinex	Pan Britannica Industries Ltd	00562	

775 Thiram + Thiabendazole

C	HY-TL	Agrichem (International) Ltd	06246	
C	HY-VIC	Agrichem (International) Ltd	06247	

776 Tolclofos-methyl

C	Basilex	The Scotts Company (UK) Limited	07494	
C	Basilex Soluble Sachets	The Scotts Company (UK) Limited	06958	31/07/2001
C	Rizolex	Sumitomo Chemical Agro Europe SA	09673	
C	Rizolex	AgrEvo UK Ltd	07271	31/07/2002
C	Rizolex 50 WP	AgrEvo UK Ltd	07272	
C	Rizolex 50 WP in Soluble Sachets	AgrEvo UK Ltd	06957	20/04/2001
C	Rizolex Flowable	Sumitomo Chemical Agro Europe SA	09358	
C	Rizolex Flowable	AgrEvo UK Ltd	07273	31/12/2001

777 Tolclofos-methyl + Prochloraz

C	Rizolex Nova	AgrEvo UK Ltd	07434	31/01/2000

778 Tolylfluanid

C	Bayer UK 456	Bayer plc	07995	
C	Elvaron M	Bayer plc	07772	

C These products are "approved for agricultural use". For further details refer to page vii.
A These products are approved for use in or near water. For further details refer to page vii.

Product Name	Marketing Company	Reg. No.	Expiry Date

779 Triadimefon

C 100-Plus	Dalgety Agriculture Ltd	05112	
C Bayleton	Bayer plc	00221	
C Bayleton 5	Bayer plc	00222	
C Landgold Triadimefon 25	Landgold & Co Ltd	06139	
C Standon Triadimefon 25	Standon Chemicals Ltd	05673	29/02/2000

780 Triadimenol

C Bayfidan	Bayer plc	02672	
C Hi-Shot	Cyanamid Agriculture Ltd	06508	
C Spinnaker	Cyanamid Agriculture Ltd	07023	

781 Triadimenol + Bitertanol + Fuberidazole

C Cereline	Bayer plc	07239	

782 Triadimenol + Bitertanol + Fuberidazole + Imidacloprid

C Cereline Secur	Bayer plc	09511	

783 Triadimenol + Ethirimol + Fuberidazole

C Bay UK 292	Bayer plc	04335	

784 Triadimenol + Fuberidazole

C Baytan	Zeneca Crop Protection	06636	31/12/2000
C Baytan	Bayer plc	00225	
C Baytan CF	Bayer plc	08193	
C Baytan Flowable	Zeneca Ltd	06635	31/12/2000
C Baytan Flowable	Bayer plc	02593	

785 Triadimenol + Fuberidazole + Imazalil

C Baytan IM	Bayer plc	00226	

786 Triadimenol + Fuberidazole + Imidacloprid

C Baytan Secur	Bayer plc	09510	

787 Triadimenol + Tebuconazole

C Garnet	Bayer plc	06391	
C Ruby	Bayer plc	06389	
C Silvacur	Bayer plc	06387	
C Silvacur 300	Bayer plc	08056	
C Veto F	Bayer plc	08057	

788 Triadimenol + Tridemorph

C Dorin	Bayer plc	08361	30/09/2001
C Dorin	Bayer plc	03292	30/09/2001

C   These products are "approved for agricultural use". For further details refer to page vii.
A   These products are approved for use in or near water. For further details refer to page vii.

Product Name	Marketing Company	Reg. No.	Expiry Date

788 Triadimenol + Tridemorph—continued

C	Jasper	BASF plc	06044	

789 Triazoxide + Imidacloprid + Tebuconazole

C	Raxil Secur	Bayer plc	08966	

790 Triazoxide + Tebuconazole

C	Raxil S	Zeneca Crop Protection	07484	31/12/2000
C	Raxil S	Bayer plc	06974	
C	Raxil S CF	Bayer plc	08192	

791 Tridemorph

C	Calixin	BASF plc	00369	
C	Standon Tridemorph 750	Standon Chemicals Ltd	05667	30/04/2002

792 Tridemorph + Carbendazim + Maneb

C	Cosmic FL	BASF plc	03473	

793 Tridemorph + Cyproconazole

C	Alto Major	Novartis Crop Protection UK Ltd	08468	30/09/2001
C	Alto Major	Sandoz Agro Ltd	06979	30/09/2001
C	Moot	Novartis Crop Protection UK Ltd	08479	30/09/2001
C	Moot	Sandoz Agro Ltd	06990	30/09/2001
C	SAN 735	Novartis Crop Protection UK Ltd	08490	30/09/2001
C	San 735	Sandoz Agro Ltd	06991	30/09/2001

794 Tridemorph + Epoxiconazole

C	BAS 47807F	BASF plc	08147	31/12/2000
C	Opus Plus	BASF plc	07363	

795 Tridemorph + Fenbuconazole

C	Unison	Pan Britannica Industries Ltd	08318	30/04/2002

796 Tridemorph + Fenpropimorph

C	BAS 46402F	BASF plc	03313	
C	Gemini	BASF plc	05684	

797 Tridemorph + Fenpropimorph + Flusilazole

C	Bingo	BASF plc	06920	
C	DUK 51	Du Pont (UK) Ltd	06764	30/09/2001
C	Justice	Du Pont (UK) Ltd	07963	30/09/2001

C   These products are "approved for agricultural use". For further details refer to page vii.
A   These products are approved for use in or near water. For further details refer to page vii.

Product Name	Marketing Company	Reg. No.	Expiry Date

798 Tridemorph + Flusilazole

C Fusion	Du Pont (UK) Ltd	04908	30/09/2001
C Meld	BASF plc	04914	
C Option	Du Pont (UK) Ltd	07951	30/09/2001

799 Tridemorph + Propiconazole

C Joust	Universal Crop Protection Ltd	08122	
C Tilt Turbo 475 EC	Novartis Crop Protection UK Ltd	08457	30/09/2001
C Tilt Turbo 475 EC	Ciba Agriculture	03476	30/09/2001

800 Tridemorph + Tebuconazole

C Allicur	Bayer plc	06468	30/09/2001
C BAS 91580F	BASF plc	06469	

801 Tridemorph + Triadimenol

C Dorin	Bayer plc	08361	30/09/2001
C Dorin	Bayer plc	03292	30/09/2001
C Jasper	BASF plc	06044	

802 Triforine

C Fairy Ring Destroyer	Vitax Ltd	05541	
C Saprol	Cyanamid Agriculture Ltd	07016	

803 Triforine + Bupirimate

C Nimrod T	The Scotts Company (UK) Limited	09268	
C Nimrod T	Miracle Professional	07865	
C Nimrod T	Zeneca Professional Products	06859	31/01/2001
C Nimrod T	ICI Agrochemicals	01499	31/05/2001

804 Vinclozolin

C Barclay Flotilla	Barclay Chemicals (UK) Ltd	07905	
C BASF Turf Protectant Fungicide	BASF plc	05457	31/08/2000
C Landgold Vinclozolin DG	Landgold & Co Ltd	06500	
C Landgold Vinclozolin SC	Landgold & Co Ltd	06459	
C Mascot Contact Turf Fungicide	Rigby Taylor Ltd	02711	31/08/2000
C Ronilan DF	BASF plc	04456	
C Ronilan FL	BASF plc	02960	
C Standon Vinclozolin	Standon Chemicals Ltd	07836	

805 Vinclozolin + Carbendazim

C Konker	BASF plc	03988	

C These products are "approved for agricultural use". For further details refer to page vii.
A These products are approved for use in or near water. For further details refer to page vii.

Product Name	Marketing Company	Reg. No.	Expiry Date

806 Vinclozolin + Chlorothalonil

| C | Curalan CL | BASF plc | 07174 | |

808 Zineb

| C | Tritoftorol | Elf Atochem Agri SA | 06872 | |
| C | Unicrop Zineb | Universal Crop Protection Ltd | 02279 | |

809 Zineb-ethylene thiuram disulphide adduct

| C | Polyram DF | BASF plc | 08234 | |

C These products are "approved for agricultural use". For further details refer to page vii.
A These products are approved for use in or near water. For further details refer to page vii.

Product Name	Marketing Company	Reg. No.	Expiry Date

PSD INSECTICIDES

1.3 Insecticides

including acaricides and nematicides

810 Abamectin

C	Dynamec	Novartis Crop Protection UK Ltd	08701	
C	Dynamec	MSD Agvet	06804	28/02/2001

811 Aldicarb

C	Aventis Temik 10G	Rhone-Poulenc Agriculture	09749
C	CleanCrop Aldicarb 10 G	United Agri Products	09694
C	Landgold Aldicarb 10G	Landgold & Co Ltd	06036
C	Me^2 Aldee	Me^2 Crop Protection Ltd	09781
C	Standon Aldicarb 10 G	Standon Chemicals Ltd	05915
C	Temik 10 G	Rhone-Poulenc Agriculture	06210

812 Alphacypermethrin

C	Acquit	Du Pont (UK) Ltd	07000	
C	Clayton Alpha-Cyper	Clayton Plant Protection (UK) Ltd	07065	29/02/2000
C	Contest	Cyanamid Agriculture Ltd	09024	
C	Contest Eco	Cyanamid Agriculture Ltd	08995	
C	Fastac	Cyanamid Agriculture Ltd	07008	
C	Fastac Dry	Cyanamid Agriculture Ltd	09025	
C	IT Alpha-cypermethrin	IT Agro Ltd	08274	
C	IT Alpha-Cypermethrin 100 EC	Whyte Agrochemicals Ltd	09372	
C	Standon Alpha-C10	Standon Chemicals Ltd	08823	

813 Aluminium phosphide

	Detia Gas Ex-T	Igrox Ltd	03792
	Detia Gas-Ex-B	Igrox Ltd	06927
	Fumitoxin	Igrox Ltd	04207
	Phostek	Terminix Peter Cox	09029
	Phostek	Killgerm Chemicals Ltd	05115
	Phostoxin I	Rentokil Initial UK Ltd	09313
	Phostoxin I	Rentokil Initial UK Ltd	05694

814 Amitraz

C	Bye Bye 20EC	Chimac-Agriphar SA	09346	
C	Mitac 20	AgrEvo UK Ltd	07265	31/05/2001
C	Mitac HF	AgrEvo UK Ltd	07358	
C	Ovasyn	AgrEvo UK Ltd	09190	
	Taktic Buildings Spray	Hoechst Roussel Vet Ltd	09136	
	Taktic Buildings Spray	Hoechst Animal Health Ltd	06504	31/07/2001

C   These products are "approved for agricultural use". For further details refer to page vii.
A   These products are approved for use in or near water. For further details refer to page vii.

Product Name	Marketing Company	Reg. No.	Expiry Date

815 Azamethiphos

Alfacron 10 WP	Novartis Animal Health UK Ltd	08587	28/02/2002
Alfacron 10WP	Ciba Agriculture	02832	28/02/2002
Alfacron 50 WP	Novartis Animal Health UK Ltd	08588	20/04/2001
Alfacron 50 WP	Ciba Agriculture	06552	31/08/2000
Alfacron Plus	Novartis Animal Health UK Ltd	09439	
Farm Fly Spray 10 WP	Rentokil Initial UK Ltd	09314	20/04/2001
Farm Fly Spray 10 WP	Rentokil Initial UK Ltd	05274	20/04/2001

816 Bendiocarb

	Ficam ULV	AgrEvo UK Ltd	07863	20/04/2001
C	Garvox 3G	AgrEvo UK Ltd	07259	20/04/2001
C	Seedox SC	Uniroyal Chemical Ltd	07591	20/04/2001

817 Benfuracarb

C	Oncol 10G	Mirfield Sales Services Ltd	08249

818 Bifenthrin

C	Starion	FMC Europe NV	09795
C	Talstar	FMC Europe NV	06913

819 Bitertanol + Fuberidazole + Imidacloprid + Triadimenol

C	Cereline Secur	Bayer plc	09511

820 Buprofezin

C	Applaud	Zeneca Crop Protection	06900

821 Carbaryl

C	Cavalier	Rhone-Poulenc Amenity	07027	20/04/2001
C	Thinsec	Zeneca Crop Protection	06710	20/04/2001

822 Carbofuran

C	Barclay Carbosect	Barclay Chemicals (UK) Ltd	05512	29/02/2000
C	Nex	Tripart Farm Chemicals Ltd	05165	
C	Rampart	Sipcam UK Ltd	05166	
C	Stefes Carbofuran	Stefes Plant Protection Ltd	06969	29/02/2000
C	Throttle	Quadrangle Agrochemicals Ltd	05204	
C	Yaltox	Bayer plc	02371	

823 Carbosulfan

C	Landgold Carbosulfan 10 G	Landgold & Co Ltd	09019	
C	Landgold Carbosulfan 10G	Landgold & Co Ltd	06046	29/02/2000
C	Marshal 10G	FMC Europe NV	08873	

C These products are "approved for agricultural use". For further details refer to page vii.
A These products are approved for use in or near water. For further details refer to page vii.

Product Name	Marketing Company	Reg. No.	Expiry Date

823 Carbosulfan—continued

C Marshal 10G	Rhone-Poulenc Agriculture	06165	20/04/2001
C Marshal Soil Insecticide suSCon CR granules	Fargro Ltd	06978	
C Standon Carbosulfan 10G	Standon Chemicals Ltd	05671	29/02/2000

824 Chlorfenvinphos

C Birlane 24	Cyanamid Agriculture Ltd	07002	
C Sapecron 240 EC	Novartis Crop Protection UK Ltd	08440	31/08/2000
C Sapecron 240 EC	Ciba Agriculture	01861	31/05/2000

825 Chlormequat

C Stay Up	Nufarm UK Ltd	09444	

826 Chlorpyrifos

C Alpha Chlorpyrifos 48EC	Makhteshim-Agan (UK) Ltd	04821	
Ballad	Headland Amenity Ltd	09775	
C Barclay Clinch	Barclay Chemicals (UK) Ltd	06148	29/02/2000
C Barclay Clinch 480	Barclay Chemicals (UK) Ltd	08080	29/02/2000
C Barclay Clinch II	Barclay Chemicals (UK) Ltd	08596	
C Choir	Whyte Agrochemicals Ltd	09778	
C Crossfire	Rhone-Poulenc Agriculture	05329	31/08/2000
C Crossfire 480	Rhone-Poulenc Amenity	08142	
C Crossfire 480	Dow AgroSciences Ltd	08141	
C CYREN	Cheminova Agro (UK) Ltd	08358	
C CYREN 4	Cheminova Agro (UK) Ltd	07964	29/02/2000
C Dispatch	Dow AgroSciences Ltd	08139	
C Dursban 4	Dow AgroSciences Ltd	07815	
C Greencrop Pontoon	Greencrop Technology Ltd	09667	
C Lorsban 480	Dow AgroSciences Ltd	08076	
C Lorsban T	Rigby Taylor Ltd	07813	
C Maraud	The Scotts Company (UK) Limited	09274	
C Pyrinex 48EC	Makhteshim-Agan (UK) Ltd	08644	
C Spannit	pbi Agrochemicals Ltd	08744	
C Spannit Granules	pbi Agrochemicals Ltd	08984	
C Spannit Granules	Pan Britannica Industries Ltd	04048	31/03/2001
C Standon Chlorpyrifos	Standon Chemicals Ltd	07633	29/02/2000
C Standon Chlorpyrifos 48	Standon Chemicals Ltd	08286	
C Suscon Green	Scotts Europe BV	09226	30/06/2001
C Suscon Green Soil Insecticide	Fargro Ltd	06312	
C Talon	Farmers Crop Chemicals Ltd	08746	
C Tripart Audax	Tripart Farm Chemicals Ltd	08745	

C These products are "approved for agricultural use". For further details refer to page vii.
A These products are approved for use in or near water. For further details refer to page vii.

Product Name	Marketing Company	Reg. No.	Expiry Date

827 Chlorpyrifos + Dimethoate

C	Atlas Sheriff	Atlas Crop Protection Ltd	07941	31/10/2000
C	Sheriff	Dow AgroSciences Ltd	05744	31/10/2000

828 Chlorpyrifos + Disulfoton

C	Twinspan	pbi Agrochemicals Ltd	08983	

829 Chlorpyrifos-methyl

C	Reldan 22	Dow AgroSciences Ltd	08191	
C	Reldan 50	Dow AgroSciences Ltd	05742	20/04/2001

830 Clofentezine

C	Apollo 50 SC	AgrEvo UK Ltd	07242	

831 Cyfluthrin

C	Baythroid	Bayer plc	04273	

832 Cypermethrin

C	Afrisect 10	Stefes UK Ltd	09114	
C	Ashlade Cypermethrin 10 EC	Cyanamid Agriculture Ltd	07412	30/06/2001
C	Ashlade Cypermethrin 10 EC	Ashlade Formulations Ltd	06229	
C	Barclay Cypersect XL	Barclay Chemicals (UK) Ltd	06509	29/02/2000
C	Chemtech Cypermethrin 10 EC	Cyanamid Agriculture Ltd	07413	30/06/2001
C	Chemtech Cypermethrin 10 EC	Chemtech (Crop Protection) Ltd	04827	
C	Clayton Cyperten	Clayton Plant Protection (UK) Ltd	07041	29/02/2000
C	Cyperkill 10	Chiltern Farm Chemicals Ltd	04119	
C	Cyperkill 25	Chiltern Farm Chemicals Ltd	09038	
C	Cyperkill 25	Mitchell Cotts Chemicals Ltd	03741	31/05/2001
C	Cyperkill 5	Mitchell Cotts Chemicals Ltd	00625	
C	Cypertox	Cyanamid Agriculture Ltd	07414	30/06/2001
C	Cypertox	Farmers Crop Chemicals Ltd	05122	
C	Luxan Cypermethrin 10	Luxan (UK) Ltd	06283	
C	Permasect C	Whyte Agrochemicals Ltd	09200	
C	Permasect C	Mirfield Sales Services Ltd	07680	31/08/2001
C	Quadrangle Cyper 10	Cyanamid Agriculture Ltd	07410	30/06/2001
C	Quadrangle Cyper 10	Quadrangle Agrochemicals	03242	
C	Ripcord	Cyanamid Agriculture Ltd	07014	30/06/2001
C	Stefes Cypermethrin	Stefes Plant Protection Ltd	05635	29/02/2000
C	Stefes Cypermethrin 2	Stefes Plant Protection Ltd	05719	29/02/2000
C	Stefes Cypermethrin 3	Stefes Plant Protection Ltd	09172	
C	Toppel 10	United Phosphorus Ltd	08772	
C	Toppel 10	Zeneca Crop Protection	06516	31/01/2001

C These products are "approved for agricultural use". For further details refer to page vii.
A These products are approved for use in or near water. For further details refer to page vii.

Product Name	Marketing Company	Reg. No.	Expiry Date

832 Cypermethrin—continued

C	Vassgro Cypermethrin Insecticide	Cyanamid Agriculture Ltd	07411	30/06/2001
C	Vassgro Cypermethrin Insecticide	L W Vass (Agricultural) Ltd	03240	

833 Cyromazine

	Neporex 2 SG	Novartis Animal Health UK Ltd	08589	
	Neporex 2 SG	Ciba Agriculture	06985	31/01/2001

834 Deltamethrin

C	ACP 625	AgrEvo UK Ltd	08619	
C	Aventis Decis	AgrEvo UK Ltd	09710	
C	Decis	AgrEvo UK Ltd	07172	
C	Decis Micro	AgrEvo UK Ltd	08618	
C	Decis Pearl	AgrEvo UK Ltd	08622	
C	Deleet	Rentokil Initial UK Ltd	09312	
C	Deleet	Rentokil Initial UK Ltd	08195	
C	Landgold Deltaland	Landgold & Co Ltd	07480	29/02/2000
C	Pearl	AgrEvo UK Ltd	08621	
C	Pearl Micro	AgrEvo UK Ltd	08620	
C	Standon Deltamethrin	Standon Chemicals Ltd	07053	29/02/2000
C	Thripstick	Aquaspersions Ltd	02134	

835 Deltamethrin + Heptenophos

C	Decisquick	AgrEvo UK Ltd	07312	20/04/2001

836 Deltamethrin + Pirimicarb

C	ACP 105	AgrEvo UK Ltd	08989	
C	Best	AgrEvo UK Ltd	08988	
C	Evidence	AgrEvo UK Ltd	06934	
C	Patriot EC	AgrEvo UK Ltd	08990	

837 Demeton-S-methyl

C	Metaphor	United Phosphorus Ltd	08017	31/10/2000

838 Diazinon

C	Basudin 5FG	Novartis Crop Protection UK Ltd	08389	30/09/2000
C	Basudin 5FG	Ciba Agriculture	00213	31/08/2000
C	Darlingtons Diazinon Granules	Darmycel	08400	30/09/2000
C	Darlingtons Diazinon Granules	Darmycel	05674	31/08/2000

839 Dicamba + Dichlorprop + MCPA

C	Cleanrun 2	The Scotts Company (UK) Limited	09271	

C These products are "approved for agricultural use". For further details refer to page vii.
A These products are approved for use in or near water. For further details refer to page vii.

Product Name	Marketing Company	Reg. No.	Expiry Date

840 1,3-Dichloropropene
C	Telone 2000	Dow AgroSciences Ltd	05748	
C	Telone II	Dow AgroSciences Ltd	05749	

841 Dichlorprop + Dicamba + MCPA
C	Cleanrun 2	The Scotts Company (UK) Limited	09271	

842 Dichlorvos
C	Darmycel Dichlorvos	Darlington Mushroom Labs	08401	30/09/2000
C	Darmycel Dichlorvos	Darlingtons Mushroom Laboratories	05699	31/08/2000
C	Luxan Dichlorvos 600	Luxan (UK) Ltd	08297	
C	Luxan Dichlorvos Aerosol 15	Luxan (UK) Ltd	08298	
	Nuvan 500 EC	Novartis Animal Health UK Ltd	08590	
	Nuvan 500 EC	Ciba Agriculture	03861	

843 Dicofol
C	Kelthane	Landseer Limited	09260	

844 Dicofol + Tetradifon
C	Childion	Hortichem Ltd	03821	

845 Dienochlor
C	Pentac Aquaflow	Novartis Crop Protection UK Ltd	08482	31/12/2000
C	Pentac Aquaflow	Sandoz Agro Ltd	05697	31/08/2000

846 Diflubenzuron
C	Dimilin 25 WP	Uniroyal Chemical Ltd	08902	
C	Dimilin Flo	Zeneca Crop Protection	08985	
C	Dimilin Flo	Uniroyal Chemical Ltd	08769	
C	Dimilin Flo	Zeneca Crop Protection	07151	31/10/2000
C	Dimilin WP	Uniroyal Chemical Ltd	08870	
C	Dimilin WP	Zeneca Crop Protection	06656	31/01/2001

847 Dimethoate
C	Barclay Dimethosect	Barclay Chemicals (UK) Ltd	08538	31/07/2002
C	BASF Dimethoate 40	BASF plc	00199	
C	Danadim	Cheminova Agro (UK) Ltd	09583	
C	Danadim Dimethoate 40	Cheminova Agro (UK) Ltd	07351	31/05/2002
C	Greencrop Pelethon	Greencrop Technology Ltd	09477	31/07/2002
C	PA Dimethoate 40	Portman Agrochemicals Ltd	01527	
C	Portman Sysdim 40	Portman Agrochemicals Ltd	06902	30/09/2000
C	Rogor L40	Sipcam UK Ltd	07611	
C	Sector	Portman Agrochemicals Ltd	08882	

C   These products are "approved for agricultural use". For further details refer to page vii.
A   These products are approved for use in or near water. For further details refer to page vii.

Product Name	Marketing Company	Reg. No.	Expiry Date

847 Dimethoate—continued

C Top Farm Dimethoate	Top Farm Formulations Ltd	05936	30/04/2000

848 Dimethoate + Chlorpyrifos

C Atlas Sheriff	Atlas Crop Protection Ltd	07941	31/10/2000
C Sheriff	Dow AgroSciences Ltd	05744	31/10/2000

849 Disulfoton

C Campbell's Disulfoton FE 10	J D Campbell & Sons Ltd	00402	20/04/2001
C Disulfoton P10	United Phosphorus Ltd	08023	
C Disyston FE-10	Bayer plc	00714	20/04/2001
C Disyston P-10	Bayer plc	00715	20/04/2001

850 Disulfoton + Chlorpyrifos

C Twinspan	pbi Agrochemicals Ltd	08983	

851 Disulfoton + Quinalphos

C Knave	Novartis Crop Protection UK Ltd	08476	31/07/2000
C Knave	Sandoz Agro Ltd	05480	31/07/2000
C Knave	Hortichem Ltd	02534	31/07/2000

852 Endosulfan

C Thiodan 20 EC	AgrEvo UK Ltd	07335	

854 Esfenvalerate

C Sumi-Alpha	Cyanamid Agriculture Ltd	07207	

855 Ethiofencarb

C Croneton	Bayer plc	00593	20/04/2001

856 Ethoprophos

C Aventis Mocap 10G	Rhone-Poulenc Agriculture	09750	
C Mocap 10G	Rhone-Poulenc Agriculture	06773	

857 Etrimfos

Satisfar	Nickerson Seed Specialists Ltd	04180	
Satisfar Dust	Sandoz Agro Ltd	04142	31/08/2000
Satisfar Dust	Nickerson Seed Specialists Ltd	04085	

858 Fatty acids

C Safer's Insecticidal Soap	Safer Ltd	07197	
C Savona	Koppert (UK) Ltd	06057	
C Savona	Koppert (UK) Ltd	03137	

C These products are "approved for agricultural use". For further details refer to page vii.
A These products are approved for use in or near water. For further details refer to page vii.

Product Name	Marketing Company	Reg. No.	Expiry Date

859 Fenazaquin

C	Matador 200 SC	Dow AgroSciences Ltd	07960	

860 Fenbutatin oxide

C	Torq	Fargro Ltd	08370	
C	Torque	Cyanamid Agriculture Ltd	07148	

861 Fenitrothion

	Antec Durakil	Antec International Ltd	05147	31/05/2000
C	Dicofen	pbi Agrochemicals Ltd	09598	
C	Dicofen	Pan Britannica Industries Ltd	00693	20/04/2001
C	Unicrop Fenitrothion 50	Universal Crop Protection Ltd	02267	20/04/2001

862 Fenitrothion + Permethrin + Resmethrin

	Turbair Grain Store Insecticide	Pan Britannica Industries Ltd	02238	20/04/2001

863 Fenoxycarb

C	Insegar	Novartis Crop Protection UK Ltd	08558	
C	Insegar	Ciba Agriculture	08329	31/08/2000

864 Fenpropathrin

C	Meothrin	Cyanamid Agriculture Ltd	07206	

865 Fenpyroximate

C	Sequel	AgrEvo UK Ltd	07624	

866 Fonofos

C	Cudgel	Zeneca Crop Protection	06648	31/07/2000
C	Fonofos Seed Treatment	Zeneca Crop Protection	06664	31/03/2000

867 Fosthiazate

C	Nemathorin 10G	Zeneca Crop Protection	08915	

868 Fuberidazole + Bitertanol + Imidacloprid + Triadimenol

C	Cereline Secur	Bayer plc	09511	

869 Fuberidazole + Imidacloprid + Triadimenol

C	Baytan Secur	Bayer plc	09510	

870 Heptenophos

C	Hostaquick	AgrEvo UK Ltd	07326	20/04/2001

C   These products are "approved for agricultural use". For further details refer to page vii.
A   These products are approved for use in or near water. For further details refer to page vii.

Product Name	Marketing Company	Reg. No.	Expiry Date

871 Heptenophos + Deltamethrin

C	Decisquick	AgrEvo UK Ltd	07312	20/04/2001

872 Imidacloprid

C	Admire	Bayer plc	07481	
C	Bayer UK 397	Bayer plc	08930	
C	Bayer UK 479	Bayer plc	08584	
C	Bayer UK368	Bayer plc	08125	
C	Gaucho	Bayer plc	06590	
C	Gaucho FS	Bayer plc	08496	
C	Intercept 5GR	The Scotts Company (UK) Limited	08126	
C	Intercept 70 WG	The Scotts Company (UK) Limited	08585	
C	Levington Professional Plus Intercept	The Scotts Company (UK) Limited	08569	
C	Rentokil Desyst	Rentokil Initial UK Ltd	09446	

873 Imidacloprid + Bitertanol + Fuberidazole + Triadimenol

C	Cereline Secur	Bayer plc	09511	

874 Imidacloprid + Fuberidazole + Triadimenol

C	Baytan Secur	Bayer plc	09510	

876 Lambda-cyhalothrin

C	Hallmark	Zeneca Crop Protection	06434	
C	Hero	Zeneca Crop Protection	07821	
C	Landgold Lambda-C	Landgold & Co Ltd	09205	
C	Landgold Lambda-C	Landgold & Co Ltd	06097	29/02/2000
C	Me^2 Lambda	Me^2 Crop Protection Ltd	09712	
C	Stampout	Phytheron 2000 SA	09777	
C	Standon Lambda-C	Standon Chemicals Ltd	08831	
C	Standon Lambda-C	Standon Chemicals Ltd	05672	29/02/2000

877 Lambda-cyhalothrin + Pirimicarb

C	Dovetail	Zeneca Crop Protection	07973	

878 Lindane

C	Atlas Steward	Atlas Crop Protection Ltd	07728	
C	Atlas Steward	Atlas Interlates Ltd	03062	29/02/2000
C	Fumite Lindane 10	Octavius Hunt Ltd	00933	
C	Fumite Lindane 40	Octavius Hunt Ltd	00934	
C	Fumite Lindane Pellets	Octavius Hunt Ltd	00937	
C	Gamma-Col	Zeneca Crop Protection	06670	30/09/2000
C	Lindane Flowable	pbi Agrochemicals Ltd	09113	
C	Lindane Flowable	Pan Britannica Industries Ltd	02610	31/08/2001

C   These products are "approved for agricultural use". For further details refer to page vii.
A   These products are approved for use in or near water. For further details refer to page vii.

215

Product Name	Marketing Company	Reg. No.	Expiry Date

879 Lindane + Thiophanate-methyl

C Castaway Plus	Rhone-Poulenc Amenity	05327	31/07/2001

880 Magnesium phosphide

Degesch Plates	Rentokil Initial UK Ltd	07603	

881 Malathion

C Malathion 60	United Phosphorus Ltd	08018	

882 MCPA + Dicamba + Dichlorprop

C Cleanrun 2	The Scotts Company (UK) Limited	09271	

883 Mefluidide

C Embark Lite	Intracrop	08749	

884 Mephosfolan

C Cytro-Lane	Cyanamid Agriculture Ltd	00626	20/04/2001

885 Methiocarb

C Bayer UK 808	Bayer plc	09513	
C Bayer UK 809	Bayer plc	09514	
C Bayer UK 892	Bayer plc	09540	
C Bayer UK 935	Bayer plc	09541	
C Club	Bayer plc	07176	
C Decoy	Bayer plc	06535	
C Decoy Plus	Bayer plc	07615	
C Decoy Wetex	Bayer plc	09707	
C Draza	Bayer plc	00765	
C Draza 2	Bayer plc	04748	
C Draza Plus	Bayer plc	06553	
C Draza Wetex	Bayer plc	09704	
C Elvitox	Bayer plc	06738	
C Epox	Bayer plc	06737	
C Exit	Bayer plc	07632	
C Karan	Bayer plc	09637	
C Lupus	Bayer plc	09638	
C Rhizopon AA Powder (0.5%)	Rhizopon UK Limited	09082	
C Rivet	Bayer plc	09512	

886 Methomyl

Golden Malrin	Novartis Animal Health UK Ltd	08821	20/04/2001

C These products are "approved for agricultural use". For further details refer to page vii.
A These products are approved for use in or near water. For further details refer to page vii.

Product Name	Marketing Company	Reg. No.	Expiry Date

887 Methoprene

C	Apex 5E	Novartis Animal Health UK Ltd	08739	
C	Apex 5E	Sandoz Speciality Pest Control Ltd	05730	31/10/2000

888 Nicotine

C	Nico Soap	United Phosphorus Ltd	07517	
C	Nicotine 40% Shreds	Dow AgroSciences Ltd	05725	
C	No-FID	Hortichem Ltd	07959	
C	XL All Insecticide	Synchemicals Ltd	02369	
C	XL All Nicotine 95%	Vitax Ltd	07402	

889 Oxamyl

C	Fielder	Du Pont (UK) Ltd	05279	
C	Vydate 10G	Du Pont (UK) Ltd	02322	

890 Permethrin

C	Darmycel Agarifume Smoke Generator	Sylvan Spawn Ltd	09564	
C	Darmycel Agarifume Smoke Generator	Darmycel	07904	30/04/2002
C	Fumite Permethrin Smoke	Octavius Hunt Ltd	00940	
	Geest Fumite MK2	Octavius Hunt Ltd	07939	
	Geestline Fumite MK2	Octavius Hunt Ltd	07218	
C	Permasect 10 EC	Mitchell Cotts Chemicals Ltd	03920	
C	Permasect 25 EC	Mitchell Cotts Chemicals Ltd	01576	
C	Permethrin 12 ED	Zeneca Crop Protection	09629	
C	Permit	pbi Agrochemicals Ltd	09609	
C	Permit	Pan Britannica Industries Ltd	01577	
C	Turbair Permethrin	pbi Agrochemicals Ltd	09616	
C	Turbair Permethrin	Pan Britannica Industries Ltd	02246	

891 Permethrin + Fenitrothion + Resmethrin

	Turbair Grain Store Insecticide	Pan Britannica Industries Ltd	02238	20/04/2001

892 Permethrin + Thiram

C	Combinex	Pan Britannica Industries Ltd	00562	

893 Phorate

C	Phorate 10G	United Phosphorus Ltd	08007	

894 Phosalone

C	Zolone Liquid	Rhone-Poulenc Agriculture	06173	20/04/2001

C   These products are "approved for agricultural use". For further details refer to page vii.
A   These products are approved for use in or near water. For further details refer to page vii.

217

Product Name	Marketing Company	Reg. No.	Expiry Date

895 Pirimicarb

C Agriguard Pirimicarb	Tronsan Ltd	09620	
C Aphox	Zeneca Crop Protection	06633	
C Barclay Pirimisect	Barclay Chemicals (UK) Ltd	09057	
C Barclay Pirimisect	Barclay Chemicals (UK) Ltd	06929	29/02/2000
C Clayton Pirimicarb 50 SG	Clayton Plant Protection (UK) Ltd	06972	29/02/2000
C Clayton Pirimicarb 50SG	Clayton Plant Protection (UK) Ltd	09221	
C Helocarb Granule 500	Helm AG	08157	20/04/2001
C Landgold Pirimicarb 50	Landgold & Co Ltd	09018	
C Landgold Pirimicarb 50	Landgold & Co Ltd	06238	29/02/2000
C Phantom	Bayer plc	04519	
C Pirimate	Portman Agrochemicals Ltd	09568	
C Pirimor	Zeneca Crop Protection	06694	20/04/2001
C Portman Pirimicarb	Portman Agrochemicals Ltd	06922	29/02/2000
C Standon Pirimicarb 50	Standon Chemicals Ltd	08878	
C Standon Pirimicarb 50	Standon Chemicals Ltd	05622	29/02/2000
C Standon Pirimicarb H	Standon Chemicals Ltd	05669	20/04/2001
C Stefes Pirimicarb	Stefes Plant Protection Ltd	05758	29/02/2000
C Unistar Pirimicarb 500	Unistar Ltd	06975	20/04/2001

896 Pirimicarb + Deltamethrin

C ACP 105	AgrEvo UK Ltd	08989	
C Best	AgrEvo UK Ltd	08988	
C Evidence	AgrEvo UK Ltd	06934	
C Patriot EC	AgrEvo UK Ltd	08990	

897 Pirimicarb + Lambda-cyhalothrin

C Dovetail	Zeneca Crop Protection	07973	

898 Pirimiphos-methyl

Actellic 2% Dust	Zeneca Crop Protection	06931	
Actellic 40 WP	Zeneca Crop Protection	06932	30/04/2000
Actellic D	Zeneca Crop Protection	06930	
Actellic Smoke Generator 10	Imperial Chemical Industries Plc	00017	30/06/2000
Actellic Smoke Generator No 10	Zeneca Public Health	07900	
Actellic Smoke Generator No 20	Zeneca Crop Protection	06627	
C Actellifog	Hortichem Ltd	08078	
C Actellifog	Zeneca Crop Protection	06628	20/04/2001
C Blex	Zeneca Crop Protection	06639	20/04/2001
C Dazzel	Zeneca Crop Protection	08024	20/04/2001
C Fumite Pirimiphos methyl Smoke	Octavius Hunt Ltd	00941	

C These products are "approved for agricultural use". For further details refer to page vii.
A These products are approved for use in or near water. For further details refer to page vii.

Product Name	Marketing Company	Reg. No.	Expiry Date

899 Propoxur

C Fumite Propoxur Smoke	Octavius Hunt Ltd	00942	31/10/2001

900 Pyrethrins

Alfadex	Novartis Animal Health UK Ltd	08591	
Alfadex	Ciba Agriculture	00074	
Turbair Flydown	pbi Agrochemicals Ltd	09613	
Turbair Flydown	Pan Britannica Industries Ltd	05482	
Turbair Kilsect Short Life Grade	pbi Agrochemicals Ltd	09615	
Turbair Kilsect Short Life Grade	Pan Britannica Industries Ltd	02240	
Turbair Super Flydown	Pan Britannica Industries Ltd	09617	
Turbair Super Flydown	Pan Britannica Industries Ltd	02249	

901 Pyrethrins + Resmethrin

C Pynosect 30 Water Miscible	Mitchell Cotts Chemicals Ltd	01653	

902 Quinalphos + Disulfoton

C Knave	Novartis Crop Protection UK Ltd	08476	31/07/2000
C Knave	Sandoz Agro Ltd	05480	31/07/2000
C Knave	Hortichem Ltd	02534	31/07/2000

903 Resmethrin

C Turbair Resmethrin Extra	pbi Agrochemicals Ltd	02247	

904 Resmethrin + Fenitrothion + Permethrin

Turbair Grain Store Insecticide	Pan Britannica Industries Ltd	02238	20/04/2001

905 Resmethrin + Pyrethrins

C Pynosect 30 Water Miscible	Mitchell Cotts Chemicals Ltd	01653	

906 Rotenone

C Devcol Liquid Derris	Devcol Ltd	06063	
C Liquid Derris	Ford Smith & Co Ltd	01213	

907 Sulphur

C Ashlade Sulphur FL	Ashlade Formulations Ltd	06478	
C Kumulus DF	BASF plc	04707	
C MSS Sulphur 80	Mirfield Sales Services Ltd	05752	

C   These products are "approved for agricultural use". For further details refer to page vii.
A   These products are approved for use in or near water. For further details refer to page vii.

Product Name	Marketing Company	Reg. No.	Expiry Date

908 Tar oils

C Sterile Tar Oil Winter Wash 60% Stock Emulsion	Coventry Chemicals Ltd	05062	
C Sterilite Tar Oil Winter Wash 80% Miscible Quality	Coventry Chemicals Ltd	05062	

909 Tau-fluvalinate

C Mavrik	Novartis Crop Protection UK Ltd	09697	
C Mavrik Aquaflow	Novartis Crop Protection UK Ltd	08347	31/08/2002

910 Tebufenpyrad

C Masai	Cyanamid Agriculture Ltd	07452	
C Masai G	Cyanamid Agriculture Ltd	07453	

911 Teflubenzuron

C Nemolt	Cyanamid Agriculture Ltd	07012	

912 Tefluthrin

C Evict	Bayer plc	09150	
C Evict	Zeneca Crop Protection	08731	31/07/2001
C Force ST	Bayer plc	09713	
C Force ST	Zeneca Crop Protection	06665	31/08/2002

913 Tetradifon

C Tedion V-18 EC	Hortichem Ltd	03820	

914 Tetradifon + Dicofol

C Childion	Hortichem Ltd	03821	

915 Thiometon

C Ekatin	Novartis Crop Protection UK Ltd	08474	30/11/2000
C Ekatin	Sandoz Agro Ltd	05281	31/08/2000

916 Thiophanate-methyl + Lindane

C Castaway Plus	Rhone-Poulenc Amenity	05327	31/07/2001

917 Thiram + Permethrin

C Combinex	Pan Britannica Industries Ltd	00562	

918 Triadimenol + Bitertanol + Fuberidazole + Imidacloprid

C Cereline Secur	Bayer plc	09511	

C These products are "approved for agricultural use". For further details refer to page vii.
A These products are approved for use in or near water. For further details refer to page vii.

Product Name	Marketing Company	Reg. No.	Expiry Date

919 Triadimenol + Fuberidazole + Imidacloprid

C   Baytan Secur   Bayer plc   09510

920 Triazamate

C   Aztec   Cyanamid Agriculture Ltd   07817

921 Triazophos

C   Hostathion   AgrEvo UK Crop Protection Ltd   07327   30/09/2000

922 Trichlorfon

C   Dipterex 80   Bayer plc   00711   20/04/2001

923 Zetacypermethrin

C Fury 10 EW	pbi Agrochemicals Ltd	09689	
C Fury 10 EW	Pan Britannica Industries Ltd	08153	31/07/2002
C Minuet EW	pbi Agrochemicals Ltd	09605	
C Minuet EW	Pan Britannica Industries Ltd	08820	31/07/2002

C   These products are "approved for agricultural use". For further details refer to page vii.
A   These products are approved for use in or near water. For further details refer to page vii.

221

Product Name	Marketing Company	Reg. No.	Expiry Date

1.4 Vertebrate Control Products
including rodenticides, mole killers and bird repellents

925 Alphachloralose

Alpha Chloralose (Pure)	Killgerm Chemicals Ltd	00082
Alphachloralose Technical	Rentokil Initial UK Ltd	09319
Alphachloralose Concentrate	Rentokil Initial UK Ltd	09317
Rentokil Alphachloralose Technical	Rentokil Initial UK Ltd	01720
Rentokil Alphachloralose Concentrate	Rentokil Initial UK Ltd	01721

926 Aluminium ammonium sulphate

Curb	Sphere Laboratories (London) Ltd	02480
Guardsman B	Chiltern Farm Chemicals Ltd	05494
Guardsman M	Chiltern Farm Chemicals Ltd	05495
Guardsman STP	Sphere Laboratories (London) Ltd	03606
Liquid Curb Crop Spray	Sphere Laboratories (London) Ltd	03164
Rezist	Barrettine Environmental Health	08576

927 Aluminium phosphide

Amos Talunex	Luxan (UK) Ltd	06567
Luxan Talunex	Luxan (UK) Ltd	06563
Phostek	Killgerm Chemicals Ltd	07921
Phostek	Killgerm Chemicals Ltd	05115
Phostoxin	Rentokil Initial UK Ltd	09315
Rentokil Phostoxin	Rentokil Initial UK Ltd	01775

928 Brodifacoum

Brodifacoum Bait	Sorex Ltd	00336	
Brodifacoum Bait Blocks	Sorex Ltd	04590	30/09/2001
Brodifacoum Paste	Rentokil Initial UK Ltd	09567	
Brodifacoum Sewer Bait	Sorex Ltd	00337	
Erasor	Sorex Ltd	09746	
Klerat	Zeneca Public Health	06869	
Klerat Mouse Tube	Zeneca Public Health	09240	28/02/2002
Klerat Mouse Tube	Zeneca Public Health	06830	31/05/2000
Klerat Wax Blocks	Zeneca Public Health	06827	
Klerat Wax Blocks	ICI Agrochemicals	04746	30/09/2001
Sorex Brodifacoum Bait Blocks	Sorex Ltd	07759	

C   These products are "approved for agricultural use". For further details refer to page vii.
A   These products are approved for use in or near water. For further details refer to page vii.

Product Name	Marketing Company	Reg. No.	Expiry Date

928 Brodifacoum—continued

C	Sorex Brodifacoum Rat and Mouse Bait	Sorex Ltd	07758	
	Sorex Brodifacoum Sewer Bait	Sorex Ltd	07760	
	Sorex Brodifacoum Sewer Bait Sachets	Sorex Ltd	07761	
	Sorexa Checkatube	Sorex Ltd	09474	
	Talon Rat And Mouse Bait (Cut Wheat)	Killgerm Chemicals Ltd	07485	
	Talon Rat And Mouse Bait (Whole Wheat)	Killgerm Chemicals Ltd	07486	

929 Bromadiolone

Product Name	Marketing Company	Reg. No.	Expiry Date
Biotrol Plus Outdoor Rat Killer	Rentokil Initial UK Ltd	09013	
Bromard	Rentokil Initial UK Ltd	09001	
Bromatrol	Rentokil Initial UK Ltd	09008	
Bromatrol Contact Dust	Rentokil Initial UK Ltd	09006	
Bromatrol Mouse Blocks	Rentokil Initial UK Ltd	09034	
Bromatrol Rat Blocks	Rentokil Initial UK Ltd	09033	
Contrac All-Weather Blox	Bell Laboratories Inc	08260	
Contrac Rodenticide	Bell Laboratories Inc	08256	
Contrac Super Blox	Bell Laboratories Inc	08262	
Deadline	Rentokil Initial UK Ltd	09037	
Deadline Contact Dust	Rentokil Initial UK Ltd	09007	
Deadline Liquid Concentrate	Rentokil Initial UK Ltd	09003	
Deadline Place Packs	Rentokil Ltd	06624	
Endorats Blue Rat Bait	Irish Drugs Ltd	08046	
Liquid Bromatrol	Rentokil Initial UK Ltd	09002	
Rentokil Biotrol Plus Outdoor Rat Killer	Rentokil Ltd	03707	31/10/2001
Rentokil Bromard	Rentokil Ltd	01727	31/10/2001
Rentokil Bromatrol	Rentokil Ltd	05077	31/10/2001
Rentokil Bromatrol Contact Dust	Rentokil Ltd	01729	31/10/2001
Rentokil Bromatrol Mouse Blocks	Rentokil Ltd	08615	31/05/2001
Rentokil Bromatrol Rat Blocks	Rentokil Ltd	08614	31/05/2001
Rentokil Deadline	Rentokil Ltd	05078	31/05/2001
Rentokil Deadline Contact Dust	Rentokil Ltd	01736	31/10/2001
Rentokil Deadline Liquid Concentrate	Rentokil Ltd	01737	31/10/2001
Rentokil Liquid Bromatrol	Rentokil Ltd	03645	31/10/2001
Slaymor	Novartis Animal Health UK Ltd	08592	

C   These products are "approved for agricultural use". For further details refer to page vii.
A   These products are approved for use in or near water. For further details refer to page vii.

Product Name	Marketing Company	Reg. No.	Expiry Date

929 Bromadiolone—continued

Slaymor	Ciba Agriculture	01958	31/05/2001
Slaymor Bait Bags	Novartis Animal Health UK Ltd	08593	
Slaymor Bait Bags	Ciba Agriculture	03183	30/11/2001
Tomcat 2 All-Weather Blox	Antec International Ltd	08261	
Tomcat 2 Rodenticide	Antec International Ltd	08257	
Tomcat 2 Super Blox	Antec International Ltd	08263	

930 Calciferol

Deerat Concentrate	Rentokil Initial UK Ltd	09011	
Rentokil Deerat Concentrate	Rentokil Ltd	01738	28/02/2002

931 Calciferol + Difenacoum

Sorexa CD Concentrate	Sorex Ltd	03513	

932 Chlorophacinone

Drat Rat and Mouse Bait	Battle Hayward & Bower Ltd	00764	
Endorats	American Products	06503	
Karate Ready to Use Rat and Mouse Bait	DiverseyLever Ltd	08658	
Karate Ready To Use Rat and Mouse Bait	Lever Industrial Ltd	05321	
Karate Ready to Use Rodenticide Sachets	DiverseyLever Ltd	08659	
Karate Ready To Use Rodenticide Sachets	Lever Industrial Ltd	05890	
Rat Eyre Rat and Mouse Bait	Vermin Eradication Supplies	03952	
Ridento Ready to use Rat Bait	Ace Chemicals Ltd	03804	
Rout	Samuel McCausland Ltd	06729	
Ruby Rat	J V Heatherington	06059	

933 Cholecalciferol + Difenacoum

Sorexa CD Mouse Bait	Sorex Ltd	08207	

934 Coumatetralyl

Racumin Contact Powder	Bayer plc	09047	
Racumin Master Mix	Bayer plc	01677	31/01/2002
Racumin Rat Bait	Bayer plc	01679	
Racumin Tracking Powder	Bayer plc	01681	31/12/2001
Townex Sachets	Town & Country Pest Services Ltd	02164	

C   These products are "approved for agricultural use". For further details refer to page vii.
A   These products are approved for use in or near water. For further details refer to page vii.

Product Name	Marketing Company	Reg. No.	Expiry Date

935 Difenacoum

Product Name	Marketing Company	Reg. No.	Expiry Date
Deosan Rataway	DiverseyLever Ltd	08803	
Deosan Rataway Bait Bags	DiverseyLever Ltd	08805	
Deosan Rataway W	DiverseyLever Ltd	08804	
Difenard	Rentokil Initial UK Ltd	09012	
Fentrol	Rentokil Initial UK Ltd	09009	
Fentrol Gel	Rentokil Initial UK Ltd	09010	
Killgerm Rat Rod	Killgerm Chemicals Ltd	05154	
Killgerm Wax Bait	Killgerm Chemicals Ltd	04096	
Neokil	Sorex Ltd	05564	
Neokil Bait Bags	Sorex Ltd	08949	
Neosorexa	Sorex Ltd	07756	
Neosorexa Bait Blocks	Sorex Ltd	07355	
Neosorexa Concentrate	Sorex Ltd	01475	
Neosorexa Liquid Concentrate	Sorex Ltd	04640	
Neosorexa Mixed Grain Bait	Sorex Ltd	08782	
Neosorexa Pellets	Sorex Ltd	08784	
Neosorexa Ratpacks	Sorex Ltd	04653	
Ratak	Zeneca Public Health	06832	
Ratak	Imperial Chemical Industries Plc	02586	
Ratak Cut Wheat	Zeneca Public Health	07392	
Ratak Wax Blocks	Zeneca Public Health	06829	
Rataway	Deosan Ltd	05560	31/05/2000
Rataway Bait Bags	Deosan Ltd	05562	31/05/2000
Rataway W	Deosan Ltd	05561	31/05/2000
Rentokil Difenard	Rentokil Initial Plc	08124	28/02/2002
Rentokil Fentrol	Rentokil Ltd	01747	28/02/2002
Rentokil Fentrol Gel	Rentokil Ltd	01749	28/02/2002
Sorexa D	Sorex Ltd	06879	
Sorexa Gel	Sorex Ltd	08315	

936 Difenacoum + Calciferol

Product Name	Marketing Company	Reg. No.	Expiry Date
Sorexa CD Concentrate	Sorex Ltd	03513	

937 Difenacoum + Cholecalciferol

Product Name	Marketing Company	Reg. No.	Expiry Date
Sorexa CD Mouse Bait	Sorex Ltd	08207	

938 Diphacinone

Product Name	Marketing Company	Reg. No.	Expiry Date
Ditrac All-Weather Bait Bar	Bell Laboratories Inc	07227	31/05/2001
Ditrac All-Weather Blox	Bell Laboratories Inc	07228	30/09/2002
Ditrac Blox	Bell Laboratories Inc	09714	
Ditrac Rat and Mouse Bait	Bell Laboratories Inc	07170	31/05/2001
Ditrac Super Blox	Bell Laboratories Inc	09715	
Ditrac Super Blox All-Weather	Bell Laboratories Inc	07229	30/09/2002

C   These products are "approved for agricultural use". For further details refer to page vii.
A   These products are approved for use in or near water. For further details refer to page vii.

Product Name	Marketing Company	Reg. No.	Expiry Date
938 Diphacinone—continued			
Tomcat All-Weather Blox	Antec International Ltd	07230	30/09/2002
Tomcat All-Weather Rodent Bar	Antec International Ltd	07231	31/05/2001
Tomcat Blox	Antec International Ltd	09716	
Tomcat Rat and Mouse Bait	Antec International Ltd	07171	31/05/2001
Tomcat Super Blox All-Weather	Antec International Ltd	07446	31/05/2001
939 Flocoumafen			
Storm	Sorex Ltd	03710	
Storm Mouse Bait Blocks	Sorex Ltd	04850	
Storm Sewer Bait	Sorex Ltd	06947	
Storm Sewer Bait Sachets	Sorex Ltd	06946	
940 Seconal			
Seconal	Killgerm Chemicals Ltd	04715	31/07/2002
941 Sodium cyanide			
Cymag	Sorex Ltd	09308	
Cymag	Zeneca Crop Protection	06651	30/11/2001
942 Sulphonated cod liver oil			
Scuttle	Fine Agrochemicals Ltd	06232	
943 Warfarin			
Grey Squirrel Liquid Concentrate	Killgerm Chemicals Ltd	06455	
Grey Squirrel Liquid Concentrate	Rodent Control Ltd	01009	31/03/2002
Killgerm Sewercide Cut Wheat Rat Bait	Killgerm Chemicals Ltd	03761	
Killgerm Sewercide Whole Wheat Rat Bait	Killgerm Chemicals Ltd	03759	
Sakarat Ready To Use (Cut Wheat)	Killgerm Chemicals Ltd	04340	
Sakarat Ready to use (Whole Wheat)	Killgerm Chemicals Ltd	01850	
Sakarat X Ready to use Warfarin Rat Bait	Killgerm Chemicals Ltd	01851	
Sewarin Extra	Killgerm Chemicals Ltd	03426	
Sewarin P	Killgerm Chemicals Ltd	01930	
Sorex Warfarin 250 ppm Rat Bait	Sorex Ltd	07371	

C These products are "approved for agricultural use". For further details refer to page vii.
A These products are approved for use in or near water. For further details refer to page vii.

Product Name	Marketing Company	Reg. No.	Expiry Date

943 Warfarin—continued

Sorex Warfarin 500 ppm Rat Bait	Sorex Ltd	07372	
Sorex Warfarin Sewer Bait	Sorex Ltd	07373	
Warfarin Concentrate	Battle Hayward & Bower Ltd	02325	
Warfarin Ready Mixed Bait	Battle Hayward & Bower Ltd	02333	

944 Zinc phosphide

Grovex Zinc phosphide	Killgerm Chemicals Ltd	06230	
RCR Zinc phosphide	Killgerm Chemicals Ltd	06231	
ZP Rodent	Bell Laboratories Inc	02822	
ZP Rodent Pellets	Antec International Ltd	07814	

945 Ziram

AAprotect	Universal Crop Protection Ltd	03784	

C   These products are "approved for agricultural use". For further details refer to page vii.
A   These products are approved for use in or near water. For further details refer to page vii.

Product Name	Marketing Company	Reg. No.	Expiry Date

1.5 Biological Pesticides

946 Bacillus thuringiensis

C	Bactospeine WP	Koppert (UK) Ltd	02913	30/09/2000
C	Bactura WP	Koppert (UK) Ltd	08732	
C	Dipel	Pan Britannica Industries Ltd	06577	30/09/2000
C	Dipel	Pan Britannica Industries Ltd	06308	31/05/2000
C	Dipel	English Woodlands Ltd	03214	30/09/2000
C	DiPel WP	English Woodlands Biocontrol	08634	
C	Novosol Flowable Concentrate	Ashlade Formulations Ltd	06566	31/03/2001
C	Thuricide HP	Novartis Crop Protection UK Ltd	08494	
C	Thuricide HP	Sandoz Agro Ltd	02136	30/06/2000

947 Bacillus thuringiensis Berliner var kurstaki

C	Novosol FC	Whyte Agrochemicals Ltd	08819

948 Peniophora gigantea

C	PG Suspension	Ecological Laboratories Ltd	08975

949 Verticillium lecanii

C	Mycotal	Koppert (UK) Ltd	04782
C	Vertalec	Koppert (UK) Ltd	04781

C These products are "approved for agricultural use". For further details refer to page vii.
A These products are approved for use in or near water. For further details refer to page vii.

Product Name	Marketing Company	Reg. No.	Expiry Date

1.6 Miscellaneous

including molluscicides, lumbricides, soil sterilants, anti-oxidants and fumigants

950 Bitertanol + Fuberidazole + Imidacloprid
C   Sibutol Secur   Bayer plc   09131

951 Carbaryl
C   Cavalier   Rhone-Poulenc Amenity   07027   20/04/2001
C   Twister Flow   Rhone-Poulenc Amenity   05712   20/04/2001

952 Carbendazim
C   Mascot Systemic   Rigby Taylor Ltd   07654   30/11/2001
C   Turfclear WDG   The Scotts Company (UK) Limited   07490
C   Twincarb   Vitax Ltd   08777

953 Chloropicrin
C   Chloropicrin Fumigant   Dewco-Lloyd Ltd   04216
C   K & S Chlorofume   K & S Fumigation Services Ltd   08722

954 Chlorpropham
   Aceto Chlorpropham 50M   Aceto Agricultural Chemicals Corporation (UK) Ltd   08929

955 Dazomet
C   Basamid   Hortichem Ltd   07204
C   Basamid   BASF plc   00192

956 1,2-Dichloropropane + 1,3-Dichloropropene
C   Cyanamid DD Soil Fumigant   Cyanamid Agriculture Ltd   07017

957 1,3-Dichloropropene + 1,2-Dichloropropane
C   Cyanamid DD Soil Fumigant   Cyanamid Agriculture Ltd   07017

958 Diphenylamine
C   No Scald DPA 31   Elf Atochem Agri SA   08312
   Shield DPA 15%   United Agri Products Limited   09550
   Shield DPA 15%   Willmot Pertwee Ltd   08313   31/05/2002

C   These products are "approved for agricultural use". For further details refer to page vii.
A   These products are approved for use in or near water. For further details refer to page vii.

229

Product Name	Marketing Company	Reg. No.	Expiry Date

959 Fuberidazole + Bitertanol + Imidacloprid

C Sibutol Secur	Bayer plc	09131	

960 Imidacloprid + Bitertanol + Fuberidazole

C Sibutol Secur	Bayer plc	09131	

961 Lindane + Thiophanate-methyl

C Castaway Plus	Rhone-Poulenc Amenity	05327	31/07/2001

962 Magnesium Phosphide

Rentokil Degesch Phostoxin Plates	Rentokil Ltd	01739	31/05/2000

963 Metaldehyde

C Aristo	De Sangosse (UK) SA	09622	
C Chiltern Hardy	Chiltern Farm Chemicals Ltd	06948	
C Clartex	pbi Agrochemicals Ltd	09213	
C Cookes 6% Metaldehyde Slug Killer	Devcol Morgan Ltd	09619	31/10/2001
C Devcol Morgan 6% Metaldehyde Slug Killer	Devcol Morgan Ltd	07422	31/10/2001
C Doff Agricultural Slug Killer with Animal Repellent	Doff Portland Ltd	06058	31/07/2001
C Doff Horticultural Slug Killer Blue Mini Pellets	Doff Portland Ltd	09666	
C Doff Horticultural Slug Killer Blue Mini Pellets	Doff Portland Ltd	05688	30/06/2002
C Doff Metaldehyde Slug Killer Mini Pellets	Doff Portland Ltd	00741	30/11/2001
C Doff New Formula Metaldehyde Slug Killer Mini Pellets	Doff Portland Ltd	09772	
C Doff New Formula Metaldehyde Slug Killer Mini Pellets	Doff Portland Ltd	09338	31/10/2002
C EM 1617/01	pbi Agrochemicals Ltd	09344	
C Escar-go 6	Chiltern Farm Chemicals Ltd	06076	
C Escar-Go Z	Lonza Ltd	08754	
C ESP	pbi Agrochemicals Ltd	09428	
C Fisons Helarion	Fisons Plc	02520	31/05/2001
C FP 107	pbi Agrochemicals Ltd	09060	
C FP 107	Zeneca Crop Protection	06666	31/05/2001
C Gastrotox 6G Slug Pellets	Truchem Ltd	04066	31/01/2000
C Hardy Z	Lonza Ltd	08755	

C These products are "approved for agricultural use". For further details refer to page vii.
A These products are approved for use in or near water. For further details refer to page vii.

Product Name	Marketing Company	Reg. No.	Expiry Date

963 Metaldehyde—continued

C	Helimax	De Sangosse (UK) SA	07350	
C	Luxan 9363	Luxan (UK) Ltd	07359	
C	Luxan Metaldehyde	Luxan (UK) Ltd	06564	
C	Lynx	De Sangosse (UK) SA	09770	
C	Lynx	De Sangosse (UK) SA	09137	31/10/2002
C	Metarex	De Sangosse (UK) SA	07752	31/10/2001
C	Metarex Green	De Sangosse (UK) SA	08131	
C	Metarex RG	De Sangosse (UK) SA	06754	
C	Metarex RGS	De Sangosse (UK) SA	09664	
C	Mifaslug	Farmers Crop Chemicals Ltd	01349	
C	Molotov	Chiltern Farm Chemicals Ltd	08295	
C	Optimol	pbi Agrochemicals Ltd	09061	
C	Optimol	Zeneca Crop Protection	06688	31/05/2001
C	pbi Slug Pellets	pbi Agrochemicals Ltd	09607	
C	Quadrangle Mini Slug Pellets	Quadrangle Agrochemicals Ltd	01670	31/01/2000
C	Regel	De Sangosse (UK) SA	08155	
C	Slug Pellets	Pan Britannica Industries Ltd	01558	28/02/2001
C	Super-Flor 6% Metaldehyde Slug Killer Mini Pellets	CMI Limited	09773	
C	Super-flor 6% Metaldehyde Slug Killer Mini Pellets	CMI Limited	05453	31/10/2002
C	Tripart Mini Slug Pellets	Tripart Farm Chemicals Ltd	02207	31/01/2000
C	Unicrop 6% Mini Slug Pellets	Universal Crop Protection Ltd	09771	
C	Unicrop 6% Mini Slug Pellets	Universal Crop Protection Ltd	02275	31/10/2002
C	Yeoman	De Sangosse (UK) SA	09623	

964 Metam-Sodium

C	Campbell's Metham Sodium	J D Campbell & Sons Ltd	00412	31/12/2000
C	Metam 510	UCB (Chem) Ltd	09796	
C	Metham Sodium 400	United Phosphorus Ltd	08051	
C	Sistan	Universal Crop Protection Ltd	01957	
C	Sistan 38	Universal Crop Protection Ltd	08646	
C	Vapam	Willmot Pertwee Ltd	09194	

965 Methiocarb

C	Barclay Poacher	Barclay Chemicals (UK) Ltd	09031	
C	Bayer UK 808	Bayer plc	09513	
C	Bayer UK 809	Bayer plc	09514	
C	Bayer UK 892	Bayer plc	09540	
C	Bayer UK 935	Bayer plc	09541	
C	Club	Bayer plc	07176	
C	Decoy	Bayer plc	06535	
C	Decoy Plus	Bayer plc	07615	
C	Decoy Wetex	Bayer plc	09707	

C   These products are "approved for agricultural use". For further details refer to page vii.
A   These products are approved for use in or near water. For further details refer to page vii.

Product Name	Marketing Company	Reg. No.	Expiry Date

965 Methiocarb—continued

C Draza	Bayer plc	00765	
C Draza 2	Bayer plc	04748	
C Draza Plus	Bayer plc	06553	
C Draza ST	Bayer plc	05315	
C Draza Wetex	Bayer plc	09704	
C Elvitox	Bayer plc	06738	
C Epox	Bayer plc	06737	
C Exit	Bayer plc	07632	
C Karan	Bayer plc	09637	
C Lupus	Bayer plc	09638	
C Rescur	Bayer plc	07942	
C Rhizopon AA Powder (0.5%)	Rhizopon UK Limited	09082	
C Rivet	Bayer plc	09512	

966 Methyl bromide

Brom-O-Gas	Great Lakes Chemical (Europe) Ltd	04508	31/01/2001
Bromomethane 100%	Brian Jones & Associates Ltd	09244	
Fumyl-O-Gas	Brian Jones & Associates Ltd	04833	
Mebrom 100	Mebrom NV	04869	
Mebrom 98	Mebrom NV	04779	31/07/2001
Methyl bromide	Rentokil Initial UK Ltd	09316	
Methyl bromide 100%	Bromine & Chemicals Ltd	01336	
Methyl bromide 98%	Bromine & Chemicals Ltd	01335	
Rentokil Methyl bromide	Rentokil Initial UK Ltd	05646	
Sobrom BM 100	Brian Jones & Associates Ltd	04381	
Sobrom BM 98	Brian Jones & Associates Ltd	04189	

967 Tar acids

C Brays Emulsion	Garden and Professional Products	00323	29/02/2000

968 Thiodicarb

C Genesis	Rhone-Poulenc Agriculture	06168	
C Genesis ST	Rhone-Poulenc Agriculture	08211	
C Judge	Rhone-Poulenc Agriculture	08163	
C Me^2 Exodus	Me^2 Crop Protection Ltd	09786	

969 Thiophanate-methyl + Lindane

C Castaway Plus	Rhone-Poulenc Amenity	05327	31/07/2001

Product Name	Marketing Company	Reg. No.	Expiry Date

970 Trinexapac-ethyl

| C | Shortcut | The Scotts Company (UK) Limited | 09254 | |

AMATEUR PRODUCTS

Amateur Products

971 Alphachloralose (Vertebrate Control)

Alphakil	Rentokil Initial UK Ltd	09318
Rentokil Alphakil	Rentokil Initial UK Ltd	01722

972 Aluminium ammonium sulphate (Vertebrate Control)

Bio Catapult	Pan Britannica Industries Ltd	07195
Curb (Garden Pack)	Sphere Laboratories (London) Ltd	03983
Scoot	William Sinclair Horticulture Ltd	07388
Stay-Off	Synchemicals Ltd	02019

973 Aluminium sulphate (Miscellaneous)

6X Slug Killer	Organic Concentrates Ltd	04702	
Fertosan Slug and Snail Powder	Growing Success Organics Ltd	08507	
Fertosan Slug and Snail Powder	Fertosan Products Ltd	00864	31/03/2000
Growing Success Slug Killer	Growing Success Organics Ltd	04386	
Septico Slug Killer	Growing Success Organics Ltd	09724	
Septico Slug Killer	Greenco	07803	30/09/2002

974 Amitrole + Atrazine (Herbicide)

Atlazin D-Weed	Nomix-Chipman Ltd	08210
Deeweed	Arable & Bulb Chemicals Ltd	00659
Do It All Path and Patio Weedkiller	Do-It-All Ltd	08156
Doff Total Path Weedkiller	Doff Portland Ltd	04632
Murphy Path Weedkiller	Fisons Plc	03630
Wilko Path Weedkiller	Wilkinson Group of Companies	06976
Woolworths Path Weedkiller	Woolworths Plc	07456

975 Amitrole + 2,4-D + Diuron (Herbicide)

B&Q Path & Patio Weedkiller	B & Q Plc	09545
Doff Long-Lasting Path Weedkiller	Doff Portland Ltd	09289
Trik	Whyte Agrochemicals Ltd	09788

976 Amitrole + 2,4-D + Diuron + Simazine (Herbicide)

Hytrol	Agrichem (International) Ltd	04540	30/10/2001

C   These products are "approved for agricultural use". For further details refer to page vii.
A   These products are approved for use in or near water. For further details refer to page vii.

Product Name	Marketing Company	Reg. No.	Expiry Date

977 Amitrole + Diquat + Paraquat + Simazine (Herbicide)

Pathclear	The Scotts Company (UK) Limited	09287	
Pathclear	Miracle Garden Care	07789	

978 Amitrole + Diuron + Simazine (Herbicide)

Hytrol Total	Agrichem (International) Ltd	08296	

979 Amitrole + Simazine (Herbicide)

Homebase Path & Drive Weed Killer	Sainsbury's Homebase	07755	
Path and Drive Weedkiller	pbi Home & Garden Ltd	05958	

980 Ammonium sulphamate (Herbicide)

Amcide	Battle Hayward & Bower Ltd	00089	
Deep Root	Growing Success Organics Ltd	08368	
Deep Root Path & Patio Weedkiller	Growing Success Organics Ltd	09365	
Root-Out	Dax Products Ltd	03510	

981 Atrazine + Amitrole (Herbicide)

Atlazin D-Weed	Nomix-Chipman Ltd	08210	
Deeweed	Arable & Bulb Chemicals Ltd	00659	
Do It All Path and Patio Weedkiller	Do-It-All Ltd	08156	
Doff Total Path Weedkiller	Doff Portland Ltd	04632	
Murphy Path Weedkiller	Fisons Plc	03630	
Wilko Path Weedkiller	Wilkinson Group of Companies	06976	
Woolworths Path Weedkiller	Woolworths Plc	07456	

982 Bacillus thuringiensis (Biological)

Bactospeine WP	Koppert (UK) Ltd	05675	30/09/2000
Bio 'BT' Caterpillar Killer	Pan Britannica Industries Ltd	06307	31/05/2000
Dipel	Pan Britannica Industries Ltd	06576	30/09/2000
Dipel	English Woodlands Ltd	04210	30/09/2000
DiPel WP	English Woodlands Biocontrol	08733	
Nature's Friends Bactospeine	Zeneca Garden Care	07070	30/09/2000

983 Bendiocarb (Insecticide)

B & Q Ant Killer	B & Q Plc	04880	
B & Q Woodlice Killer	B & Q Plc	07155	
Camco Insect Powder	AgrEvo UK Ltd	07576	
Delta Insect Powder	AgrEvo UK Ltd	07547	20/04/2001
Do It All Ant Killer	Do-It-All Ltd	04854	
Do-It-All Woodlice Killer	Do-It-All Ltd	08105	

C   These products are "approved for agricultural use". For further details refer to page vii.
A   These products are approved for use in or near water. For further details refer to page vii.

Product Name	Marketing Company	Reg. No.	Expiry Date

983 Bendiocarb (Insecticide)—continued

Product Name	Marketing Company	Reg. No.	Expiry Date
Doff Ant Control Powder	Doff Portland Ltd	04881	
Doff Wasp Nest Killer	Doff Portland Ltd	06114	
Doff Woodlice Killer	Doff Portland Ltd	06081	
Ficam Insect Powder	Agrevo UK Ltd	07546	20/04/2001
Focus Ant Killer	Focus DIY Ltd	06235	
Great Mills Ant Killer	Great Mills (Retail) Ltd	04852	
Great Mills Woodlice Killer	Great Mills (Retail) Ltd	07066	
Homebase Ant Killer	Homebase Ltd	06236	
Homebase Wasp Nest Killer	Homebase Ltd	08281	
Homebase Woodlice Killer	Homebase Ltd	06994	
Portland Brand Ant Killer	Doff Portland Ltd	07068	
Premier Ant Killer	Premier Way Ltd	07644	
Secto Ant and Crawling Insect Powder	Sinclair Animal & Household Care Ltd	07895	
Secto Wasp Killer Powder	Sinclair Animal & Household Care Ltd	07896	
Wilko Ant Destroyer	Wilkinson Group of Companies	04853	
Wilko Woodlice Killer	Wilkinson Group of Companies	07156	
Woolworths Ant Killer	Woolworths Plc	07067	
Woolworths Woodlice Killer	Woolworths Plc	07057	

984 Benzalkonium chloride + Copper sulphate (Herbicide)

Product Name	Marketing Company	Reg. No.	Expiry Date
Algizin P	Waterlife Research Industries Ltd	08851	
Lotus Algicide	Waterlife Research Industries Ltd	08852	
Lotus Fungicide	Waterlife Research Industries Ltd	08853	

985 Bifenthrin (Insecticide)

Product Name	Marketing Company	Reg. No.	Expiry Date
Bifenthrin 0.5% Tablets	FMC Corporation NV	08603	
Bio New Greenfly Killer Plus	pbi Home & Garden Ltd	09043	
Bio New Sprayday	pbi Home & Garden Ltd	09042	31/07/2002
Bio Sprayday Greenfly Killer Plus	pbi Home & Garden Ltd	09669	
Blitz Bug Gun	Miracle Garden Care Ltd	09040	28/02/2002
Bug Gun! For Gardens	The Scotts Company (UK) Limited	09463	
Bug-Free Concentrate	Phostrogen Ltd	09762	
Bug-Free Ready To Use	Phostrogen Ltd	09766	
Polysect Insecticide	The Scotts Company (UK) Limited	09650	
Polysect Insecticide	Solaris (Garden Div Of Monsanto)	08625	
Polysect Insecticide	Solaris (Garden Div Of Monsanto)	08624	30/06/2002
Polysect Insecticide	Monsanto Agricultural Co	06909	31/07/2000
Polysect Insecticide	Monsanto Agricultural Co	06908	31/07/2000
Polysect Insecticide Ready To Use	The Scotts Company (UK) Limited	09651	

C   These products are "approved for agricultural use". For further details refer to page vii.
A   These products are approved for use in or near water. For further details refer to page vii.

Product Name	Marketing Company	Reg. No.	Expiry Date

985 Bifenthrin (Insecticide)—continued

Polysect Insecticide Ready To Use	Solaris (Garden Div Of Monsanto)	08627	
Polysect Insecticide Ready To Use	Solaris (Garden Div Of Monsanto)	08626	30/06/2002
Sybol Extra	The Scotts Company (UK) Limited	09723	
Sybol Extra	Miracle Garden Care Ltd	09041	

986 Bioallethrin + Permethrin (Insecticide)

Bio Spraydex Greenfly Killer	pbi Home & Garden Ltd	07404	31/01/2001
Floracid	Perycut Insectengun Ltd	06798	
Longer Lasting Bug Gun	Miracle Garden Care	08021	

987 Bone oil (Vertebrate Control)

Renardine	Roebuck-Eyot Ltd	06378	
Renardine 72-2	Roebuck-Eyot Ltd	06769	

988 Borax (Insecticide)

Nippon Ant Killer Liquid	Vitax Ltd	05270	

989 Brodifacoum (Vertebrate Control)

Mouser	Zeneca Public Health	06831	
Mouser	Imperial Chemical Industries Plc	03213	

990 Bromadiolone (Vertebrate Control)

Biotrol Plus Outdoor Rat Killer	Rentokil Initial UK Ltd	09036	
Mouse Killer	Rentokil Initial UK Ltd	09662	
Murphy Mouse Killer	Levington Horticulture Ltd	08815	
Rentokil Biotrol Plus Outdoor Rat Killer	Rentokil Ltd	02936	31/05/2001
Rentokil Rodine C	Rentokil Ltd	03318	31/10/2001
Rentokil Total Mouse Killer System	Rentokil Ltd	02994	30/11/2001
Rodine C	Rentokil Initial UK Ltd	09004	
Slaymor Bait Bags	Novartis Animal Health UK Ltd	09264	
Slaymor Mouse Bait	Novartis Animal Health UK Ltd	09657	
Tomcat 2 Mouse Bait Station	Antec International Ltd	08642	
Tomcat 2 Rat and Mouse Blox	Antec International Ltd	08259	
Tomcat 2 Rat and Mouse Killer	Antec International Ltd	08338	
Total Mouse Killer System	Rentokil Initial UK Ltd	09005	

991 Bupirimate + Pirimicarb + Triforine (Fungicide) (Insecticide)

'Roseclear' 2	The Scotts Company (UK) Limited	09498	
'Roseclear' 2	Miracle Garden Care Ltd	08736	

C   These products are "approved for agricultural use". For further details refer to page vii.

A   These products are approved for use in or near water. For further details refer to page vii.

Product Name	Marketing Company	Reg. No.	Expiry Date

992 Bupirimate + Triforine (Fungicide)

Nimrod T	The Scotts Company (UK) Limited	09502	
Nimrod T	Miracle Garden Care	07876	
Nimrod T	Zeneca Garden Care	06843	31/01/2001
Nimrod T	Imperial Chemical Industries Plc	03982	31/05/2001

993 Buprofezin (Insecticide)

Whitefly Gun!	Miracle Garden Care	08235	

994 Butoxycarboxim (Insecticide)

Systemic Insecticide Pins	Monsanto Plc, Solaris - Garden Division	09330	
Systemic Insecticide Pins	Phostrogen Ltd	08209	30/11/2001

995 Captan + 1-Naphthylacetic acid (Fungicide) (Herbicide)

Doff Hormone Rooting Powder	Doff Portland Ltd	01065	
Homebase Rooting Hormone Powder	Sainsbury's Homebase	07630	
Murphy Hormone Rooting Powder	The Scotts Company (UK) Limited	07923	
Murphy Hormone Rooting Powder	Fisons Plc	03618	
New Strike	pbi Home & Garden Ltd	05956	31/07/2002
Rooting Powder	Vitax Ltd	06334	

996 Captan + 1-Naphthylacetic acid (Herbicide)

Bio Strike	pbi Home & Garden Ltd	09674	

997 Carbendazim (Fungicide)

AgrEvo Garden Systemic Fungicide	AgrEvo UK Ltd	08133	
B & Q Systemic Fungicide Concentrate	B & Q Plc	07417	
Do It All Plant Disease Control	Do-It-All Ltd	08132	
Doff Plant Disease Control	Doff Portland Ltd	07159	
Murphy Systemic Action Fungicide	Levington Horticulture Ltd	07558	
PBI Supercarb	pbi Home & Garden Ltd	03981	31/07/2001
Spotless	The Scotts Company (UK) Limited	09534	
Spotless	AgrEvo UK Ltd	08167	
Wilko Plant Disease Control	Wilkinson Group of Companies	08963	

998 Chlorpyrifos (Insecticide)

EM 1340/8	Pan Britannica Industries Ltd	08190	31/10/2000

C   These products are "approved for agricultural use". For further details refer to page vii.
A   These products are approved for use in or near water. For further details refer to page vii.

Product Name	Marketing Company	Reg. No.	Expiry Date

998 Chlorpyrifos (Insecticide)—continued

| New Chlorophos | pbi Home & Garden Ltd | 08773 | |

999 Cholecalciferol + Difenacoum (Vertebrate Control)

| Sorexa CD Mouse Killer | Sorex Ltd | 08208 | |

1000 Citronella oil + Methyl nonyl ketone (Vertebrate Control)

| Secto Keep Off | Sinclair Animal & Household Care Ltd | 07890 | |

1001 Citrus Extract (Herbicide)

Algae Control No 3 Anti-Slime Algae	Interpet Ltd	09071	
Biotal Bio-Clear	Interpet Ltd	09072	
Blagdon Pondsafe Bio Clear	Biotal Industrial Products Ltd	09073	
Duckweed Control	Interpet Ltd	09070	

1002 Citrus Extract (Miscellaneous)

| Bio-Active Algaway | Interpet Ltd | 09076 | |

1003 Clopyralid + Fluroxypyr + MCPA (Herbicide)

Weed-B-Gon	Solaris (Garden Div Of Monsanto)	08926	
Weed-B-Gon	Dow AgroSciences Ltd	08925	
Weed-B-Gon Ready To Use	Solaris (Garden Div Of Monsanto)	09451	
Weed-B-Gon Ready To Use	Dow AgroSciences Ltd	09450	

1004 Copper (Fungicide)

| Bordeaux Mixture | Vitax Ltd | 07162 | |

1006 Copper chloride + Metaldehyde (Miscellaneous)

| Sera Snailpur | Sera GmbH | 09587 | |
| Sera Snailpur | Sera Werke Heimtierbedarf | 09064 | 30/04/2001 |

1007 Copper hydrate (Miscellaneous)

| Algimycin PLL | Certikin International Ltd | 09096 | |
| Algimycin PLL | Primemix Ltd | 08859 | 30/06/2001 |

1008 Copper oxychloride (Fungicide)

| Murphy Traditional Copper Fungicide | Levington Horticulture Ltd | 07574 | |
| Murphy Traditional Copper Fungicide | Fisons Plc | 04585 | |

1009 Copper sulphate (Fungicide)

| B & Q Garden Fungicide Spray | B & Q Plc | 08543 | 31/08/2000 |

C   These products are "approved for agricultural use". For further details refer to page vii.
A   These products are approved for use in or near water. For further details refer to page vii.

Product Name	Marketing Company	Reg. No.	Expiry Date

1009 Copper sulphate (Fungicide)—continued

B & Q Garden Fungicide Spray	B & Q Plc	06285	
Bio Spraydex General Purpose Fungicide	pbi Home & Garden Ltd	07419	31/08/2000
PBI Cheshunt Compound	pbi Home & Garden Ltd	00485	

1010 Copper sulphate (Herbicide)

Algalit	Symbionics	09077	30/06/2002
Aqualife Pond Algicide	Rosewood Pet Products Ltd	08838	
Betta Algae Clear	J & K Aquatics	08843	
Green Algae Control	King British Aquarium Accessories	08855	
Hydroperfect Algalit HP-352	Symbionics Sarl	09519	
Hydroperfect Algalit HP-355	Symbionics Sarl	09520	
Hydroperfect Algalit HP-357	Symbionics Sarl	09521	
Hydroperfect Lemnalit HP-372	Symbionics Sarl	09528	
Hydroperfect Lemnalit HP-375	Symbionics Sarl	09529	
Hydroperfect Lemnalit HP-377	Symbionics Sarl	09530	
Lemnalit	Symbionics	09078	30/04/2002
Mister Green Algalit MG-352	Symbionics Sarl	09522	
Mister Green Algalit MG-355	Symbionics Sarl	09523	
Mister Green Algalit MG-357	Symbionics Sarl	09524	
Mister Green Lemnalit MG-372	Symbionics Sarl	09531	
Mister Green Lemnalit MG-375	Symbionics Sarl	09532	
Mister Green Lemnalit MG-377	Symbionics Sarl	09533	
Pond Pride Green Algae Control	Sinclair Animal & Household Care Ltd	09100	
Sera Algopur	Sera Werke Heimtierbedarf	09065	
Waterscene Coldwater - Algae Control	Waterscene Enterprises Co. Ltd	08869	

1011 Copper sulphate (Herbicide) (Miscellaneous)

| Aqualife Algae and Snail Killer | Rosewood Pet Products Ltd | 08839 | |

1012 Copper sulphate (Miscellaneous)

Betta Snail Clear	J & K Aquatics	08842	
BWG Anti Snails	Technical Aquatic Products	08847	
King British Formula KB7	King British Aquarium Accessories	08854	30/06/2001
Molluzin	Waterlife Research Industries Ltd	08849	
Snail Control	Sinclair Animal & Household Care Ltd	09101	
Snail Smasher	Technical Aquatic Products	08840	

C These products are "approved for agricultural use". For further details refer to page vii.
A These products are approved for use in or near water. For further details refer to page vii.

Product Name	Marketing Company	Reg. No.	Expiry Date

1013 Copper sulphate + Benzalkonium chloride (Herbicide)

Algizin P	Waterlife Research Industries Ltd	08851	
Lotus Algicide	Waterlife Research Industries Ltd	08852	
Lotus Fungicide	Waterlife Research Industries Ltd	08853	

1014 Copper sulphate + Ferrous sulphate (Miscellaneous)

Snailaway	Interpet Ltd	02457	

1015 Copper sulphate + Ferrous sulphate + Magnesium sulphate (Miscellaneous)

Snail Away	Interpet Ltd	08868	

1016 Copper sulphate + Simazine (Herbicide)

PPI Clarity Excel	Pet Products International Ltd	09095	
Tap Algasan	Technical Aquatic Products	08845	

1017 Coumatetralyl (Vertebrate Control)

Bio Racumin Rat Bait	pbi Home & Garden Ltd	09158	
PBI Racumin Mouse Bait	pbi Home & Garden Ltd	01678	31/01/2002
PBI Racumin Rat Bait	pbi Home & Garden Ltd	01680	

1018 Cresylic acid (Miscellaneous)

Armillatox	Armillatox Ltd	00115	30/09/2000

1019 Cypermethrin + Propiconazole (Fungicide) (Insecticide)

Murphy Roseguard	Levington Horticulture Ltd	08564	

1020 2,4-D + Amitrole + Diuron (Herbicide)

B Patio Weedkiller	B & Q Plc	09545	
Doff Long-Lasting Path Weedkiller	Doff Portland Ltd	09289	
Trik	Whyte Agrochemicals Ltd	09788	

1021 2,4-D + Amitrole + Diuron + Simazine (Herbicide)

Hytrol	Agrichem (International) Ltd	04540	30/10/2001

1022 2,4-D + Dicamba (Herbicide)

B & Q Granular Weed and Feed	B & Q Plc	05294	
B & Q Lawnweed Spray	B & Q Plc	05804	
Betterware Spot Weedkiller Spray	Betterware UK Ltd	08377	
Bio Lawn Weedkiller	pbi Home & Garden Ltd	00268	
Elliott Easyweeder	Thomas Elliott Ltd	07693	31/07/2001

C These products are "approved for agricultural use". For further details refer to page vii.
A These products are approved for use in or near water. For further details refer to page vii.

Product Name	Marketing Company	Reg. No.	Expiry Date

1022 2,4-D + Dicamba (Herbicide)—continued

Elliott's Touchweeder	Thomas Elliott Ltd	09160	
Green Up Lawn Feed and Weed	Vitax Ltd	06419	
Green Up Weedfree Lawn Weedkiller	Vitax Ltd	05322	
Green Up Weedfree Spot Weedkiller for Lawns	Vitax Ltd	06321	
Lawn Builder Plus Weed Control	Miracle Garden Care Ltd	07936	
Lawn Weed Gun	Zeneca Garden Care	06838	31/05/2001
Lawn Weed Gun	ICI Garden & Professional Products	04407	31/05/2001
Lawnsman Weed and Feed	Miracle Garden Care	07785	31/07/2000
Lawnsman Weed and Feed	Zeneca Garden Care	06842	31/07/2000
PBI Toplawn	Pan Britannica Industries Ltd	02145	31/07/2000
Tesco Lawn Feed 'n' Weed	Tesco Stores Ltd	05617	31/12/2000
Toplawn Feed with Weedkiller	Pan Britannica Industries Ltd	08289	
Verdone Gun!	The Scotts Company (UK) Limited	09643	
Verdone Gun!	Miracle Garden Care Ltd	07671	

1023 2,4-D + Dicamba + Ferrous sulphate (Herbicide)

B & Q Triple Action Lawn Care	B & Q Plc	05282	31/10/2001
Bio Supergreen 123	pbi Home & Garden Ltd	07962	31/01/2001
Greensward	Miracle Garden Care	07775	31/12/2000
Toplawn Feed with Weed & Mosskiller	pbi Home & Garden Ltd	08120	31/12/2001
Triple Action 'Grasshopper'	Miracle Garden Care	07791	31/12/2000
Triple Action Grasshopper	Zeneca Garden Care	06852	31/12/2000

1024 2,4-D + Dicamba + Ferrous sulphate heptahydrate (Herbicide)

Toplawn Feed, Weed & Mosskiller	pbi Home & Garden Ltd	09350	

1025 2,4-D + Dicamba + Mecoprop (Herbicide)

B & Q Tree Stump and Brushwood Killer	B & Q Plc	07645	
New Formulation SBK Brushwood Killer	Vitax Ltd	05043	31/10/2000

1026 2,4-D + Dicamba + Mecoprop-P (Herbicide)

New Formulation SBK Brushwood Killer	Vitax Ltd	08737	

C   These products are "approved for agricultural use". For further details refer to page vii.
A   These products are approved for use in or near water. For further details refer to page vii.

Product Name	Marketing Company	Reg. No.	Expiry Date

1027 2,4-D + Dichlorprop (Herbicide)

Product Name	Marketing Company	Reg. No.	Expiry Date
AgrEvo Lawn Weed Killer	AgrEvo UK Ltd	08137	
AgrEvo Lawn Weedkiller Concentrate	AgrEvo UK Ltd	08912	
B & Q Lawn Feed and Weed	B & Q Plc	05487	
B & Q Lawn Spot Weeder	B & Q Plc	05486	
B & Q Lawn Weedkiller	B & Q Plc	05324	
B & Q Nettle Gun	B & Q Plc	06897	
Do It All Lawn Feed & Weed	Do-It-All Ltd	09311	
Do It All Wipeout! Lawn Weed Killer	Do-It-All Ltd	08197	
Do-It-All Lawn Weedkiller	Do-It-All Ltd	08106	
Doff Lawn Feed and Weed	Doff Portland Ltd	05117	
Doff Lawn Spot Weeder	Doff Portland Ltd	03995	
Doff Nettle Gun	Doff Portland Ltd	07158	
Doff New Formula Lawn Weedkiller	Doff Portland Ltd	05666	
Fisons Clover-kil	Fisons Plc	05808	29/02/2000
Fisons Ready to Use Cloverkil	Fisons Plc	05585	31/05/2000
Fisons Ready-to-Use Lawn Weedkiller	Fisons Plc	05868	
Fisons Water-on Lawn Weedkiller	Fisons Plc	05807	
Focus Lawn Spot Weeder	Focus DIY Ltd	07468	
Focus Lawn Weedkiller	Focus DIY Ltd	06096	
Focus Nettle Gun	Focus DIY Ltd	06896	
Great Mills Lawn Feed & Weed	Great Mills (Retail) Ltd	08367	
Great Mills Lawn Spot Weeder	Great Mills (Retail) Ltd	05014	
Great Mills Lawn Weedkiller	Great Mills (Retail) Ltd	09296	
Homebase Lawn Weedkiller Liquid	Homebase Ltd	07539	
Homebase Lawn Weedkiller Ready To Use Sprayer	Sainsbury's Homebase House and Garden Centres	07536	
Homebase Liquid Lawn Feed and Weed	Homebase Ltd	07989	
Homebase Nettle Gun	Sainsbury's Homebase House and Garden Centres	07538	
Lawn Weedkiller	L C Solutions Ltd	09736	
Lawn Weedkiller Concentrate	L C Solutions Ltd	09740	
Levington Clover-kil	Levington Horticulture Ltd	07501	29/02/2000
Levington Ready to Use Clover-kil	Levington Horticulture Ltd	07498	31/05/2000
Levington Ready to Use Lawn Weedkiller	Levington Horticulture Ltd	07497	

C   These products are "approved for agricultural use". For further details refer to page vii.
A   These products are approved for use in or near water. For further details refer to page vii.

Product Name	Marketing Company	Reg. No.	Expiry Date

1027 2,4-D + Dichlorprop (Herbicide)—continued

Levington Water-on Lawn Weedkiller	Levington Horticulture Ltd	07495	
Murphy Clover-Kil	Fisons Plc	05271	
Tumbleweed Clover	Levington Horticulture Ltd	08325	
Tumbleweed Clover Ready to Use	Levington Horticulture Ltd	08546	
Tumbleweed Lawns	Levington Horticulture Ltd	08088	
Tumbleweed Lawns Ready to Use	Levington Horticulture Ltd	08089	
Wilko Lawn Spot Weeder	Wilkinson Home & Garden Stores	05130	
Wilko Lawn Weed and Feed	Wilkinson Group of Companies	07473	
Wilko Lawn Weedkiller	Wilkinson Group of Companies	03749	
Wilko Nettle Gun	Wilkinson Group of Companies	06898	
Woolworths Lawn Spot Weeder	Woolworths Plc	07105	
Woolworths Lawn Weed and Feed	Woolworths Plc	07474	
Woolworths Lawn Weedkiller	Woolworths Plc	07064	

1028 2,4-D + Ferrous sulphate + Mecoprop (Herbicide)

Asda Lawn Weed and Feed and Mosskiller	Asda Stores Ltd	03819	
Green Up Feed and Weed Plus Mosskiller	Vitax Ltd	06513	31/05/2001
Supergreen Feed, Weed & Mosskiller	Pan Britannica Industries Ltd	08118	31/01/2001
Weed 'N' Feed Extra	Vitax Ltd	06506	31/05/2001

1029 2,4-D + Ferrous sulphate + Mecoprop-P (Herbicide)

ASB Greenworld Lawn Weed, Feed and Mosskiller Granular	ASB Greenworld Ltd	09354	
B & Q Lawn Feed, Weed and Mosskiller Granular	B & Q Plc	09355	
Doff Granular Lawn Feed, Weed and Mosskiller	Doff Portland Ltd	08876	
Elliott's Lawn and Fine Turf Feed and Weed plus Moss Killer	Thomas Elliott Ltd	09464	
Gem Lawn Weed and Feed + Mosskiller	Gem Gardening	07087	
Green Up Feed and Weed Plus Mosskiller	Vitax Ltd	09046	

C   These products are "approved for agricultural use". For further details refer to page vii.
A   These products are approved for use in or near water. For further details refer to page vii.

Product Name	Marketing Company	Reg. No.	Expiry Date

1029 2,4-D + Ferrous sulphate + Mecoprop-P (Herbicide)—continued

Proctors Lawn Feed Weed and Mosskiller	H & T Proctor (Division of Willett & Son (Bristol) Ltd)	07808	
Supergreen Feed, Weed and Mosskiller 2	pbi Home & Garden Ltd	09224	
Weed 'N' Feed Extra	Vitax Ltd	09045	
Westland Triple Action Weed, Feed And Mosskill	Westland Horticulture	09640	

1030 2,4-D + Mecoprop (Herbicide)

Bio Spraydex Lawn Spot Weeder	Pan Britannica Industries Ltd	07418	30/09/2000
Homebase Lawn Feed and Weed Soluble	Sainsbury's Homebase	07690	30/09/2001
J Arthur Bower's Lawn Food with Weedkiller	Sinclair Horticulture & Leisure Ltd	04301	30/04/2000
Lawn Feed and Weed Granules	pbi Home & Garden Ltd	05902	30/09/2001
Lawn Spot Weed Granules	Pan Britannica Industries Ltd	05903	
Supergreen and Weed	Pan Britannica Industries Agrochemicals Ltd	08911	30/09/2001
Supergreen and Weed	Rhone-Poulenc Agriculture	05348	30/04/2000
Supergreen Double (Feed and Weed)	Pan Britannica Industries Ltd	05949	28/02/2001
Supertox	Rhone-Poulenc Agriculture	05350	30/04/2000
Supertox Lawn Weedkiller	Pan Britannica Industries Ltd	05948	30/09/2000
Supertox Spot	pbi Home & Garden Ltd	05951	30/11/2001
Toplawn Lawn Weedkiller	pbi Home & Garden Ltd	08694	31/01/2002
Toplawn Ready to Use Lawn Weedkiller	pbi Home & Garden Ltd	08695	31/12/2001
Verdone	Miracle Garden Care	08657	30/06/2001
Verdone 2	Miracle Garden Care	07792	31/08/2000
Verdone 2	Zeneca Garden Care	06853	31/08/2000
Verdone 2	Imperial Chemical Industries Plc	03271	31/05/2001
Wilko Lawn Feed 'N' Weed	Wilkinson Home & Garden Stores	04403	30/04/2000
Wilko Lawn Food with Weedkiller	Wilkinson Home & Garden Stores	04302	31/12/2000

1031 2,4-D + Mecoprop-P (Herbicide)

ASB Greenworld Lawn Weed, And Feed Granular	ASB Greenworld Ltd	09356	
B & Q Lawn Weed and Feed Granular	B & Q Plc	09357	
Bio Toplawn Spot Weeder	pbi Home & Garden Ltd	09672	
Gem Lawn Weed & Feed	Gem Gardening	07086	

C   These products are "approved for agricultural use". For further details refer to page vii.

A   These products are approved for use in or near water. For further details refer to page vii.

Product Name	Marketing Company	Reg. No.	Expiry Date

1031 2,4-D + Mecoprop-P (Herbicide)—continued

Green Up Spot Lawn Weedkiller	Vitax Ltd	06028	
Homebase Lawn Feed and Weed Soluble	Sainsbury's Homebase	09217	
Lawn Feed and Weed Granules	pbi Home & Garden Ltd	09225	
Supergreen and Weed	pbi Home & Garden Ltd	09218	
Supertox Spot	pbi Home & Garden Ltd	09306	
Toplawn Lawn Weedkiller	pbi Home & Garden Ltd	09589	
Toplawn Lawn Weedkiller	pbi Home & Garden Ltd	09408	31/05/2002
Toplawn Ready to Use Lawn Weedkiller	pbi Home & Garden Ltd	09348	31/07/2002
Verdone	The Scotts Company (UK) Limited	09642	
Verdone	Miracle Garden Care Ltd	09097	
Vitax 'Green Up' Granular Lawn Feed and Weed	Vitax Ltd	06158	
Wilko Lawn Feed & Weed Liquid	Wilkinson Home & Garden Stores	08789	

1032 2,4-D + 2,3,6-TBA (Herbicide)

Touchweeder	Thomas Elliott Ltd	02864	

1033 Dicamba + 2,4-D (Herbicide)

B & Q Granular Weed and Feed	B & Q Plc	05294	
B & Q Lawnweed Spray	B & Q Plc	05804	
Betterware Spot Weedkiller Spray	Betterware UK Ltd	08377	
Bio Lawn Weedkiller	pbi Home & Garden Ltd	00268	
Elliott Easyweeder	Thomas Elliott Ltd	07693	31/07/2001
Elliott's Touchweeder	Thomas Elliott Ltd	09160	
Green Up Lawn Feed and Weed	Vitax Ltd	06419	
Green Up Weedfree Lawn Weedkiller	Vitax Ltd	05322	
Green Up Weedfree Spot Weedkiller for Lawns	Vitax Ltd	06321	
Lawn Builder Plus Weed Control	Miracle Garden Care Ltd	07936	
Lawn Weed Gun	Zeneca Garden Care	06838	31/05/2001
Lawn Weed Gun	ICI Garden & Professional Products	04407	31/05/2001
Lawnsman Weed and Feed	Miracle Garden Care	07785	31/07/2000
Lawnsman Weed and Feed	Zeneca Garden Care	06842	31/07/2000

C   These products are "approved for agricultural use". For further details refer to page vii.
A   These products are approved for use in or near water. For further details refer to page vii.

Product Name	Marketing Company	Reg. No.	Expiry Date

1033 Dicamba + 2,4-D (Herbicide)—continued

PBI Toplawn	Pan Britannica Industries Ltd	02145	31/07/2000
Tesco Lawn Feed 'n' Weed	Tesco Stores Ltd	05617	31/12/2000
Toplawn Feed with Weedkiller	Pan Britannica Industries Ltd	08289	
Verdone Gun!	The Scotts Company (UK) Limited	09643	
Verdone Gun!	Miracle Garden Care Ltd	07671	

1034 Dicamba + 2,4-D + Ferrous sulphate (Herbicide)

B & Q Triple Action Lawn Care	B & Q Plc	05282	31/10/2001
Bio Supergreen 123	pbi Home & Garden Ltd	07962	31/01/2001
Greensward	Miracle Garden Care	07775	31/12/2000
Toplawn Feed with Weed & Mosskiller	pbi Home & Garden Ltd	08120	31/12/2001
Triple Action 'Grasshopper'	Miracle Garden Care	07791	31/12/2000
Triple Action Grasshopper	Zeneca Garden Care	06852	31/12/2000

1035 Dicamba + 2,4-D + Ferrous sulphate heptahydrate (Herbicide)

Toplawn Feed, Weed & Mosskiller	pbi Home & Garden Ltd	09350	

1036 Dicamba + 2,4-D + Mecoprop (Herbicide)

B & Q Tree Stump and Brushwood Killer	B & Q Plc	07645	
New Formulation SBK Brushwood Killer	Vitax Ltd	05043	31/10/2000

1037 Dicamba + 2,4-D + Mecoprop-P (Herbicide)

New Formulation SBK Brushwood Killer	Vitax Ltd	08737	

1038 Dicamba + Dichlorprop + Ferrous sulphate + MCPA (Herbicide)

'Grasshopper' Triple Action	Miracle Garden Care	08643	
'Greensward' 2	Miracle Garden Care	08380	30/11/2001
B & Q Granular Weed and Feed and Mosskiller For Lawns	B & Q Plc	08381	
Evergreen Grasshopper	The Scotts Company (UK) Limited	09639	
Miracle-Gro Lawn Food with Weed and Moss Control	Miracle Garden Care	08609	

1039 Dicamba + Dichlorprop + MCPA (Herbicide)

'Verdone Plus'	Miracle Garden Care	08689	

C   These products are "approved for agricultural use". For further details refer to page vii.

A   These products are approved for use in or near water. For further details refer to page vii.

Product Name	Marketing Company	Reg. No.	Expiry Date

1039 Dicamba + Dichlorprop + MCPA (Herbicide)—continued

B & Q Granular Weed & Feed for Lawns	B & Q Plc	07850	
B & Q Liquid Weed and Feed	B & Q Plc	05293	
Boots Nettle and Bramble Weedkiller	The Boots Company Plc	03455	
Grasshopper Weed & Feed	Miracle Garden Care Ltd	07849	
Greensward Triple Action	The Scotts Company (UK) Limited	09324	
Groundclear	pbi Home & Garden Ltd	05953	
Groundclear Spot	pbi Home & Garden Ltd	05950	30/11/2000
Lawnsman Liquid Weed and Feed	Imperial Chemical Industries Plc	03610	31/05/2001
Liquid Weed and Feed	Miracle Garden Care	07786	31/07/2000
Liquid Weed and Feed	Zeneca Garden Care	06887	31/07/2000
Liquid Weed and Feed	ICI Garden & Professional Products	05869	31/07/2000
Miracle-Gro Weed & Feed	Miracle Garden Care	08550	
Woolworths Liquid Lawn Feed and Weed	Woolworths Plc	07060	

1040 Dicamba + Dichlorprop + Mecoprop (Herbicide)

New Supertox	pbi Home & Garden Ltd	06128	30/09/2001

1041 Dicamba + Dichlorprop + Mecoprop-P (Herbicide)

New Supertox	pbi Home & Garden Ltd	09209	

1042 Dicamba + MCPA (Herbicide)

Green Up Liquid Lawn Feed'n Weed	Vitax Ltd	08196	

1043 Dicamba + MCPA + Mecoprop (Herbicide)

Bio Spot	pbi Home & Garden Ltd	05071	31/12/2001
Bio Weed Pencil	pbi Home & Garden Ltd	04054	31/12/2001
Evergreen Feed and Weed Liquid	Levington Horticulture Ltd	07492	30/04/2000
Fisons Evergreen Feed and Weed Liquid	Fisons Plc	05664	30/04/2000
Homebase Lawn Feed and Weed Liquid	Homebase Ltd	06086	30/04/2000
Homebase Weed Pen	Sainsbury's Homebase	07648	31/12/2001

1044 Dicamba + MCPA + Mecoprop-P (Herbicide)

Bio Lawn Weed Pencil	pbi Home & Garden Ltd	09656	
Bio Spot	pbi Home & Garden Ltd	09210	

C   These products are "approved for agricultural use". For further details refer to page vii.
A   These products are approved for use in or near water. For further details refer to page vii.

Product Name	Marketing Company	Reg. No.	Expiry Date

1044 Dicamba + MCPA + Mecoprop-P (Herbicide)—continued

Bio Weed Pencil	pbi Home & Garden Ltd	09212	30/06/2002
Evergreen Feed and Weed Liquid	Levington Horticulture Ltd	07766	
Homebase Lawn Feed and Weed Liquid	Homebase Ltd	07765	
Homebase Weed Pen	Sainsbury's Homebase	09211	

1045 Dicamba + Mecoprop (Herbicide)

Elliott Touchweeder	Thomas Elliott Ltd	07694	30/04/2000

1046 Dichlobenil (Herbicide)

Bio Weed Ban	pbi Home & Garden Ltd	09705	
Casoron G4	Vitax Ltd	06866	
Casoron G4 Weed Block	Vitax Ltd	09371	
Path and Shrub Guard	Miracle Garden Care	08055	

1047 Dichlorophen (Fungicide) (Herbicide)

Bio Mosskiller Ready-to-Use	pbi Home & Garden Ltd	07734	

1048 Dichlorophen (Herbicide)

AgrEvo Moss Killer	AgrEvo UK Ltd	08143	31/05/2000
Bio Mossclear	pbi Home & Garden Ltd	07673	31/12/2000
Do It All De-Moss	Do-It-All Ltd	08240	31/05/2000
Doff Lawn Moss Killer	Doff Portland Ltd	08748	31/05/2000
Doff Moss Killer	Doff Portland Ltd	07433	31/05/2000
Homebase Lawn Mosskiller Ready to Use Sprayer	Sainsbury's Homebase House and Garden Centres	07664	31/05/2000
Moss Gun	Zeneca Garden Care	06821	31/07/2000
Moss Gun!	Miracle Garden Care	07787	28/02/2001
Mossgun	Imperial Chemical Industries Plc	03326	31/07/2000
Tumbleweed Moss Ready to Use	Levington Horticulture Ltd	08081	

1049 Dichlorophen + 1-Naphthylacetic acid (Fungicide) (Herbicide)

Homebase Rooting Hormone Liquid	Sainsbury's Homebase	07698	

1050 Dichlorophen + 1-Naphthylacetic acid (Fungicide) (Miscellaneous)

Baby Bio Roota	pbi Home & Garden Ltd	09305	

1051 Dichlorophen + 1-Naphthylacetic acid (Herbicide)

Bio Roota	pbi Home & Garden Ltd	00271	30/11/2001

C   These products are "approved for agricultural use". For further details refer to page vii.
A   These products are approved for use in or near water. For further details refer to page vii.

Product Name	Marketing Company	Reg. No.	Expiry Date

1052 Dichlorprop + 2,4-D (Herbicide)

Product Name	Marketing Company	Reg. No.	Expiry Date
AgrEvo Lawn Weed Killer	AgrEvo UK Ltd	08137	
AgrEvo Lawn Weedkiller Concentrate	AgrEvo UK Ltd	08912	
B & Q Lawn Feed and Weed	B & Q Plc	05487	
B & Q Lawn Spot Weeder	B & Q Plc	05486	
B & Q Lawn Weedkiller	B & Q Plc	05324	
B & Q Nettle Gun	B & Q Plc	06897	
Do It All Lawn Feed & Weed	Do-It-All Ltd	09311	
Do It All Wipeout! Lawn Weed Killer	Do-It-All Ltd	08197	
Do-It-All Lawn Weedkiller	Do-It-All Ltd	08106	
Doff Lawn Feed and Weed	Doff Portland Ltd	05117	
Doff Lawn Spot Weeder	Doff Portland Ltd	03995	
Doff Nettle Gun	Doff Portland Ltd	07158	
Doff New Formula Lawn Weedkiller	Doff Portland Ltd	05666	
Fisons Clover-kil	Fisons Plc	05808	29/02/2000
Fisons Ready to Use Cloverkil	Fisons Plc	05585	31/05/2000
Fisons Ready-to-Use Lawn Weedkiller	Fisons Plc	05868	
Fisons Water-on Lawn Weedkiller	Fisons Plc	05807	
Focus Lawn Spot Weeder	Focus DIY Ltd	07468	
Focus Lawn Weedkiller	Focus DIY Ltd	06096	
Focus Nettle Gun	Focus DIY Ltd	06896	
Great Mills Lawn Feed & Weed	Great Mills (Retail) Ltd	08367	
Great Mills Lawn Spot Weeder	Great Mills (Retail) Ltd	05014	
Great Mills Lawn Weedkiller	Great Mills (Retail) Ltd	09296	
Homebase Lawn Weedkiller Liquid	Homebase Ltd	07539	
Homebase Lawn Weedkiller Ready To Use Sprayer	Sainsbury's Homebase House and Garden Centres	07536	
Homebase Liquid Lawn Feed and Weed	Homebase Ltd	07989	
Homebase Nettle Gun	Sainsbury's Homebase House and Garden Centres	07538	
Lawn Weedkiller	L C Solutions Ltd	09736	
Lawn Weedkiller Concentrate	L C Solutions Ltd	09740	
Levington Clover-kil	Levington Horticulture Ltd	07501	29/02/2000
Levington Ready to Use Clover-kil	Levington Horticulture Ltd	07498	31/05/2000
Levington Ready to Use Lawn Weedkiller	Levington Horticulture Ltd	07497	

C   These products are "approved for agricultural use". For further details refer to page vii.
A   These products are approved for use in or near water. For further details refer to page vii.

Product Name	Marketing Company	Reg. No.	Expiry Date

1052 Dichlorprop + 2,4-D (Herbicide)—continued

Product Name	Marketing Company	Reg. No.	Expiry Date
Levington Water-on Lawn Weedkiller	Levington Horticulture Ltd	07495	
Murphy Clover-Kil	Fisons Plc	05271	
Tumbleweed Clover	Levington Horticulture Ltd	08325	
Tumbleweed Clover Ready to Use	Levington Horticulture Ltd	08546	
Tumbleweed Lawns	Levington Horticulture Ltd	08088	
Tumbleweed Lawns Ready to Use	Levington Horticulture Ltd	08089	
Wilko Lawn Spot Weeder	Wilkinson Home & Garden Stores	05130	
Wilko Lawn Weed and Feed	Wilkinson Group of Companies	07473	
Wilko Lawn Weedkiller	Wilkinson Group of Companies	03749	
Wilko Nettle Gun	Wilkinson Group of Companies	06898	
Woolworths Lawn Spot Weeder	Woolworths Plc	07105	
Woolworths Lawn Weed and Feed	Woolworths Plc	07474	
Woolworths Lawn Weedkiller	Woolworths Plc	07064	

1053 Dichlorprop + Dicamba + Ferrous sulphate + MCPA (Herbicide)

Product Name	Marketing Company	Reg. No.	Expiry Date
'Grasshopper' Triple Action	Miracle Garden Care	08643	
'Greensward' 2	Miracle Garden Care	08380	30/11/2001
B & Q Granular Weed and Feed and Mosskiller For Lawns	B & Q Plc	08381	
Evergreen Grasshopper	The Scotts Company (UK) Limited	09639	
Miracle-Gro Lawn Food with Weed and Moss Control	Miracle Garden Care	08609	

1054 Dichlorprop + Dicamba + MCPA (Herbicide)

Product Name	Marketing Company	Reg. No.	Expiry Date
'Verdone Plus'	Miracle Garden Care	08689	
B & Q Granular Weed & Feed for Lawns	B & Q Plc	07850	
B & Q Liquid Weed and Feed	B & Q Plc	05293	
Boots Nettle and Bramble Weedkiller	The Boots Company Plc	03455	
Grasshopper Weed & Feed	Miracle Garden Care Ltd	07849	
Greensward Triple Action	The Scotts Company (UK) Limited	09324	
Groundclear	pbi Home & Garden Ltd	05953	
Groundclear Spot	pbi Home & Garden Ltd	05950	30/11/2000
Lawnsman Liquid Weed and Feed	Imperial Chemical Industries Plc	03610	31/05/2001
Liquid Weed and Feed	Miracle Garden Care	07786	31/07/2000
Liquid Weed and Feed	Zeneca Garden Care	06887	31/07/2000

C   These products are "approved for agricultural use". For further details refer to page vii.

A   These products are approved for use in or near water. For further details refer to page vii.

Product Name	Marketing Company	Reg. No.	Expiry Date

1054 Dichlorprop + Dicamba + MCPA (Herbicide)—continued

Liquid Weed and Feed	ICI Garden & Professional Products	05869	31/07/2000
Miracle-Gro Weed & Feed	Miracle Garden Care	08550	
Woolworths Liquid Lawn Feed and Weed	Woolworths Plc	07060	

1055 Dichlorprop + Dicamba + Mecoprop (Herbicide)

| New Supertox | pbi Home & Garden Ltd | 06128 | 30/09/2001 |

1056 Dichlorprop + Dicamba + Mecoprop-P (Herbicide)

| New Supertox | pbi Home & Garden Ltd | 09209 | |

1057 Dichlorprop + Ferrous sulphate + MCPA (Herbicide)

J Arthur Bower's Granular Feed, Weed and Mosskiller	William Sinclair Horticulture Ltd	07042	
J Arthur Bower's Lawn Feed, Weed and Mosskiller	William Sinclair Horticulture Ltd	09459	
J Arthur Bower's Lawn Weed and Mosskiller.	William Sinclair Horticulture Ltd	09460	
Wilko Granular Feed, Weed and Mosskiller	Wilkinson Group of Companies	07471	
Wilko Granular Feed, Weed and Mosskiller for Lawns	Wilkinson Group of Companies	09731	
Wilko Lawn Feed, Weed and Mosskiller	Wilkinson Home & Garden Stores	04602	

1058 Dichlorprop + MCPA (Herbicide)

Doff Lawn Weed and Feed Soluble Powder	Doff Portland Ltd	05708	
J Arthur Bower's Granular Feed and Weed	William Sinclair Horticulture Ltd	07164	
J Arthur Bower's Lawn Feed and Weed	William Sinclair Horticulture Ltd	09458	
J Arthur Bower's Lawn Weedkiller	William Sinclair Horticulture Ltd	09461	
Wilko Granular Feed and Weed for Lawns	Wilkinson Group of Companies	09732	
Wilko Lawn Feed and Weed	Wilkinson Group of Companies	08816	
Wilko Soluble Lawn Food and Weedkiller	Wilkinson Home & Garden Stores	04391	

1059 Difenacoum (Vertebrate Control)

| Bio Mouse Killer | pbi Home & Garden Ltd | 09548 | |

C   These products are "approved for agricultural use". For further details refer to page vii.
A   These products are approved for use in or near water. For further details refer to page vii.

Product Name	Marketing Company	Reg. No.	Expiry Date

1059 Difenacoum (Vertebrate Control)—continued

Bio Rat Killer	pbi Home & Garden Ltd	09549	
Endem	Downland Marketing Company	08951	
Endem Bait Sachets	Downland Marketing Company	08952	
Neokil	Sorex Ltd	08948	
Neokil Bait Bags	Sorex Ltd	08950	
Neosorexa	Sorex Ltd	07757	
Neosorexa Bait Blocks	Sorex Ltd	07356	
Neosorexa Mixed Grain Bait	Sorex Ltd	08783	
Neosorexa Pellets	Sorex Ltd	08785	
Neosorexa Ratpacks	Sorex Ltd	09276	
Ratak	Zeneca Public Health	06828	
Sorexa D Mouse Killer	Sorex Ltd	06901	

1060 Difenacoum + Cholecalciferol (Vertebrate Control)

Sorexa CD Mouse Killer	Sorex Ltd	08208	

1061 Dikegulac (Herbicide)

Cutlass	The Scotts Company (UK) Limited	09543	
Cutlass	Miracle Garden Care	07783	
Cutlass	Zeneca Garden Care	06839	31/05/2001
Cutlass	Imperial Chemical Industries Plc	00617	31/05/2001

1062 Dimethoate (Insecticide)

Doff Systemic Insecticide	Doff Portland Ltd	02658	

1063 Diphacinone (Vertebrate Control)

Tomcat Rat and Mouse Blox	Antec International Ltd	07794	31/05/2001
Tomcat Rat and Mouse Pellets	Antec International Ltd	07825	31/05/2001

1064 Diquat (Herbicide)

Weedol Gun!	The Scotts Company (UK) Limited	09279	
Weedol Gun!	Miracle Garden Care Ltd	09079	30/09/2001

1065 Diquat + Amitrole + Paraquat + Simazine (Herbicide)

Pathclear	The Scotts Company (UK) Limited	09287	
Pathclear	Miracle Garden Care	07789	

1066 Diquat + Paraquat (Herbicide)

MF00296A	Miracle Garden Care	09152	
Weedol	The Scotts Company (UK) Limited	09280	
Weedol	Miracle Garden Care	07750	30/09/2001
Weedol	Zeneca Garden Care	06863	31/05/2001
Weedol Gun!	Miracle Garden Care	08254	31/07/2001

C   These products are "approved for agricultural use". For further details refer to page vii.
A   These products are approved for use in or near water. For further details refer to page vii.

Product Name	Marketing Company	Reg. No.	Expiry Date

1067 Diquat + Paraquat + Simazine (Herbicide)

| 'Pathclear' S | The Scotts Company (UK) Limited | 09285 | |
| 'Pathclear' S | Miracle Garden Care Ltd | 08992 | 31/10/2001 |

1068 Disodium hydrogen phosphate (Miscellaneous)

| Algae Control No 2 Anti-hair Algae | Interpet Ltd | 09067 | |

1069 Diuron + Amitrole + 2,4-D (Herbicide)

B&Q Path & Patio Weedkiller	B & Q Plc	09545	
Doff Long-Lasting Path Weedkiller	Doff Portland Ltd	09289	
Trik	Whyte Agrochemicals Ltd	09788	

1070 Diuron + Amitrole + 2,4-D + Simazine (Herbicide)

| Hytrol | Agrichem (International) Ltd | 04540 | 30/10/2001 |

1071 Diuron + Amitrole + Simazine (Herbicide)

| Hytrol Total | Agrichem (International) Ltd | 08296 | |

1072 Diuron + Glufosinate-ammonium (Herbicide)

AgrEvo Path Weed Killer	AgrEvo Environmental Health Ltd	08616	
Tumbleweed Paths and Patios	Levington Horticulture Ltd	08613	
Tumbleweed Paths and Patios	Levington Horticulture Ltd	08107	31/07/2000

1073 Diuron + Glyphosate (Herbicide)

| Pathclear Gun! | The Scotts Company (UK) Limited | 09492 | |
| Weedatak Path and Drive | Monsanto Plc | 07659 | |

1074 Fatty acids (Herbicide)

| Bio SpeedWeed | pbi Home & Garden Ltd | 06134 | |

1075 Fatty acids (Insecticide)

AgrEvo Rose & Flower Greenfly Killer	AgrEvo UK Ltd	09508	
AgrEvo Rose & Flower Insect Killer	AgrEvo UK Ltd	09507	
Agrevo Rose PestKiller	AgrEvo UK Ltd	09509	
B&Q Insect Spray for Houseplants	B & Q Plc	09416	
B&Q Insect Spray for Roses and Flowers	B & Q Plc	09417	
Bio Pest Pistol	pbi Home & Garden Ltd	07233	

C   These products are "approved for agricultural use". For further details refer to page vii.
A   These products are approved for use in or near water. For further details refer to page vii.

Product Name	Marketing Company	Reg. No.	Expiry Date

1075 Fatty acids (Insecticide)—continued

Product Name	Marketing Company	Reg. No.	Expiry Date
Bug Gun! For Roses & Flowers	The Scotts Company (UK) Limited	09453	
Do It All De-Bug!2	Focus DIY Ltd	09506	
Doff Houseplant Pest Spray	Doff Portland Ltd	09419	
Doff Rose & Flower Pest Spray	Doff Portland Ltd	09418	
General Insect Killer	Growing Success Organics Ltd	08808	
General Insect Killer	Nehra Cooke's Chemicals Ltd	07462	30/11/2000
Get Off Insect	Pet and Garden Manufacturing Plc	09162	
Great Mills Insect Spray for Roses	Doff Portland Ltd	09420	
Greenco GR1	Doff Portland Ltd	09455	
Greenco GR1	Doff Portland Ltd	07544	
Greenco GR3 Pest Jet	Doff Portland Ltd	09178	
Greenco GR3 Pest Jet	Greenco	07621	31/08/2001
Greenco GR3 Pest Jet	Greenco	07545	
Greenfly & Blackfly Killer	Growing Success Organics Ltd	08807	
Greenfly and Blackfly Killer	Nehra Cooke's Chemicals Ltd	07461	30/11/2000
Home Base Pest Gun 2	Homebase Ltd	09422	
Homebase Houseplant Insecticide Spray 2	Homebase Ltd	09421	
Houseplant Pest Killer	L C Solutions Ltd	09774	
Nature's Answer Organic Insecticide Ready to Use	Levington Horticulture Ltd	07627	28/02/2002
Phostrogen House Plant Insecticide	Phostrogen Ltd	06538	
Phostrogen Safer's Garden Insecticide Concentrate	Monsanto Plc	09332	
Phostrogen Safer's Garden Insecticide Concentrate	Phostrogen Ltd	05499	30/11/2001
Phostrogen Safer's Ready-to-Use Fruit & Vegetable Insecticide	Monsanto Plc	09329	
Phostrogen Safer's Ready-To-Use Fruit & Vegetable Insecticide	Phostrogen Ltd	04329	30/11/2001
Phostrogen Safer's Ready-To-Use House Plant Insecticide	Monsanto Plc	09333	
Phostrogen Safer's Ready-To-Use House Plant Insecticide	Phostrogen Ltd	04328	30/11/2001
Phostrogen Safer's RTU Rose & Flower Insecticide	Monsanto Plc, Solaris - Garden Division	09331	
Phostrogen Safer's RTU Rose & Flower Insecticide	Phostrogen Ltd	04341	30/11/2001
Rose & Flower Insect Killer	L C Solutions Ltd	09739	

C These products are "approved for agricultural use". For further details refer to page vii.
A These products are approved for use in or near water. For further details refer to page vii.

Product Name	Marketing Company	Reg. No.	Expiry Date

1075 Fatty acids (Insecticide)—continued

Rose & Flower Insecticide	Growing Success Organics Ltd	08809	
Rose and Flower Insecticide	Nehra Cooke's Chemicals Ltd	07620	30/11/2000
Safer's Insecticidal Soap Ready-to-Use	Safer Ltd	06573	
Savona Rose Spray	Safer Ltd	06297	
Wilko Flower and Rose Insecticide Spray	Wilkinson Group of Companies	09424	
Wilko Insecticidal Houseplant Spray	Wilkinson Group of Companies	09423	
Woolworths Greenfly Killer for Flowers	Woolworths Plc	09427	
Woolworths Insecticide Spray for Flowers	Woolworths Plc	09426	
Woolworths Insecticide Spray for Houseplants	Woolworths Plc	09425	

1076 Fatty acids + Glufosinate-ammonium (Herbicide)

Bio Kills Weeds Dead Fast	pbi Home & Garden Ltd	09173	
Bio Kills Weeds Dead Fast Concentrate	pbi Home & Garden Ltd	09752	
Bio Speedweed Ultra	pbi Home & Garden Ltd	09186	

1077 Fatty acids + Pyrethrins (Insecticide)

Phostrogen Safer's All Purpose Insecticide Ready to Use	Phostrogen Ltd	08093	
Safer's Indoor Trounce Insecticide Concentrate	Safer Ltd	08097	
Safer's Indoor Trounce Insecticide Ready to Use	Safer Ltd	08094	
Safer's Trounce Insecticide Concentrate	Safer Ltd	08098	
Safer's Trounce Insecticide Ready to Use	Safer Ltd	08095	

1078 Fatty acids + Sulphur (Fungicide) (Insecticide)

Nature's Answer Fungicide and Insect Killer	The Scotts Company (UK) Limited	07628	

1079 Fenitrothion (Insecticide)

Bio Fruit Spray	pbi Home & Garden Ltd	07994	20/04/2001

C   These products are "approved for agricultural use". For further details refer to page vii.
A   These products are approved for use in or near water. For further details refer to page vii.

Product Name	Marketing Company	Reg. No.	Expiry Date

1080 Ferrous sulphate (Herbicide)

Product Name	Marketing Company	Reg. No.	Expiry Date
Asda Lawn Sand	Asda Stores Ltd	06062	
Asda Lawn Sand	Asda Stores Ltd	04520	
B & Q Granular Autumn Lawn Food with Mosskiller	B & Q Plc	07887	
B & Q Granular Lawn Feed and Mosskiller	B & Q Plc	07915	
B & Q Mosskiller for Lawns	B & Q Plc	05827	
Bio Velvas	pbi Home & Garden Ltd	08545	
Boots Lawn Moss Killer and Fertiliser	The Boots Company Plc	02494	
Chempak Lawn Sand	Chempak Ltd	05723	
Cooke's Lawn Mosskiller	Nehra Cooke's Chemicals Ltd	08114	
Country Gardens Lawn Sand	Country Garden Centres Ltd	07969	
Doff Lawn Mosskiller and Fertilizer	Doff Portland Ltd	05689	
Evergreen Autumn	Levington Horticulture Ltd	08586	
Evergreen Lawn Sand	Levington Horticulture Ltd	08629	
Evergreen Mosskil	Levington Horticulture Ltd	08632	
Evergreen Mosskil Extra	Levington Horticulture Ltd	08669	
Fisons Lawn Sand	Fisons Plc	00885	
Fisons Mosskil Extra	Fisons Plc	03267	30/11/2000
Gardenstore, Lawn Feed and Mosskiller	Texas Homecare Ltd	07445	31/05/2000
Gem Lawn Sand	Gem Gardening	04555	
Green Up Mossfree	Vitax Ltd	05639	
Homebase Autumn Lawn Feed and Moss Killer	Homebase Ltd	08607	
Homebase Lawn Feed and Moss killer	Homebase Ltd	09334	
Homebase Lawn Feed and Moss killer	Homebase Ltd	07476	30/11/2001
Homebase Lawn Feed and Mosskiller	Homebase Ltd	06107	
Homebase Lawn Sand	Sainsbury's Homebase	07667	
J Arthur Bower's Lawn Sand	William Sinclair Horticulture Ltd	07028	
J Arthur Bowers Lawn Mosskiller	William Sinclair Horticulture Ltd	09457	
Lawn Builder Plus Moss Control	Miracle Garden Care Ltd	08612	
Lawn Sand	Wessex Horticultural Products Ltd	08936	
Levington Autumn Extra	Levington Horticulture Ltd	07477	
Levington Lawn Sand	Levington Horticulture Ltd	07499	
Maxicrop Mosskiller and Lawn Tonic	Maxicrop International Ltd	04661	

C   These products are "approved for agricultural use". For further details refer to page vii.
A   These products are approved for use in or near water. For further details refer to page vii.

Product Name	Marketing Company	Reg. No.	Expiry Date

1080 Ferrous sulphate (Herbicide)—continued

Product Name	Marketing Company	Reg. No.	Expiry Date
Moss Control Plus Lawn Fertilizer	Miracle Garden Care Ltd	07911	31/07/2000
Moss Control Plus Lawn Fertilizer	OM Scott & Sons Ltd	05613	31/01/2000
Mosskil Extra	Levington Horticulture Ltd	07478	
Mosskiller for Lawns	Miracle Garden Care	07869	
Murphy Lawn Feed and Mosskiller	Murphy Home and Garden Ltd	07513	
PBI Velvas	Pan Britannica Industries Ltd	02291	31/12/2000
Phostrogen Soluble Mosskiller and Lawn Tonic	Monsanto Plc, Solaris - Garden Division	09335	
Phostrogen Soluble Mosskiller and Lawn Tonic	Phostrogen Ltd	07112	30/11/2001
Premier Autumn Lawn Feed with Mosskiller	Premier Way Ltd	07049	
Toplawn Feed with Mosskiller	pbi Home & Garden Ltd	08104	
Vitax Lawn Sand	Vitax Ltd	04352	
Wilko Lawn Sand	Wilkinson Home & Garden Stores	04084	

1081 Ferrous sulphate + Copper sulphate (Miscellaneous)

Product Name	Marketing Company	Reg. No.	Expiry Date
Snailaway	Interpet Ltd	02457	

1082 Ferrous sulphate + Copper sulphate + Magnesium sulphate (Miscellaneous)

Product Name	Marketing Company	Reg. No.	Expiry Date
Snail Away	Interpet Ltd	08868	

1083 Ferrous sulphate + 2,4-D + Dicamba (Herbicide)

Product Name	Marketing Company	Reg. No.	Expiry Date
B & Q Triple Action Lawn Care	B & Q Plc	05282	31/10/2001
Bio Supergreen 123	pbi Home & Garden Ltd	07962	31/01/2001
Greensward	Miracle Garden Care	07775	31/12/2000
Toplawn Feed with Weed & Mosskiller	pbi Home & Garden Ltd	08120	31/12/2001
Triple Action 'Grasshopper'	Miracle Garden Care	07791	31/12/2000
Triple Action Grasshopper	Zeneca Garden Care	06852	31/12/2000

1084 Ferrous sulphate + 2,4-D + Mecoprop (Herbicide)

Product Name	Marketing Company	Reg. No.	Expiry Date
Asda Lawn Weed and Feed and Mosskiller	Asda Stores Ltd	03819	
Green Up Feed and Weed Plus Mosskiller	Vitax Ltd	06513	31/05/2001
Supergreen Feed, Weed & Mosskiller	Pan Britannica Industries Ltd	08118	31/01/2001
Weed 'N' Feed Extra	Vitax Ltd	06506	31/05/2001

C   These products are "approved for agricultural use". For further details refer to page vii.
A   These products are approved for use in or near water. For further details refer to page vii.

Product Name	Marketing Company	Reg. No.	Expiry Date

1085 Ferrous sulphate + 2,4-D + Mecoprop-P (Herbicide)

Product Name	Marketing Company	Reg. No.	Expiry Date
ASB Greenworld Lawn Weed, Feed and Mosskiller Granular	ASB Greenworld Ltd	09354	
B & Q Lawn Feed, Weed and Mosskiller Granular	B & Q Plc	09355	
Doff Granular Lawn Feed, Weed and Mosskiller	Doff Portland Ltd	08876	
Elliott's Lawn and Fine Turf Feed and Weed plus Moss Killer	Thomas Elliott Ltd	09464	
Gem Lawn Weed and Feed + Mosskiller	Gem Gardening	07087	
Green Up Feed and Weed Plus Mosskiller	Vitax Ltd	09046	
Proctors Lawn Feed Weed and Mosskiller	H & T Proctor (Division of Willett & Son (Bristol) Ltd)	07808	
Supergreen Feed, Weed and Mosskiller 2	pbi Home & Garden Ltd	09224	
Weed 'N' Feed Extra	Vitax Ltd	09045	
Westland Triple Action Weed, Feed And Mosskill	Westland Horticulture	09640	

1086 Ferrous sulphate + Dicamba + Dichlorprop + MCPA (Herbicide)

Product Name	Marketing Company	Reg. No.	Expiry Date
'Grasshopper' Triple Action	Miracle Garden Care	08643	
'Greensward' 2	Miracle Garden Care	08380	30/11/2001
B & Q Granular Weed and Feed and Mosskiller For Lawns	B & Q Plc	08381	
Evergreen Grasshopper	The Scotts Company (UK) Limited	09639	
Miracle-Gro Lawn Food with Weed and Moss Control	Miracle Garden Care	08609	

1087 Ferrous sulphate + Dichlorprop + MCPA (Herbicide)

Product Name	Marketing Company	Reg. No.	Expiry Date
J Arthur Bower's Granular Feed, Weed and Mosskiller	William Sinclair Horticulture Ltd	07042	
J Arthur Bower's Lawn Feed, Weed and Mosskiller	William Sinclair Horticulture Ltd	09459	
J Arthur Bower's Lawn Weed and Mosskiller.	William Sinclair Horticulture Ltd	09460	
Wilko Granular Feed, Weed and Mosskiller	Wilkinson Group of Companies	07471	
Wilko Granular Feed, Weed and Mosskiller for Lawns	Wilkinson Group of Companies	09731	
Wilko Lawn Feed, Weed and Mosskiller	Wilkinson Home & Garden Stores	04602	

C   These products are "approved for agricultural use". For further details refer to page vii.
A   These products are approved for use in or near water. For further details refer to page vii.

Product Name	Marketing Company	Reg. No.	Expiry Date

1088 Ferrous sulphate + MCPA + Mecoprop (Herbicide)

Evergreen Extra	Levington Horticulture Ltd	07493	30/04/2000
Fisons Evergreen Extra	Fisons Plc	03890	
Gardenstore Lawn Feed, Weed and Mosskiller	Texas Homecare Ltd	07448	31/05/2000

1089 Ferrous sulphate + MCPA + Mecoprop-P (Herbicide)

B & Q Granular Lawn Feed, Weed and Mosskiller	B & Q Plc	07762	
Do-It-All Complete Lawn Care	Do-It-All Ltd	08123	
Evergreen Easy	Levington Horticulture Ltd	08103	
Evergreen Extra	Levington Horticulture Ltd	07549	
Evergreen Triple Action	Levington Horticulture Ltd	09199	
Gardenstore Lawn Feed, Weed and Mosskiller	Texas Homecare Ltd	07550	31/10/2001
Great Mills Lawn Feed, Weed and Mosskiller	Great Mills (Retail) Ltd	07552	
Homebase Lawn Feed, Weed and Mosskiller	Homebase Ltd	07551	
Levington Gold	Levington Horticulture Ltd	07548	
Murphy Ultra Lawn Feed, Weed and Mosskiller	Murphy Home and Garden Ltd	07553	
Wickes Lawn Feed, Weed and Mosskiller	Wickes Building Supplies Ltd	08800	
Wilko Granular Lawn Feed, Weed and Mosskiller	Wilkinson Group of Companies	07981	
Woolworths Lawn Feed, Weed and Mosskiller	Woolworths Plc	07554	

1090 Ferrous sulphate heptahydrate (Herbicide)

Autumn Toplawn Feed and Mosskiller	pbi Home & Garden Ltd	09161	

1091 Ferrous sulphate heptahydrate + 2,4-D + Dicamba (Herbicide)

Toplawn Feed, Weed & Mosskiller	pbi Home & Garden Ltd	09350	

1092 Fluroxypyr + Clopyralid + MCPA (Herbicide)

Weed-B-Gon	Solaris (Garden Div Of Monsanto)	08926	
Weed-B-Gon	Dow AgroSciences Ltd	08925	
Weed-B-Gon Ready To Use	Solaris (Garden Div Of Monsanto)	09451	
Weed-B-Gon Ready To Use	Dow AgroSciences Ltd	09450	

C   These products are "approved for agricultural use". For further details refer to page vii.
A   These products are approved for use in or near water. For further details refer to page vii.

Product Name	Marketing Company	Reg. No.	Expiry Date

1093 Fluroxypyr + Mecoprop-P (Herbicide)

Verdone Extra	Miracle Garden Care Ltd	08999
Verdone Extra	Dow AgroSciences Ltd	08998

1094 Garlic oil + Orange peel oil + Orange pith oil (Vertebrate Control)

6X Cat Repellent	Organic Concentrates Ltd	07979
Growing Success Cat Repellent	Growing Success Organics Ltd	06262

1095 Glufosinate-ammonium (Herbicide)

AgrEvo Garden Weedkiller Concentrate	AgrEvo UK Ltd	08200
AgrEvo Garden Weedkiller Spray	AgrEvo UK Ltd	08176
AgrEvo Patio Weedkiller	AgrEvo UK Ltd	08177
Do It All Liquid Weedkiller Concentrate	Do-It-All Ltd	08201
Do It All Nettle Gun	Do-It-All Ltd	08180
Do It All Topple Weed!	Do-It-All Ltd	08130
Doff Knockdown Weedkiller Concentrate	Doff Portland Ltd	08199
Doff Knockdown Weedkiller Spray	Doff Portland Ltd	07769
Doff New Formula Nettle Gun	Doff Portland Ltd	08178
Doff Path and Patio Weedkiller	Doff Portland Ltd	08181
Doff Rose Weedkiller	Doff Portland Ltd	09342
Garden Weedkiller	L C Solutions Ltd	09735
Garden Weedkiller Concentrate	L C Solutions Ltd	09741
Great Mills Fast Action Weedkiller Ready to Use	Great Mills (Retail) Ltd	07114
Homebase Contact Action Weedkiller	Homebase Ltd	06735
Homebase Contact Action WeedkillerReady to Use Sprayer	Homebase Ltd	06736
Homebase Nettle Spray	Homebase Ltd	08179
Patio Weedkiller	L C Solutions Ltd	09734
Tumbleweed General Purpose	Levington Horticulture Ltd	07992
Tumbleweed General Purpose Ready To Use	Levington Horticulture Ltd	07993
Wilko Complete Weedkiller Concentrate	Wilkinson Group of Companies	08968
Wilko Complete Weedkiller Spray	Wilkinson Home & Garden Stores	07771

C   These products are "approved for agricultural use". For further details refer to page vii.
A   These products are approved for use in or near water. For further details refer to page vii.

Product Name	Marketing Company	Reg. No.	Expiry Date

1095 Glufosinate-ammonium (Herbicide)—continued

Product Name	Marketing Company	Reg. No.	Expiry Date
Woolworths Complete Weedkiller Concentrate	Woolworths Plc	08001	
Woolworths Complete Weedkiller Spray	Woolworths Plc	07770	

1096 Glufosinate-ammonium + Diuron (Herbicide)

Product Name	Marketing Company	Reg. No.	Expiry Date
AgrEvo Path Weed Killer	AgrEvo Environmental Health Ltd	08616	
Tumbleweed Paths and Patios	Levington Horticulture Ltd	08613	
Tumbleweed Paths and Patios	Levington Horticulture Ltd	08107	31/07/2000

1097 Glufosinate-ammonium + Fatty acids (Herbicide)

Product Name	Marketing Company	Reg. No.	Expiry Date
Bio Kills Weeds Dead Fast	pbi Home & Garden Ltd	09173	
Bio Kills Weeds Dead Fast Concentrate	pbi Home & Garden Ltd	09752	
Bio Speedweed Ultra	pbi Home & Garden Ltd	09186	

1098 Glyphosate (Herbicide)

Product Name	Marketing Company	Reg. No.	Expiry Date
B & Q Complete Weedkiller	B & Q Plc	05290	
B & Q Complete Weedkiller Ready To Use	B & Q Plc	06722	
Biactive Roundup Brushkiller	Solaris (Garden Div Of Monsanto)	08220	31/07/2001
Biactive Roundup Brushkiller	Monsanto Plc	07589	31/07/2001
Biactive Roundup GC	Solaris (Garden Div Of Monsanto)	08219	31/07/2001
Biactive Roundup GC	Monsanto Plc	07590	31/07/2001
Bio Glyphosate	pbi Home & Garden Ltd	09242	
Bio Glyphosate Pen	pbi Home & Garden Ltd	09663	
Bio Glyphosate Ready-to-Use	pbi Home & Garden Ltd	09241	
Bio WeedEasy	Pan Britannica Industries Ltd	08567	
Boots Systemic Weed & Grass Killer Ready To Use	The Boots Company Plc	05028	
Glypho	Miracle Garden Care	07737	
Glypho	Zeneca Garden Care	06441	25/07/2003
Glypho Gun!	Miracle Garden Care	07738	
Glypho Gun!	Zeneca Garden Care	06443	31/08/2000
Great Mills Systemic Action Weedkiller Ready to Use	Great Mills (Retail) Ltd	07080	
Greenscape Ready to Use	Solaris (Garden Div Of Monsanto)	08218	
Greenscape Ready To Use Weed Killer	Monsanto Plc	04676	31/08/2001
Greenscape Weedkiller	Solaris (Garden Div Of Monsanto)	08216	
Greenscape Weedkiller	Monsanto Plc	04321	

C   These products are "approved for agricultural use". For further details refer to page vii.
A   These products are approved for use in or near water. For further details refer to page vii.

Product Name	Marketing Company	Reg. No.	Expiry Date

1098 Glyphosate (Herbicide)—continued

High Strength Tough Weed Gun!	Miracle Garden Care Ltd	08967	
Homebase Systemic Action Weedkiller	Homebase Ltd	06622	
Homebase Systemic Action Weedkiller Ready to Use	Homebase Ltd	06623	
Knock Out	Premier Way Ltd	07846	
Knock Out Weedkiller Ready To Use	Premier Way Ltd	07948	
MON 44068 Garden Weedkiller	Solaris (Garden Div Of Monsanto)	08222	
Mon 44068 Garden Weedkiller	Monsanto Plc	07367	31/12/2001
MON 77020 Garden Weedkiller	Solaris (Garden Div Of Monsanto)	08223	
MON 77020 Garden Weedkiller	Monsanto Plc	07368	30/04/2002
MON 77021 Garden Weedkiller	Solaris (Garden Div Of Monsanto)	08224	
MON 77021 Garden Weedkiller	Monsanto Plc	07369	30/04/2002
Murphy Tumbleweed Gel	Murphy Home & Garden Products	04009	
New Improved Leaf Action Roundup Brushkiller RTU	Solaris (Garden Div Of Monsanto)	08265	30/06/2002
New Improved Leaf Action Roundup Weedkiller RTU	Monsanto Roundup Lawn & Garden	08264	
Nomix Weedkiller	Nomix-Chipman Ltd	09591	
Roundup Brushkiller	Solaris (Garden Div Of Monsanto)	09166	30/06/2002
Roundup Brushkiller	Solaris (Garden Div Of Monsanto)	08225	
Roundup Brushkiller	Monsanto Plc	05755	
Roundup Estate	Monsanto Plc	09399	
Roundup GC	Monsanto Roundup Lawn & Garden	09167	
Roundup GC	Solaris (Garden Div Of Monsanto)	08217	
Roundup GC	Monsanto Plc	05538	
Roundup Micro	Solaris (Garden Div Of Monsanto)	08226	
Roundup Ready-to-Use Faster Acting Formula	Monsanto Roundup Lawn & Garden	09277	
Roundup Tab	Monsanto Plc	05917	29/02/2000
Roundup Tough Weedkiller	Monsanto Roundup Lawn & Garden	09627	
Roundup Tough Weedkiller Ready-To-Use	Monsanto Roundup Lawn & Garden	09628	

C   These products are "approved for agricultural use". For further details refer to page vii.
A   These products are approved for use in or near water. For further details refer to page vii.

Product Name	Marketing Company	Reg. No.	Expiry Date

1098 Glyphosate (Herbicide)—continued

Roundup Ultra 3000	Monsanto Roundup Lawn & Garden	08172	
SBK Ready to Use Tough Weedkiller	Vitax Ltd	08937	
SBK Tough Weed and Bramble killer	Vitax Ltd	08938	
SBK Tough Weed killer	Vitax Ltd	08939	
Tough Weed Gun!	Miracle Garden Care	07748	
Tough Weed Gun!	Zeneca Garden Care	06442	31/08/2000
Tough Weed Killer	Miracle Garden Care	07749	
Tough Weed Killer	Zeneca Garden Care	06440	31/08/2000
Tumbleweed Original	Levington Horticulture Ltd	07974	
Tumbleweed Original Extra Strong	Levington Horticulture Ltd	07975	
Tumbleweed Original Extra Strong Gel	Levington Horticulture Ltd	08090	
Tumbleweed Original Extra Strong Ready To Use	Levington Horticulture Ltd	08091	
Tumbleweed Original Ultra Concentrated	Levington Horticulture Ltd	07976	
Weedclear	Miracle Garden Care Ltd	08672	
Wickes General Purpose Weedkiller Ready To Use	Wickes Building Supplies Ltd	08825	
Woolworths Weedkiller	Woolworths Plc	07108	

1099 Glyphosate + Diuron (Herbicide)

| Pathclear Gun! | The Scotts Company (UK) Limited | 09492 | |
| Weedatak Path and Drive | Monsanto Plc | 07659 | |

1100 Heptenophos + Permethrin (Insecticide)

| Murphy Systemic Action Insecticide | Levington Horticulture Ltd | 07557 | 30/04/2001 |
| Murphy Tumblebug | Levington Horticulture Ltd | 07571 | 30/04/2001 |

1101 Imidacloprid (Insecticide)

Bio Provado Complete Pest Killer	pbi Home & Garden Ltd	09691	
Bio Provado Vine Weevil Killer	pbi Home & Garden Ltd	09660	
Levington Plant Protection	Levington Horticulture Ltd	09377	
Plant Protection Compost	Levington Horticulture Ltd	08365	

1102 Indol-3-ylbutyric-4 acid (Herbicide)

| Clearcut II | SupaPlants Ltd | 08901 | |

C   These products are "approved for agricultural use". For further details refer to page vii.
A   These products are approved for use in or near water. For further details refer to page vii.

Product Name	Marketing Company	Reg. No.	Expiry Date

1102 Indol-3-ylbutyric-4 acid (Herbicide)—continued

Clearcut II	Levington Horticulture Ltd	07827	31/01/2001
Clonex	Growth Technology	09441	

1103 4-Indol-3-ylbutyric acid + 1-Naphthylacetic acid + Thiram (Fungicide) (Herbicide)

Boots Hormone Rooting Powder	The Boots Company Plc	01067	

1104 Lindane (Insecticide)

Doff Ant Killer	Doff Portland Ltd	00739	
Doff Gamma BHC Dust	Doff Portland Ltd	04868	
Doff Weevil Killer	Doff Portland Ltd	08138	
Murphy Gamma BHC Dust	Levington Horticulture Ltd	08896	31/05/2000
Murphy Gamma BHC Dust	Fisons Plc	04006	31/05/2000

1105 Magnesium sulphate + Copper sulphate + Ferrous sulphate (Miscellaneous)

Snail Away	Interpet Ltd	08868	

1106 Malathion (Insecticide)

Duramitex	Harkers Ltd	08664	
Duramitex	Harkers Ltd	02512	31/08/2000
Malathion Greenfly Killer	Pan Britannica Industries Ltd	01247	31/08/2000
Murphy Liquid Malathion	Levington Horticulture Ltd	07881	
Murphy Malathion Dust	Levington Horticulture Ltd	07880	30/04/2001

1107 Malathion + Permethrin (Insecticide)

Bio Crop Saver	Pan Britannica Industries Ltd	03969	30/04/2000

1108 Maleic hydrazide (Herbicide)

Stop Gro G8	Botanical Developments	05923	
Stop Gro G8	Synchemicals Ltd	02029	

1109 Mancozeb (Fungicide)

PBI Dithane 945	pbi Home & Garden Ltd	00718	

1110 MCPA + Clopyralid + Fluroxypyr (Herbicide)

Weed-B-Gon	Solaris (Garden Div Of Monsanto)	08926	
Weed-B-Gon	Dow AgroSciences Ltd	08925	
Weed-B-Gon Ready To Use	Solaris (Garden Div Of Monsanto)	09451	
Weed-B-Gon Ready To Use	Dow AgroSciences Ltd	09450	

C These products are "approved for agricultural use". For further details refer to page vii.
A These products are approved for use in or near water. For further details refer to page vii.

Product Name	Marketing Company	Reg. No.	Expiry Date

1111 MCPA + Dicamba (Herbicide)

| Green Up Liquid Lawn Feed 'n Weed | Vitax Ltd | 08196 | - |

1112 MCPA + Dicamba + Dichlorprop (Herbicide)

'Verdone Plus'	Miracle Garden Care	08689	
B & Q Granular Weed & Feed for Lawns	B & Q Plc	07850	
B & Q Liquid Weed and Feed	B & Q Plc	05293	
Boots Nettle and Bramble Weedkiller	The Boots Company Plc	03455	
Grasshopper Weed & Feed	Miracle Garden Care Ltd	07849	
Greensward Triple Action	The Scotts Company (UK) Limited	09324	
Groundclear	pbi Home & Garden Ltd	05953	
Groundclear Spot	pbi Home & Garden Ltd	05950	30/11/2000
Lawnsman Liquid Weed and Feed	Imperial Chemical Industries Plc	03610	31/05/2001
Liquid Weed and Feed	Miracle Garden Care	07786	31/07/2000
Liquid Weed and Feed	Zeneca Garden Care	06887	31/07/2000
Liquid Weed and Feed	ICI Garden & Professional Products	05869	31/07/2000
Miracle-Gro Weed & Feed	Miracle Garden Care	08550	
Woolworths Liquid Lawn Feed and Weed	Woolworths Plc	07060	

1113 MCPA + Dicamba + Dichlorprop + Ferrous sulphate (Herbicide)

'Grasshopper' Triple Action	Miracle Garden Care	08643	
'Greensward' 2	Miracle Garden Care	08380	30/11/2001
B & Q Granular Weed and Feed and Mosskiller For Lawns	B & Q Plc	08381	
Evergreen Grasshopper	The Scotts Company (UK) Limited	09639	
Miracle-Gro Lawn Food with Weed and Moss Control	Miracle Garden Care	08609	

1114 MCPA + Dicamba + Mecoprop (Herbicide)

Bio Spot	pbi Home & Garden Ltd	05071	31/12/2001
Bio Weed Pencil	pbi Home & Garden Ltd	04054	31/12/2001
Evergreen Feed and Weed Liquid	Levington Horticulture Ltd	07492	30/04/2000
Fisons Evergreen Feed and Weed Liquid	Fisons Plc	05664	30/04/2000
Homebase Lawn Feed and Weed Liquid	Homebase Ltd	06086	30/04/2000
Homebase Weed Pen	Sainsbury's Homebase	07648	31/12/2001

C These products are "approved for agricultural use". For further details refer to page vii.
A These products are approved for use in or near water. For further details refer to page vii.

269

Product Name	Marketing Company	Reg. No.	Expiry Date

1115 MCPA + Dicamba + Mecoprop-P (Herbicide)

Bio Lawn Weed Pencil	pbi Home & Garden Ltd	09656	
Bio Spot	pbi Home & Garden Ltd	09210	
Bio Weed Pencil	pbi Home & Garden Ltd	09212	30/06/2002
Evergreen Feed and Weed Liquid	Levington Horticulture Ltd	07766	
Homebase Lawn Feed and Weed Liquid	Homebase Ltd	07765	
Homebase Weed Pen	Sainsbury's Homebase	09211	

1116 MCPA + Dichlorprop (Herbicide)

Doff Lawn Weed and Feed Soluble Powder	Doff Portland Ltd	05708	
J Arthur Bower's Granular Feed and Weed	William Sinclair Horticulture Ltd	07164	
J Arthur Bower's Lawn Feed and Weed	William Sinclair Horticulture Ltd	09458	
J Artur Bower's Lawn Weedkiller	William Sinclair Horticulture Ltd	09461	
Wilko Granular Feed and Weed for Lawns	Wilkinson Group of Companies	09732	
Wilko Lawn Feed and Weed	Wilkinson Group of Companies	08816	
Wilko Soluble Lawn Food and Weedkiller	Wilkinson Home & Garden Stores	04391	

1117 MCPA + Dichlorprop + Ferrous sulphate (Herbicide)

J Arthur Bower's Granular Feed, Weed and Mosskiller	William Sinclair Horticulture Ltd	07042	
J Arthur Bower's Lawn Feed, Weed and Mosskiller	William Sinclair Horticulture Ltd	09459	
J Arthur Bower's Lawn Weed and Mosskiller.	William Sinclair Horticulture Ltd	09460	
Wilko Granular Feed, Weed and Mosskiller	Wilkinson Group of Companies	07471	
Wilko Granular Feed, Weed and Mosskiller for Lawns	Wilkinson Group of Companies	09731	
Wilko Lawn Feed, Weed and Mosskiller	Wilkinson Home & Garden Stores	04602	

1118 MCPA + Ferrous sulphate + Mecoprop (Herbicide)

Evergreen Extra	Levington Horticulture Ltd	07493	30/04/2000
Fisons Evergreen Extra	Fisons Plc	03890	
Gardenstore Lawn Feed, Weed and Mosskiller	Texas Homecare Ltd	07448	31/05/2000

C   These products are "approved for agricultural use". For further details refer to page vii.
A   These products are approved for use in or near water. For further details refer to page vii.

Product Name	Marketing Company	Reg. No.	Expiry Date

1119 MCPA + Ferrous sulphate + Mecoprop-P (Herbicide)

B & Q Granular Lawn Feed, Weed and Mosskiller	B & Q Plc	07762	
Do-It-All Complete Lawn Care	Do-It-All Ltd	08123	
Evergreen Easy	Levington Horticulture Ltd	08103	
Evergreen Extra	Levington Horticulture Ltd	07549	
Evergreen Triple Action	Levington Horticulture Ltd	09199	
Gardenstore Lawn Feed, Weed and Mosskiller	Texas Homecare Ltd	07550	31/10/2001
Great Mills Lawn Feed, Weed and Mosskiller	Great Mills (Retail) Ltd	07552	
Homebase Lawn Feed, Weed and Mosskiller	Homebase Ltd	07551	
Levington Gold	Levington Horticulture Ltd	07548	
Murphy Ultra Lawn Feed, Weed and Mosskiller	Murphy Home and Garden Ltd	07553	
Wickes Lawn Feed, Weed and Mosskiller	Wickes Building Supplies Ltd	08800	
Wilko Granular Lawn Feed, Weed and Mosskiller	Wilkinson Group of Companies	07981	
Woolworths Lawn Feed, Weed and Mosskiller	Woolworths Plc	07554	

1120 MCPA + Mecoprop (Herbicide)

Fisons Evergreen 90	Fisons Plc	03131	30/04/2000
Fisons Evergreen Feed and Weed	Fisons Plc	05906	30/04/2000
Gardenstore Lawn Feed and Weed	Texas Homecare Ltd	07360	30/04/2000
Homebase Lawn Feed and Weed	Homebase Ltd	05614	30/04/2000

1121 MCPA + Mecoprop-P (Herbicide)

B & Q Granular Lawn Feed and Weed	B & Q Plc	07793	
Do-It-All Granular Lawn Feed and Weed	Do-It-All Ltd	07971	
Evergreen Feed and Weed	Levington Horticulture Ltd	07595	
Gardenstore Lawn Feed and Weed	Texas Homecare Ltd	07593	30/11/2000
Homebase Lawn Feed and Weed	Homebase Ltd	07592	
Murphy Lawn Feed and Weed	Murphy Home and Garden Ltd	07596	

C   These products are "approved for agricultural use". For further details refer to page vii.
A   These products are approved for use in or near water. For further details refer to page vii.

Product Name	Marketing Company	Reg. No.	Expiry Date
1122 Mecoprop + 2,4-D (Herbicide)			
Bio Spraydex Lawn Spot Weeder	Pan Britannica Industries Ltd	07418	30/09/2000
Homebase Lawn Feed and Weed Soluble	Sainsbury's Homebase	07690	30/09/2001
J Arthur Bower's Lawn Food with Weedkiller	Sinclair Horticulture & Leisure Ltd	04301	30/04/2000
Lawn Feed and Weed Granules	pbi Home & Garden Ltd	05902	30/09/2001
Lawn Spot Weed Granules	Pan Britannica Industries Ltd	05903	
Supergreen and Weed	Pan Britannica Industries Agrochemicals Ltd	08911	30/09/2001
Supergreen and Weed	Rhone-Poulenc Agriculture	05348	30/04/2000
Supergreen Double (Feed and Weed)	Pan Britannica Industries Ltd	05949	28/02/2001
Supertox	Rhone-Poulenc Agriculture	05350	30/04/2000
Supertox Lawn Weedkiller	Pan Britannica Industries Ltd	05948	30/09/2000
Supertox Spot	pbi Home & Garden Ltd	05951	30/11/2001
Toplawn Lawn Weedkiller	pbi Home & Garden Ltd	08694	31/01/2002
Toplawn Ready to Use Lawn Weedkiller	pbi Home & Garden Ltd	08695	31/12/2001
Verdone	Miracle Garden Care	08657	30/06/2001
Verdone 2	Miracle Garden Care	07792	31/08/2000
Verdone 2	Zeneca Garden Care	06853	31/08/2000
Verdone 2	Imperial Chemical Industries Plc	03271	31/05/2001
Wilko Lawn Feed 'N' Weed	Wilkinson Home & Garden Stores	04403	30/04/2000
Wilko Lawn Food with Weedkiller	Wilkinson Home & Garden Stores	04302	31/12/2000
1123 Mecoprop + 2,4-D + Dicamba (Herbicide)			
B & Q Tree Stump and Brushwood Killer	B & Q Plc	07645	
New Formulation SBK Brushwood Killer	Vitax Ltd	05043	31/10/2000
1124 Mecoprop + 2,4-D + Ferrous sulphate (Herbicide)			
Asda Lawn Weed and Feed and Mosskiller	Asda Stores Ltd	03819	
Green Up Feed and Weed Plus Mosskiller	Vitax Ltd	06513	31/05/2001
Supergreen Feed, Weed & Mosskiller	Pan Britannica Industries Ltd	08118	31/01/2001
Weed 'N' Feed Extra	Vitax Ltd	06506	31/05/2001

C   These products are "approved for agricultural use". For further details refer to page vii.

A   These products are approved for use in or near water. For further details refer to page vii.

Product Name	Marketing Company	Reg. No.	Expiry Date

1125 Mecoprop + Dicamba (Herbicide)

Elliott Touchweeder	Thomas Elliott Ltd	07694	30/04/2000

1126 Mecoprop + Dicamba + Dichlorprop (Herbicide)

New Supertox	pbi Home & Garden Ltd	06128	30/09/2001

1127 Mecoprop + Dicamba + MCPA (Herbicide)

Bio Spot	pbi Home & Garden Ltd	05071	31/12/2001
Bio Weed Pencil	pbi Home & Garden Ltd	04054	31/12/2001
Evergreen Feed and Weed Liquid	Levington Horticulture Ltd	07492	30/04/2000
Fisons Evergreen Feed and Weed Liquid	Fisons Plc	05664	30/04/2000
Homebase Lawn Feed and Weed Liquid	Homebase Ltd	06086	30/04/2000
Homebase Weed Pen	Sainsbury's Homebase	07648	31/12/2001

1128 Mecoprop + Ferrous sulphate + MCPA (Herbicide)

Evergreen Extra	Levington Horticulture Ltd	07493	30/04/2000
Fisons Evergreen Extra	Fisons Plc	03890	
Gardenstore Lawn Feed, Weed and Mosskiller	Texas Homecare Ltd	07448	31/05/2000

1129 Mecoprop + MCPA (Herbicide)

Fisons Evergreen 90	Fisons Plc	03131	30/04/2000
Fisons Evergreen Feed and Weed	Fisons Plc	05906	30/04/2000
Gardenstore Lawn Feed and Weed	Texas Homecare Ltd	07360	30/04/2000
Homebase Lawn Feed and Weed	Homebase Ltd	05614	30/04/2000

1130 Mecoprop-P + 2,4-D (Herbicide)

ASB Greenworld Lawn Weed, And Feed Granular	ASB Greenworld Ltd	09356	
B & Q Lawn Weedand Feed Granular	B & Q Plc	09357	
Bio Toplawn Spot Weeder	pbi Home & Garden Ltd	09672	
Gem Lawn Weed & Feed	Gem Gardening	07086	
Green Up Spot Lawn Weedkiller	Vitax Ltd	06028	
Homebase Lawn Feed and Weed Soluble	Sainsbury's Homebase	09217	

C   These products are "approved for agricultural use". For further details refer to page vii.
A   These products are approved for use in or near water. For further details refer to page vii.

Product Name	Marketing Company	Reg. No.	Expiry Date

1130 Mecoprop-P + 2,4-D (Herbicide)—continued

Lawn Feed and Weed Granules	pbi Home & Garden Ltd	09225	
Supergreen and Weed	pbi Home & Garden Ltd	09218	
Supertox Spot	pbi Home & Garden Ltd	09306	
Toplawn Lawn Weedkiller	pbi Home & Garden Ltd	09589	
Toplawn Lawn Weedkiller	pbi Home & Garden Ltd	09408	31/05/2002
Toplawn Ready to Use Lawn Weedkiller	pbi Home & Garden Ltd	09348	31/07/2002
Verdone	The Scotts Company (UK) Limited	09642	
Verdone	Miracle Garden Care Ltd	09097	
Vitax 'Green Up' Granular Lawn Feed and Weed	Vitax Ltd	06158	
Wilko Lawn Feed & Weed Liquid	Wilkinson Home & Garden Stores	08789	

1131 Mecoprop-P + 2,4-D + Dicamba (Herbicide)

New Formulation SBK Brushwood Killer	Vitax Ltd	08737	

1132 Mecoprop-P + 2,4-D + Ferrous sulphate (Herbicide)

ASB Greenworld Lawn Weed, Feed and Mosskiller Granular	ASB Greenworld Ltd	09354	
B & Q Lawn Feed, Weed and Mosskiller Granular	B & Q Plc	09355	
Doff Granular Lawn Feed, Weed and Mosskiller	Doff Portland Ltd	08876	
Elliott's Lawn and Fine Turf Feed and Weed plus Moss Killer	Thomas Elliott Ltd	09464	
Gem Lawn Weed and Feed + Mosskiller	Gem Gardening	07087	
Green Up Feed and Weed Plus Mosskiller	Vitax Ltd	09046	
Proctors Lawn Feed Weed and Mosskiller	H & T Proctor (Division of Willett & Son (Bristol) Ltd)	07808	
Supergreen Feed, Weed and Mosskiller 2	pbi Home & Garden Ltd	09224	
Weed 'N' Feed Extra	Vitax Ltd	09045	
Westland Triple Action Weed, Feed And Mosskill	Westland Horticulture	09640	

C   These products are "approved for agricultural use". For further details refer to page vii.
A   These products are approved for use in or near water. For further details refer to page vii.

Product Name	Marketing Company	Reg. No.	Expiry Date

1133 Mecoprop-P + Dicamba + Dichlorprop (Herbicide)

| New Supertox | pbi Home & Garden Ltd | 09209 | |

1134 Mecoprop-P + Dicamba + MCPA (Herbicide)

Bio Lawn Weed Pencil	pbi Home & Garden Ltd	09656	
Bio Spot	pbi Home & Garden Ltd	09210	
Bio Weed Pencil	pbi Home & Garden Ltd	09212	30/06/2002
Evergreen Feed and Weed Liquid	Levington Horticulture Ltd	07766	
Homebase Lawn Feed and Weed Liquid	Homebase Ltd	07765	
Homebase Weed Pen	Sainsbury's Homebase	09211	

1135 Mecoprop-P + Ferrous sulphate + MCPA (Herbicide)

B & Q Granular Lawn Feed, Weed and Mosskiller	B & Q Plc	07762	
Do-It-All Complete Lawn Care	Do-It-All Ltd	08123	
Evergreen Easy	Levington Horticulture Ltd	08103	
Evergreen Extra	Levington Horticulture Ltd	07549	
Evergreen Triple Action	Levington Horticulture Ltd	09199	
Gardenstore Lawn Feed, Weed and Mosskiller	Texas Homecare Ltd	07550	31/10/2001
Great Mills Lawn Feed, Weed and Mosskiller	Great Mills (Retail) Ltd	07552	
Homebase Lawn Feed, Weed and Mosskiller	Homebase Ltd	07551	
Levington Gold	Levington Horticulture Ltd	07548	
Murphy Ultra Lawn Feed, Weed and Mosskiller	Murphy Home and Garden Ltd	07553	
Wickes Lawn Feed, Weed and Mosskiller	Wickes Building Supplies Ltd	08800	
Wilko Granular Lawn Feed, Weed and Mosskiller	Wilkinson Group of Companies	07981	
Woolworths Lawn Feed, Weed and Mosskiller	Woolworths Plc	07554	

1136 Mecoprop-P + Fluroxypyr (Herbicide)

| Verdone Extra | Miracle Garden Care Ltd | 08999 | |
| Verdone Extra | Dow AgroSciences Ltd | 08998 | |

1137 Mecoprop-P + MCPA (Herbicide)

| B & Q Granular Lawn Feed and Weed | B & Q Plc | 07793 | |

C   These products are "approved for agricultural use". For further details refer to page vii.

A   These products are approved for use in or near water. For further details refer to page vii.

Product Name	Marketing Company	Reg. No.	Expiry Date

1137 Mecoprop-P + MCPA (Herbicide)—continued

Do-It-All Granular Lawn Feed and Weed	Do-It-All Ltd	07971	
Evergreen Feed and Weed	Levington Horticulture Ltd	07595	
Gardenstore Lawn Feed and Weed	Texas Homecare Ltd	07593	30/11/2000
Homebase Lawn Feed and Weed	Homebase Ltd	07592	
Murphy Lawn Feed and Weed	Murphy Home and Garden Ltd	07596	

1138 Metaldehyde (Miscellaneous)

AgrEvo Slug Killer Pellets	AgrEvo UK Ltd	09551	
Aro Slug Killer Blue Mini Pellets	Makro Self Serv Wholesalers Ltd	06289	31/10/2001
B & Q Slug Killer Blue Mini Pellets	B & Q Plc	09678	
B & Q Slug Killer Blue Mini Pellets	B & Q Plc	05607	31/07/2002
Bio Slug Mini Pellets	pbi Home & Garden Ltd	08595	
Cookes 3% Metaldehyde Slug Killer	Devcol Morgan Ltd	09618	31/10/2001
Devcol Morgan 3% Metaldehyde Slug Killer	Devcol Morgan Ltd	07113	31/10/2001
Do-It-All Slug Killer Pellets	Do-It-All Ltd	09679	
Do-It-All Slug Killer Pellets	Do-It-All Ltd	04895	31/07/2002
Doff Slugoids Slug Killer Blue Mini Pellets	Doff Portland Ltd	09665	
Doff Slugoids Slug Killer Blue Mini Pellets	Doff Portland Ltd	00744	30/06/2002
Focus Slug Killer Pellets	Focus DIY Ltd	09680	
Focus Slug Killer Pellets	Focus DIY Ltd	06027	31/07/2002
Great Mills Slug Killer Blue Mini Pellets	Great Mills (Retail) Ltd	09681	
Great Mills Slug Killer Blue Mini Pellets	Great Mills (Retail) Ltd	06088	31/07/2002
Homebase Slug Killer Blue Mini Pellets	Sainsbury's Homebase House and Garden Centres	09682	
Homebase Slug Killer Blue Mini Pellets	Sainsbury's Homebase House and Garden Centres	06410	31/07/2002
Murphy Dilute Slugit Liquid	Levington Horticulture Ltd	09239	
Murphy Slugit Liquid	Levington Horticulture Ltd	07560	
Murphy Slugit Liquid	Fisons Plc	03633	31/05/2001
Murphy Slugits	The Scotts Company (UK) Limited	09576	
Murphy Slugits	The Scotts Company (UK) Limited	07559	31/05/2002
Murphy Slugits	Murphy Home & Garden Products	03634	31/05/2001

C   These products are "approved for agricultural use". For further details refer to page vii.
A   These products are approved for use in or near water. For further details refer to page vii.

Product Name	Marketing Company	Reg. No.	Expiry Date

1138 Metaldehyde (Miscellaneous)—continued

PBI Slug Mini Pellets	Pan Britannica Industries Ltd	02611	31/05/2001
Portland Brand Slug Killer Blue Mini Pellets	Doff Portland Ltd	09677	
Portland Brand Slug Killer Blue Mini Pellets	Doff Portland Ltd	06411	31/07/2002
Slug Xtra	Miracle Garden Care	07858	31/05/2001
Slugit Xtra	The Scotts Company (UK) Limited	09465	
Slugit Xtra	Miracle Garden Care Ltd	09195	
Super Slug and Snail Killer	Chiltern Farm Chemicals Ltd	09575	
Wilko Slug Killer Blue Mini Pellets	Wilkinson Home & Garden Stores	09683	
Wilko Slug Killer Blue Mini Pellets	Wilkinson Home & Garden Stores	05608	31/07/2002
Wilko Slug Killer Blue Mini Pellets	Wilkinson Home & Garden Stores	05032	31/01/2000
Woolworths Slug Killer Pellets	Woolworths Plc	09684	
Woolworths Slug Killer Pellets	Woolworths Plc	06924	31/07/2002
Woolworths Slug Pellets	Woolworths Plc	07038	

1140 Metaldehyde + Copper chloride (Miscellaneous)

Sera Snailpur	Sera GmbH	09587	
Sera Snailpur	Sera Werke Heimtierbedarf	09064	30/04/2001

1141 Methiocarb (Miscellaneous)

Bio Slug Guard	pbi Home & Garden Ltd	09593	
Slug Gard	pbi Home & Garden Ltd	01963	

1142 Methyl nonyl ketone (Vertebrate Control)

Get Off My Garden	Pet and Garden Manufacturing Plc	06614	
Get Off Spray	Pet and Garden Manufacturing Plc	08919	
Wash and Get Off Spray	Pet and Garden Manufacturing Plc	08918	

1143 Methyl nonyl ketone + Citronella oil (Vertebrate Control)

Secto Keep Off	Sinclair Animal & Household Care Ltd	07890	

1144 Monolinuron (Herbicide)

AlgoFin	Tetra	08862	

C   These products are "approved for agricultural use". For further details refer to page vii.
A   These products are approved for use in or near water. For further details refer to page vii.

Product Name	Marketing Company	Reg. No.	Expiry Date

1145 Myclobutanil (Fungicide)

Bio Fungus Fighter	pbi Home & Garden Ltd	09624	
Bio Systhane	pbi Home & Garden Ltd	08570	
Systhane	Pan Britannica Industries Ltd	04523	30/06/2000
Systhane	Pan Britannica Industries Ltd	04522	

1146 Naphthalene (Vertebrate Control)

Scent Off Buds	Synchemicals Ltd	02907	
Scent off Gel	Vitax Ltd	07366	
Scent Off Pellets	Vitax Ltd	01888	

1147 1-Naphthylacetic acid (Herbicide)

Strike 2	pbi Home & Garden Ltd	05952	

1148 1-Naphthylacetic acid + Captan (Fungicide) (Herbicide)

Doff Hormone Rooting Powder	Doff Portland Ltd	01065	
Homebase Rooting Hormone Powder	Sainsbury's Homebase	07630	
Murphy Hormone Rooting Powder	The Scotts Company (UK) Limited	07923	
Murphy Hormone Rooting Powder	Fisons Plc	03618	
New Strike	pbi Home & Garden Ltd	05956	31/07/2002
Rooting Powder	Vitax Ltd	06334	

1149 1-Naphthylacetic acid + Captan (Herbicide)

Bio Strike	pbi Home & Garden Ltd	09674	

1150 1-Naphthylacetic acid + Dichlorophen (Fungicide) (Herbicide)

Homebase Rooting Hormone Liquid	Sainsbury's Homebase	07698	

1151 1-Naphthylacetic acid + Dichlorophen (Fungicide) (Miscellaneous)

Baby Bio Roota	pbi Home & Garden Ltd	09305	

1152 1-Naphthylacetic acid + Dichlorophen (Herbicide)

Bio Roota	pbi Home & Garden Ltd	00271	30/11/2001

1153 1-Naphthylacetic acid + 4-Indol-3-ylbutyric acid + Thiram (Fungicide) (Herbicide)

Boots Hormone Rooting Powder	The Boots Company Plc	01067	

C   These products are "approved for agricultural use". For further details refer to page vii.
A   These products are approved for use in or near water. For further details refer to page vii.

Product Name	Marketing Company	Reg. No.	Expiry Date

1154 Orange peel oil + Garlic oil + Orange pith oil (Vertebrate Control)

6X Cat Repellent	Organic Concentrates Ltd	07979	
Growing Success Cat Repellent	Growing Success Organics Ltd	06262	

1155 Orange pith oil + Garlic oil + Orange peel oil (Vertebrate Control)

6X Cat Repellent	Organic Concentrates Ltd	07979	
Growing Success Cat Repellent	Growing Success Organics Ltd	06262	

1156 P-[(Diiodomethyl)sulfonyl]toluol (Herbicide)

Barrel Feature Clear	Interpet Ltd	09074	

1157 Paraquat + Amitrole + Diquat + Simazine (Herbicide)

Pathclear	The Scotts Company (UK) Limited	09287	
Pathclear	Miracle Garden Care	07789	

1158 Paraquat + Diquat (Herbicide)

MF00296A	Miracle Garden Care	09152	
Weedol	The Scotts Company (UK) Limited	09280	
Weedol	Miracle Garden Care	07750	30/09/2001
Weedol	Zeneca Garden Care	06863	31/05/2001
Weedol Gun!	Miracle Garden Care	08254	31/07/2001

1159 Paraquat + Diquat + Simazine (Herbicide)

'Pathclear' S	The Scotts Company (UK) Limited	09285	
'Pathclear' S	Miracle Garden Care Ltd	08992	31/10/2001

1160 Penconazole (Fungicide)

Murphy Tumbleblite II	Levington Horticulture Ltd	07573	
Murphy Tumbleblite II Ready to Use	Levington Horticulture Ltd	07572	

1161 Pepper (Vertebrate Control)

PBI Pepper Dust	pbi Home & Garden Ltd	01569	
Pepper Dust	Vitax Ltd	09635	
Pepper Dust	Synchemicals Ltd	01570	
Secto Pepper Dust	Sinclair Animal & Household Care Ltd	07891	

1162 Permethrin (Herbicide)

Miracle-Gro Bug Spray	Miracle Garden Care Ltd	09176	

C   These products are "approved for agricultural use". For further details refer to page vii.
A   These products are approved for use in or near water. For further details refer to page vii.

Product Name	Marketing Company	Reg. No.	Expiry Date

1163 Permethrin (Insecticide)

Product Name	Marketing Company	Reg. No.	Expiry Date
Bio Flydown	pbi Home & Garden Ltd	00267	
Bio Kill	Jesmond Ltd	07735	31/01/2001
Bio Sprayday	pbi Home & Garden Ltd	00272	
Fumite Whitefly Smoke Cone	The Scotts Company (UK) Limited	09535	
Fumite Whitefly Smoke Cone	Miracle Garden Care	07839	
Homebase All-In-One Insecticide	Sainsbury's Homebase	07946	
Levington Insect Spray for Houseplants	Levington Horticulture Ltd	07466	
Picket	Miracle Garden Care	07740	
Picket	Zeneca Garden Care	06846	30/11/2000

1164 Permethrin + Bioallethrin (Insecticide)

Product Name	Marketing Company	Reg. No.	Expiry Date
Bio Spraydex Greenfly Killer	pbi Home & Garden Ltd	07404	31/01/2001
Floracid	Perycut Insectengun Ltd	06798	
Longer Lasting Bug Gun	Miracle Garden Care	08021	

1165 Permethrin + Heptenophos (Insecticide)

Product Name	Marketing Company	Reg. No.	Expiry Date
Murphy Systemic Action Insecticide	Levington Horticulture Ltd	07557	30/04/2001
Murphy Tumblebug	Levington Horticulture Ltd	07571	30/04/2001

1166 Permethrin + Malathion (Insecticide)

Product Name	Marketing Company	Reg. No.	Expiry Date
Bio Crop Saver	Pan Britannica Industries Ltd	03969	30/04/2000

1167 Permethrin + Sulphur + Triforine (Fungicide) (Insecticide)

Product Name	Marketing Company	Reg. No.	Expiry Date
Bio Multirose	pbi Home & Garden Ltd	05716	
Homebase Rose Care	Sainsbury's Homebase	07754	

1168 Phenothrin + Tetramethrin (Insecticide)

Product Name	Marketing Company	Reg. No.	Expiry Date
Pesguard House and Plant Spray	Sumitomo Chemical (UK) Plc	07873	30/04/2001

1169 Pirimicarb (Insecticide)

Product Name	Marketing Company	Reg. No.	Expiry Date
Rapid Aerosol	Miracle Garden Care	07741	20/04/2001
Rapid Aerosol	Zeneca Garden Care	06847	31/12/2000
Rapid Aerosol	Imperial Chemical Industries Plc	01689	20/04/2001
Rapid Greenfly Killer	The Scotts Company (UK) Limited	09500	
Rapid Greenfly Killer	Miracle Garden Care	07742	30/09/2001
Rapid Greenfly Killer	Zeneca Garden Care	06848	20/04/2001
Rapid Greenfly Killer	Imperial Chemical Industries Plc	01690	20/04/2001

C   These products are "approved for agricultural use". For further details refer to page vii.
A   These products are approved for use in or near water. For further details refer to page vii.

Product Name	*Marketing Company*	*Reg. No.*	*Expiry Date*

1170 Pirimicarb + Bupirimate + Triforine (Fungicide) (Insecticide)

'Roseclear' 2	The Scotts Company (UK) Limited	09498	
'Roseclear' 2	Miracle Garden Care Ltd	08736	

1171 Pirimiphos-methyl (Insecticide)

Ant Powder	Woolworths Plc	06977	31/01/2000
Antkiller Dust	Miracle Garden Care	07736	31/10/2000
Antkiller Dust	Zeneca Garden Care	06865	31/10/2000
Antkiller Dust	Imperial Chemical Industries Plc	00101	30/06/2000
B & Q Antkiller Dust	B & Q Plc	05830	20/04/2001
Fumite General Purpose Insecticide Smoke Cone	The Scotts Company (UK) Limited	09501	
Fumite General Purpose Insecticide Smoke Cone	Miracle Garden Care	07838	
Sybol	Miracle Garden Care	07745	20/04/2001
Sybol	Zeneca Garden Care	06888	31/03/2000
Sybol Dust	Miracle Garden Care	07746	30/09/2001
Sybol Dust	Zeneca Garden Care	06851	20/04/2001
Sybol Dust	Imperial Chemical Industries Plc	05665	30/06/2000

1172 Pirimiphos-methyl + Resmethrin + Tetramethrin (Insecticide)

'Sybol' Aerosol	Miracle Garden Care	07790	
Miracle Gro Insect Spray	Miracle Garden Care	07670	31/08/2000
Miracle-Gro Bug Spray	Miracle Garden Care	08690	20/04/2001

1173 Polymeric quaternary ammonium chloride (Herbicide)

Feature Clear	Interpet Ltd	09075	

1174 Propiconazole (Fungicide)

Murphy Tumbleblite	Levington Horticulture Ltd	09327	
Tumbleblite	Levington Horticulture Ltd	07567	
Tumbleblite	Fisons Plc	04691	

1175 Propiconazole + Cypermethrin (Fungicide) (Insecticide)

Murphy Roseguard	Levington Horticulture Ltd	08564	

1176 Pyrethrins (Insecticide)

AgrEvo Garden Insect Killer	AgrEvo UK Ltd	08096	
AgrEvo House Plant Insect Killer	AgrEvo UK Ltd	08099	
Aquablast Bug Spray	Agropharm Ltd	03461	
B & Q Complete Insecticide Spray	B & Q Plc	06964	

C These products are "approved for agricultural use". For further details refer to page vii.
A These products are approved for use in or near water. For further details refer to page vii.

Product Name	Marketing Company	Reg. No.	Expiry Date

1176 Pyrethrins (Insecticide)—continued

Product Name	Marketing Company	Reg. No.	Expiry Date
B & Q Fruit and Vegetable Insecticide Spray	B & Q Plc	05766	
B & Q House Plant Insecticide Spray	B & Q Plc	05828	
B & Q Houseplant Insecticide Spray	B & Q Plc	07032	
B & Q Insecticide Spray for Fruit and Vegetable	B & Q Plc	05829	
B & Q Insecticide Spray For Roses and Flowers	B & Q Plc	05875	
B & Q Rose and Flower Insecticide Spray	B & Q Plc	05769	
Bio Friendly Anti-Ant Duster	pbi Home & Garden Ltd	05100	
Bug Gun!	The Scotts Company (UK) Limited	09538	
Bug Gun!	The Scotts Company (UK) Limited	09536	
Bug Gun!	Miracle Garden Care	07782	
Bug Gun!	Miracle Garden Care	07781	
Bug Gun!	Zeneca Garden Care	06837	31/05/2001
Bug Gun!	Zeneca Garden Care	06836	31/05/2001
Bug Gun!	Imperial Chemical Industries Plc	05966	31/05/2001
Bug Gun!	Imperial Chemical Industries Plc	05965	31/05/2001
Bug Gun! Attack	The Scotts Company (UK) Limited	09590	
Co-op Garden Maker Rose & Flower Insecticide Spray	Co-Operative Wholesale Society Ltd	05609	
Devcol All Purpose Natural Insecticide Spray	Devcol Ltd	05802	
Do It All De-Bug! Insect Killer	Do-It-All Ltd	08184	
Do It All Fruit and Vegetable Insecticide Spray	Do-It-All Ltd	05767	
Do It All Rose and Flower Insecticide Spray	Do-It-All Ltd	05765	
Doff 'All in One' Insecticide Spray	Doff Portland Ltd	06069	
Doff Fruit and Vegetable Insecticide Spray	Doff Portland Ltd	04040	
Doff Greenfly Killer	Doff Portland Ltd	07030	
Doff Houseplant Insecticide Spray	Doff Portland Ltd	06066	
Doff Rose and Flower Insecticide Spray	Doff Portland Ltd	04041	
Doff Rose Insecticide Spray	Doff Portland Ltd	08747	
Fellside Green Fruit and Vegetable Insect Spray	Doff Portland Ltd	05008	

C   These products are "approved for agricultural use". For further details refer to page vii.
A   These products are approved for use in or near water. For further details refer to page vii.

Product Name	Marketing Company	Reg. No.	Expiry Date

1176 Pyrethrins (Insecticide)—continued

Product Name	Marketing Company	Reg. No.	Expiry Date
Fellside Green Rose and Flower Insect Spray	Doff Portland Ltd	05007	
Fisons Nature's Answer to Insect Pests	Fisons Plc	06936	30/06/2000
Focus All-In-One Insecticide Spray	Focus DIY Ltd	07510	
Focus Fruit and Vegetable Insecticide Spray	Focus DIY Ltd	06067	
Focus Rose and Flower Insecticide Spray	Focus DIY Ltd	06068	
Garden Insect Killer	L C Solutions Ltd	09738	
Great Mills Complete Insecticide Spray	Great Mills (Retail) Ltd	07031	
Great Mills Fruit and Vegetable Insecticide Spray	Great Mills (Retail) Ltd	05772	
Great Mills Rose and Flower Insecticide Spray	Great Mills (Retail) Ltd	05771	
Homebase Houseplant Insecticide	Sainsbury's Homebase	07692	
Homebase Pest Gun	Sainsbury's Homebase House and Garden Centres	06962	
Houseplant Insect Killer	L C Solutions Ltd	09737	
Keri Insect Spray	The Scotts Company (UK) Limited	09539	
Keri Insect Spray	Miracle Garden Care	07784	
Keri Insect Spray	Zeneca Garden Care	06885	31/05/2001
Keri Insect Spray	ICI Garden & Professional Products	06155	31/05/2001
Levington Natural Houseplant Insect Spray	The Scotts Company (UK) Limited + Levington Horticulture Ltd	07511	
Murphy Bugmaster	The Scotts Company (UK) Limited	07575	
Nature's Answer to Insect Pests	The Scotts Company (UK) Limited + Levington Horticulture Ltd	07504	
Nature's Answer to Insect Pests Concentrate	The Scotts Company (UK) Limited + Levington Horticulture Ltd	08683	
Natures "Bug Gun!"	The Scotts Company (UK) Limited	09537	
Natures 'Bug Gun!'	Miracle Garden Care	07788	
Natures Bug Gun	Zeneca Garden Care	07077	31/05/2001
PBI Anti-Ant Duster	pbi Home & Garden Ltd	00098	
Premier Multi-Pest Insecticide Spray	Premier Group Ltd	07424	
Py Powder	Vitax Ltd	05542	
Py Spray Garden Insect Killer	Vitax Ltd	06085	
Py Spray Insect Killer	Vitax Ltd	05543	
Rentokil Garden Insect Killer	Rentokil Ltd	07852	

C   These products are "approved for agricultural use". For further details refer to page vii.
A   These products are approved for use in or near water. For further details refer to page vii.

Product Name	Marketing Company	Reg. No.	Expiry Date

1176 Pyrethrins (Insecticide)—continued

Product Name	Marketing Company	Reg. No.	Expiry Date
Secto Nature Care Garden Insect Powder	Sinclair Animal & Household Care Ltd	07894	
Secto Nature Care Garden Insect Spray	Sinclair Animal & Household Care Ltd	07892	
Secto Nature Care Houseplant Insect Spray	Sinclair Animal & Household Care Ltd	07893	
Texas All-In-One Insecticide Spray	Texas Homecare Ltd	07426	
Trappit	Agrisense -BCS Limited	09452	
Vapona All-In-One	Sara Lee Household and Personal Care	06965	
Vapona House and Garden Plant Insect Killer	Sara Lee Household and Personal Care	05996	
Wickes General Purpose Insecticide Ready to Use	Wickes Building Supplies Ltd	08787	
Wilko Fruit and Vegetable Insecticide Spray	Wilkinson Group of Companies	09577	
Wilko Fruit and Vegetable Insecticide Spray	Wilkinson Home & Garden Stores	05770	31/10/2000
Wilko Greenfly Killer	Wilkinson Home & Garden Stores	08727	
Wilko Houseplant Insecticide Spray	Wilkinson Home & Garden Stores	07425	
Wilko Multi-Purpose Insecticide Spray	Wilkinson Home & Garden Stores	06963	
Wilko Rose and Flower Insecticide Spray	Wilkinson Home & Garden Stores	05768	
Woolworths Complete Insecticide Spray	Woolworths Plc	06961	

1177 Pyrethrins + Fatty acids (Insecticide)

Product Name	Marketing Company	Reg. No.	Expiry Date
Phostrogen Safer's All Purpose Insecticide Ready to Use	Phostrogen Ltd	08093	
Safer's Indoor Trounce Insecticide Concentrate	Safer Ltd	08097	
Safer's Indoor Trounce Insecticide Ready to Use	Safer Ltd	08094	
Safer's Trounce Insecticide Concentrate	Safer Ltd	08098	
Safer's Trounce Insecticide Ready to Use	Safer Ltd	08095	

1178 Pyrethrins + Resmethrin (Insecticide)

Product Name	Marketing Company	Reg. No.	Expiry Date
House Plant Pest Killer	Vitax Ltd	06432	

C These products are "approved for agricultural use". For further details refer to page vii.
A These products are approved for use in or near water. For further details refer to page vii.

Product Name	Marketing Company	Reg. No.	Expiry Date

1179 Resmethrin + Pirimiphos-methyl + Tetramethrin (Insecticide)
'Sybol' Aerosol	Miracle Garden Care	07790	
Miracle Gro Insect Spray	Miracle Garden Care	07670	31/08/2000
Miracle-Gro Bug Spray	Miracle Garden Care	08690	20/04/2001

1180 Resmethrin + Pyrethrins (Insecticide)
| House Plant Pest Killer | Vitax Ltd | 06432 | |

1181 Rotenone (Insecticide)
Bio Friendly Insect Spray	pbi Home & Garden Ltd	05148	31/03/2000
Bio Friendly Pest Duster	pbi Home & Garden Ltd	06811	
Bio Liquid Derris Plus	pbi Home & Garden Ltd	07059	
BioSect	pbi Home & Garden Ltd	07807	
Derris Dust	Vitax Ltd	05452	
Doff Derris Dust	Doff Portland Ltd	00740	
Murphy Derris Dust	Fisons Plc	04005	
Nature's Answer Derris Dust	Levington Horticulture Ltd	09015	
PBI Liquid Derris	pbi Home & Garden Ltd	01214	30/11/2000
Stirling Rescue Wasp Nest Killer	STV International Ltd	08205	
Wasp Exterminator	Battle Hayward & Bower Ltd	06333	

1182 Rotenone + Sulphur (Fungicide) (Insecticide)
| Bio Back to Nature Pest & Disease Duster | Pan Britannica Industries Ltd | 00265 | |

1183 Simazine (Herbicide)
Algae Control No 1 Anti-Algae	Interpet Ltd	09068	
Algae Destroyer For Freshwater Aquariums	Aquarium Pharmaceuticals E.C. Inc	08856	
Algae Destroyer For Ponds	Aquarium Pharmaceuticals E.C. Inc	08858	
Algae Destroyer Liquid	Aquarium Pharmaceuticals E.C. Inc	08857	
Aquarium Algicide	Technical Aquatic Products	08841	
BWG Aquarium Anti Algae	Balgdon/KFS	08848	
BWG Pool Clinic Algicide	Blagdon Garden Products	08846	
Feature Algae Control	Interpet Ltd	09069	
Pond Doctor Algicide	Technical Aquatic Products	08810	
PPI Clarity Plus	Technical Aquatic Products	08837	
Tetra Algimin	Tetra	08861	

C These products are "approved for agricultural use". For further details refer to page vii.
A These products are approved for use in or near water. For further details refer to page vii.

Product Name	Marketing Company	Reg. No.	Expiry Date

1184 Simazine + Amitrole (Herbicide)

Homebase Path & Drive Weed Killer	Sainsbury's Homebase	07755	
Path and Drive Weedkiller	pbi Home & Garden Ltd	05958	

1185 Simazine + Amitrole + 2,4-D + Diuron (Herbicide)

Hytrol	Agrichem (International) Ltd	04540	30/10/2001

1186 Simazine + Amitrole + Diquat + Paraquat (Herbicide)

Pathclear	The Scotts Company (UK) Limited	09287	
Pathclear	Miracle Garden Care	07789	

1187 Simazine + Amitrole + Diuron (Herbicide)

Hytrol Total	Agrichem (International) Ltd	08296	

1188 Simazine + Copper sulphate (Herbicide)

PPI Clarity Excel	Pet Products International Ltd	09095	
Tap Algasan	Technical Aquatic Products	08845	

1189 Simazine + Diquat + Paraquat (Herbicide)

'Pathclear' S	The Scotts Company (UK) Limited	09285	
'Pathclear' S	Miracle Garden Care Ltd	08992	31/10/2001

1190 Sodium chlorate (Herbicide)

Barrettine Sodium Chlorate (Fire Suppressed) Weedkiller	Cooke's Chemicals (Sales) Ltd	06617	
Battle, Hayward and Bower Sodium Chlorate Weedkiller with Fire Depressant	Battle Hayward & Bower Ltd	05876	
Blanchard's Sodium Chlorate Weedkiller (Fire Suppressed)	Blanchard Martin and Simmonds Ltd	05649	
Cooke's Liquid Sodium Chlorate Weedkiller	Cooke's Chemicals (Sales) Ltd	04280	
Cooke's Sodium Chlorate Weedkiller with Fire Depressant	Cooke's Chemicals (Sales) Ltd	04281	
Devcol Path Weedkiller	Devcol Ltd	06580	
Devcol-Sodium Chlorate Weedkiller	Devcol Ltd	05656	
Doff Path Weedkiller	Doff Portland Ltd	07044	
Doff Sodium Chlorate Weedkiller	Doff Portland Ltd	00500	

C   These products are "approved for agricultural use". For further details refer to page vii.
A   These products are approved for use in or near water. For further details refer to page vii.

Product Name	Marketing Company	Reg. No.	Expiry Date

1190 Sodium chlorate (Herbicide)—continued

Focus Sodium Chlorate Weedkiller	Focus DIY Ltd	06000	
Gem Sodium Chlorate Weedkiller	Joseph Metcalf Ltd	04159	
Great Mills Sodium Chlorate Weedkiller	Great Mills (Retail) Ltd	07078	
Homebase Sodium Chlorate Weedkiller	Sainsbury's Homebase House and Garden Centres	06620	
Morgan's Sodium Chlorate Weedkiller	David Morgan (Nottingham) Ltd	08928	
Premier Sodium Chlorate	Premier Way Ltd	07469	
Strathclyde Sodium Chlorate Weedkiller	Strathclyde Chemical Co Ltd	07421	
Wilko Sodium Chlorate Weedkiller	Wilkinson Home & Garden Stores	06281	

1191 Sulphur (Fungicide)

Green Sulphur	Vitax Ltd	05782	
Safer's Natural Garden Fungicide	Safer Ltd	06298	
Sulphur Candle	Growing Success Organics Ltd	08688	
Sulphur Candles	Battle Hayward & Bower Ltd	02039	
Yellow Sulphur	Vitax Ltd	05783	

1192 Sulphur (Vertebrate Control)

Murphy Mole Smokes	Levington Horticulture Ltd	07562	31/05/2000
Murphy Mole Smokes	Fisons Plc	03615	

1193 Sulphur + Fatty acids (Fungicide) (Insecticide)

Nature's Answer Fungicide and Insect Killer	The Scotts Company (UK) Limited	07628	

1194 Sulphur + Permethrin + Triforine (Fungicide) (Insecticide)

Bio Multirose	pbi Home & Garden Ltd	05716	
Homebase Rose Care	Sainsbury's Homebase	07754	

1195 Sulphur + Rotenone (Fungicide) (Insecticide)

Bio Back to Nature Pest & Disease Duster	Pan Britannica Industries Ltd	00265	

1196 Tar acids (Fungicide) (Herbicide)

Armillatox	Armillatox Ltd	06234	

C   These products are "approved for agricultural use". For further details refer to page vii.
A   These products are approved for use in or near water. For further details refer to page vii.

Product Name	Marketing Company	Reg. No.	Expiry Date

1197 Tar acids (Miscellaneous)

| Jeyes Fluid | Jeyes Ltd | 04606 | |

1198 Tar oils (Insecticide)

| Murphy Mortegg | The Scotts Company (UK) Limited | 07879 | |

1199 2,3,6-TBA + 2,4-D (Herbicide)

| Touchweeder | Thomas Elliott Ltd | 02864 | |

1200 Terbutryn (Herbicide)

Algae-Kit	Intercel UK	08383	
Algae-Kit	Intercel UK	04545	31/08/2000
Algizin A	Waterlife Research Industries Ltd	08850	
Blanc-Kit	Intercel UK	08391	
Blanc-Kit	Intercel UK	04546	31/08/2000

1201 Tetramethrin + Phenothrin (Insecticide)

| Pesguard House and Plant Spray | Sumitomo Chemical (UK) Plc | 07873 | 30/04/2001 |

1202 Tetramethrin + Pirimiphos-methyl + Resmethrin (Insecticide)

'Sybol' Aerosol	Miracle Garden Care	07790	
Miracle Gro Insect Spray	Miracle Garden Care	07670	31/08/2000
Miracle-Gro Bug Spray	Miracle Garden Care	08690	20/04/2001

1203 Thiophanate-methyl (Fungicide)

| Liquid Club Root Control | pbi Home & Garden Ltd | 05957 | |

1204 Thiram + 4-Indol-3-ylbutyric acid + 1-Naphthylacetic acid (Fungicide) (Herbicide)

| Boots Hormone Rooting Powder | The Boots Company Plc | 01067 | |

1205 Triclopyr (Herbicide)

Murphy Nettlemaster	Levington Horticulture Ltd	07561	
Murphy Nettlemaster	Fisons Plc	06757	31/10/2000
New Garlon RTU	Dow AgroSciences Ltd	06758	
Tumbleweed Brushwood Ready to Use	Levington Horticulture Ltd	08645	

1206 Triforine + Bupirimate (Fungicide)

Nimrod T	The Scotts Company (UK) Limited	09502	
Nimrod T	Miracle Garden Care	07876	
Nimrod T	Zeneca Garden Care	06843	31/01/2001
Nimrod T	Imperial Chemical Industries Plc	03982	31/05/2001

C   These products are "approved for agricultural use". For further details refer to page vii.
A   These products are approved for use in or near water. For further details refer to page vii.

Product Name	Marketing Company	Reg. No.	Expiry Date

1207 Triforine + Bupirimate + Pirimicarb (Fungicide) (Insecticide)

'Roseclear' 2	The Scotts Company (UK) Limited	09498	
'Roseclear' 2	Miracle Garden Care Ltd	08736	

1208 Triforine + Permethrin + Sulphur (Fungicide) (Insecticide)

Bio Multirose	pbi Home & Garden Ltd	05716	
Homebase Rose Care	Sainsbury's Homebase	07754	

C   These products are "approved for agricultural use". For further details refer to page vii.
A   These products are approved for use in or near water. For further details refer to page vii.

3

PSD PRODUCT TRADE NAME INDEX

The number after the Trade Name gives the Active Ingredient Code Number under which the Active Ingredient in the product occurs in the Professional or Amateur Sections.

B & Q Insecticide Spray For Roses and Flowers *1176*
B & Q Insecticide Spray for Fruit and Vegetable *1176*
B & Q Lawn Feed and Weed *1027*
B & Q Lawn Feed, Weed and Mosskiller Granular *1029*
B & Q Lawn Spot Weeder *1027*
B & Q Lawn Feed and Weed Granular *1031*
B & Q Lawn Weedkiller *1027*
B & Q Lawnweed Spray *1022*
B & Q Liquid Weed and Feed *1039*
B & Q Mosskiller for Lawns *1080*
B & Q Nettle Gun *1027*
B & Q Path & Patio Weedkiller *975*
B & Q Rose and Flower Insecticide Spray *1176*
B & Q Slug Killer Blue Mini Pellets *1138*
B & Q Systemic Fungicide Concentrate *997*
B & Q Tree Stump and Brushwood Killer *1025*
B & Q Triple Action Lawn Care *1023*
B & Q Woodlice Killer *983*
B-Nine *126*
BAS 03729H *333*
BAS 421F *586*
BAS 438 *523*
BAS 46402F *601*
BAS 47807F *568*
BAS 48500F *594*
BAS 91580F *750*
BASF 3C Chlormequat 600 *63*
BASF 3C Chlormequat 720 *63*
BASF 3C Chlormequat 750 *63*
BASF CMPP Amine 60 *322*
BASF Dimethoate 40 *847*
BASF MCPA Amine 50 *293*
BASF Turf Protectant Fungicide *804*
BASF Turf Systemic Fungicide *495*
BB Chlorothalonil *518*
BH CMPP Extra *322*
BH CMPP/2,4-D *123*
BL500 *79*
BR Destral *4*
BUK 84000F *600*
BWG Anti Snails *1012*
BWG Aquarium Anti Algae *1183*
BWG Pool Clinic Algicide *1183*
Baby Bio Roota *1050*
Bacara *176*
Bactospeine WP *946*
Bactura WP *946*
Ballad *826*
Banco *24*
Banlene Super *146*
Banvel M *144*
Banvel P *147*
Banvel T *151*

Banweed *372*
Barbarian *236*
Barclay Amtrak *550*
Barclay Avalon *566*
Barclay Banshee *73*
Barclay Banshee XL *64*
Barclay Bezant *483*
Barclay Bolt *711*
Barclay Busker *739*
Barclay Canter *104*
Barclay Carbosect *822*
Barclay Champion *58*
Barclay Claddagh *58*
Barclay Cleanup *236*
Barclay Cleave *1*
Barclay Clinch *826*
Barclay Clinch 480 *826*
Barclay Clinch II *826*
Barclay Cluster *559*
Barclay Cobbler *605*
Barclay Coolmore *69*
Barclay Corrib *518*
Barclay Corrib 500 *518*
Barclay Cypersect XL *832*
Barclay Dart *236*
Barclay Desiquat *181*
Barclay Dimethosect *847*
Barclay Dingo *16*
Barclay Dodex *562*
Barclay Eyetak *697*
Barclay Eyetak 40 *697*
Barclay Eyetak 450 *697*
Barclay Fentin Flow *603*
Barclay Fentin Flow 532 *603*
Barclay Flotilla *804*
Barclay Flumen *363*
Barclay Foran *728*
Barclay Gallup *236*
Barclay Gallup 360 *236*
Barclay Gallup Amenity *236*
Barclay Garryowen *236*
Barclay Goalpost *196*
Barclay Goldpost *196*
Barclay Guideline *263*
Barclay Guideline 500 *263*
Barclay Hat-Trick *145*
Barclay Haybob *113*
Barclay Haybob II *113*
Barclay Holdup *63*
Barclay Holdup 600 *63*
Barclay Holdup 640 *63*
Barclay Hurler *216*
Barclay Karaoke *90*
Barclay Keeper *190*
Barclay Keeper 200 *190*
Barclay Keeper 500 FL *190*

PSD PRODUCT INDEX

PSD **PRODUCT INDEX**

Bug Gun! Attack *1176*
Bug Gun! For Gardens *985*
Bug Gun! For Roses & Flowers *1075*
Bug-Free Concentrate *985*
Bug-Free Ready To Use *985*
Buggy SG *236*
Bullet *109*
Bulwark Flo *408*
Bumper 250 EC *711*
Bumper P *705*
Burex 430 SC *58*
Butisan S *353*
Buzzard *177*
Bye Bye 20EC *814*
Bygran F *429*
Bygran S *751*
CDA Rovral *643*
CDA Spasor *236*
CDA Supertox 30 *123*
CDA Vanquish *236*
CMPP 60% *322*
CX 171 *195*
CYREN *826*
CYREN 4 *826*
Cadence *135*
Calibre *450*
Calidan *502*
Calirus *484*
Calixin *791*
Calypso *63*
Camco Insect Powder *983*
Campbell's CMPP *322*
Campbell's Dioweed 50 *113*
Campbell's Disulfoton FE 10 *849*
Campbell's Field Marshall *145*
Campbell's Field Marshall *145*
Campbell's Grassland Herbicide *145*
Campbell's MCPA 50 *293*
Campbell's Metham Sodium *964*
Camppex *121*
Capture *36*
Carbate Flowable *495*
Carbetamex *49*
Cardel Egret *236*
Cardel Glyphosate *236*
Casoron G *152*
Casoron G4 *152*
Casoron G4 Weed Block *1046*
Casoron GSR *152*
Castaway Plus *879*
Cavalier *821*
Celest *606*
Centex *424*
Centra *344*
Cercobin Liquid *766*
Cereline *488*

Cereline Secur *487*
Cerone *69*
Challenge *235*
Challenge 2 *235*
Challenge 60 *235*
Chandor *290*
Charger *69*
Check Turf II *346*
Checkmate *417*
Cheetah R *197*
Cheetah Super *198*
Chempak Lawn Sand *1080*
Chemtech Cypermethrin 10 EC *832*
Cherokee 318.5 EC *574*
Chieftain *550*
Childion *844*
Chiltern Hardy *963*
Chipko Diuron 80 *184*
Chipman Diuron 80 *184*
Chipman Diuron Flowable *184*
Chipman Garlon 4 *460*
Chloronil *518*
Chloropicrin Fumigant *953*
Chlortoluron 500 *74*
Choir *826*
Chryzoplus Grey 0.8% *245*
Chryzopon Rose 0.1% *245*
Chryzosan White 0.6% *245*
Chryzotek Beige *245*
Chryzotop Green *245*
Ciba Chlormequat 460 *63*
Ciba Chlormequat 5C 460:320 *63*
Ciba Chlormequat 730 *63*
Cirrus *392*
Cirsium *293*
Citadel *209*
Citation *360*
Citation 70 *360*
Clarion *236*
Clarosan *438*
Clarosan 1FG *438*
Clartex *963*
Claymore *383*
Clayton Alpha-Cyper *812*
Clayton Benson *355*
Clayton Benzeb *483*
Clayton CCC 750 *63*
Clayton Chizm *495*
Clayton Chloron *74*
Clayton Cyperten *832*
Clayton Cyprocon *541*
Clayton Diquat *181*
Clayton Fencer *197*
Clayton Fenican 625 *178*
Clayton Fenican-IPU *178*
Clayton Fluroxypyr *216*

Clayton Flutriafol *615*
Clayton Gantry *566*
Clayton Glyphosate *236*
Clayton Glyphosate - 360 *236*
Clayton IPU *263*
Clayton Krypton *539*
Clayton Lenacil 80W *282*
Clayton Metazachlor *353*
Clayton Metsulfuron *363*
Clayton Pendalin *383*
Clayton Pendalin-IPU 37 *274*
Clayton Pendalin-IPU 47 *274*
Clayton Pirimicarb 50 SG *895*
Clayton Pirimicarb 50SG *895*
Clayton Propel *408*
Clayton Propiconazole *711*
Clayton Quatrow *181*
Clayton Rhizeup *236*
Clayton Siptu 50 FL *263*
Clayton Siptu 500 *263*
Clayton Standup *63*
Clayton Stobik *480*
Clayton Swath *236*
Clayton Tebucon *739*
Clayton Turret *518*
CleanCrop Aldicarb 10 G *811*
CleanCrop Cyprodinil *550*
CleanCrop Cyprothal *522*
CleanCrop Diquat *181*
CleanCrop Fenpropimorph *586*
CleanCrop Fonic *M 73*
CleanCrop Kresoxazole *566*
CleanCrop Kresoxazole Plus *565*
CleanCrop Metribuzin *360*
CleanCrop MTZ 500 *353*
CleanCrop Phenmedipham *392*
CleanCrop PropaQ *407*
Cleanrun 2 *142*
Clean-Up-360 *236*
Clearcut II *1102*
Cleaval *108*
Clenecorn *322*
Clenecorn II *322*
Clenecorn Super *333*
Clinic *236*
Clinic Pro TMF *236*
Cliophar *90*
Clonex *1102*
Clortosip *518*
Clortosip 500 *518*
Clovotox *333*
Club *885*
Co-op Garden Maker Rose & Flower Insecticide
    Spray *1176*
CoPilot *415*
Cogito *720*

Colstar *594*
Combinex *696*
Comet *353*
Commando *208*
Comodor 600 *428*
Compass *645*
Compete Forte *210*
Competitor *210*
Compitox Extra *322*
Compitox Extra P *333*
Compitox Plus *333*
Condox *147*
Contact 75 *518*
Contest *812*
Contest Eco *812*
Contrac All-Weather Blox *929*
Contrac Rodenticide *929*
Contrac Super Blox *929*
Contrast *500*
Controller Fungicide *711*
Convoy *460*
Cooke's Lawn Mosskiller *1080*
Cooke's Liquid Sodium Chlorate Weedkiller *1190*
Cooke's Professional Sodium Chlorate Weedkiller
    with Fire Depressant *424*
Cooke's Sodium Chlorate Weedkiller with Fire
    Depressant *1190*
Cooke's Weedclear *424*
Cookes 3% Metaldehyde Slug Killer *1138*
Cookes 6% Metaldehyde Slug Killer *963*
Corbel *586*
Corbel CL *523*
Corniche *169*
Corral *209*
Cosmic FL *506*
Cougar *178*
Country Gardens Lawn Sand *1080*
Croneton *855*
Cropsafe 5C Chlormequat *63*
Croptex Bronze *390*
Croptex Chrome *80*
Croptex Fungex *532*
Croptex Pewter *79*
Croptex Steel *426*
Crossfire *826*
Crossfire 480 *826*
Crusader *S 33*
Cudgel *866*
Cultar *376*
Cuprokylt *534*
Cuprokylt FL *534*
Cuprokylt L *534*
Cuprosana H *534*
Curalan CL *531*
Curb *926*
Curb (Garden Pack) *972*

Guardsman STP *926*
HY-CARB *495*
HY-D *113*
HY-MCPA *293*
HY-TL *765*
HY-VIC *765*
Hallmark *876*
Halo *525*
Halophen RE 49 *553*
Halt *739*
Hardy Z *963*
Harlequin 500 SC *275*
Harmony Express *56*
Harmony M *366*
Harvest *235*
Hawk *89*
Headland Addstem *495*
Headland Addstem DF *495*
Headland Archer *19*
Headland Charge *322*
Headland Dual *504*
Headland Inorganic Liquid Copper *534*
Headland Judo *408*
Headland Maple *503*
Headland Regain *495*
Headland Relay *145*
Headland Relay P *146*
Headland Relay Turf *146*
Headland Spear *293*
Headland Spirit *664*
Headland Staff *113*
Headland Sulphur *734*
Headland Sword *235*
Headland Venus *734*
Headland Zebra Flo *651*
Headland Zebra WP *651*
Helimax *963*
Helm 75 WG Newtec *651*
Helm 75WG *651*
Helmet *550*
Helmsman *50*
Helocarb Granule 500 *895*
Helosate *236*
Herbasan *392*
Herboxone *113*
Herboxone 60 *113*
Hero *876*
Herrisol *145*
Herrisol New *146*
Hi-Shot *780*
Hickson Phenmedipham *392*
Hickstor 10 *429*
Hickstor 3 *429*
Hickstor 5 *429*
Hickstor 6 *429*
Hickstor 6 Plus MBC *48*

High Strength Tough Weed Gun! *1098*
Hilite *236*
Hilite 120 *236*
Hispor 45 WP *509*
Hispor 45WP *509*
Hoegrass *168*
Holdfast D *150*
Home Base Pest Gun 2 *1075*
Homebase All-In-One Insecticide *1163*
Homebase Ant Killer *983*
Homebase Autumn Lawn Feed and Moss
   Killer *1080*
Homebase Contact Action Weedkiller *1095*
Homebase Contact Action Weedkiller Ready to
   Use Sprayer *1095*
Homebase Houseplant Insecticide *1176*
Homebase Houseplant Insecticide Spray 2 *1075*
Homebase Lawn Feed and Moss killer *1080*
Homebase Lawn Feed and Weed *1120*
Homebase Lawn Feed and Weed Liquid *1043*
Homebase Lawn Feed and Weed Soluble *1030*
Homebase Lawn Feed, Weed and Mosskiller *1089*
Homebase Lawn Mosskiller Ready to Use
   Sprayer *1048*
Homebase Lawn Sand *1080*
Homebase Lawn Weedkiller Liquid *1027*
Homebase Lawn Weedkiller Ready To Use
   Sprayer *1027*
Homebase Liquid Lawn Feed and Weed *1027*
Homebase Nettle Gun *1027*
Homebase Nettle Spray *1095*
Homebase Path & Drive Weed Killer *979*
Homebase Pest Gun *1176*
Homebase Rooting Hormone Liquid *1049*
Homebase Rooting Hormone Powder *995*
Homebase Rose Care *1167*
Homebase Slug Killer Blue Mini Pellets *1138*
Homebase Sodium Chlorate Weedkiller *1190*
Homebase Systemic Action Weedkiller *1098*
Homebase Systemic Action Weedkiller Ready to
   Use *1098*
Homebase Wasp Nest Killer *983*
Homebase Weed Pen *1043*
Homebase Woodlice Killer *983*
Hortag Tecnacarb *48*
Hortag Tecnazene 10% Granules *751*
Hortag Tecnazene Double Dust *751*
Hortag Tecnazene Potato Dust *751*
Hortag Tecnazene Potato Granules *751*
Hortichem 2-Aminobutane *478*
Hostaquick *870*
Hostathion *921*
Hotspur *97*
House Plant Pest Killer *1178*
Houseplant Insect Killer *1176*
Houseplant Pest Killer *1075*

PSD PRODUCT INDEX

Legumex Extra *15*
Lektan *350*
Lemnalit *1010*
Lentagran EC *412*
Lentagran WP *412*
Lentipur CL 500 *74*
Levi *181*
Levington Autumn Extra *1080*
Levington Clover-kil *1027*
Levington Gold *1089*
Levington Insect Spray for Houseplants *1163*
Levington Lawn Sand *1080*
Levington Natural Houseplant Insect Spray *1176*
Levington Octave *697*
Levington Plant Protection *1101*
Levington Professional Plus Intercept *872*
Levington Ready to Use Clover-kil *1027*
Levington Ready to Use Lawn Weedkiller *1027*
Levington Water-on Lawn Weedkiller *1027*
Lexone 70 DF *360*
Lexus 50 DF *211*
Lexus Class *52*
Lexus Class WSB *52*
Lexus Millenium *215*
Lexus Millenium WSB *215*
Lexus XPE *212*
Lexus XPE WSB *212*
Leyclene *37*
Lindane Flowable *878*
Linnet *290*
Linurex 50SC *286*
Linuron Flowable *286*
Liquid Bromatrol *929*
Liquid Club Root Control *1203*
Liquid Curb Crop Spray *926*
Liquid Derris *906*
Liquid Linuron *286*
Liquid Weed and Feed *1039*
Lo-Gran *453*
Lo-Gran 20WG *453*
Loft *3*
Longer Lasting Bug Gun *986*
Longlife Clearun 2 *142*
Longlife Plus *116*
Longlife Renovator *117*
Longlife Renovator 2 *141*
Lonpar *94*
Lontrel 100 *90*
Lontrel Plus *95*
Lorate 20DF *363*
Lorsban 480 *826*
Lorsban T *826*
Lotus *85*
Lotus Algicide *984*
Lotus Fungicide *984*
Lucifer *87*

Lupus *885*
Luxan 2,4-D *113*
Luxan 9363 *963*
Luxan CMPP *322*
Luxan CMPP 600 *322*
Luxan Chloridazon *58*
Luxan Chlorotoluron 500 Flowable *74*
Luxan Cypermethrin 10 *832*
Luxan Dichlobenil Granules *152*
Luxan Dichlorvos 600 *842*
Luxan Dichlorvos Aerosol 15 *842*
Luxan Gro-Stop 300 EC *79*
Luxan Gro-Stop Basis *79*
Luxan Gro-Stop Fog *79*
Luxan Gro-Stop HN *79*
Luxan Isoproturon 500 Flowable *263*
Luxan MCPA 500 *293*
Luxan Mancozeb Flowable *651*
Luxan Maneb 80 *664*
Luxan Maneb 800 WP *664*
Luxan Metaldehyde *963*
Luxan Micro-Sulphur *734*
Luxan Phenmedipham *392*
Luxan Talunex *927*
Lynx *963*
Lyric *610*
MC Flowable *504*
MCC 25 EC *112*
MCPA 25% *293*
MCPA 500 *293*
MF00296A *1066*
MM70 *350*
MON 10EC *754*
MON 240 *236*
MON 44068 Garden Weedkiller *1098*
MON 44068 Pro *236*
MON 52276 *236*
MON 77020 Garden Weedkiller *1098*
MON 77021 Garden Weedkiller *1098*
MSS 2,4-D Amine *113*
MSS 2,4-D Ester *113*
MSS 2,4-DB + MCPA *131*
MSS 2,4-DP *155*
MSS 2,4-DP + MCPA *164*
MSS Aminotriazole Technical *3*
MSS Atrazine 50 FL *10*
MSS Atrazine 80 WP *10*
MSS Betaren Flow *392*
MSS CIPC 40 EC *79*
MSS CIPC 50 LF *79*
MSS CIPC 50M *79*
MSS CIPC 5G *79*
MSS CMPP *322*
MSS Chlormequat 40 *63*
MSS Chlormequat 460 *63*
MSS Chlormequat 60 *63*

MSS Chlormequat 70 *63*
MSS Chlorothalonil *518*
MSS Chlortoluron 500 *74*
MSS Diuron 500 FL *184*
MSS Diuron 50FL *184*
MSS Flotin 480 *603*
MSS Glyfield *236*
MSS Iprofile *263*
MSS Linuron 50 *286*
MSS Linuron 500 *286*
MSS MCPA 50 *293*
MSS MCPB + MCPA *315*
MSS Mircam *149*
MSS Mircam Plus *146*
MSS Mircarb *495*
MSS Mircell *63*
MSS Mirprop *333*
MSS Mirquat *63*
MSS Optica *333*
MSS Primasan *190*
MSS Protrum G *392*
MSS Simazine 50 FL *418*
MSS Sulphur 80 *734*
MSS Thor *190*
MSS Trifluralin 48 EC *468*
MTM CIPC 40 *79*
MTM Eminent *108*
MTM Grassland Herbicide *146*
MTM Trifluralin *468*
Magnum *60*
Mainstay *518*
Malathion 60 *881*
Malathion Greenfly Killer *1106*
Mallard *580*
Manconex *651*
Mandate 75 WDG *651*
Mandate 80 WP *651*
Mandolin *392*
Mandops Barleyquat *B 63*
Mandops Bettaquat *B 63*
Mandops Chlormequat 460 *63*
Mandops Chlormequat 700 *63*
Maneb 80 *664*
Manex *664*
Manex II *651*
Manipulator *63*
Mantis *711*
Mantis 250EC *711*
Mantle *598*
Mantle 425 EC *598*
Mantra *565*
Manzate 200 DF *651*
Manzate 200 PI *651*
Maraud *826*
Marks 2,4-D-A *113*
Marks 2,4-DB *127*

Marks 2,4-DB Extra *131*
Marks MCPA 50A *293*
Marks MCPA P30 *293*
Marks MCPA S25 *293*
Marks MCPA SP *293*
Marks MCPB *318*
Marks Mecoprop K *322*
Marks Mecoprop-E *322*
Marks Polytox-K *155*
Marks Polytox-M *120*
Marnoch Chlorothalonil *518*
Marnoch Glyphosate *236*
Marnoch Metazachlor *353*
Marquise *350*
Marshal 10G *823*
Marshal Soil Insecticide suSCon CR granules *823*
Masai *910*
Masai G *910*
Mascot Clover Killer *322*
Mascot Cloverkiller-P *333*
Mascot Contact Turf Fungicide *804*
Mascot Mosskiller *153*
Mascot Selective Weedkiller *123*
Mascot Selective-P *124*
Mascot Super Selective-P *146*
Mascot Systemic *495*
Master Sward *15*
Masterspray *45*
Matador 200 SC *859*
Match *556*
Match SL *556*
Matrikerb *101*
Mavrik *909*
Mavrik Aquaflow *909*
Maxicrop Mosskiller and Conditioner *202*
Maxicrop Mosskiller and Lawn Tonic *1080*
Mazide 25 *291*
Mazide Selective *143*
Mazin *664*
Me^2 Aldee *811*
Me^2 Azoxystrobin *480*
Me^2 Cymoxeb *539*
Me^2 Exodus *968*
Me^2 KME *566*
Me^2 Lambda *876*
Me^2 Tebuconazole *739*
Mebrom 100 *966*
Mebrom 98 *966*
Meld *614*
Menace 80 EDF *408*
Menara *547*
Meothrin *864*
Merit *389*
Merlin *528*
Metam 510 *964*
Metaphor *837*

PSD PRODUCT INDEX

SAN 710 *546*
SAN 710 HF *546*
SAN 735 *549*
SBK Ready to Use Tough Weedkiller *1098*
SBK Tough Weed and Bramble killer *1098*
SBK Tough Weed killer *1098*
SHL Lawn Sand *202*
SHL Lawn Sand Plus *154*
SHL Turf Feed & Weed *164*
SHL Turf Feed and Weed and Mosskiller *163*
SL 471 200 EC *725*
SL 520 *88*
SL 552A *582*
SL 556 500 EC *585*
SL 567A *679*
SL 600 *410*
SL-556 500 EC *585*
SL571A *597*
Sabre *263*
Sabre 500 *263*
Safer's Indoor Trounce Insecticide
 Concentrate *1077*
Safer's Indoor Trounce Insecticide Ready to
 Use *1077*
Safer's Insecticidal Soap *858*
Safer's Insecticidal Soap Ready-to-Use *1075*
Safer's Natural Garden Fungicide *1191*
Safer's Trounce Insecticide Concentrate *1077*
Safer's Trounce Insecticide Ready to Use *1077*
Safran *604*
Sage *733*
Sakarat Ready To Use (Cut Wheat) *943*
Sakarat Ready to use (Whole Wheat) *943*
Sakarat X Ready to use Warfarin Rat Bait *943*
Salute *190*
Salvo *605*
Sambarin 312.5 SC *529*
San 735 *549*
San 845H *135*
Sanaphen M *293*
Sanction *610*
Sandofan Pepite *662*
Sanuron *184*
Sapecron 240 EC *824*
Saprol *802*
Saracen *557*
Saracen WDG *557*
Satellite *65*
Satisfar *857*
Satisfar Dust *857*
Savona *858*
Savona Rose Spray *1075*
Scala *726*
Scarab 5EW *573*
Scent Off Buds *1146*
Scent Off Pellets *1146*

Scent off Gel *1146*
Sceptre *415*
Scoot *972*
Scotts Octave *697*
Sculptor *58*
Scuttle *942*
Scythe LC *378*
Seconal *940*
Secto Keep Off *1000*
Secto Nature Care Garden Insect Powder *1176*
Secto Nature Care Garden Insect Spray *1176*
Secto Nature Care Houseplant Insect Spray *1176*
Secto Pepper Dust *1161*
Secto Wasp Killer Powder *983*
Sector *847*
Seedox SC *816*
Selective Weedkiller *123*
Semeron *134*
Semeron 25 WP *134*
Senate *442*
Sencorex WG *360*
Sentinel 2 *404*
Septico Slug Killer *973*
Sequel *865*
Sera Algopur *1010*
Sera Snailpur *1006*
Seradix 1 *245*
Seradix 2 *245*
Seradix 3 *245*
Seritox 50 *164*
Seritox Turf *164*
Setter 33 *15*
Sewarin Extra *943*
Sewarin P *943*
Sheen *583*
Sheriff *827*
Shield DPA 15% *958*
Shire *178*
Shirlan *605*
Shirlan Programme *605*
Shogun *407*
Shogun 100EC *407*
Shortcut *970*
Sibutol *486*
Sibutol CF *486*
Sibutol LS *486*
Sibutol New Formula *486*
Sibutol Secur *950*
Sickle *38*
Sierraron G *152*
Sigma PCT *63*
Silvacur *748*
Silvacur 300 *748*
Silvapron D *113*
Simba 20 DF *363*
Sinbar *432*

Teldor *578*
Telone 2000 *840*
Telone II *840*
Temik 10 G *811*
Terbel Triple *119*
Tern *580*
Terpal *73*
Terpal C *64*
Terpitz *73*
Terraclor 20D *731*
Terraclor Flo *731*
Terset *41*
Tesco Lawn Feed 'n' Weed *1022*
Tetra 5634 *754*
Tetra Algimin *1183*
Texas All-In-One Insecticide Spray *1176*
Thinsec *821*
Thiodan 20 EC *852*
Thiovit *734*
Thiraflo *768*
Thripstick *834*
Throttle *822*
Thuricide HP *946*
Tigress *169*
Tigress Ultra *169*
Tilt *711*
Tilt 250EC *711*
Tilt Turbo 475 EC *721*
Timbrel *460*
Tipoff *369*
Tiptor *546*
Titus *416*
Tolkan *263*
Tolkan Liquid *263*
Tolkan Turbo *178*
Tolkan WDG *263*
Tolugan 700 *74*
Tolugan Extra *76*
Tomahawk *216*
Tomahawk 2000 *216*
Tomcat 2 All-Weather Blox *929*
Tomcat 2 Mouse Bait Station *990*
Tomcat 2 Rat and Mouse Blox *990*
Tomcat 2 Rat and Mouse Killer *990*
Tomcat 2 Rodenticide *929*
Tomcat 2 Super Blox *929*
Tomcat All-Weather Blox *938*
Tomcat All-Weather Rodent Bar *938*
Tomcat Blox *938*
Tomcat Rat and Mouse Bait *938*
Tomcat Rat and Mouse Blox *1063*
Tomcat Rat and Mouse Pellets *1063*
Tomcat Super Blox All-Weather *938*
Top Farm Bentazone *16*
Top Farm CMPP *322*
Top Farm Carbendazim - 435 *495*

Top Farm Chlormequat 640 *63*
Top Farm Chlorothalonil 500 *518*
Top Farm Dimethoate *847*
Top Farm Diquat-200 *181*
Top Farm Ethofumesate 200 *190*
Top Farm Fluazinam *605*
Top Farm IPU 500 *263*
Top Farm MCPA 500 *293*
Top Farm Paraquat *378*
Top Farm Paraquat 200 *378*
Top Farm Propiconazole 250 *711*
Top Farm Propyzamide 500 *408*
Top Farm Toluron 500 *74*
Topas *691*
Topas 100 EC *691*
Topas C 50 WP *494*
Topas D275 SC *560*
Topik *86*
Toplawn Feed with Mosskiller *1080*
Toplawn Feed with Weed & Mosskiller *1023*
Toplawn Feed with Weedkiller *1022*
Toplawn Feed, Weed & Mosskiller *1024*
Toplawn Lawn Weedkiller *1030*
Toplawn Ready to Use Lawn Weedkiller *1030*
Toppel 10 *832*
Topshot *17*
Torch *732*
Tordon 101 *125*
Tordon 22K *398*
Torq *860*
Torque *860*
Total *188*
Total Mouse Killer System *990*
Totem *77*
Totril *247*
Touchdown *236*
Touchdown LA *236*
Touche *188*
Touchweeder *1032*
Tough Weed Gun! *1098*
Tough Weed Killer *1098*
Townex Sachets *934*
Trappit *1176*
Treflan *468*
Tribel 480 *460*
Tribunil *355*
Tribunil WG *355*
Tribute *146*
Tribute Plus *146*
Trident *524*
Tridex *651*
Triflur *468*
Triflurex 48EC *468*
Trifluron *290*
Trifolex-Tra *315*
Trifsan *468*

PSD
PRODUCT INDEX

4
PSD ACTIVE INGREDIENT INDEX

The number after the Active Ingredient gives the Ingredient Code Number for the Professional or Amateur Sections.

PSD
ACTIVE INGREDIENT INDEX

PSD
ACTIVE INGREDIENT INDEX

PART C

HSE Registered Products

1
ANTIFOULING PRODUCTS

Antifouling Products

1 Copper

Avonclad	Avon Technical Products	Professional	6396
Copperbot	Copperbot 98 Ltd	Amateur Professional	6860
Copperbot 2000	Wessex Resins and Adhesives Ltd	Amateur Professional	6680
Copperguard	Synthetic Solutions Ltd	Amateur Professional	6670
VC 17M-EP	International Coatings Ltd	Amateur Professional Professional (Aquaculture)	6102

2 Copper + Dichlorophenyl dimethylurea

Coppercoat	Aquarius Marine Coatings Ltd	Amateur Professional	6428

3 Copper + Zinc pyrithione + 2-Methylthio-4-tertiary-butylamino-6-cyclopropylamino-S-triazine

VC17M-HS	International Coatings Ltd	Amateur Professional	5960

4 Copper metal

W Amercoat 67E	Ameron BV	Amateur Professional	3201
W Amercoat 70E	Ameron BV	Amateur Professional Professional (Aquaculture)	3202
W Amercoat 70ESP	Ameron BV	Amateur Professional Professional (Aquaculture)	3203
W Copperbot	C-Defence International Ltd	Amateur Professional	5262
D Crystic Copperclad 70PA	Scott Bader Co Ltd	Amateur Professional	3479
CU15	Hippo Marine Products Ltd	Amateur Professional	5872
Miricoat A.F. Coating	Miricoat Ltd	Professional	5587
D VC 17M EP-Antifouling	Extensor AB	Amateur Professional	3318

Product Name	Marketing Company	Use	HSE No.

4 Copper metal—continued

VC17M — International Coatings Ltd — Amateur / Professional / Professional (Aquaculture) — 4780

5 Copper metal + 2-Methylthio-4-tertiary-butylamino-6-cyclopropylamino-S-triazine

VC 17M Tropicana — International Coatings Ltd — Amateur / Professional — 4218

6 Copper naphthenate + Cuprous oxide + Dichlofluanid + Zinc naphthenate

W Teamac Killa Copper Plus — Teal and Mackrill Ltd — Amateur / Professional — 4659

7 Copper resinate + Cuprous oxide

Double Shield Antifouling — Indestructible Paint Company Ltd — Amateur — 6040

W Double Shield Antifouling — Llewellyn Ryland Ltd — Amateur / Professional — 3471

8 Copper resinate + Cuprous oxide + 2-Methylthio-4-tertiary-butylamino-6-cyclopropylamino-S-triazine

D Sigmaplane Ecol Antifouling — Sigma Coatings BV — Amateur / Professional — 4348

9 Copper resinate + Cuprous oxide + Zineb

Sigma Pilot Ecol Antifouling — Sigma Coatings Ltd — Amateur / Professional — 4933

10 Copper sulphate + 2,4,5,6-Tetrachloro isophthalonitrile

W Flexgard VI Waterbase Preservative — Flexabar Aquatech Corporation — Professional / Professional (Aquaculture) — 3319

11 Cuprous oxide

D (ATMC 129) Amercoat 279 — Ameron (UK) Ltd — Amateur / Professional — 3942

D Admiralty Antifouling (to TS10240) — Sigma Coatings BV — Amateur / Professional — 3490

Algicide Antifouling — Blakes Marine Paints — Amateur / Professional — 3219

AMC Sport Antifouling — Aquarius Marine Coatings Ltd — Amateur / Professional — 6395

Product Name	Marketing Company	Use	HSE No.

11 Cuprous oxide—continued

Product Name	Marketing Company	Use	HSE No.
W Amercoat 275	Ameron BV	Amateur Professional	3204
W Amercoat 277	Ameron BV	Amateur Professional	3205
W Amercoat 279	Ameron BV	Amateur Professional	3206
Anti-Fouling Paint 161P (Red and Chocolate to TS10240)	Witham Oil & Paint (Lowestoft) Ltd	Amateur Professional	3503
Antifouling Paint 161P (Red and Chocolate to TS10240)	International Coatings Ltd	Professional	3401
Antifouling Sargasso	Jotun-Henry Clark Ltd	Amateur Professional	6073
W Antifouling Sargasso Non-Tin	Jotun-Henry Clark Ltd	Amateur Professional	3418
Antifouling Seaguardian	Jotun-Henry Clark Ltd	Amateur Professional	3856
Antifouling Seaguardian (Black and Blue)	Jotun-Henry Clark Ltd	Professional	4273
Antifouling Seavictor 40	Jotun-Henry Clark Ltd	Amateur Professional	4957
Antifouling Super	Jotun-Henry Clark Ltd	Amateur Professional	5812
Antifouling Super Tropic	Jotun-Henry Clark Ltd	Amateur Professional	3413
Antifouling Supertropic	Jotun-Henry Clark Ltd	Amateur Professional	6470
Aquagard (Flexgard XI)	Marineware Ltd	Amateur Professional	6589
Aquasafe	GJOCO A/S	Professional Professional (Aquaculture)	5983
Aquasafe W	GJOCO A/S	Professional Professional (Aquaculture)	6353
Armachlor AF275	Johnstone's Paints Plc	Amateur Professional	5929
Armacote AF259	Johnstone's Paints Plc	Amateur Professional	5928
Armarine AF259	Johnstone's Paints Plc	Amateur Professional	5926
Armarine AF275	Johnstone's Paints Plc	Amateur Professional	5927
Awlgrip Awlstar Gold Label Antifouling	NOF Europe NV	Amateur Professional	5065
Blueline Copper SBA100	International Paint Ltd	Amateur Professional	5140

Product Name	Marketing Company	Use	HSE No.

11 Cuprous oxide—continued

Product Name	Marketing Company	Use	HSE No.
Boatguard	International Coatings Ltd	Amateur Professional Professional (Aquaculture)	3399
Bottomkote	International Coatings Ltd	Amateur Professional Professional (Aquaculture)	5903
Bradite CA21	Bradite Ltd	Amateur Professional	6147
Bradite MA55	Bradite Ltd	Amateur Professional	6178
Broads Freshwater	Blakes Marine Paints	Amateur Professional	3220
C-Clean 100	Camrex Holdings BV	Amateur Professional	5946
C-Clean 200	Camrex Holdings BV	Amateur Professional	5943
C-Worthy	Benfleet Marine Wholesale	Amateur Professional	5476
W Classica 3786/093 Red	Ernesto Stoppani SPA	Amateur Professional	4797
Cobra V	Valiant Marine	Amateur Professional	5194
Cooper's Copolymer Antifouling	Coopers Marine Paints	Amateur Professional	5609
Copperpaint	International Paint Ltd	Amateur Professional	4119
Cupron Plus T.F	New Guard Coatings Ltd	Amateur Professional	5661
D Cupron T.F.	Veneziani SPA	Amateur Professional	4935
D Devoe ABC 3 Black Antifouling	Ameron BV	Amateur Professional	3315
D Devoe ABC 3 Red Antifouling	Ameron BV	Amateur Professional	3317
Envoy TF100	W and J Leigh and Company	Amateur Professional	3951
Flagspeed Antifouling	C W Wastnage Ltd	Amateur Professional	5825
Flexgard VI-II Waterbase Preservative	Aquatess Ltd	Professional Professional (Aquaculture)	6543
Hard Racing	International Paint Ltd	Amateur Professional	3393
Hempel's Antifouling 761GB	Hempel Paints Ltd	Amateur Professional	3339

341

Product Name	Marketing Company	Use	HSE No.

11 Cuprous oxide—continued

Product Name	Marketing Company	Use	HSE No.
Hempel's Antifouling Nordic 7133	Hempel Paints Ltd	Amateur Professional	3325
Hempel's Antifouling Paint 161P (Red and Chocolate to TS10240)	Hempel Paints Ltd	Amateur Professional	3355
Hempel's Copper Bottom Paint 7116	Hempel Paints Ltd	Amateur Professional	4274
Hempel's Net Antifouling 715GB	Hempel Paints Ltd	Professional Professional (Aquaculture)	6342
Hempel's Tin Free Antifouling 744GB	Hempel Paints Ltd	Amateur Professional	3330
Hempel's Tin Free Antifouling 7660	Hempel Paints Ltd	Amateur Professional	3338
Inflatable Boat Antifouling	Polymarine Ltd	Amateur Professional	6647
Interclene Extra BAA100 Series	International Paint Ltd	Amateur Professional	3371
Interclene Premium BCA300 Series	International Coatings Ltd	Amateur Professional Professional (Aquaculture)	3372
Interclene Super BCA400 Series (BCA400 Red)	International Coatings Ltd	Amateur Professional	4084
Interclene Underwater Premium BCA468 Red	International Coatings Ltd	Amateur Professional	5059
International TBT Free Copolymer Antifouling BQA100 Series	International Coatings Ltd	Amateur Professional	3375
Marclear Full Strength EU45 Antifouling	Marclear Marine Products Ltd	Amateur Professional	5987
Marclear High Strength Antifouling	Marclear Marine Products Ltd	Amateur Professional	5264
Micron 400 Series	International Coatings Ltd	Amateur Professional	5728
Micron CSC 100 Series	International Coatings Ltd	Amateur Professional	5731
Netrex AF	Tulloch Enterprises	Professional Professional (Aquaculture)	5684
Noa-Noa Rame	Ernesto Stoppani SPA	Amateur Professional	4795
Nordrift Antifouling	Norland Distributors	Amateur Professional	5993
Norimp 2000 Black	Jotun-Henry Clark Ltd	Professional Professional (Aquaculture)	3404

Product Name	Marketing Company	Use	HSE No.

11 Cuprous oxide—continued

Product Name	Marketing Company	Use	HSE No.
Penguin Racing	Marine and Industrial Sealants	Amateur Professional	5673
Pilot Antifouling	Blakes Marine Paints	Amateur Professional	3226
Professional	Colorificio Attiva SRL	Amateur Professional	5981
D Puma Antifouling	Blakes Marine Paints	Amateur Professional	3227
Ravax AF	Camrex Chugoku Ltd	Amateur Professional	5319
Seatender 10	Camrex Chugoku Ltd	Amateur Professional	5321
Seatender 7	Camrex Chugoku Ltd	Amateur Professional	5320
Shearwater Racing Antifouling	Blakes Marine Paints	Amateur Professional	3228
Shiprite Sailing	Bradite Ltd	Amateur Professional	6302
Speedclean Antifouling	Mariner	Amateur Professional	5077
Standard Antifouling	Skipper (UK) Ltd	Amateur Professional	6194
Super Tropical Antifouling	Blakes Marine Paints	Amateur Professional	3229
Super Tropical Extra Antifouling	Blakes Marine Paints	Amateur Professional	3230
W Superspeed	D R Margetson	Amateur Professional	5191
Superspeed	Mariner	Amateur Professional	6210
Teamac Tropical Copper Antifouling (C/260/65)	Teal and Mackrill Ltd	Amateur Professional	3496
Tiger Cruising	Blakes Marine Paints	Amateur Professional	5099
D Titan Tin Free Antifouling	Blakes Marine Paints	Amateur Professional	3232
Trawler	International Coatings Ltd	Amateur Professional	3398
W Tropical Super Service Antifouling Paint	Devoe Coatings BV	Amateur Professional	3316
TS 10240 Antifouling ADA160 Series	International Coatings Ltd	Amateur Professional	3386
Unitas Antifouling Paint Chocolate	Witham Oil & Paint (Lowestoft) Ltd	Amateur Professional	3499
Viniline	Skipper (UK) Ltd	Amateur Professional	6193
Vinilstop 9926 Red.	Ernesto Stoppani SPA	Professional	4798

HSE ANTIFOULING PRODUCTS

Product Name	Marketing Company	Use	HSE No.

11 Cuprous oxide—continued

Vinyl Antifouling 2000	Akzo Coatings BV	Amateur Professional	5633

12 Cuprous oxide + 2,3,5,6-Tetrachloro-4-(methyl sulphonyl)pyridine

Grassline TF Anti-Fouling Type M396	W and J Leigh and Company	Amateur Professional	3462

13 Cuprous oxide + 2,4,5,6-Tetrachloro isophthalonitrile

Flexgard VI	Flexabar Aquatech Corporation	Professional Professional (Aquaculture)	6035
D Hempel's Antifouling Classic 7654C	Hempel Paints Ltd	Amateur Professional	5298
D Hempel's Antifouling Combic 7199C	Hempel Paints Ltd	Amateur Professional	5296
D Hempel's Antifouling Nautic 7190C	Hempel Paints Ltd	Amateur Professional	5294
TFA 10 LA	Camrex Chugoku Ltd	Amateur Professional	5361

14 Cuprous oxide + 2,4,5,6-Tetrachloro isophthalonitrile + 4,5-Dichloro-2-N-octyl-4-isothiazolin-3-one

C-Clean 400	Camrex Holdings BV	Amateur Professional	5947
Seatender 15	Camrex Chugoku Ltd	Amateur Professional	5348
W TFA 10 HG	Camrex Chugoku Ltd	Amateur Professional	5366
W TFA 10G	Camrex Chugoku Ltd	Amateur Professional	5347

15 Cuprous oxide + 2-(Thiocyanomethylthio) benzothiazole

A3 Antifouling	Nautix SA	Amateur Professional	4367
W Titan Tin Free	Blakes Marine Paints	Amateur Professional	3556

16 Cuprous oxide + 2-Methylthio-4-tertiary-butylamino-6-cyclopropylamino-S-triazine

Algicide	Blakes Marine Paints	Amateur Professional	5738
Aquaspeed	Blakes Marine Paints	Amateur Professional	4511
Aquaspeed Antifouling	Blakes Marine Paints	Amateur Professional	5817

16 Cuprous oxide + 2-Methylthio-4-tertiary-butylamino-6-cyclopropylamino-S-triazine—continued

Product Name	Marketing Company	Use	HSE No.
Boat Paint General Purpose Antifouling	Boat Paint Direct	Amateur	6198
Boat Paint Hard Antifouling	Boat Paint Direct	Amateur	6197
Boat Paint High Copper Anti-Fouling	Boat Paint Direct	Amateur	6199
Broads Black Antifouling	Blakes Marine Paints	Amateur Professional	5739
Broads Freshwater Red	Blakes Marine Paints	Amateur Professional	5736
Challenger Antifouling	Blakes Marine Paints	Amateur Professional	4099
W Even TF	Veneziani SPA	Amateur Professional	5166
Even Tin Free	Mark Dowland Marine Ltd	Amateur Professional	5841
Hard Racing Antifouling	Blakes Marine Paints	Amateur Professional	5704
Hempel's Antifouling Classic 7611 Red (Tin Free) 5000	Hempel Paints Ltd	Amateur Professional	5064
Hempel's Antifouling Classic 76540	Hempel Paints Ltd	Amateur Professional	5291
Hempel's Antifouling Combic 71990	Hempel Paints Ltd	Amateur Professional	5600
Hempel's Antifouling Combic 71992	Hempel Paints Ltd	Amateur Professional	5601
D Hempel's Antifouling Combic Tin Free 71990	Hempel Paints Ltd	Amateur Professional	5266
Hempel's Antifouling Forte Tin Free 7625	Hempel Paints Ltd	Amateur Professional	3351
Hempel's Antifouling Mille Dynamic	Hempel Paints Ltd	Amateur	4440
Hempel's Antifouling Nautic 71900	Hempel Paints Ltd	Amateur Professional	5290
Hempel's Antifouling Nautic 71902	Hempel Paints Ltd	Amateur Professional	5605
Hempel's Antifouling Nautic 8190c	Hempel Paints Ltd	Amateur Professional	6043
Hempel's Antifouling Olympic Hi-7661	Hempel Paints Ltd	Amateur Professional	4898
Hempel's Antifouling Olympic Tin Free 7154	Hempel Paints Ltd	Amateur Professional	3718
Hempel's Antifouling Tin Free 745GB	Hempel Paints Ltd	Amateur Professional	3331
Hempel's Antifouling Tin Free 750GB	Hempel Paints Ltd	Amateur Professional	3332
Hempel's Hard Racing 76480	Hempel Paints Ltd	Amateur Professional	5538

Product Name	Marketing Company	Use	HSE No.

16 Cuprous oxide + 2-Methylthio-4-tertiary-butylamino-6-cyclopropylamino-S-triazine—continued

Hempel's Mille Dynamic 71700	Hempel Paints Ltd	Amateur Professional	5574
Hempel's Spinnaker Antifouling 78200	Hempel Paints Ltd	Amateur Professional	6026
Hempels Antifouling Bravo Tin Free 7610	Hempel Paints Ltd	Amateur Professional	4482
Interspeed System 2 BRA143 Brown	International Paint Ltd	Amateur Professional	4301
Interviron BQA450 Series	International Paint Ltd	Amateur Professional	4657
W Long Life T.F.	Veneziani SPA	Amateur Professional	4936
Longlife Antifouling	Skipper (UK) Ltd	Amateur Professional	6195
Micron CSC 200 Series	International Coatings Ltd	Amateur Professional	5732
Patente Laxe	Teais SA	Amateur Professional	5497
Pilot	Blakes Marine Paints	Amateur Professional	5959
W Raffaello Plus Tin Free	Veneziani SPA	Amateur Professional	3642
Shiprite Speed	Bradite Ltd	Amateur Professional	6603
Titan FGA Antifouling	Blakes Marine Paints	Amateur Professional	5681
VC Offshore	International Coatings Ltd	Amateur Professional	4777
XM Anti-Fouling C2000 Cruising Self Eroding	X M Yachting	Amateur	6176
XM Anti-Fouling P4000 Hard	X M Yachting	Amateur	6175
XM Antifouling HS3000 High Performance Self Eroding	X M Yachting	Amateur	6124

17 Cuprous oxide + 2-Methylthio-4-tertiary-butylamino-6-cyclopropylamino-S-triazine + Copper resinate

D Sigmaplane Ecol Antifouling	Sigma Coatings BV	Amateur Professional	4348

18 Cuprous oxide + 4,5-Dichloro-2-N-octyl-4-isothiazolin-3-one

Antifouling Seavictor 50	Jotun-Henry Clark Ltd	Professional	4958
Hempel's Antifouling Classic 7654E	Hempel Paints Ltd	Amateur Professional	5286
Hempel's Antifouling Combic 7199E	Hempel Paints Ltd	Amateur Professional	5277

Product Name	Marketing Company	Use	HSE No.

18 Cuprous oxide + 4,5-Dichloro-2-N-octyl-4-isothiazolin-3-one—continued

Hempel's Antifouling Globic SP-ECO 81900	Hempel Paints Ltd	Professional	6531
Hempel's Antifouling Globic SP-ECO 81990	Hempel Paints Ltd	Professional	6532
Hempel's Antifouling Nautic 7190E	Hempel Paints Ltd	Amateur Professional	5283
Hempel's Antifouling Nautic Tin Free 7190E	Hempel Paints Ltd	Amateur Professional	5145
Hempel's Antifouling Tin Free 743GB	Hempel Paints Ltd	Amateur Professional	3329
Hempel's Antifouling Tin Free 751GB	Hempel Paints Ltd	Amateur Professional	3333
Hempel's Antifouling Tin Free 7662	Hempel Paints Ltd	Amateur Professional	3336
Hempel's Tin Free Antifouling 7626	Hempel Paints Ltd	Amateur Professional	3337
Hempels Antifouling Globic SP-ECO 81920	Hempel Paints Ltd	Professional	6877
Micron CSC 300 Series	International Coatings Ltd	Amateur Professional	5724
Titan FGA-E Antifouling	Blakes Marine Paints	Amateur Professional	5768

19 Cuprous oxide + Copper resinate

Double Shield Antifouling	Indestructible Paint Company Ltd	Amateur	6040
W Double Shield Antifouling	Llewellyn Ryland Ltd	Amateur Professional	3471

20 Cuprous oxide + Cuprous sulphide

D Admiralty Antifouling Black (to TS10239)	Sigma Coatings BV	Amateur Professional	3489
D Black Anti-Fouling Paint 317 (to TS10239)	Witham Oil & Paint (Lowestoft) Ltd	Amateur Professional	3502
Black Antifouling Paint 317 (to TS10239)	Hempel Paints Ltd	Amateur Professional	3370
Hempel's Antifouling 762GB	Hempel Paints Ltd	Amateur Professional	3340
D Unitas Antifouling Paint Black	Witham Oil & Paint (Lowestoft) Ltd	Amateur Professional	3501

21 Cuprous oxide + Cuprous thiocyanate + 2,3,5,6-Tetrachloro-4-(methyl sulphonyl)pyridine + 2-Methylthio-4-tertiary-butylamino-6-cyclopropylamino-S-triazine

Envoy TF 400	W and J Leigh and Company	Amateur Professional	4432

Product Name	Marketing Company	Use	HSE No.

21 Cuprous oxide + Cuprous thiocyanate + 2,3,5,6-Tetrachloro-4-(methyl sulphonyl)pyridine + 2-Methylthio-4-tertiary-butylamino-6-cyclopropylamino-S-triazine—continued

Envoy TF 500	W and J Leigh and Company	Amateur Professional	5599
Meridian MP40 Antifouling	Deangate Marine Products	Amateur Professional	5027
Micro-Tech	Marine (Chemi-Technics) Ltd	Amateur Professional	6316

22 Cuprous oxide + Cuprous thiocyanate + Dichlorophenyl dimethylurea + Zineb + 2,3,5,6-Tetrachloro-4-(methyl sulphonyl)pyridine

Even Extreme	New Guard Coatings Ltd	Amateur Professional	6625

23 Cuprous oxide + Dichlofluanid

Aqua-Net	Steen-Hansen Maling AS	Professional Professional (Aquaculture)	5845
Copper Net	Steen-Hansen Maling AS	Professional Professional (Aquaculture)	6034
Halcyon 5000 (Base)	Waterline	Amateur Professional	5396
Hempel's Antifouling Rennot 7150	Hempel Paints Ltd	Professional Professional (Aquaculture)	3364
Hempel's Antifouling Rennot 7177	Hempel Paints Ltd	Professional Professional (Aquaculture)	3365
Hempel's Antifouling Tin Free 742GB	Hempel Paints Ltd	Amateur Professional	3328
Interspeed Ultra	International Coatings Ltd	Amateur Professional	6660
Micron Extra	International Coatings Ltd	Amateur Professional	6663
Net-Guard	Steen-Hansen Maling AS	Professional (Aquaculture)	5657
Penguin Non-Stop	Marine and Industrial Sealants	Amateur Professional	5671
Seashield (Base)	Waterline	Amateur Professional	5438
Slipstream Antifouling	Witham Oil & Paint (Lowestoft) Ltd	Amateur Professional	3721
D Tiger Tin Free Antifouling	Blakes Marine Paints	Amateur Professional	3231

Product Name	Marketing Company	Use	HSE No.

24 Cuprous oxide + Dichlofluanid + 2-Methylthio-4-tertiary-butylamino-6-cyclopropylamino-S-triazine

Antifouling 1.2	Plastimo International	Amateur Professional	6159

25 Cuprous oxide + Dichlofluanid + Dichlorophenyl dimethylurea

Antifouling 1.1	Plastimo International	Amateur Professional	6158

26 Cuprous oxide + Dichlofluanid + Dichlorophenyl dimethylurea + 2-Methylthio-4-tertiary-butylamino-6-cyclopropylamino-S-triazine

Antifouling 1.3 Black and Red (400-03/ind.3)	Plastimo International	Amateur Professional	6350

27 Cuprous oxide + Dichlofluanid + Zinc naphthenate + Copper naphthenate

W Teamac Killa Copper Plus	Teal and Mackrill Ltd	Amateur Professional	4659

28 Cuprous oxide + Dichlorophen

Aquacleen	Mariner	Amateur Professional	5667

29 Cuprous oxide + Dichlorophenyl dimethylurea

A4 Antifouling	Nautix SA	Amateur Professional	4369
Aquarius Extra Strong	International Paint Ltd	Amateur Professional	4280
Blakes Hard Racing	Blakes Marine Paints	Amateur Professional	5764
Blueline Tropical SBA300	International Paint Ltd	Amateur Professional	5139
Boatgard Antifouling	International Coatings Ltd	Amateur Professional	6854
Cruiser Premium	International Coatings Ltd	Amateur Professional	5127
Drake Antifouling	Colorificio Attiva SRL	Amateur Professional	6671
Even Tin Free Light Grey	Mark Dowland Marine Ltd	Amateur Professional	5839
Giraglia	Colorificio Attiva SRL	Amateur Professional	5980
W Grafo Anti-Foul SW	Grafo Coatings Ltd	Amateur Professional	5380
Hempel's Antifouling 8199D	Hempel Paints Ltd	Amateur Professional	6177
Hempel's Antifouling 8199F	Hempel Paints Ltd	Amateur Professional	6179

HSE
ANTIFOULING PRODUCTS

349

Product Name	Marketing Company	Use	HSE No.

29 Cuprous oxide + Dichlorophenyl dimethylurea—continued

Product Name	Marketing Company	Use	HSE No.
Hempel's Antifouling Classic 7654B	Hempel Paints Ltd	Amateur Professional	5285
Hempel's Antifouling Combic 7199B	Hempel Paints Ltd	Amateur Professional	5274
Hempel's Antifouling Nautic 7190B	Hempel Paints Ltd	Amateur Professional	5273
Hempel's Antifouling Nautic 8190H	Hempel Paints Ltd	Amateur Professional	6042
Hempel's Bravo 7610A	Hempel Paints Ltd	Amateur Professional	5603
Hempel's Hard Racing 7648A	Hempel Paints Ltd	Amateur Professional	5535
Hempel's Mille Dynamic 7170A	Hempel Paints Ltd	Amateur Professional	5536
Hempel's Seatech Antifouling 7820A	Hempel Paints Ltd	Amateur Professional	6098
International Tin Free SPC BNA100 Series	International Paint Ltd	Amateur Professional	5186
Intersmooth Hisol Tin Free BGA620 Series	International Paint Ltd	Amateur Professional	4787
Intersmooth Tin Free BGA530 Series	International Paint Ltd	Amateur Professional	4611
Interspeed Extra BWA500 Red	International Coatings Ltd	Amateur Professional	4303
Interspeed Extra Strong	International Coatings Ltd	Amateur Professional	4819
Interspeed Super BWA900 Red	International Coatings Ltd	Amateur Professional	4884
Interspeed Super BWA909 Black	International Coatings Ltd	Amateur Professional	5058
Interspeed System 2 BRA142 Brown	International Paint Ltd	Amateur Professional	4302
Interswift Tin Free BQA400 Series	International Coatings Ltd	Amateur Professional	4842
Interswift Tin-Free SPC BTA540 Series	International Paint Ltd	Amateur Professional	5078
Interviron Super BQA400 Series	International Coatings Ltd	Amateur Professional	5409
Long Life Tin Free	Mark Dowland Marine Ltd	Amateur Professional	5840
Micron 500 Series	International Coatings Ltd	Amateur Professional	5729
Micron CSC	International Paint Ltd	Amateur Professional	4775
Micron CSC Extra	International Coatings Ltd	Amateur Professional	6263

29 Cuprous oxide + Dichlorophenyl dimethylurea—continued

Product Name	Marketing Company	Use	HSE No.
Micron Plus Antifouling	International Paint Ltd	Amateur Professional	5133
Noa-Noa Rame	Ernesto Stoppani SPA	Amateur Professional	4796
Prima	International Coatings Ltd	Amateur Professional	6683
Seajet 033	Camrex Chugoku Ltd	Amateur Professional	5331
Sigmaplane Ecol Antifouling	Sigma Coatings Ltd	Amateur Professional	5670
Slippy Bottom	Thincoat Technology International Ltd	Amateur Professional	6156
Soft Antifouling	International Coatings Ltd	Amateur Professional	6067
Teamac Antifouling "D" (C/258/65)	Teal and Mackrill Ltd	Amateur Professional	6418
Teamac Killa Copper Premium Antifouling (C/262/24)	Teal and Mackrill Ltd	Amateur Professional	6410
Titan FGA	Hempel Paints Ltd	Amateur Professional	5767

30 Cuprous oxide + Dichlorophenyl dimethylurea + 2,3,5,6-Tetrachloro-4-(methyl sulphonyl)pyridine

Product Name	Marketing Company	Use	HSE No.
Raffaello 3	New Guard Coatings Ltd	Amateur Professional	5826

31 Cuprous oxide + Dichlorophenyl dimethylurea + 2,4,5,6-Tetrachloro isophthalonitrile

Product Name	Marketing Company	Use	HSE No.
C-Clean 300	Camrex Holdings BV	Amateur Professional	5942
Seatender 12	Camrex Chugoku Ltd	Amateur Professional	5324
TFA 10	Camrex Chugoku Ltd	Amateur Professional	5346
W TFA 10 H	Camrex Chugoku Ltd	Amateur Professional	5358

32 Cuprous oxide + Dichlorophenyl dimethylurea + 2,4,5,6-Tetrachloro isophthalonitrile + 2-Methylthio-4-tertiary-butylamino-6-cyclopropylamino-S-triazine

Product Name	Marketing Company	Use	HSE No.
Sigmaplane Ecol HA 120 Antifouling	Sigma Coatings Ltd	Amateur Professional	5788

Product Name	Marketing Company	Use	HSE No.

33 Cuprous oxide + Dichlorophenyl dimethylurea + 2-Methylthio-4-tertiary-butylamino-6-cyclopropylamino-S-triazine

A3 Antifouling 072015	Nautix SA	Amateur Professional	5224
A4 Antifouling 072017	Nautix SA	Amateur Professional	5226
Le Marin	Nautix SA	Amateur Professional	5480

34 Cuprous oxide + Dichlorophenyl dimethylurea + 4,5-Dichloro-2-N-octyl-4-isothiazolin-3-one

TFA-10 LA Q	Camrex Chugoku Ltd	Professional	6581

35 Cuprous oxide + Dichlorophenyl dimethylurea + Tributyltin methacrylate + Tributyltin oxide

A11 Antifouling 072014	Nautix SA	Professional	5376
Interswift BKO000/700 Series	International Coatings Ltd	Professional	5640

36 Cuprous oxide + Maneb + Tributyltin methacrylate

D Nu Wave A/F Flat Bottom	Kansai Paint Europe Ltd	Professional	3447
D Nu Wave A/F Vertical Bottom	Kansai Paint Europe Ltd	Professional	3446
D Rabamarine A/F No 2500 HS	Kansai Paint Europe Ltd	Professional	3444
D Rabamarine A/F No 2500M HS	Kansai Paint Europe Ltd	Professional	3445

37 Cuprous oxide + Maneb + Tributyltin methacrylate + Tributyltin oxide

D AF Seaflo Mark 2	Camrex Chugoku Ltd	Professional	5944
D C-Clean 5000	Camrex Holdings BV	Professional	5955

38 Cuprous oxide + Thiram

D Grassline TF Anti-Fouling Type M395	W and J Leigh and Company	Amateur Professional	3461

39 Cuprous oxide + Tributyltin acrylate

ABC # 1 Antifouling	Ameron BV	Professional	3214
W Amercoat 697	Ameron BV	Professional	3513

40 Cuprous oxide + Tributyltin methacrylate

Devran MCP Antifouling Red	Devoe Coatings BV	Professional	3311
Devran MCP Antifouling Red Brown	Devoe Coatings BV	Professional	3310
W Takata LLL Antifouling	NOF Europe NV	Professional	4050
W Takata LLL Antifouling Hi-Solid	NOF Europe NV	Professional	4052
W Takata LLL Antifouling LS	NOF Europe NV	Professional	4051
W Takata LLL Antifouling LS Hi-Solid	NOF Europe NV	Professional	4053
W Takata LLL Antifouling No 2001	NOF Europe NV	Professional	4054

Product Name	Marketing Company	Use	HSE No.

41 Cuprous oxide + Tributyltin methacrylate + Tributyltin oxide

AF Seaflo Mark 2-1	Camrex Chugoku Ltd	Professional	5316
W AF Seaflo Z-100 HS-1	Camrex Chugoku Ltd	Professional	5318
Antifouling Seaconomy	Jotun-Henry Clark Ltd	Professional	5791
Antifouling Seaconomy	Jotun-Henry Clark Ltd	Professional	6261
W Antifouling Seaconomy 200	Jotun-Henry Clark Ltd	Professional	4436
W Antifouling Seaconomy 300	Jotun-Henry Clark Ltd	Professional	4272
Antifouling Seamate FB30	Jotun-Henry Clark Ltd	Professional	6506
W Antifouling Seamate HB 33	Jotun-Henry Clark Ltd	Professional	4270
Antifouling Seamate HB 99 Black	Jotun-Henry Clark Ltd	Professional	3410
Antifouling Seamate HB 99 Dark Red	Jotun-Henry Clark Ltd	Professional	3412
Antifouling Seamate HB 99 Light Red	Jotun-Henry Clark Ltd	Professional	3411
Antifouling Seamate HB22	Jotun-Henry Clark Ltd	Professional	4271
Antifouling Seamate HB22	Jotun-Henry Clark Ltd	Professional	6301
Antifouling Seamate HB33	Jotun-Henry Clark Ltd	Professional	5811
Antifouling Seamate HB66	Jotun-Henry Clark Ltd	Professional	5698
Antifouling Seamate HB66	Jotun-Henry Clark Ltd	Professional	6475
Antifouling Seamate SB33	Jotun-Henry Clark Ltd	Professional	6517
C-Clean 6000	Camrex Holdings BV	Professional	5945
Grassline ABL Antifouling Type M349	W and J Leigh and Company	Professional	5374
W Hempel's Antifouling Combic 76990	Hempel Paints Ltd	Professional	3348
Hempel's Antifouling Nautic HI 76900	Hempel Paints Ltd	Professional	3360
W Hempel's Antifouling Nautic HI 76910	Hempel Paints Ltd	Professional	3361
Hempel's Antifouling Nautic SP-ACE 79031	Hempel Paints Ltd	Professional	6452
Hempel's Antifouling Nautic SP-ACE 79051	Hempel Paints Ltd	Professional	6453
Intersmooth SPC BFA090/BFA190 Series	International Paint Ltd	Professional	3377

42 Cuprous oxide + Tributyltin methacrylate + Tributyltin oxide + 4,5-Dichloro-2-N-octyl-4-isothiazolin-3-one

Intersmooth Hisol BFO270/950/970 Series	Courtaulds Coatings (Holdings) Ltd	Professional	5641

43 Cuprous oxide + Tributyltin methacrylate + Tributyltin oxide + Zineb

Blueline SBA900 Series	International Paint Ltd	Professional	5138
W Hempel's Antifouling 7690D	Hempel Paints Ltd	Professional	6056
Hempel's Antifouling Economic SP-SEA 74030	Hempel Paints Ltd	Professional	5581
Hempel's Antifouling Nautic 7690B	Hempel Paints Ltd	Professional	5580

Product Name	Marketing Company	Use	HSE No.

43 Cuprous oxide + Tributyltin methacrylate + Tributyltin oxide + Zineb—continued

W Hempel's Antifouling Nautic 7691B	Hempel Paints Ltd	Professional	5586
Hempel's Antifouling Nautic SP-ACE 79030	Hempel Paints Ltd	Professional	6331
Hempel's Antifouling Nautic SP-ACE 79050	Hempel Paints Ltd	Professional	6325
Hempel's Antifouling Nautic SP-ACE 79070	Hempel Paints Ltd	Professional	6328
Hempel's Economic SP-SEA 74010	Hempel Paints Ltd	Professional	6330
Hempel's Economic SP-SEA 74040	Hempel Paints Ltd	Professional	6329
Intersmooth 110 Standard	International Coatings Ltd	Professional	5907
Intersmooth 130 Ultra	International Paint Ltd	Professional	5914
Intersmooth 210 Standard	International Coatings Ltd	Professional	5915
Intersmooth 220 Premium	International Coatings Ltd	Professional	5911
Intersmooth 230 Ultra	International Paint Ltd	Professional	5912
Intersmooth 320 Premium	International Paint Ltd	Professional	5910
Intersmooth 330 Ultra	International Coatings Ltd	Professional	5924
Intersmooth Hisol 2000 BFA270 Series	International Coatings Ltd	Professional	3844
Intersmooth Hisol 9000 BFA970 Series	International Coatings Ltd	Professional	5461
Intersmooth Hisol BFA250/BFA900 series	International Coatings Ltd	Professional	3848
Interswift BKA000/700 Series	International Coatings Ltd	Professional	3384
Seaflo 10	Camrex Chugoku Ltd	Professional	6283

44 Cuprous oxide + Tributyltin methacrylate + Zineb

Antifouling Seamate HB22	Jotun-Henry Clark Ltd	Professional	5362
Antifouling Seamate HB33	Jotun-Henry Clark Ltd	Professional	5363
Sigmaplane HA Antifouling	Sigma Coatings BV	Professional	4345
W Sigmaplane HB	Sigma Coatings BV	Professional	3487
Sigmaplane HB Antifouling	Sigma Coatings Ltd	Professional	5721
Sigmaplane TA Antifouling	Sigma Coatings BV	Professional	4346

45 Cuprous oxide + Zinc naphthenate

W Teamac Killa Copper	Teal and Mackrill Ltd	Amateur Professional	3494
W Teamac Super Tropical	Teal and Mackrill Ltd	Amateur Professional	3497

46 Cuprous oxide + Zinc pyrithione

Blakes Antifouling 87910	Blakes Marine Paints	Amateur Professional	6450
Intersmooth 360 Ecoloflex	International Coatings Ltd	Amateur Professional	6057

Product Name	Marketing Company	Use	HSE No.

46 Cuprous oxide + Zinc pyrithione—continued

Product Name	Marketing Company	Use	HSE No.
Intersmooth 365 Ecoloflex	International Coatings Ltd	Amateur Professional	6557
Intersmooth 465 Ecoloflex	International Coatings Ltd	Amateur Professional	6556
W Micron 600 Series	International Paint Ltd	Amateur Professional	5733
Micron Optima	International Coatings Ltd	Amateur Professional	5941
VC Offshore Extra 100 Series	International Coatings Ltd	Amateur Professional	5730
VC Offshore SP Antifouling	International Coatings Ltd	Amateur Professional	3507

47 Cuprous oxide + Zineb

Product Name	Marketing Company	Use	HSE No.
Blueline SPC Tin Free SBA700 Series	International Coatings Ltd	Amateur Professional	5214
Equatorial	International Paint Ltd	Amateur Professional	4121
Hempel's Antifouling Combic 7199f	Hempel Paints Ltd	Amateur Professional	5744
Inter 100	International Coatings Ltd	Amateur Professional	4120
Interclene 245	International Coatings Ltd	Amateur Professional	6233
Interspeed 340	International Coatings Ltd	Amateur Professional	6089
Interspeed System 2 BRA140/ BRA240 Series	International Coatings Ltd	Amateur Professional	3847
Interviron BQA200 Series	International Coatings Ltd	Amateur Professional	3846

48 Cuprous oxide + Zineb + Copper resinate

Product Name	Marketing Company	Use	HSE No.
Sigma Pilot Ecol Antifouling	Sigma Coatings Ltd	Amateur Professional	4933

49 Cuprous oxide + Ziram

Product Name	Marketing Company	Use	HSE No.
D Awlgrip Awlstar Antifouling ABC 3 Black	Devoe Coatings BV	Amateur Professional	6036
D Awlgrip Awlstar Antifouling ABC 3 Blue	Devoe Coatings BV	Amateur Professional	6044
D Awlgrip Awlstar Antifouling ABV 3 Red	Devoe Coatings BV	Amateur Professional	6038
D Awlstar Gold Label Anti-fouling	Grow Group Incorporated	Amateur Professional	3674
D Cheetah Antifouling	Blakes Marine Paints	Amateur Professional	3223

Product Name	Marketing Company	Use	HSE No.

49 Cuprous oxide + Ziram—continued

D Devoe ABV 3 Antifouling Blue	Devoe Coatings BV	Amateur Professional	6037
D Hempel's Antifouling Classic 7654A	Hempel Paints Ltd	Amateur Professional	5297
D Hempel's Antifouling Combic 7199A	Hempel Paints Ltd	Amateur Professional	5295
D Hempel's Antifouling Nautic 7190A	Hempel Paints Ltd	Amateur Professional	5293

50 Cuprous sulphide + Cuprous oxide

D Admiralty Antifouling Black (to TS10239)	Sigma Coatings BV	Amateur Professional	3489
D Black Anti-Fouling Paint 317 (to TS10239)	Witham Oil & Paint (Lowestoft) Ltd	Amateur Professional	3502
Black Antifouling Paint 317 (to TS10239)	Hempel Paints Ltd	Amateur Professional	3370
Hempel's Antifouling 762GB	Hempel Paints Ltd	Amateur Professional	3340
D Unitas Antifouling Paint Black	Witham Oil & Paint (Lowestoft) Ltd	Amateur Professional	3501

51 Cuprous thiocyanate

Antifouling Broken White DL-2253	International Paint Ltd	Amateur Professional	3400
Boot Top	International Coatings Ltd	Amateur Professional	3388
Hempel's Antifouling 763GB	Hempel Paints Ltd	Amateur Professional	3537
Interspeed 2001	International Coatings Ltd	Amateur Professional	5727
Unitas Antifouling Paint White	Witham Oil & Paint (Lowestoft) Ltd	Amateur Professional	3500

52 Cuprous thiocyanate + 2,3,5,6-Tetrachloro-4-(methyl sulphonyl)pyridine + 2-Methylthio-4-tertiary-butylamino-6-cyclopropylamino-S-triazine + Cuprous oxide

Envoy TF 400	W and J Leigh and Company	Amateur Professional	4432
Envoy TF 500	W and J Leigh and Company	Amateur Professional	5599
Meridian MP40 Antifouling	Deangate Marine Products	Amateur Professional	5027
Micro-Tech	Marine (Chemi-Technics) Ltd	Amateur Professional	6316

53 Cuprous thiocyanate + 2,4,5,6-Tetrachloro isophthalonitrile

D Gummipaint A/F	Veneziani SPA	Amateur Professional	4934

54 Cuprous thiocyanate + 2-Methylthio-4-tertiary-butylamino-6-cyclopropylamino-S-triazine

AL-27	International Paint Ltd	Amateur Professional	3613
Aquaspeed White	Blakes Marine Paints	Amateur Professional	4513
Boot Top Plus (Gull White)	International Paint Ltd	Amateur Professional	3389
D Broads Sweetwater Antifouling	Blakes Marine Paints	Amateur Professional	3222
Cruiser Superior 100 Series	International Coatings Ltd	Amateur Professional	5723
W Even TF Light Grey	Veneziani SPA	Amateur Professional	5167
Hard Racing Antifouling White	Blakes Marine Paints	Amateur Professional	5705
Hempel's Hard Racing 76380	Hempel Paints Ltd	Amateur Professional	5540
Hempel's Mille Dynamic 71600	Hempel Paints Ltd	Amateur Professional	5576
Hempel's Tin Free Hard Racing 7648	Hempel Paints Ltd	Amateur Professional	3367
Interspeed 2000	International Coatings Ltd	Amateur Professional	4148
Marclear Powerboat Antifouling	Marclear Marine Products Ltd	Amateur Professional	6586
MPX	International Coatings Ltd	Amateur Professional	4818
Propeller T.F.	New Guard Coatings Ltd	Amateur Professional	5660
Quicksilver Antifouling	International Coatings Ltd	Amateur Professional	5938
Raffaello Alloy	Mark Dowland Marine Ltd	Amateur Professional	5492
Tiger Cruising White	Blakes Marine Paints	Amateur Professional	5712
Tiger White	Blakes Marine Paints	Amateur Professional	4512
Tigerline Antifouling	Blakes Marine Paints	Amateur Professional	3842
Tigerline Boottop	Blakes Marine Paints	Amateur Professional	5763
Titan FGA Antifouling White	Blakes Marine Paints	Amateur Professional	5680

Product Name	Marketing Company	Use	HSE No.

54 Cuprous thiocyanate + 2-Methylthio-4-tertiary-butylamino-6-cyclopropylamino-S-triazine—continued

VC Offshore	International Coatings Ltd	Amateur Professional	4779
VC Prop-O-Drev	International Coatings Ltd	Amateur Professional	4217

55 Cuprous thiocyanate + 4,5-Dichloro-2-N-octyl-4-isothiazolin-3-one

Interspeed 2003	International Coatings Ltd	Professional	5726

56 Cuprous thiocyanate + Dichlofluanid

Interspeed 2000 White	International Coatings Ltd	Amateur Professional	6631
Penguin Non-Stop White	Marine and Industrial Sealants	Amateur Professional	5672
Penguin Racing White	Marine and Industrial Sealants	Amateur Professional	5674
Selfpolishing Antifouling 2000	Akzo Nobel Coatings BV	Amateur Professional	6521
Trilux	International Coatings Ltd	Amateur Professional	6661

57 Cuprous thiocyanate + Dichlofluanid + 2-Methylthio-4-tertiary-butylamino-6-cyclopropylamino-S-triazine

Antifouling 1.2 (W)	Plastimo International	Amateur Professional	6160
Hempel's Antifouling Mille Dynamic 717 GB	Hempel Paints Ltd	Amateur Professional	4437

58 Cuprous thiocyanate + Dichlofluanid + Dichlorophenyl dimethylurea + 2-Methylthio-4-tertiary-butylamino-6-cyclopropylamino-S-triazine

Antifouling 1.3	Plastimo International	Amateur Professional	6161
Antifouling 1.3 White (327-04)	Plastimo International	Amateur Professional	6349

59 Cuprous thiocyanate + Dichlofluanid + Zinc pyrithione + 2-Methylthio-4-tertiary-butylamino-6-cyclopropylamino-S-triazine

Antifouling 1.3 (W)	Plastimo International	Amateur Professional	6162

60 Cuprous thiocyanate + Dichlorophenyl dimethylurea

Antialga	New Guard Coatings Ltd	Amateur Professional	5662
Aqua 12	International Coatings Ltd	Amateur Professional	4802

60 Cuprous thiocyanate + Dichlorophenyl dimethylurea—continued

Aquarius AL	International Paint Ltd	Amateur Professional	4295
Cruiser Superior	International Paint Ltd	Amateur Professional	4776
Hempel's Hard Racing 7638A	Hempel Paints Ltd	Amateur Professional	5539
Hempel's Mille ALU 71602	Hempel Paints Ltd	Amateur Professional	5578
Hempel's Mille Dynamic 7160A	Hempel Paints Ltd	Amateur Professional	5575
Seatech Antifouling White	Blakes Marine Paints	Amateur Professional	6140
Tiger Cruising-A Antifouling	Blakes Marine Paints	Amateur Professional	5765

61 Cuprous thiocyanate + Dichlorophenyl dimethylurea + 2,3,5,6-Tetrachloro-4-(methyl sulphonyl)pyridine

Raffaello Racing	New Guard Coatings Ltd	Amateur Professional	6451

62 Cuprous thiocyanate + Dichlorophenyl dimethylurea + 2-Methylthio-4-tertiary-butylamino-6-cyclopropylamino-S-triazine

A3 Antifouling 072016	Nautix SA	Amateur Professional	5225
A3 Teflon Antifouling 062015	Nautix SA	Amateur Professional	5481
A4 Antifouling 072018	Nautix SA	Amateur Professional	5227
A4 Teflon Antifouling 062017	Nautix SA	Amateur Professional	5482

63 Cuprous thiocyanate + Dichlorophenyl dimethylurea + 4,5-Dichloro-2-N-octyl-4-isothiazolin-3-one

Sprint	New Guard Coatings Ltd	Professional	6449

64 Cuprous thiocyanate + Dichlorophenyl dimethylurea + Zineb + 2,3,5,6-Tetrachloro-4-(methyl sulphonyl)pyridine + Cuprous oxide

Even Extreme	New Guard Coatings Ltd	Amateur Professional	6625

65 Cuprous thiocyanate + Tributyltin methacrylate + Tributyltin oxide

Antifouling HB 66 Ocean Green	Jotun-Henry Clark Ltd	Professional	3408
Antifouling Seamate HB 66 STD 4037	Jotun-Henry Clark Ltd	Professional	4269
W Antifouling Seamate HB Green	Jotun-Henry Clark Ltd	Professional	3430

65 Cuprous thiocyanate + Tributyltin methacrylate + Tributyltin oxide—continued

Product Name	Marketing Company	Use	HSE No.
Antifouling Seamate HB22 Roundel Blue	Jotun-Henry Clark Ltd	Professional	4872
Antifouling Seamate HB66 Black	Jotun-Henry Clark Ltd	Professional	4871
Intersmooth Hisol SPC Antifouling BFA949 Red	International Coatings Ltd	Professional	4949
Intersmooth SPC BFA040/BFA050 Series	International Coatings Ltd	Professional	3376

66 Cuprous thiocyanate + Tributyltin methacrylate + Tributyltin oxide + Zineb

Product Name	Marketing Company	Use	HSE No.
Cruiser Copolymer	International Paint Ltd	Professional	3514
Intersmooth Hisol BFA948 Orange	International Coatings Ltd	Professional	4281
Micron 25 Plus	International Coatings Ltd	Professional	3402
Superyacht 800 Antifouling	International Coatings Ltd	Professional	4778

67 Cuprous thiocyanate + Zinc pyrithione

Product Name	Marketing Company	Use	HSE No.
Hempel's Antifouling Combic 7199C	Hempel Paints Ltd	Amateur Professional	5742
Interspeed 2002	International Coatings Ltd	Amateur Professional	5725
Titan FGA-C Antifouling	Blakes Marine Paints	Amateur Professional	5766

68 Dichlofluanid

Product Name	Marketing Company	Use	HSE No.
Bayer AFC	Bayer Plc	Amateur Professional	5163

69 Dichlofluanid + 2-Methylthio-4-tertiary-butylamino-6-cyclopropylamino-S-triazine + Cuprous oxide

Product Name	Marketing Company	Use	HSE No.
Antifouling 1.2	Plastimo International	Amateur Professional	6159

70 Dichlofluanid + 2-Methylthio-4-tertiary-butylamino-6-cyclopropylamino-S-triazine + Cuprous thiocyanate

Product Name	Marketing Company	Use	HSE No.
Antifouling 1.2 (W)	Plastimo International	Amateur Professional	6160
Hempel's Antifouling Mille Dynamic 717 GB	Hempel Paints Ltd	Amateur Professional	4437

71 Dichlofluanid + Cuprous oxide

Product Name	Marketing Company	Use	HSE No.
Aqua-Net	Steen-Hansen Maling AS	Professional Professional (Aquaculture)	5845
Copper Net	Steen-Hansen Maling AS	Professional Professional (Aquaculture)	6034

Product Name	Marketing Company	Use	HSE No.

71 Dichlofluanid + Cuprous oxide—continued

Product Name	Marketing Company	Use	HSE No.
Halcyon 5000 (Base)	Waterline	Amateur Professional	5396
Hempel's Antifouling Rennot 7150	Hempel Paints Ltd	Professional Professional (Aquaculture)	3364
Hempel's Antifouling Rennot 7177	Hempel Paints Ltd	Professional Professional (Aquaculture)	3365
Hempel's Antifouling Tin Free 742GB	Hempel Paints Ltd	Amateur Professional	3328
Interspeed Ultra	International Coatings Ltd	Amateur Professional	6660
Micron Extra	International Coatings Ltd	Amateur Professional	6663
Net-Guard	Steen-Hansen Maling AS	Professional (Aquaculture)	5657
Penguin Non-Stop	Marine and Industrial Sealants	Amateur Professional	5671
Seashield (Base)	Waterline	Amateur Professional	5438
Slipstream Antifouling	Witham Oil & Paint (Lowestoft) Ltd	Amateur Professional	3721
D Tiger Tin Free Antifouling	Blakes Marine Paints	Amateur Professional	3231

72 Dichlofluanid + Cuprous thiocyanate

Product Name	Marketing Company	Use	HSE No.
Interspeed 2000 White	International Coatings Ltd	Amateur Professional	6631
Penguin Non-Stop White	Marine and Industrial Sealants	Amateur Professional	5672
Penguin Racing White	Marine and Industrial Sealants	Amateur Professional	5674
Selfpolishing Antifouling 2000	Akzo Nobel Coatings BV	Amateur Professional	6521
Trilux	International Coatings Ltd	Amateur Professional	6661

73 Dichlofluanid + Dichlorophenyl dimethylurea + 2-Methylthio-4-Tertiary-butylamino-6-cyclopropylamino-S-triazine + Cuprous oxide

Product Name	Marketing Company	Use	HSE No.
Antifouling 1.3 Black and Red (400-03/Ind.3)	Plastimo International	Amateur Professional	6350

74 Dichlofluanid + Dichlorophenyl dimethylurea + 2-Methylthio-4-tertiary-butylamino-6-cyclopropylamino-S-triazine + Cuprous thiocyanate

Product Name	Marketing Company	Use	HSE No.
Antifouling 1.3	Plastimo International	Amateur Professional	6161

HSE

Product Name	Marketing Company	Use	HSE No.

74 Dichlofluanid + Dichlorophenyl dimethylurea + 2-Methylthio-4-tertiary-butylamino-6-cyclopropylamino-S-triazine + Cuprous thiocyanate—continued

Antifouling 1.3 white (327-04)	Plastimo International	Amateur Professional	6349

75 Dichlofluanid + Dichlorophenyl dimethylurea + Cuprous oxide

Antifouling 1.1	Plastimo International	Amateur Professional	6158

76 Dichlofluanid + Dichlorophenyl dimethylurea + Zinc pyrithione

Antifouling 1.4	Plastimo International	Amateur Professional	6163

77 Dichlofluanid + Zinc naphthenate + Copper naphthenate + Cuprous oxide

W Teamac Killa Copper Plus	Teal and Mackrill Ltd	Amateur Professional	4659

78 Dichlofluanid + Zinc pyrithione + 2-Methylthio-4-tertiary-butylamino-6-cyclopropylamino-S-triazine + Cuprous thiocyanate

Antifouling 1.3 (W)	Plastimo International	Amateur Professional	6162

79 4,5-Dichloro-2-N-octyl-4-isothiazolin-3-one + Cuprous oxide

Antifouling Seavictor 50	Jotun-Henry Clark Ltd	Professional	4958
Hempel's Antifouling Classic 7654E	Hempel Paints Ltd	Amateur Professional	5286
Hempel's Antifouling Combic 7199E	Hempel Paints Ltd	Amateur Professional	5277
Hempel's Antifouling Globic SP-ECO 81900	Hempel Paints Ltd	Professional	6531
Hempel's Antifouling Globic SP-ECO 81990	Hempel Paints Ltd	Professional	6532
Hempel's Antifouling Nautic 7190E	Hempel Paints Ltd	Amateur Professional	5283
Hempel's Antifouling Nautic Tin Free 7190E	Hempel Paints Ltd	Amateur Professional	5145
Hempel's Antifouling Tin Free 743GB	Hempel Paints Ltd	Amateur Professional	3329
Hempel's Antifouling Tin Free 751GB	Hempel Paints Ltd	Amateur Professional	3333
Hempel's Antifouling Tin Free 7662	Hempel Paints Ltd	Amateur Professional	3336
Hempel's Tin Free Antifouling 7626	Hempel Paints Ltd	Amateur Professional	3337
Hempels Antifouling Globic SP-ECO 81920	Hempel Paints Ltd	Professional	6877
Micron CSC 300 Series	International Coatings Ltd	Professional	5724

79 4,5-Dichloro-2-N-octyl-4-isothiazolin-3-one + Cuprous oxide—continued

Titan FGA-E Antifouling	Blakes Marine Paints	Amateur Professional	5768

80 4,5-Dichloro-2-N-octyl-4-isothiazolin-3-one + Cuprous oxide + 2,4,5,6-Tetrachloro isophthalonitrile

C-Clean 400	Camrex Holdings BV	Amateur Professional	5947
Seatender 15	Camrex Chugoku Ltd	Professional	5348
W TFA 10 HG	Camrex Chugoku Ltd	Amateur Professional	5366
W TFA 10G	Camrex Chugoku Ltd	Amateur Professional	5347

81 4,5-Dichloro-2-N-octyl-4-isothiazolin-3-one + Cuprous oxide + Dichlorophenyl dimethylurea

TFA-10 LA Q	Camrex Chugoku Ltd	Professional	6581

82 4,5-Dichloro-2-N-octyl-4-isothiazolin-3-one + Cuprous oxide + Tributyltin methacrylate + Tributyltin oxide

Intersmooth Hisol BFO270/950/970 Series	Courtaulds Coatings (Holdings) Ltd	Professional	5641

83 4,5-Dichloro-2-N-octyl-4-isothiazolin-3-one + Cuprous thiocyanate

Interspeed 2003	International Coatings Ltd	Professional	5726

84 4,5-Dichloro-2-N-octyl-4-isothiazolin-3-one + Cuprous thiocyanate + Dichlorophenyl dimethylurea

Sprint	New Guard Coatings Ltd	Professional	6449

85 4,5-Dichloro-2-N-octyl-4-isothiazolin-3-one + Tributyltin methacrylate + Tributyltin oxide

Antifouling Alusea Turbo	Jotun-Henry Clark Ltd	Professional	6464

86 4,5-Dichloro-2-N-octyl-4-isothiazolin-3-one + Zinc pyrithione

Hempel's Antifouling Nautic 7180E	Hempel Paints Ltd	Amateur Professional	5771
Lynx-E	Blakes Marine Paints	Amateur Professional	5793

87 Dichlorophen + Cuprous oxide

Aquacleen	Mariner Ltd	Amateur Professional	5667

88 Dichlorophenyl dimethylurea + 2,3,5,6-Tetrachloro-4-(methyl sulphonyl)pyridine + Cuprous oxide

Raffaello 3	New Guard Coatings Ltd	Amateur Professional	5826

89 Dichlorophenyl dimethylurea + 2,3,5,6-Tetrachloro-4-(methyl sulphonyl)pyridine + Cuprous thiocyanate

Raffaello Racing	New Guard Coatings Ltd	Amateur Professional	6451

90 Dichlorophenyl dimethylurea + 2,4,5,6-Tetrachloro isophthalonitrile + 2-Methylthio-4-tertiary-butylamino-6-cyclopropylamino-S-triazine + Cuprous oxide

Sigmaplane Ecol HA 120 Antifouling	Sigma Coatings Ltd	Amateur Professional	5788

91 Dichlorophenyl dimethylurea + 2,4,5,6-Tetrachloro isophthalonitrile + Cuprous oxide

C-Clean 300	Camrex Holdings BV	Amateur Professional	5942
Seatender 12	Camrex Chugoku Ltd	Amateur Professional	5324
TFA 10	Camrex Chugoku Ltd	Amateur Professional	5346
W TFA 10 H	Camrex Chugoku Ltd	Amateur Professional	5358

92 Dichlorophenyl dimethylurea + 2-Methylthio-4-tertiary-butylamino-6-cyclopropylamino-S-triazine + Cuprous oxide

A3 Antifouling 072015	Nautix SA	Amateur Professional	5224
A4 Antifouling 072017	Nautix SA	Amateur Professional	5226
Le Marin	Nautix SA	Amateur Professional	5480

93 Dichlorophenyl dimethylurea + 2-Methylthio-4-tertiary-butylamino-6-cyclopropylamino-S-triazine + Cuprous oxide + Dichlofluanid

Antifouling 1.3 Black and Red (400-03/Ind.3)	Plastimo International	Amateur Professional	6350

94 Dichlorophenyl dimethylurea + 2-Methylthio-4-tertiary-butylamino-6-cyclopropylamino-S-triazine + Cuprous thiocyanate

A3 Antifouling 072016	Nautix SA	Amateur Professional	5225
A3 Teflon Antifouling 062015	Nautix SA	Amateur Professional	5481

Product Name	Marketing Company	Use	HSE No.

94 Dichlorophenyl dimethylurea + 2-Methylthio-4-tertiary-butylamino-6-cyclopropylamino-S-triazine + Cuprous thiocyanate—continued

A4 Antifouling 072018	Nautix SA	Amateur Professional	5227
A4 Teflon Antifouling 062017	Nautix SA	Amateur Professional	5482

95 Dichlorophenyl dimethylurea + 2-Methylthio-4-tertiary-butylamino-6-cyclopropylamino-S-triazine + Cuprous thiocyanate + Dichlofluanid

Antifouling 1.3	Plastimo International	Amateur Professional	6161
Antifouling 1.3 White (327-04)	Plastimo International	Amateur Professional	6349

96 Dichlorophenyl dimethylurea + 4,5-Dichloro-2-N-octyl-4-isothiazolin-3-one + Cuprous oxide

TFA-10 LA Q	Camrex Chugoku Ltd	Professional	6581

97 Dichlorophenyl dimethylurea + 4,5-Dichloro-2-N-octyl-4-isothiazolin-3-one + Cuprous thiocyanate

Sprint	New Guard Coatings Ltd	Professional	6449

98 Dichlorophenyl dimethylurea + Copper

Coppercoat	Aquarius Marine Coatings Ltd	Amateur Professional	6428

99 Dichlorophenyl dimethylurea + Cuprous oxide

A4 Antifouling	Nautix SA	Amateur Professional	4369
Aquarius Extra Strong	International Paint Ltd	Amateur Professional	4280
Blakes Hard Racing	Blakes Marine Paints	Amateur Professional	5764
Blueline Tropical SBA300	International Paint Ltd	Amateur Professional	5139
Boatgard Antifouling	International Coatings Ltd	Amateur Professional	6854
Cruiser Premium	International Coatings Ltd	Amateur Professional	5127
Drake Antifouling	Colorificio Attiva SRL	Amateur Professional	6671
Even Tin Free Light Grey	Mark Dowland Marine Ltd	Amateur Professional	5839
Giraglia	Colorificio Attiva SRL	Amateur Professional	5980
W Grafo Anti-Foul SW	Grafo Coatings Ltd	Amateur Professional	5380

Product Name	Marketing Company	Use	HSE No.

99 Dichlorophenyl dimethylurea + Cuprous oxide—continued

Product Name	Marketing Company	Use	HSE No.
Hempel's Antifouling 8199D	Hempel Paints Ltd	Amateur Professional	6177
Hempel's Antifouling 8199F	Hempel Paints Ltd	Amateur Professional	6179
Hempel's Antifouling Classic 7654B	Hempel Paints Ltd	Amateur Professional	5285
Hempel's Antifouling Combic 7199B	Hempel Paints Ltd	Amateur Professional	5274
Hempel's Antifouling Nautic 7190B	Hempel Paints Ltd	Amateur Professional	5273
Hempel's Antifouling Nautic 8190H	Hempel Paints Ltd	Amateur Professional	6042
Hempel's Bravo 7610A	Hempel Paints Ltd	Amateur Professional	5603
Hempel's Hard Racing 7648A	Hempel Paints Ltd	Amateur Professional	5535
Hempel's Mille Dynamic 7170A	Hempel Paints Ltd	Amateur Professional	5536
Hempel's Seatech Antifouling 7820A	Hempel Paints Ltd	Amateur Professional	6098
International Tin Free SPC BNA100 Series	International Paint Ltd	Amateur Professional	5186
Intersmooth Hisol Tin Free BGA620 Series	International Paint Ltd	Amateur Professional	4787
Intersmooth Tin Free BGA530 Series	International Paint Ltd	Amateur Professional	4611
Interspeed Extra BWA500 Red	International Coatings Ltd	Amateur Professional	4303
Interspeed Extra Strong	International Coatings Ltd	Amateur Professional	4819
Interspeed Super BWA900 Red	International Coatings Ltd	Amateur Professional	4884
Interspeed Super BWA909 Black	International Coatings Ltd	Amateur Professional	5058
Interspeed System 2 BRA142 Brown	International Paint Ltd	Amateur Professional	4302
Interswift Tin Free BQA400 Series	International Coatings Ltd	Amateur Professional	4842
Interswift Tin-Free SPC BTA540 Series	International Paint Ltd	Amateur Professional	5078
Interviron Super BQA400 Series	International Coatings Ltd	Amateur Professional	5409
Long Life Tin Free	Mark Dowland Marine Ltd	Amateur Professional	5840
Micron 500 Series	International Coatings Ltd	Amateur Professional	5729

Product Name	Marketing Company	Use	HSE No.

99 Dichlorophenyl dimethylurea + Cuprous oxide—continued

Micron CSC	International Paint Ltd	Amateur Professional	4775
Micron CSC Extra	International Coatings Ltd	Amateur Professional	6263
Micron Plus Antifouling	International Paint Ltd	Amateur Professional	5133
Noa-Noa Rame	Ernesto Stoppani SPA	Amateur Professional	4796
Prima	International Coatings Ltd	Amateur Professional	6683
Seajet 033	Camrex Chugoku Ltd	Amateur Professional	5331
Sigmaplane Ecol Antifouling	Sigma Coatings Ltd	Amateur Professional	5670
Slippy Bottom	Thincoat Technology International Ltd	Amateur Professional	6156
Soft Antifouling	International Coatings Ltd	Amateur Professional	6067
Teamac Antifouling "D" (C/258/65)	Teal and Mackrill Ltd	Amateur Professional	6418
Teamac Killa Copper Premium Antifouling (C/262/24)	Teal and Mackrill Ltd	Amateur Professional	6410
Titan FGA	Hempel Paints Ltd	Amateur Professional	5767

100 Dichlorophenyl dimethylurea + Cuprous oxide + Dichlofluanid

Antifouling 1.1	Plastimo International	Amateur Professional	6158

101 Dichlorophenyl dimethylurea + Cuprous thiocyanate

Antialga	New Guard Coatings Ltd	Amateur Professional	5662
Aqua 12	International Coatings Ltd	Amateur Professional	4802
Aquarius AL	International Paint Ltd	Amateur Professional	4295
Cruiser Superior	International Paint Ltd	Amateur Professional	4776
Hempel's Hard Racing 7638A	Hempel Paints Ltd	Amateur Professional	5539
Hempel's Mille ALU 71602	Hempel Paints Ltd	Amateur Professional	5578
Hempel's Mille Dynamic 7160A	Hempel Paints Ltd	Amateur Professional	5575
Seatech Antifouling White	Blakes Marine Paints	Amateur Professional	6140

Product Name	Marketing Company	Use	HSE No.

101 Dichlorophenyl dimethylurea + Cuprous thiocyanate—continued

Tiger Cruising-A Antifouling	Blakes Marine Paints	Amateur Professional	5765

102 Dichlorophenyl dimethylurea + Tributyltin methacrylate + Tributyltin oxide + Cuprous oxide

A11 antifouling 072014	Nautix SA	Professional	5376
Interswift BKO000/700 Series	International Coatings Ltd	Professional	5640

103 Dichlorophenyl dimethylurea + Zinc pyrithione

Hempel's Antifouling Nautic 7180B	Hempel Paints Ltd	Amateur Professional	5770
Lynx-A	Blakes Marine Paints	Amateur Professional	5794

104 Dichlorophenyl dimethylurea + Zinc pyrithione + 2-(Thiocyanomethylthio)benzothiazole + 2-Methylthio-4-tertiary-butylamino-6-cyclopropylamino-S-triazine

A2 Antifouling	Nautix SA	Amateur Professional	4832
A2 Teflon Antifouling	Nautix SA	Amateur Professional	4833

105 Dichlorophenyl dimethylurea + Zinc pyrithione + 2-Methylthio-4-tertiary-butylamino-6-cyclopropylamino-S-triazine

A6 Antifouling	Nautix SA	Amateur Professional	4944
A7 Teflon Antifouling	Nautix SA	Amateur Professional	4945
A7 Teflon Antifouling 072019	Nautix SA	Amateur Professional	5228
A8 Antifouling	Nautix SA	Amateur Professional	5982

106 Dichlorophenyl dimethylurea + Zinc pyrithione + Dichlofluanid

Antifouling 1.4	Plastimo International	Amateur Professional	6163

107 Dichlorophenyl dimethylurea + Zineb + 2,3,5,6-Tetrachloro-4-(methyl sulphonyl)pyridine + Cuprous oxide + Cuprous thiocyanate

Even Extreme	New Guard Coatings Ltd	Amateur Professional	6625

108 Maneb + Tributyltin methacrylate + Cuprous oxide

D Nu Wave A/F Flat Bottom	Kansai Paint Europe Ltd	Professional	3447
D Nu Wave A/F Vertical Bottom	Kansai Paint Europe Ltd	Professional	3446

Product Name	Marketing Company	Use	HSE No.

108 Maneb + Tributyltin methacrylate + Cuprous oxide—continued

D Rabamarine A/F No 2500 HS	Kansai Paint Europe Ltd	Professional	3444
D Rabamarine A/F No 2500M HS	Kansai Paint Europe Ltd	Professional	3445

109 Maneb + Tributyltin methacrylate + Tributyltin oxide + Cuprous oxide

D AF Seaflo Mark 2	Camrex Chugoku Ltd	Professional	5944
D C-Clean 5000	Camrex Holdings BV	Professional	5955

110 2-Methylthio-4-tertiary-butylamino-6-cyclopropylamino-S-triazine + 2,3,5,6-Tetrachloro-4-(methyl sulphonyl)pyridine

Envoy TCF 600 Copper/Tin Free Antifouling	W and J Leigh and Company	Professional	6482

111 2-Methylthio-4-tertiary-butylamino-6-cyclopropylamino-S-triazine + Copper + Zinc pyrithione

VC17M-HS	International Coatings Ltd	Amateur Professional	5960

112 2-Methylthio-4-tertiary-butylamino-6-cyclopropylamino-S-triazine + Copper metal

VC 17M Tropicana	International Coatings Ltd	Amateur Professional	4218

113 2-Methylthio-4-tertiary-butylamino-6-cyclopropylamino-S-triazine + Copper resinate + Cuprous oxide

D Sigmaplane Ecol Antifouling	Sigma Coatings BV	Amateur Professional	4348

114 2-Methylthio-4-tertiary-butylamino-6-cyclopropylamino-S-triazine + Cuprous oxide

Algicide	Blakes Marine Paints	Amateur Professional	5738
Aquaspeed	Blakes Marine Paints	Amateur Professional	4511
Aquaspeed Antifouling	Blakes Marine Paints	Amateur Professional	5817
Boat Paint General Purpose Antifouling	Boat Paint Direct	Amateur	6198
Boat Paint Hard Antifouling	Boat Paint Direct	Amateur	6197
Boat Paint High Copper Anti-Fouling	Boat Paint Direct	Amateur	6199
Broads Black Antifouling	Blakes Marine Paints	Amateur Professional	5739
Broads Freshwater Red	Blakes Marine Paints	Amateur Professional	5736
Challenger Antifouling	Blakes Marine Paints	Amateur Professional	4099

Product Name	Marketing Company	Use	HSE No.

114 2-Methylthio-4-tertiary-butylamino-6-cyclopropylamino-S-triazine + Cuprous oxide—continued

Product Name	Marketing Company	Use	HSE No.
W Even TF	Veneziani SPA	Amateur Professional	5166
Even Tin Free	Mark Dowland Marine Ltd	Amateur Professional	5841
Hard Racing Antifouling	Blakes Marine Paints	Amateur Professional	5704
Hempel's Antifouling Classic 7611 Red (Tin Free) 5000	Hempel Paints Ltd	Amateur Professional	5064
Hempel's Antifouling Classic 76540	Hempel Paints Ltd	Amateur Professional	5291
Hempel's Antifouling Combic 71990	Hempel Paints Ltd	Amateur Professional	5600
Hempel's Antifouling Combic 71992	Hempel Paints Ltd	Amateur Professional	5601
D Hempel's Antifouling Combic Tin Free 71990	Hempel Paints Ltd	Amateur Professional	5266
Hempel's Antifouling Forte Tin Free 7625	Hempel Paints Ltd	Amateur Professional	3351
Hempel's Antifouling Mille Dynamic	Hempel Paints Ltd	Amateur	4440
Hempel's Antifouling Nautic 71900	Hempel Paints Ltd	Amateur Professional	5290
Hempel's Antifouling Nautic 71902	Hempel Paints Ltd	Amateur Professional	5605
Hempel's Antifouling Nautic 8190C	Hempel Paints Ltd	Amateur Professional	6043
Hempel's Antifouling Olympic Hi-7661	Hempel Paints Ltd	Amateur Professional	4898
Hempel's Antifouling Olympic Tin Free 7154	Hempel Paints Ltd	Amateur Professional	3718
Hempel's Antifouling Tin Free 745GB	Hempel Paints Ltd	Amateur Professional	3331
Hempel's Antifouling Tin Free 750GB	Hempel Paints Ltd	Amateur Professional	3332
Hempel's Hard Racing 76480	Hempel Paints Ltd	Amateur Professional	5538
Hempel's Mille Dynamic 71700	Hempel Paints Ltd	Amateur Professional	5574
Hempel's Spinnaker Antifouling 78200	Hempel Paints Ltd	Amateur Professional	6026
Hempels Antifouling Bravo Tin Free 7610	Hempel Paints Ltd	Amateur Professional	4482
Interspeed System 2 BRA143 Brown	International Paint Ltd	Amateur Professional	4301
Interviron BQA450 Series	International Paint Ltd	Amateur Professional	4657

Product Name	Marketing Company	Use	HSE No.

114 2-Methylthio-4-tertiary-butylamino-6-cyclopropylamino-S-triazine + Cuprous oxide—continued

W Long Life T.F.	Veneziani SPA	Amateur Professional	4936
Longlife Antifouling	Skipper (UK) Ltd	Amateur Professional	6195
Micron CSC 200 Series	International Coatings Ltd	Amateur Professional	5732
Patente Laxe	Teais SA	Amateur Professional	5497
Pilot	Blakes Marine Paints	Amateur Professional	5959
W Raffaello Plus Tin Free	Veneziani SPA	Amateur Professional	3642
Shiprite Speed	Bradite Ltd	Amateur Professional	6603
Titan FGA Antifouling	Blakes Marine Paints	Amateur Professional	5681
VC Offshore	International Coatings Ltd	Amateur Professional	4777
XM Anti-Fouling C2000 Cruising Self Eroding	X M Yachting	Amateur	6176
XM Anti-Fouling P4000 Hard	X M Yachting	Amateur	6175
XM Antifouling HS3000 High Performance Self Eroding	X M Yachting	Amateur	6124

115 2-Methylthio-4-tertiary-butylamino-6-cyclopropylamino-S-triazine + Cuprous oxide + Cuprous thiocyanate + 2,3,5,6-Tetrachloro-4-(methyl sulphonyl)pyridine

Envoy TF 400	W and J Leigh and Company	Amateur Professional	4432
Envoy TF 500	W and J Leigh and Company	Amateur Professional	5599
Meridian MP40 Antifouling	Deangate Marine Products	Amateur Professional	5027
Micro-Tech	Marine (Chemi-Technics) Ltd	Amateur Professional	6316

116 2-Methylthio-4-tertiary-butylamino-6-cyclopropylamino-S-triazine + Cuprous oxide + Dichlofluanid

Antifouling 1.2	Plastimo International	Amateur Professional	6159

Product Name	Marketing Company	Use	HSE No.

117 2-Methylthio-4-tertiary-butylamino-6-cyclopropylamino-S-triazine + Cuprous oxide + Dichlofluanid + Dichlorophenyl dimethylurea

Antifouling 1.3 Black and Red (400-03/Ind.3)	Plastimo International	Amateur Professional	6350

118 2-Methylthio-4-tertiary-butylamino-6-cyclopropylamino-S-triazine + Cuprous oxide + Dichlorophenyl dimethylurea

A3 Antifouling 072015	Nautix SA	Amateur Professional	5224
A4 Antifouling 072017	Nautix SA	Amateur Professional	5226
Le Marin	Nautix SA	Amateur Professional	5480

119 2-Methylthio-4-tertiary-butylamino-6-cyclopropylamino-S-triazine + Cuprous oxide + Dichlorophenyl dimethylurea + 2,4,5,6-Tetrachloro isophthalonitrile

Sigmaplane Ecol HA 120 Antifouling	Sigma Coatings Ltd	Amateur Professional	5788

120 2-Methylthio-4-tertiary-butylamino-6-cyclopropylamino-S-triazine + Cuprous thiocyanate

AL-27	International Paint Ltd	Amateur Professional	3613
Aquaspeed White	Blakes Marine Paints	Amateur Professional	4513
Boot Top Plus (Gull White)	International Paint Ltd	Amateur Professional	3389
D Broads Sweetwater Antifouling	Blakes Marine Paints	Amateur Professional	3222
Cruiser Superior 100 Series	International Coatings Ltd	Amateur Professional	5723
W Even TF Light Grey	Veneziani SPA	Amateur Professional	5167
Hard Racing Antifouling White	Blakes Marine Paints	Amateur Professional	5705
Hempel's Hard Racing 76380	Hempel Paints Ltd	Amateur Professional	5540
Hempel's Mille Dynamic 71600	Hempel Paints Ltd	Amateur Professional	5576
Hempel's Tin Free Hard Racing 7648	Hempel Paints Ltd	Amateur Professional	3367
Interspeed 2000	International Coatings Ltd	Amateur Professional	4148
Marclear Powerboat Antifouling	Marclear Marine Products Ltd	Amateur Professional	6586

Product Name	Marketing Company	Use	HSE No.

120 2-Methylthio-4-tertiary-butylamino-6-cyclopropylamino-S-triazine + Cuprous thiocyanate—continued

MPX	International Coatings Ltd	Amateur Professional	4818
Propeller T.F.	New Guard Coatings Ltd	Amateur Professional	5660
Quicksilver Antifouling	International Coatings Ltd	Amateur Professional	5938
Raffaello Alloy	Mark Dowland Marine Ltd	Amateur Professional	5492
Tiger Cruising White	Blakes Marine Paints	Amateur Professional	5712
Tiger White	Blakes Marine Paints	Amateur Professional	4512
Tigerline Antifouling	Blakes Marine Paints	Amateur Professional	3842
Tigerline Boottop	Blakes Marine Paints	Amateur Professional	5763
Titan FGA Antifouling White	Blakes Marine Paints	Amateur Professional	5680
VC Offshore	International Coatings Ltd	Amateur Professional	4779
VC Prop-O-Drev	International Coatings Ltd	Amateur Professional	4217

121 2-Methylthio-4-tertiary-butylamino-6-cyclopropylamino-S-triazine + Cuprous thiocyanate + Dichlofluanid

Antifouling 1.2 (W)	Plastimo International	Amateur Professional	6160
Hempel's Antifouling Mille Dynamic 717 GB	Hempel Paints Ltd	Amateur Professional	4437

122 2-Methylthio-4-tertiary-butylamino-6-cyclopropylamino-S-triazine + Cuprous thiocyanate + Dichlofluanid + Dichlorophenyl dimethylurea

Antifouling 1.3	Plastimo International	Amateur Professional	6161
Antifouling 1.3 White (327-04)	Plastimo International	Amateur Professional	6349

123 2-Methylthio-4-tertiary-butylamino-6-cyclopropylamino-S-triazine + Cuprous thiocyanate + Dichlofluanid + Zinc pyrithione

Antifouling 1.3 (W)	Plastimo International	Amateur Professional	6162

Product Name	Marketing Company	Use	HSE No.

**124 2-Methylthio-4-tertiary-butylamino-6-cyclopropylamino-S-triazine +
Cuprous thiocyanate + Dichlorophenyl dimethylurea**

A3 Antifouling 072016	Nautix SA	Amateur Professional	5225
A3 Teflon Antifouling 062015	Nautix SA	Amateur Professional	5481
A4 Antifouling 072018	Nautix SA	Amateur Professional	5227
A4 Teflon Antifouling 062017	Nautix SA	Amateur Professional	5482

**125 2-Methylthio-4-tertiary-butylamino-6-cyclopropylamino-S-triazine +
Dichlorophenyl dimethylurea + Zinc pyrithione**

A6 Antifouling	Nautix SA	Amateur Professional	4944
A7 Teflon Antifouling	Nautix SA	Amateur Professional	4945
A7 Teflon Antifouling 072019	Nautix SA	Amateur Professional	5228
A8 Antifouling	Nautix SA	Amateur Professional	5982

**126 2-Methylthio-4-tertiary-butylamino-6-cyclopropylamino-S-triazine +
Dichlorophenyl dimethylurea + Zinc pyrithione + 2-
(Thiocyanomethylthio)benzothiazole**

A2 Antifouling	Nautix SA	Amateur Professional	4832
A2 Teflon Antifouling	Nautix SA	Amateur Professional	4833

127 2-Methylthio-4-tertiary-butylamino-6-cyclopropylamino-S-triazine + Thiram

D Broads Freshwater Antifouling	Blakes Marine Paints	Amateur Professional	3221
D Lynx Metal Free Antifouling	Blakes Marine Paints	Amateur Professional	3555
D Trawler Tin-Free	Blakes Marine Paints	Amateur Professional	3235

**128 2-Methylthio-4-tertiary-butylamino-6-cyclopropylamino-S-triazine + Thiram
+ 2-(Thiocyanomethylthio)benzothiazole**

D Lynx Antifouling	Blakes Marine Paints	Amateur Professional	3618
D Lynx Antifouling Plus	Blakes Marine Paints	Amateur Professional	5865

Product Name	Marketing Company	Use	HSE No.

129 2-Methylthio-4-tertiary-butylamino-6-cyclopropylamino-S-triazine + Tributyltin methacrylate

| Antifouling Alusea | Jotun-Henry Clark Ltd | Professional | 4569 |

130 2-Methylthio-4-tertiary-butylamino-6-cyclopropylamino-S-triazine + Tributyltin methacrylate + Tributyltin oxide

| Antifouling Alusea Classic | Jotun-Henry Clark Ltd | Professional | 6463 |

131 2-Methylthio-4-tertiary-butylamino-6-cyclopropylamino-S-triazine + Zinc pyrithione

Hempel's Antifouling Nautic 71800	Hempel Paints Ltd	Amateur Professional	5769
Lynx Plus	Blakes Marine Paints	Amateur Professional	5792
White Tiger Cruising	Blakes Marine Paints	Amateur Professional	6873

132 No Recognised Active Ingredient

W Bioclean DX	Camrex Chugoku Ltd	Amateur Professional Professional (Aquaculture)	5317
Interclene AQ HZA700 Series (Base)	International Paint Ltd	Amateur Professional Professional (Aquaculture)	4765
Intersleek BXA 810/820 (Base)	International Coatings Ltd	Amateur Professional Professional (Aquaculture)	3403
Intersleek BXA560 Series (Base)	International Coatings Ltd	Amateur Professional Professional (Aquaculture)	4785
Intersleek BXA580 Series (Base)	International Coatings Ltd	Amateur Professional	4786
Intersleek FCS HKA560 Series (Base)	International Paint Ltd	Amateur Professional Professional (Aquaculture)	4767
Intersleek FCS HKA580 Series (Base)	International Paint Ltd	Amateur Professional Professional (Aquaculture)	4766
Penguin Aqualine	Jotun-Henry Clark Ltd	Amateur Professional	6196

Product Name	Marketing Company	Use	HSE No.

133 2,4,5,6-Tetrachloro isophthalonitrile

W Flexgard IV Waterbase Preservative	Flexabar Aquatech Corporation	Professional Professional (Aquaculture)	3321
W Flexgard V Waterbase Preservative	Flexabar Aquatech Corporation	Professional Professional (Aquaculture)	3320

134 2,4,5,6-Tetrachloro isophthalonitrile + 2-Methylthio-4-tertiary-butylamino-6-cyclopropylamino-S-triazine + Cuprous oxide + Dichlorophenyl dimethylurea

Sigmaplane Ecol HA 120 Antifouling	Sigma Coatings Ltd	Amateur Professional	5788

135 2,4,5,6-Tetrachloro isophthalonitrile + 4,5-Dichloro-2-N-octyl-4-isothiazolin-3-one + Cuprous oxide

C-Clean 400	Camrex Holdings BV	Amateur Professional	5947
Seatender 15	Camrex Chugoku Ltd	Professional	5348
W TFA 10 HG	Camrex Chugoku Ltd	Amateur Professional	5366
W TFA 10G	Camrex Chugoku Ltd	Amateur Professional	5347

136 2,4,5,6-Tetrachloro isophthalonitrile + Copper sulphate

W Flexgard VI Waterbase Preservative	Flexabar Aquatech Corporation	Professional Professional (Aquaculture)	3319

137 2,4,5,6-Tetrachloro isophthalonitrile + Cuprous oxide

Flexgard VI	Flexabar Aquatech Corporation	Professional Professional (Aquaculture)	6035
D Hempel's Antifouling Classic 7654C	Hempel Paints Ltd	Amateur Professional	5298
D Hempel's Antifouling Combic 7199C	Hempel Paints Ltd	Amateur Professional	5296
D Hempel's Antifouling Nautic 7190C	Hempel Paints Ltd	Amateur Professional	5294
TFA 10 LA	Camrex Chugoku Ltd	Amateur Professional	5361

138 2,4,5,6-Tetrachloro isophthalonitrile + Cuprous oxide + Dichlorophenyl dimethylurea

C-Clean 300	Camrex Holdings BV	Amateur Professional	5942

Product Name	Marketing Company	Use	HSE No.

138 2,4,5,6-Tetrachloro isophthalonitrile + Cuprous oxide + Dichlorophenyl dimethylurea—continued

Seatender 12	Camrex Chugoku Ltd	Amateur Professional	5324
TFA 10	Camrex Chugoku Ltd	Amateur Professional	5346
W TFA 10 H	Camrex Chugoku Ltd	Amateur Professional	5358

139 2,4,5,6-Tetrachloro isophthalonitrile + Cuprous thiocyanate

D Gummipaint A/F	Veneziani SPA	Amateur Professional	4934

140 2,3,5,6-Tetrachloro-4-(methyl sulphonyl)pyridine + 2-Methylthio-4-tertiary-butylamino-6-cyclopropylamino-S-triazine

Envoy TCF 600 Copper/Tin Free Antifouling	W and J Leigh and Company	Professional	6482

141 2,3,5,6-Tetrachloro-4-(methyl sulphonyl)pyridine + 2-Methylthio-4-tertiary-butylamino-6-cyclopropylamino-S-triazine + Cuprous oxide + Cuprous thiocyanate

Envoy TF 400	W and J Leigh and Company	Amateur Professional	4432
Envoy TF 500	W and J Leigh and Company	Amateur Professional	5599
Meridian MP40 Antifouling	Deangate Marine Products	Amateur Professional	5027
Micro-tech	Marine (Chemi-Technics) Ltd	Amateur Professional	6316

142 2,3,5,6-Tetrachloro-4-(methyl sulphonyl)pyridine + Cuprous oxide

Grassline TF Anti-Fouling Type M396	W and J Leigh and Company	Amateur Professional	3462

143 2,3,5,6-Tetrachloro-4-(methyl sulphonyl)pyridine + Cuprous oxide + Cuprous thiocyanate + Dichlorophenyl dimethylurea + Zineb

Even Extreme	New Guard Coatings Ltd	Amateur Professional	6625

144 2,3,5,6-Tetrachloro-4-(methyl sulphonyl)pyridine + Cuprous oxide + Dichlorophenyl dimethylurea

Raffaello 3	New Guard Coatings Ltd	Amateur Professional	5826

Product Name	Marketing Company	Use	HSE No.

145 2,3,5,6-Tetrachloro-4-(methyl sulphonyl)pyridine + Cuprous thiocyanate + Dichlorophenyl dimethylurea

Raffaello Racing	New Guard Coatings Ltd	Amateur Professional	6451

146 2-(Thiocyanomethylthio)benzothiazole + 2-Methylthio-4-tertiary-butylamino-6-cyclopropylamino-S-triazine + Dichlorophenyl dimethylurea + Zinc pyrithione

A2 Antifouling	Nautix SA	Amateur Professional	4832
A2 Teflon Antifouling	Nautix SA	Amateur Professional	4833

147 2-(Thiocyanomethylthio)benzothiazole + 2-Methylthio-4-tertiary-butylamino-6-cyclopropylamino-S-triazine + Thiram

D Lynx Antifouling	Blakes Marine Paints	Amateur Professional	3618
D Lynx Antifouling Plus	Blakes Marine Paints	Amateur Professional	5865

148 2-(Thiocyanomethylthio)benzothiazole + Cuprous oxide

A3 Antifouling	Nautix SA	Amateur Professional	4367
W Titan Tin Free	Blakes Marine Paints	Amateur Professional	3556

149 Thiram + 2-(Thiocyanomethylthio)benzothiazole + 2-Methylthio-4-tertiary-butylamino-6-cyclopropylamino-S-triazine

D Lynx Antifouling	Blakes Marine Paints	Amateur Professional	3618
D Lynx Antifouling Plus	Blakes Marine Paints	Amateur Professional	5865

150 Thiram + 2-Methylthio-4-tertiary-butylamino-6-cyclopropylamino-S-triazine

D Broads Freshwater Antifouling	Blakes Marine Paints	Amateur Professional	3221
D Lynx Metal Free Antifouling	Blakes Marine Paints	Amateur Professional	3555
D Trawler Tin-Free	Blakes Marine Paints	Amateur Professional	3235

151 Thiram + Cuprous oxide

D Grassline TF Anti-Fouling Type M395	W and J Leigh and Company	Amateur Professional	3461

Product Name	Marketing Company	Use	HSE No.

152 Tributyltin acrylate + Cuprous oxide

ABC # 1 Antifouling	Ameron BV	Professional	3214
W Amercoat 697	Ameron BV	Professional	3513

153 Tributyltin methacrylate + 2-Methylthio-4-tertiary-butylamino-6-cyclopropylamino-S-triazine

Antifouling Alusea	Jotun-Henry Clark Ltd	Professional	4569

154 Tributyltin methacrylate + Cuprous oxide

Devran MCP Antifouling Red	Devoe Coatings BV	Professional	3311
Devran MCP Antifouling Red Brown	Devoe Coatings BV	Professional	3310
W Takata LLL Antifouling	NOF Europe NV	Professional	4050
W Takata LLL Antifouling Hi-Solid	NOF Europe NV	Professional	4052
Takata LLL Antifouling LS	NOF Europe NV	Professional	4051
W Takata LLL Antifouling LS Hi-Solid	NOF Europe NV	Professional	4053
W Takata LLL Antifouling No 2001	NOF Europe NV	Professional	4054

155 Tributyltin methacrylate + Cuprous oxide + Maneb

D Nu Wave A/F Flat Bottom	Kansai Paint Europe Ltd	Professional	3447
D Nu Wave A/F Vertical Bottom	Kansai Paint Europe Ltd	Professional	3446
D Rabamarine A/F No 2500 HS	Kansai Paint Europe Ltd	Professional	3444
D Rabamarine A/F No 2500M HS	Kansai Paint Europe Ltd	Professional	3445

156 Tributyltin methacrylate + Tributyltin oxide

Hempel's Antifouling Nautic 76800	Hempel Paints Ltd	Professional	6025

157 Tributyltin methacrylate + Tributyltin oxide + 2-Methylthio-4-tertiary-butylamino-6-cyclopropylamino-S-triazine

Antifouling Alusea Classic	Jotun-Henry Clark Ltd	Professional	6463

158 Tributyltin methacrylate + Tributyltin oxide + 4,5-Dichloro-2-N-octyl-4-isothiazolin-3-one

Antifouling Alusea Turbo	Jotun-Henry Clark Ltd	Professional	6464

159 Tributyltin methacrylate + Tributyltin oxide + 4,5-Dichloro-2-N-octyl-4-isothiazolin-3-one + Cuprous oxide

Intersmooth Hisol BFO270/950/970 Series	Courtaulds Coatings (Holdings) Ltd	Professional	5641

160 Tributyltin methacrylate + Tributyltin oxide + Cuprous oxide

AF Seaflo Mark 2-1	Camrex Chugoku Ltd	Professional	5316
W AF Seaflo Z-100 HS-1	Camrex Chugoku Ltd	Professional	5318
Antifouling Seaconomy	Jotun-Henry Clark Ltd	Professional	5791
Antifouling Seaconomy	Jotun-Henry Clark Ltd	Professional	6261
W Antifouling Seaconomy 200	Jotun-Henry Clark Ltd	Professional	4436
W Antifouling Seaconomy 300	Jotun-Henry Clark Ltd	Professional	4272

Product Name	Marketing Company	Use	HSE No.

160 Tributyltin methacrylate + Tributyltin oxide + Cuprous oxide—continued

Product Name	Marketing Company	Use	HSE No.
Antifouling Seamate FB30	Jotun-Henry Clark Ltd	Professional	6506
W Antifouling Seamate HB 33	Jotun-Henry Clark Ltd	Professional	4270
Antifouling Seamate HB 99 Black	Jotun-Henry Clark Ltd	Professional	3410
Antifouling Seamate HB 99 Dark Red	Jotun-Henry Clark Ltd	Professional	3412
Antifouling Seamate HB 99 Light Red	Jotun-Henry Clark Ltd	Professional	3411
Antifouling Seamate HB22	Jotun-Henry Clark Ltd	Professional	4271
Antifouling Seamate HB22	Jotun-Henry Clark Ltd	Professional	6301
Antifouling Seamate HB33	Jotun-Henry Clark Ltd	Professional	5811
Antifouling Seamate HB66	Jotun-Henry Clark Ltd	Professional	5698
Antifouling Seamate HB66	Jotun-Henry Clark Ltd	Profeessional	6475
Antifouling Seamate SB33	Jotun-Henry Clark Ltd	Professional	6517
C-Clean 6000	Camrex Holdings BV	Professional	5945
Grassline ABL Antifouling Type M349	W and J Leigh and Company	Professional	5374
W Hempel's Antifouling Combic 76990	Hempel Paints Ltd	Professional	3348
Hempel's Antifouling Nautic Hi 76900	Hempel Paints Ltd	Professional	3360
W Hempel's Antifouling Nautic Hi 76910	Hempel Paints Ltd	Professional	3361
Hempel's Antifouling Nautic SP-ACE 79031	Hempel Paints Ltd	Professional	6452
Hempel's Antifouling Nautic SP-ACE 79051	Hempel Paints Ltd	Professional	6453
Intersmooth SPC BFA090/BFA190 Series	International Paint Ltd	Professional	3377

161 Tributyltin methacrylate + Tributyltin oxide + Cuprous oxide + Dichlorophenyl dimethylurea

Product Name	Marketing Company	Use	HSE No.
A11 Antifouling 072014	Nautix SA	Professional	5376
Interswift BKO000/700 Series	International Coatings Ltd	Professional	5640

162 Tributyltin methacrylate + Tributyltin oxide + Cuprous oxide + Maneb

Product Name	Marketing Company	Use	HSE No.
D AF Seaflo Mark 2	Camrex Chugoku Ltd	Professional	5944
D C-Clean 5000	Camrex Holdings BV	Professional	5955

163 Tributyltin methacrylate + Tributyltin oxide + Cuprous thiocyanate

Product Name	Marketing Company	Use	HSE No.
Antifouling HB 66 Ocean Green	Jotun-Henry Clark Ltd	Professional	3408
Antifouling Seamate HB 66 Std 4037	Jotun-Henry Clark Ltd	Professional	4269
W Antifouling Seamate HB Green	Jotun-Henry Clark Ltd	Professional	3430
Antifouling Seamate HB22 Roundel Blue	Jotun-Henry Clark Ltd	Professional	4872
Antifouling Seamate HB66 Black	Jotun-Henry Clark Ltd	Professional	4871

Product Name	Marketing Company	Use	HSE No.

163 Tributyltin methacrylate + Tributyltin oxide + Cuprous thiocyanate—continued

Intersmooth Hisol SPC Antifouling BFA949 Red	International Coatings Ltd	Professional	4949
Intersmooth SPC BFA040/BFA050 Series	International Coatings Ltd	Professional	3376

164 Tributyltin methacrylate + Tributyltin oxide + Zinc pyrithione

Hempel's Antifouling Nautic 7680B	Hempel Paints Ltd	Professional	6172

165 Tributyltin methacrylate + Tributyltin oxide + Zineb + Cuprous oxide

Blueline SBA900 Series	International Paint Ltd	Professional	5138
W Hempel's Antifouling 7690D	Hempel Paints Ltd	Professional	6056
Hempel's Antifouling Economic SP-SEA 74030	Hempel Paints Ltd	Professional	5581
Hempel's Antifouling Nautic 7690B	Hempel Paints Ltd	Professional	5580
W Hempel's Antifouling Nautic 7691B	Hempel Paints Ltd	Professional	5586
Hempel's Antifouling Nautic SP-ACE 79030	Hempel Paints Ltd	Professional	6331
Hempel's Antifouling Nautic SP-ACE 79050	Hempel Paints Ltd	Professional	6325
Hempel's Antifouling Nautic SP-ACE 79070	Hempel Paints Ltd	Professional	6328
Hempel's Economic SP-SEA 74010	Hempel Paints Ltd	Professional	6330
Hempel's Economic SP-SEA 74040	Hempel Paints Ltd	Professional	6329
Intersmooth 110 Standard	International Coatings Ltd	Professional	5907
Intersmooth 130 Ultra	International Paint Ltd	Professional	5914
Intersmooth 210 Standard	International Coatings Ltd	Professional	5915
Intersmooth 220 Premium	International Coatings Ltd	Professional	5911
Intersmooth 230 Ultra	International Paint Ltd	Professional	5912
Intersmooth 320 Premium	International Paint Ltd	Professional	5910
Intersmooth 330 Ultra	International Coatings Ltd	Professional	5924
Intersmooth Hisol 2000 BFA270 Series	International Coatings Ltd	Professional	3844
Intersmooth Hisol 9000 BFA970 Series	International Coatings Ltd	Professional	5461
Intersmooth Hisol BFA250/BFA900 Series	International Coatings Ltd	Professional	3848
Interswift BKA000/700 Series	International Coatings Ltd	Professional	3384
Seaflo 10	Camrex Chugoku Ltd	Professional	6283

166 Tributyltin methacrylate + Tributyltin oxide + Zineb + Cuprous thiocyanate

Cruiser Copolymer	International Paint Ltd	Professional	3514
Intersmooth Hisol BFA948 Orange	International Coatings Ltd	Professional	4281
Micron 25 Plus	International Coatings Ltd	Professional	3402
Superyacht 800 Antifouling	International Coatings Ltd	Professional	4778

HSE ANTIFOULING PRODUCTS

Product Name	Marketing Company	Use	HSE No.

167 Tributyltin methacrylate + Zineb + Cuprous oxide

Antifouling Seamate HB22	Jotun-Henry Clark Ltd	Professional	5362
Antifouling Seamate HB33	Jotun-Henry Clark Ltd	Professional	5363
Sigmaplane HA Antifouling	Sigma Coatings BV	Professional	4345
W Sigmaplane HB	Sigma Coatings BV	Professional	3487
Sigmaplane HB Antifouling	Sigma Coatings Ltd	Professional	5721
Sigmaplane TA Antifouling	Sigma Coatings BV	Professional	4346

168 Tributyltin oxide + 2-Methylthio-4-tertiary-butylamino-6-cyclopropylamino-S-triazine + Tributyltin methacrylate

Antifouling Alusea Classic	Jotun-Henry Clark Ltd	Professional	6463

169 Tributyltin oxide + 4,5-Dichloro-2-N-octyl-4-isothiazolin-3-one + Cuprous oxide + Tributyltin methacrylate

Intersmooth Hisol BFO270/950/970 Series	Courtaulds Coatings (Holdings) Ltd	Professional	5641

170 Tributyltin oxide + 4,5-Dichloro-2-N-octyl-4-isothiazolin-3-one + Tributyltin methacrylate

Antifouling Alusea Turbo	Jotun-Henry Clark Ltd	Professional	6464

171 Tributyltin oxide + Cuprous oxide + Dichlorophenyl dimethylurea + Tributyltin methacrylate

A11 Antifouling 072014	Nautix SA	Professional	5376
Interswift BKO000/700 Series	International Coatings Ltd	Professional	5640

172 Tributyltin oxide + Cuprous oxide + Maneb + Tributyltin methacrylate

D AF Seaflo Mark 2	Camrex Chugoku Ltd	Professional	5944
D C-Clean 5000	Camrex Holdings BV	Professional	5955

173 Tributyltin oxide + Cuprous oxide + Tributyltin methacrylate

AF Seaflo Mark 2-1	Camrex Chugoku Ltd	Professional	5316
W AF Seaflo Z-100 HS-1	Camrex Chugoku Ltd	Professional	5318
Antifouling Seaconomy	Jotun-Henry Clark Ltd	Professional	5791
Antifouling Seaconomy	Jotun-Henry Clark Ltd	Professional	6261
W Antifouling Seaconomy 200	Jotun-Henry Clark Ltd	Professional	4436
W Antifouling Seaconomy 300	Jotun-Henry Clark Ltd	Professional	4272
Antifouling Seamate FB30	Jotun-Henry Clark Ltd	Professional	6506
W Antifouling Seamate HB 33	Jotun-Henry Clark Ltd	Professional	4270
Antifouling Seamate HB 99 Black	Jotun-Henry Clark Ltd	Professional	3410
Antifouling Seamate HB 99 Dark Red	Jotun-Henry Clark Ltd	Professional	3412
Antifouling Seamate HB 99 Light Red	Jotun-Henry Clark Ltd	Professional	3411
Antifouling Seamate HB22	Jotun-Henry Clark Ltd	Professional	4271
Antifouling Seamate HB22	Jotun-Henry Clark Ltd	Professional	6301
Antifouling Seamate HB33	Jotun-Henry Clark Ltd	Professional	5811

Product Name	Marketing Company	Use	HSE No.

173 Tributyltin oxide + Cuprous oxide + Tributyltin methacrylate—continued

Antifouling Seamate HB66	Jotun-Henry Clark Ltd	Professional	5698
Antifouling Seamate HB66	Jotun-Henry Clark Ltd	Professional	6475
Antifouling Seamate SB33	Jotun-Henry Clark Ltd	Professional	6517
C-Clean 6000	Camrex Holdings BV	Professional	5945
Grassline ABL Antifouling Type M349	W and J Leigh and Company	Professional	5374
W Hempel's Antifouling Combic 76990	Hempel Paints Ltd	Professional	3348
Hempel's Antifouling Nautic Hi 76900	Hempel Paints Ltd	Professional	3360
W Hempel's Antifouling Nautic Hi 76910	Hempel Paints Ltd	Professional	3361
Hempel's Antifouling Nautic SP-ACE 79031	Hempel Paints Ltd	Professional	6452
Hempel's Antifouling Nautic SP-ACE 79051	Hempel Paints Ltd	Professional	6453
Intersmooth SPC BFA090/BFA190 Series	International Paint Ltd	Professional	3377

174 Tributyltin oxide + Cuprous thiocyanate + Tributyltin methacrylate

Antifouling HB 66 Ocean Green	Jotun-Henry Clark Ltd	Professional	3408
Antifouling Seamate HB 66 STD 4037	Jotun-Henry Clark Ltd	Professional	4269
W Antifouling Seamate HB Green	Jotun-Henry Clark Ltd	Professional	3430
Antifouling Seamate HB22 Roundel Blue	Jotun-Henry Clark Ltd	Professional	4872
Antifouling Seamate HB66 Black	Jotun-Henry Clark Ltd	Professional	4871
Intersmooth Hisol SPC Antifouling BFA949 Red	International Coatings Ltd	Professional	4949
Intersmooth SPC BFA040/BFA050 Series	International Coatings Ltd	Professional	3376

175 Tributyltin oxide + Tributyltin methacrylate

Hempel's Antifouling Nautic 76800	Hempel Paints Ltd	Professional	6025

176 Tributyltin oxide + Zinc pyrithione + Tributyltin methacrylate

Hempel's Antifouling Nautic 7680B	Hempel Paints Ltd	Professional	6172

177 Tributyltin oxide + Zineb + Cuprous oxide + Tributyltin methacrylate

Blueline SBA900 Series	International Paint Ltd	Professional	5138
W Hempel's Antifouling 7690D	Hempel Paints Ltd	Professional	6056
Hempel's Antifouling Economic SP-SEA 74030	Hempel Paints Ltd	Professional	5581
Hempel's Antifouling Nautic 7690B	Hempel Paints Ltd	Professional	5580
W Hempel's Antifouling Nautic 7691B	Hempel Paints Ltd	Professional	5586
Hempel's Antifouling Nautic SP-ACE 79030	Hempel Paints Ltd	Professional	6331

Product Name	Marketing Company	Use	HSE No.

177 Tributyltin oxide + Zineb + Cuprous oxide + Tributyltin methacrylate—
continued

Hempel's Antifouling Nautic SP-ACE 79050	Hempel Paints Ltd	Professional	6325
Hempel's Antifouling Nautic SP-ACE 79070	Hempel Paints Ltd	Professional	6328
Hempel's Economic SP-SEA 74010	Hempel Paints Ltd	Professional	6330
Hempel's Economic SP-SEA 74040	Hempel Paints Ltd	Professional	6329
Intersmooth 110 Standard	International Coatings Ltd	Professional	5907
Intersmooth 130 Ultra	International Paint Ltd	Professional	5914
Intersmooth 210 Standard	International Coatings Ltd	Professional	5915
Intersmooth 220 Premium	International Coatings Ltd	Professional	5911
Intersmooth 230 Ultra	International Paint Ltd	Professional	5912
Intersmooth 320 Premium	International Paint Ltd	Professional	5910
Intersmooth 330 Ultra	International Coatings Ltd	Professional	5924
Intersmooth Hisol 2000 BFA270 Series	International Coatings Ltd	Professional	3844
Intersmooth Hisol 9000 BFA970 Series	International Coatings Ltd	Professional	5461
Intersmooth Hisol BFA250/BFA900 Series	International Coatings Ltd	Professional	3848
Interswift BKA000/700 Series	International Coatings Ltd	Professional	3384
Seaflo 10	Camrex Chugoku Ltd	Professional	6283

178 Tributyltin oxide + Zineb + Cuprous thiocyanate + Tributyltin methacrylate

Cruiser Copolymer	International Paint Ltd	Professional	3514
Intersmooth Hisol BFA948 Orange	International Coatings Ltd	Professional	4281
Micron 25 Plus	International Coatings Ltd	Professional	3402
Superyacht 800 Antifouling	International Coatings Ltd	Professional	4778

179 Zinc naphthenate + Copper naphthenate + Cuprous oxide + Dichlofluanid

W Teamac Killa Copper Plus	Teal and Mackrill Ltd	Amateur Professional	4659

180 Zinc naphthenate + Cuprous oxide

W Teamac Killa Copper	Teal and Mackrill Ltd	Amateur Professional	3494
W Teamac Super Tropical	Teal and Mackrill Ltd	Amateur Professional	3497

181 Zinc pyrithione

No Foul-Zo	E Paint Company Inc	Amateur Professional	5789

Product Name	Marketing Company	Use	HSE No.

182 Zinc pyrithione + 2-(Thiocyanomethylthio)benzothiazole + 2-Methylthio-4-tertiary-butylamino-6-cyclopropylamino-S-triazine + Dichlorophenyl dimethylurea

| A2 Antifouling | Nautix SA | Amateur Professional | 4832 |
| A2 Teflon Antifouling | Nautix SA | Amateur Professional | 4833 |

183 Zinc pyrithione + 2-Methylthio-4-tertiary-butylamino-6-cyclopropylamino-S-triazine

Hempel's Antifouling Nautic 71800	Hempel Paints Ltd	Amateur Professional	5769
Lynx Plus	Blakes Marine Paints	Amateur Professional	5792
White Tiger Cruising	Blakes Marine Paints	Amateur Professional	6873

184 Zinc pyrithione + 2-Methylthio-4-tertiary-butylamino-6-cyclopropylamino-S-triazine + Copper

| VC17M-HS | International Coatings Ltd | Amateur Professional | 5960 |

185 Zinc pyrithione + 2-Methylthio-4-tertiary-butylamino-6-cyclopropylamino-S-triazine + Cuprous thiocyanate + Dichlofluanid

| Antifouling 1.3 (W) | Plastimo International | Amateur Professional | 6162 |

186 Zinc pyrithione + 2-Methylthio-4-tertiary-butylamino-6-cyclopropylamino-S-triazine + Dichlorophenyl dimethylurea

A6 Antifouling	Nautix SA	Amateur Professional	4944
A7 Teflon Antifouling	Nautix SA	Amateur Professional	4945
A7 Teflon Antifouling 072019	Nautix SA	Amateur Professional	5228
A8 Antifouling	Nautix SA	Amateur Professional	5982

187 Zinc pyrithione + 4,5-Dichloro-2-N-octyl-4-isothiazolin-3-one

| Hempel's Antifouling Nautic 7180E | Hempel Paints Ltd | Amateur Professional | 5771 |
| Lynx-E | Blakes Marine Paints | Amateur Professional | 5793 |

188 Zinc pyrithione + Cuprous oxide

| Blakes Antifouling 87910 | Blakes Marine Paints | Amateur Professional | 6450 |

Product Name	Marketing Company	Use	HSE No.

188 Zinc pyrithione + Cuprous oxide—continued

Intersmooth 360 Ecoloflex	International Coatings Ltd	Amateur Professional	6057
Intersmooth 365 Ecoloflex	International Coatings Ltd	Amateur Professional	6557
Intersmooth 465 Ecoloflex	International Coatings Ltd	Amateur Professional	6556
W Micron 600 Series	International Paint Ltd	Amateur Professional	5733
Micron Optima	International Coatings Ltd	Amateur Professional	5941
VC Offshore Extra 100 Series	International Coatings Ltd	Amateur Professional	5730
VC Offshore SP Antifouling	International Coatings Ltd	Amateur Professional	3507

189 Zinc pyrithione + Cuprous thiocyanate

Hempel's Antifouling Combic 7199C	Hempel Paints Ltd	Amateur Professional	5742
Interspeed 2002	International Coatings Ltd	Amateur Professional	5725
Titan FGA-C Antifouling	Blakes Marine Paints	Amateur Professional	5766

190 Zinc pyrithione + Dichlofluanid + Dichlorophenyl dimethylurea

Antifouling 1.4	Plastimo International	Amateur Professional	6163

191 Zinc pyrithione + Dichlorophenyl dimethylurea

Hempel's Antifouling Nautic 7180B	Hempel Paints Ltd	Amateur Professional	5770
Lynx-A	Blakes Marine Paints	Amateur Professional	5794

192 Zinc pyrithione + Tributyltin methacrylate + Tributyltin oxide

Hempel's Antifouling Nautic 7680B	Hempel Paints Ltd	Professional	6172

193 Zineb + 2,3,5,6-Tetrachloro-4-(methyl sulphonyl)pyridine + Cuprous oxide + Cuprous thiocyanate + Dichlorophenyl dimethylurea

Even Extreme	New Guard Coatings Ltd	Amateur Professional	6625

194 Zineb + Copper resinate + Cuprous oxide

Sigma Pilot Ecol Antifouling	Sigma Coatings Ltd	Amateur Professional	4933

Product Name	Marketing Company	Use	HSE No.

195 Zineb + Cuprous oxide

Blueline SPC Tin Free SBA700 Series	International Coatings Ltd	Amateur Professional	5214
Equatorial	International Paint Ltd	Amateur Professional	4121
Hempel's Antifouling Combic 7199F	Hempel Paints Ltd	Amateur Professional	5744
Inter 100	International Coatings Ltd	Amateur Professional	4120
Interclene 245	International Coatings Ltd	Amateur Professional	6233
Interspeed 340	International Coatings Ltd	Amateur Professional	6089
Interspeed System 2 BRA140/ BRA240 Series	International Coatings Ltd	Amateur Professional	3847
Interviron BQA200 Series	International Coatings Ltd	Amateur Professional	3846

196 Zineb + Cuprous oxide + Tributyltin methacrylate

Antifouling Seamate HB22	Jotun-Henry Clark Ltd	Professional	5362
Antifouling Seamate HB33	Jotun-Henry Clark Ltd	Professional	5363
Sigmaplane HA Antifouling	Sigma Coatings BV	Professional	4345
W Sigmaplane HB	Sigma Coatings BV	Professional	3487
Sigmaplane HB Antifouling	Sigma Coatings Ltd	Professional	5721
Sigmaplane TA Antifouling	Sigma Coatings BV	Professional	4346

197 Zineb + Cuprous oxide + Tributyltin methacrylate + Tributyltin oxide

Blueline SBA900 Series	International Paint Ltd	Professional	5138
W Hempel's Antifouling 7690D	Hempel Paints Ltd	Professional	6056
Hempel's Antifouling Economic SP-SEA 74030	Hempel Paints Ltd	Professional	5581
Hempel's Antifouling Nautic 7690B	Hempel Paints Ltd	Professional	5580
W Hempel's Antifouling Nautic 7691B	Hempel Paints Ltd	Professional	5586
Hempel's Antifouling Nautic SP-ACE 79030	Hempel Paints Ltd	Professional	6331
Hempel's Antifouling Nautic SP-ACE 79050	Hempel Paints Ltd	Professional	6325
Hempel's Antifouling Nautic SP-ACE 79070	Hempel Paints Ltd	Professional	6328
Hempel's Economic SP-SEA 74010	Hempel Paints Ltd	Professional	6330
Hempel's Economic SP-SEA 74040	Hempel Paints Ltd	Professional	6329
Intersmooth 110 Standard	International Coatings Ltd	Professional	5907
Intersmooth 130 Ultra	International Paint Ltd	Professional	5914
Intersmooth 210 Standard	International Coatings Ltd	Professional	5915
Intersmooth 220 Premium	International Coatings Ltd	Professional	5911
Intersmooth 230 Ultra	International Paint Ltd	Professional	5912
Intersmooth 320 Premium	International Paint Ltd	Professional	5910

Product Name	Marketing Company	Use	HSE No.

197 Zineb + Cuprous oxide + Tributyltin methacrylate + Tributyltin oxide—continued

Intersmooth 330 Ultra	International Coatings Ltd	Professional	5924
Intersmooth Hisol 2000 BFA270 Series	International Coatings Ltd	Professional	3844
Intersmooth Hisol 9000 BFA970 Series	International Coatings Ltd	Professional	5461
Intersmooth Hisol BFA250/BFA900 Series	International Coatings Ltd	Professional	3848
Interswift BKA000/700 Series	International Coatings Ltd	Professional	3384
Seaflo 10	Camrex Chugoku Ltd	Professional	6283

198 Zineb + Cuprous thiocyanate + Tributyltin methacrylate + Tributyltin oxide

Cruiser Copolymer	International Paint Ltd	Professional	3514
Intersmooth Hisol BFA948 Orange	International Coatings Ltd	Professional	4281
Micron 25 Plus	International Coatings Ltd	Professional	3402
Superyacht 800 Antifouling	International Coatings Ltd	Professional	4778

199 Ziram + Cuprous oxide

D Awlgrip Awlstar Antifouling ABC 3 Black	Devoe Coatings BV	Amateur Professional	6036
D Awlgrip Awlstar Antifouling ABC 3 Blue	Devoe Coatings BV	Amateur Professional	6044
D Awlgrip Awlstar Antifouling ABC 3 Red	Devoe Coatings BV	Amateur Professional	6038
D Awlstar Gold Label Anti-Fouling	Grow Group Incorporated	Amateur Professional	3674
D Cheetah Antifouling	Blakes Marine Paints	Amateur Professional	3223
D Devoe ABC 3 Antifouling Blue	Devoe Coatings BV	Amateur Professional	6037
D Hempel's Antifouling Classic 7654A	Hempel Paints Ltd	Amateur Professional	5297
D Hempel's Antifouling Combic 7199A	Hempel Paints Ltd	Amateur Professional	5295
D Hempel's Antifouling Nautic 7190A	Hempel Paints Ltd	Amateur Professional	5293

2
BIOCIDAL PAINTS

Biocidal Paints

200 Carbendazim

Glidden Fungicidal Acrylic Eggshell	Imperial Chemical Industries Plc	Amateur Professional	6662
Glidden Fungicidal Vinyl Matt	Imperial Chemical Industries Plc	Amateur Professional	6658
Glidden Fungicidal Vinyl Silk	Imperial Chemical Industries Plc	Amateur Professional	6659

201 Carbendazim + Dichlorophenyl dimethylurea + 2-Octyl-2h-isothiazolin-3-one

Anti-Mould Gloss	Plascon International	Amateur	6524
Aquaguard Anti-Mould Acrylic Matt Finish	Leyland Paint Company Ltd	Amateur Professional	6338
Biocheck Water Based Gloss	Mould Growth Consultants Ltd	Amateur Professional	6573
Manders Stop Mould	Manders Paints Ltd	Amateur Professional	6459
Sterashield	Johnstone's Paints Plc	Amateur Professional	6333
Wickes Mouldkill Emulsion	Wickes Building Supplies Ltd	Amateur Professional	6458

202 Carbendazim + Dichlorophenyl dimethylurea + Dithio-2,2'-bis(benzmethylamide) + 2-Octyl-2h-isothiazolin-3-one

Biocheck 01	Mould Growth Consultants Ltd	Professional	6536
Biocheck SP	Mould Growth Consultants Ltd	Amateur Professional	6600
Biocure Emulsion	Biotech Environmental (UK) Ltd	Amateur Professional	6850

203 Dichlofluanid

Dulux Trade Mouldshield Cellar Paint	ICI Paints	Amateur Professional	6481

204 Dichlorophen

Roofcoat TP15	Mould Growth Consultants Ltd	Professional	6578

205 Dichlorophenyl dimethylurea + 2-Octyl-2h-isothiazolin-3-one + Carbendazim

Anti-Mould Gloss	Plascon International	Amateur	6524
Aquaguard Anti-Mould Acrylic Matt Finish	Leyland Paint Company Ltd	Amateur Professional	6338

205 Dichlorophenyl dimethylurea + 2-Octyl-2h-isothiazolin-3-one + Carbendazim—continued

Product Name	Marketing Company	Use	HSE No.
Biocheck Water Based Gloss	Mould Growth Consultants Ltd	Amateur Professional	6573
Manders Stop Mould	Manders Paints Ltd	Amateur Professional	6459
Sterashield	Johnstone's Paints Plc	Amateur Professional	6333
Wickes Mouldkill Emulsion	Wickes Building Supplies Ltd	Amateur Professional	6458

206 Dichlorophenyl dimethylurea + Dithio-2,2'-bis(benzmethylamide) + 2-Octyl-2h-isothiazolin-3-one + Carbendazim

Product Name	Marketing Company	Use	HSE No.
Biocheck 01	Mould Growth Consultants Ltd	Professional	6536
Biocheck SP	Mould Growth Consultants Ltd	Amateur Professional	6600
Biocure Emulsion	Biotech Environmental (UK) Ltd	Amateur Professional	6850

207 Dithio-2,2'-bis(benzmethylamide)

Product Name	Marketing Company	Use	HSE No.
Biocheck 2507	Mould Growth Consultants Ltd	Professional	6610
Biocheck C	Mould Growth Consultants Ltd	Amateur Professional	6596
Biocheck Matt	Mould Growth Consultants Ltd	Amateur Professional	6599
Biocheck Silk	Mould Growth Consultants Ltd	Amateur Professional	6597
Bioshield	Mould Growth Consultants Ltd	Amateur Professional	6604

208 Dithio-2,2'-Bis(benzmethylamide) + 2-Octyl-2h-isothiazolin-3-one + Carbendazim + Dichlorophenyl dimethylurea

Product Name	Marketing Company	Use	HSE No.
Biocheck 01	Mould Growth Consultants Ltd	Professional	6536
Biocheck SP	Mould Growth Consultants Ltd	Amateur Professional	6600
Biocure Emulsion	Biotech Environmental (UK) Ltd	Amateur Professional	6850

Product Name	Marketing Company	Use	HSE No.

209 3-Iodo-2-propynyl-n-butyl carbamate

Product Name	Marketing Company	Use	HSE No.
ACS Dry Rot Paint	Advanced Chemical Specialties Ltd	Amateur Amateur (Wood Treatment) Professional Professional (Wood Treatment)	6310
Anti-Mould Emulsion	Plascon International	Amateur	6473
Biotech Fungicidal Additive	Biotech Environmental (UK) Ltd	Amateur Professional	6650
Dulux Trade Mouldshield Fungicidal Eggshell	ICI Paints	Amateur Professional	6480
Dulux Trade Mouldshield Fungicidal Matt	ICI Paints	Amateur Professional	6494
Lectros Dry Rot Paint	Advanced Chemical Specialties Ltd	Amateur Amateur (Wood Treatment) Professional Professional (Wood Treatment)	6520
MGC Fungicidal Additive	Mould Growth Consultants Ltd	Amateur Professional	6605

210 3-Iodo-2-propynyl-n-butyl Carbamate + Propiconazole

Product Name	Marketing Company	Use	HSE No.
Silexine Anticon	Remtox (Chemicals) Ltd	Amateur Professional	6476
Silexine Fungi-Chek Emulsion	Remtox (Chemicals) Ltd	Amateur Professional	6477
Silexine Fungi-Chek Oil Matt Paint	Remtox (Chemicals) Ltd	Amateur Professional	6478

211 2-Octyl-2h-isothiazolin-3-one

Product Name	Marketing Company	Use	HSE No.
ACS Antimould Microporous Replastering Paint	Advanced Chemical Specialties Ltd	Amateur Professional	6355
Azygo Antimould Coating (New Works)	Azygo International Direct Ltd	Amateur Professional	6483
Azygo Antimould Paint	Azygo International Direct Ltd	Amateur Professional	6406

Product Name	Marketing Company	Use	HSE No.

211 2-Octyl-2h-isothiazolin-3-one—continued

Halocell 221	Mould Growth Consultants Ltd	Professional	6587
Lectros Antimould Breathable Paint	Lectros International Ltd	Amateur Professional	6496
Lectros Antimould Emulsion	Lectros International Ltd	Amateur Professional	6505

212 2-Octyl-2h-isothiazolin-3-one + Carbendazim + Dichlorophenyl dimethylurea

Anti-mould Gloss	Plascon International	Amateur	6524
Aquaguard Anti-mould Acrylic Matt Finish	Leyland Paint Company Ltd	Amateur Professional	6338
Biocheck Water Based Gloss	Mould Growth Consultants Ltd	Amateur Professional	6573
Manders Stop Mould	Manders Paints Ltd	Amateur Professional	6459
Sterashield	Johnstone's Paints Plc	Amateur Professional	6333
Wickes Mouldkill Emulsion	Wickes Building Supplies Ltd	Amateur Professional	6458

213 2-Octyl-2h-isothiazolin-3-one + Carbendazim + Dichlorophenyl dimethylurea + Dithio-2,2'-bis(benzmethylamide)

Biocheck 01	Mould Growth Consultants Ltd	Professional	6536
Biocheck SP	Mould Growth Consultants Ltd	Amateur Professional	6600
Biocure Emulsion	Biotech Environmental (UK) Ltd	Amateur Professional	6850

214 Permethrin + Pyrethrins + 2,3,5,6-tetrachloro-4-(methyl sulphonyl)pyridine

Stericide AM	Resin Surfaces Ltd	Amateur Professional	6513

215 Propiconazole + 3-Iodo-2-propynyl-n-butyl carbamate

Silexine Anticon	Remtox (Chemicals) Ltd	Amateur Professional	6476
Silexine Fungi-Chek Emulsion	Remtox (Chemicals) Ltd	Amateur Professional	6477
Silexine Fungi-Chek Oil Matt Paint	Remtox (Chemicals) Ltd	Amateur Professional	6478

216 Pyrethrins + 2,3,5,6-Tetrachloro-4-(methyl sulphonyl)pyridine + Permethrin

Stericide AM	Resin Surfaces Ltd	Amateur Professional	6513

Product Name	Marketing Company	Use	HSE No.

217 2,3,5,6-Tetrachloro-4-(methyl sulphonyl)pyridine

Haloseal	Mould Growth Consultants Ltd	Amateur Professional	6572
Stericide	Resin Surfaces Ltd	Amateur Professional	6454

218 2,3,5,6-Tetrachloro-4-(methyl sulphonyl)pyridine + Permethrin + Pyrethrins

Stericide AM	Resin Surfaces Ltd	Amateur Professional	6513

3
INSECT (REPELLENT)

HSE
INSECT (REPELLENT)

Product Name	Marketing Company	Use	HSE No.

Insect (repellent)

219 Citronella oil

Aztec B.B.Q. Patio Candles	Chartan Aldred Ltd	Amateur	5613
B.B.Q Fly Repellent Terracotta Pot Candle	Eclipse Candles Ltd	Amateur	5627
B.B.Q. Fly Repellant Candle	Eclipse Candles Ltd	Amateur	5625
B.B.Q. Patio Candles	Sherwood Promark Ltd	Amateur	5556
Betterware Adjustable Insect Repellent	Betterware UK Ltd	Amateur	6883
W Citromax Citronella Insect Repellent Liquid Candle Oil	Lamplight Farms Inc	Amateur	5423
Citronella Bug Spray	STV International Ltd	Amateur	6369
Citronella Candle (Code D10)	Colony Gift Corporation Ltd	Amateur	6005
Citronella Floater Candle	Colony Gift Corporation Ltd	Amateur	6004
Citronella Scented Candle	Colony Gift Corporation Ltd	Amateur	6006
Floral Bouquet (Oil of Citronella)	Parlour Products Plc	Amateur	5886
Killgerm Repel	Killgerm Chemicals Ltd	Amateur Professional	5790
Killgerm Repel RTU	Killgerm Chemicals Ltd	Amateur Professional	5971
Vapona Citronella Candle	Ashe Ltd	Amateur	6861

220 Deltamethrin

K-Othrine Moustiquaire S.C.	Agrevo UK Limited	Amateur Professional	6129

221 1,4-Dichlorobenzene

Bouchard Anti Moth Proofer Pouches	IBA UK Ltd	Amateur	6403
W Moth Repellent	Boots the Chemist Ltd	Amateur	5079

222 1,4-Dichlorobenzene + Naphthalene

Mothaks	Sara Lee Household and Body Care UK Ltd	Amateur	5124
Vapona Mothaks	Ashe Consumer Products Ltd	Amateur	5682

223 Naphthalene

Dragon Brand Moth Balls	R A Davies and Partners Ltd	Amateur	5385
Jertox Moth Balls	Thornton and Ross Ltd	Amateur	5057
Phernal Brand Moth Balls	Harrow Drug Company	Amateur	5740

224 Naphthalene + 1,4-Dichlorobenzene

Mothaks	Sara Lee Household and Body Care UK Ltd	Amateur	5124

Product Name	Marketing Company	Use	HSE No.

224 Naphthalene + 1,4-Dichlorobenzene—continued

Vapona Mothaks	Ashe Consumer Products Ltd	Amateur	5682

225 Permethrin

Peripel	Agrevo UK Limited	Professional	5265
Raid Mothproofer	Johnson Wax Ltd	Amateur	5646

226 Pyrethrins

Repel Clothing Spray	Ibis Products Ltd	Amateur	6212
X-Gnat Advanced Insect Repellent - Fabric Spray	X-Gnat Laboratories Ltd	Amateur	6155

HSE INSECT (REPELLENT)

397

4
INSECTICIDES

Product Name	Marketing Company	Use	HSE No.

Insecticides

227 Allethrin

Spira 'No-Bite' Outdoor Mosquito Coils	Masters London Ltd	Amateur	4421

228 d-Allethrin

W Baygon	Scholl Consumer Products Ltd	Amateur	5752
Gelert Mosquito Repellent	Bryncir Products Ltd	Amateur	5465
Jeyes Expel Plug-In Flying Insect Killer	Jeyes Group Ltd	Amateur	5653
Jeyes Kontrol Plug-In Flying Insect Killer	Jeyes Group Ltd	Amateur	5652
Moskil Mosquito Repellent	I & M Steiner Ltd	Amateur	4342
Pynamin Forte 40MG Mat	Sumitomo Chemical (UK) Plc	Amateur	5852
Pynamin Forte Mat 120	Sumitomo Chemical (UK) Plc	Amateur	4417
Shelltox Mat 2	Temana International Ltd	Amateur	4742
Vapona Plug-In Flying Insect Killer	Ashe Consumer Products Ltd	Amateur	5708

229 d-Allethrin + Cypermethrin

Shelltox Crawling Insect Killer Liquid Spray	Temana International Ltd	Amateur Professional	3565
Vapona Ant and Crawling Insect Spray	Ashe Consumer Products Ltd	Amateur	3593

230 d-Allethrin + Cypermethrin + Permethrin + Tetramethrin

Elf Insecticide	Elf Oil UK Ltd	Amateur	6669

231 d-Allethrin + Cypermethrin + Tetramethrin

Bop Flying and Crawling Insect Killer	Robert McBride Ltd - Exports	Amateur Professional	4136
Bop Flying and Crawling Insect Killer (Water Based)	Robert McBride Ltd - Exports	Amateur Professional	4137
Bop Flying Insect Killer	Robert McBride Ltd - Exports	Amateur Professional	4141
Bop Flying Insect Killer (Mc)	Robert McBride Ltd - Exports	Amateur Professional	4140
Bop Flying and Crawling Insect Killer (Mc)	Robert Mcbride Ltd - Exports	Amateur Professional	4135

232 d-Allethrin + Dichlorvos + Permethrin + Tetramethrin

Bop	Robert McBride Ltd - Exports	Amateur	4773

Product Name	Marketing Company	Use	HSE No.

233 d-Allethrin + Permethrin

New Vapona Fly and Wasp Killer	Ashe Consumer Products Ltd	Amateur	4201
Raid Outdoor Insectguard	Johnson Wax Ltd	Amateur	5459

234 d-Allethrin + Permethrin + Tetramethrin

Hail Plus Crawling Insect Killer	Robert McBride Ltd - Exports	Amateur	5567
New Super Raid Insecticide	Johnson Wax Ltd	Amateur Professional	4583

235 d-Allethrin + Tetramethrin

Boots Dry Fly and Wasp Killer	Boots the Chemist Ltd	Amateur	4037
W Boots Dry Fly and Wasp Killer 3	Boots the Chemist Ltd	Amateur	4241
Fly, Wasp and Mosquito Killer	Rentokil Initial UK Ltd	Amateur	4882
New Vapona Fly Killer Dry Formulation	Ashe Consumer Products Ltd	Amateur	4202
Raid Fly and Wasp Killer	Johnson Wax Ltd	Amateur	3000
Sainsbury's Fly, Wasp and Mosquito Killer	J Sainsbury Plc	Amateur	5164

236 d-Allethrin + d-Phenothrin

Boots New Improved Fly and Wasp Killer	Boots the Chemist Ltd	Amateur	6432
Fly and Ant Killer	Rentokil Initial UK Ltd	Amateur	5561
Jeyes Crawling Insect Spray	Jeyes Group Ltd	Amateur	5270
Kleeneze New Improved Fly Killer	Kleeneze Ltd	Amateur	6431
Kleenoff Crawling Insect Spray	Jeyes Group Ltd	Amateur	5269
Kwik Insecticide	Hand Associates Ltd	Amateur	4290
Nippon Ready For Use Fly Killer Spray	Vitax Ltd	Amateur	4631
Pedigree Bedding Spray	Thomas's Europe	Amateur	5427
Pesguard PS 102	Sumitomo Chemical (UK) Plc	Amateur	5312
Pesguard PS 102A	Sumitomo Chemical (UK) Plc	Amateur	4173
Pesguard PS 102B	Sumitomo Chemical (UK) Plc	Amateur	4174
Pesguard PS 102C	Sumitomo Chemical (UK) Plc	Amateur	4175
Pesguard PS 102D	Sumitomo Chemical (UK) Plc	Amateur	4176
W Secto Ant and Crawling Insect Spray.	Sinclair Animal and Household Care Ltd	Amateur	5714
W Target Flying Insect Killer	Reckitt and Colman Products Ltd	Amateur	4178
Whiskas Bedding Spray	Thomas's Europe	Amateur	4837
Whiskas Bedding Pest Control Spray	Thomas's Europe	Amateur	4596

237 Alphacypermethrin

Cytrol Alpha 25 EC	Pelgar International Ltd	Professional	6677
Cytrol Alpha 5 SC	Pelgar International Ltd	Professional	6423

237 Alphacypermethrin—continued

Fendona 1.5 SC	Sorex Ltd	Professional	4092
Fendona 6SC	Sorex Ltd	Professional	4455
Fendona ASC	Rentokil Initial UK Ltd	Professional	4946
Fendona Lacquer	Sorex Ltd	Amateur Professional	4278
Littac	Sorex Ltd	Professional Professional (Animal Husbandry)	5176

238 Alphacypermethrin + Flufenoxuron

Tenopa	Sorex Ltd	Professional	6206

239 Alphacypermethrin + Tetramethrin

Cytrol Alpha Super 50/50 SE	Pelgar International Ltd	Professional	6678

240 Azamethiphos

AL63 Crawling Insect Killer	Reckitt and Colman Products Ltd	Professional	5450
Raid Flyguard	Johnson Wax Ltd	Amateur	5713

241 Bacillus thuringiensis var israelensis

Bactimos Flowable Concentrate	Abbott Laboratories	Professional	4792
Bactimos Wettable Powder	Abbott Laboratories	Professional	4793
Skeetal Flowable Concentrate	Industrial Pesticides (N.W)	Professional	5706
D Skeetal Flowable Concentrate	Novo Enzyme Products Ltd	Professional	3603
Teknar HP-D	Killgerm Chemicals Ltd	Professional	5355
Vectobac 12AS	Univar Plc	Professional	6205

242 Bendiocarb

Agrevo Ant Killer Powder	Agrevo UK Limited	Amateur	6071
Agrevo Woodlice Killer	Agrevo UK Limited	Amateur	6069
Ant Bait	Rentokil Initial UK Ltd	Amateur	6896
Ant Killer Powder	LC Solutions Ltd	Amateur	6846
Bendiocarb Dusting Powder	Rentokil Initial UK Ltd	Professional	5125
Bendiocarb Wettable Powder	Rentokil Initial UK Ltd	Professional	5416
Bio Ant Kill Plus Powder	PBI Home & Garden Ltd	Amateur	6512
Bob Martin Home Flea Powder	The Bob Martin Company	Amateur	6656
Camco Ant Powder	Agrevo UK Limited	Amateur	5218
Doff Ant Control Powder	Doff Portland Ltd	Amateur	6868
Ficam 20W	Agrevo UK Limited	Professional	3682
Ficam D	Agrevo UK Limited	Professional	4829
Ficam W	Agrevo UK Limited	Professional	5390
D Ficam Wasp and Hornet Spray	Agrevo UK Limited	Professional	5253
Murphy Kil-Ant Powder	The Scotts Company (UK) Limited	Amateur	5237

Product Name	Marketing Company	Use	HSE No.

242 Bendiocarb—continued

Rentokil Ant and Insect Killer	Rentokil Initial UK Ltd	Amateur	5810
Rentokil Ant and Insect Powder Professional	Rentokil Initial UK Ltd	Professional	5386
Rentokil Wasp Killer	Rentokil Initial UK Ltd	Amateur	5813
Secto Ant Bait	Sinclair Animal and Household Care Ltd	Amateur	5088
Wasp Nest Destroyer	LC Solutions Ltd	Amateur	6844
Wasp Nest Killer Professional	Rentokil Initial UK Ltd	Professional	5651
Woodlice Killer	LC Solutions Ltd	Amateur	6845

243 Bendiocarb + Pyrethrins

Ficam Plus	Agrevo UK Limited	Professional	4830

244 Bendiocarb + Tetramethrin

Camco Insect Spray	Agrevo UK Limited	Amateur	4222
Devcol Household Insect Spray	Devcol Ltd	Amateur	5512

245 Benzalkonium chloride + Naphthalene + Permethrin + Pyrethrins

W Roxem D	Horton Hygiene Company	Professional	5107

246 Benzalkonium chloride + Natamycin

D Tymasil	Brocades (Great Britain) Ltd	Amateur	3625

247 Benzyl benzoate

W Acarosan Foam	Crawford Pharmaceuticals	Amateur	5750
W Acarosan Moist Powder	Crawford Pharmaceuticals	Amateur	5751

248 Bioallethrin

Betterware Flying Insect Killer	Betterware UK Ltd	Amateur	6137
Boots Electric Mosquito Killer	Boots the Chemist Ltd	Amateur	3871
Boots UK Flying Insect Killer	Boots the Chemist Ltd	Amateur	4145
Buzz Off 1	Volex Accessories Ltd	Amateur	5823
W Buzz-Off	Ross Consumer Electronics	Amateur	3854
D Buzz-Off 2	Ross Consumer Electronics	Amateur	3855
Far and Away	Dencon Accessories Ltd	Amateur	5753
Globol Pyrethrum Electrical Evaporator	Globol Chemicals (UK) Ltd	Amateur	4439
Haden Mosquito and Flying Insect Killer	D H Haden Plc	Amateur	5716
Lyvia Mosquito Killer	Lyvia Electrical Ltd	Amateur	4212
Mosqui - Go Electric	Jack Rogers and Co Ltd	Amateur	4083
Mosquito Killer Travel Pack	Culmstock Ltd	Amateur	4926
Mosquito Repellent	Forward Chemicals Ltd	Amateur	5746
D Mosquito Repellent	Shopfield Ltd	Amateur	4653
Nippon Flying Insect Killer Tablets	Vitax Ltd	Amateur	5199

HSE **INSECTICIDES**

Product Name	Marketing Company	Use	HSE No.

248 Bioallethrin—continued

Pif Paf Mosquito Mats	Reckitt and Colman Products Ltd	Amateur	5156
Rentokil Flying Insect Killer	Rentokil Initial UK Ltd	Amateur	5976
Shelltox Mat 1	Temana International Ltd	Amateur	4740
Spira 'No-Bite' Mosquito Killer	Masters London Ltd	Amateur	3727
Stradz Mosquito Killer	Stradz Ltd	Amateur	4144
Superdrug Mosquito Killer	Superdrug Stores Plc	Amateur	4078
Terminator Flying Insect Killer	Sanoda Ltd	Amateur	6139
Vapona Plug-In	Ashe Consumer Products Ltd	Amateur	5773
Wahl Envoyage Mosquito Killer	Wahl Europe Ltd	Amateur	5108
Woolworth Mosquito Killer	F W Woolworth Ltd	Amateur	4210

249 S-Bioallethrin + Bioallethrin

Actomite	Ceuta Healthcare Ltd	Amateur	4182

250 Bioallethrin + Bioresmethrin

Days Fly Spray	Day and Sons (Crewe) Ltd	Professional Professional (Animal Husbandry)	6457
Flyclear BHB	Battle, Hayward and Bower Ltd	Professional	6435
Pybuthrin 33 BB	Agrevo UK Limited	Professional	5162
Trilanco Fly Spray	Trilanco	Professional Professional (Animal Husbandry)	6456

251 S-Bioallethrin + Bioresmethrin

Procontrol FIK H20	Agrevo UK Limited	Amateur Professional	6579
Procontrol FIK Super	Agrevo UK Limited	Amateur Professional	6580
Wasp Killer	LC Solutions Ltd	Amateur	6853

252 S-Bioallethrin + Deltamethrin

Crackdown Rapide	Agrevo UK Limited	Professional	5996

253 Bioallethrin + Dichlorvos + Permethrin

Farco Rapid Kill	F A Richard & Company Ltd	Amateur	6152

254 S-Bioallethrin + Dichlorvos + Permethrin

Pif Paf Insecticide	Reckitt and Colman Products Ltd	Amateur Professional	5150

Product Name	Marketing Company	Use	HSE No.

255 Bioallethrin + Methoprene + Permethrin

Arrest	Arnolds Veterinary Products Ltd	Amateur	6389
Bob Martin Home Flea Fogger Plus	The Bob Martin Company	Amateur	6630
Bob Martin Home Flea Spray Plus	The Bob Martin Company	Amateur	6223
R.I.P. Fleas	The Bob Martin Company	Amateur	6207

256 Bioallethrin + Permethrin

Agrevo Procontrol Crawling Insect Killer	Agrevo UK Limited	Amateur Professional	6116
Agrevo Procontrol Flying Insect Killer	Agrevo UK Limited	Amateur Professional	6117
Agrevo Procontrol Flying Insect Killer Aerosol	Agrevo UK Limited	Professional	6230
Ant Gun! 2	Miracle Garden Care Ltd	Amateur Professional	5629
D Ant Gun! 2	Zeneca Garden Care	Amateur Professional	5184
AntKiller Spray 2	B & Q Plc	Amateur Professional	5185
Baby Bio Houseplant Spray	PBI Home & Garden Ltd	Amateur Professional	5828
Bio Spraydex Ant and Insect Killer	PBI Home & Garden Ltd	Amateur Professional	5829
Crawling Insect and Ant Killer	Rentokil Initial UK Ltd	Amateur	4650
Creepy Crawly Gun!	Miracle Garden Care Ltd	Amateur Professional	5183
Defest II	Sherley's Ltd	Amateur	6002
Farco Flying Insect Killer	F A Richard & Company Ltd	Amateur	6151
Fleegard	Bayer Plc	Amateur	5564
Homebase Ant and Crawling Insect Killer	Sainsbury's Homebase House and Garden Centres	Amateur	5846
Insectrol	Rentokil Initial UK Ltd	Amateur	4568
Insectrol Professional	Rentokil Initial UK Ltd	Professional	4624
Kayo	Nebulous Ltd	Amateur Professional	5881
Nippon Fly Killer Pads	Vitax Ltd	Amateur	5426
PC Insect Killer	Perycut Chemie AG	Amateur Professional	4415
Pif Paf Crawling Insect Killer	Reckitt and Colman Products Ltd	Amateur Professional	5153
Pif Paf Flying Insect Killer	Reckitt and Colman Products Ltd	Amateur Professional	5151
Pif Paf Flying Insect Killer Aerosol	Reckitt and Colman Products Ltd	Amateur	5152
Pybuthrin Fly Killer	Agrevo UK Limited	Amateur Professional	5141

405

Product Name	Marketing Company	Use	HSE No.

256 Bioallethrin + Permethrin—continued

Product Name	Marketing Company	Use	HSE No.
Pybuthrin Fly Spray.	Agrevo UK Limited	Amateur	5116
Sainsbury's Crawling Insect and Ant Killer	J Sainsbury Plc	Amateur	5231
W Spraydex Ant & Insect Killer	Pan Britannica Industries Ltd	Amateur	5645
W Spraydex Houseplant Spray	Spraydex Ltd	Amateur	3540
W Spraydex Insect Killer	Spraydex Ltd	Amateur	3834
Vapona Ant and Crawling Insect Killer	Ashe Consumer Products Ltd	Amateur Professional	5206
Vapona Fly and Wasp Killer	Ashe Consumer Products Ltd	Amateur	5205
D Zap Pest Control	Delsol Ltd	Amateur Professional	5068

257 S-Bioallethrin + Permethrin

Product Name	Marketing Company	Use	HSE No.
Agrevo Procontrol Yellow Flying Insect Killer Aerosol	Agrevo UK Limited	Professional	6232
Aqua Reslin Premium	Agrevo UK Limited	Professional	5172
Aqua Reslin Super	Agrevo UK Limited	Professional	5175
Atta-x Flying Insect Killer	Sanderson Curtis Ltd	Amateur	6348
Bug Wars	Jack Rogers and Co Ltd	Amateur	6361
W Coopex S25% EC	Agrevo UK Limited	Professional	5146
W Imperator Fog S 18/6	Zeneca Public Health	Professional	4960
W Imperator Fog Super S 30/15	Zeneca Public Health	Professional	4955
W Imperator ULV S 6/3	Zeneca Public Health	Professional	4954
W Imperator ULV Super S 10/4	Zeneca Public Health	Professional	4953
W Neopybuthrin Premium	Agrevo UK Limited	Professional	5149
W Neopybuthrin Super	Agrevo UK Limited	Professional	5148
Pif Paf Yellow Flying Insect Killer	Reckitt and Colman Products Ltd	Amateur	5937
Purge	Aztec Aerosols Limited	Amateur	6634
Resigen	Agrevo UK Limited	Professional	5143
Reslin Premium	Agrevo UK Limited	Professional	5174
W Reslin Super	Agrevo UK Limited	Professional	5173

258 Bioallethrin + Permethrin + Pyrethrins

Product Name	Marketing Company	Use	HSE No.
Supabug Crawling Insect Killer	Sharpstow International Homecare Products Ltd	Amateur	4794
Tyrax	DBC	Amateur	6829

259 Bioallethrin + S-Bioallethrin

Product Name	Marketing Company	Use	HSE No.
Actomite	Ceuta Healthcare Ltd	Amateur	4182

260 S-Bioallethrin + Tetramethrin

Product Name	Marketing Company	Use	HSE No.
Vapona Wasp Killer Aerosol	Ashe Consumer Products Ltd	Amateur	6063

Product Name	Marketing Company	Use	HSE No.

261 S-Bioallethrin + Tetramethrin + Permethrin

Agrevo Procontrol Crawling Insect Killer Aerosol	Agrevo UK Limited	Professional	6231
Pif Paf Crawling Insect Killer Aerosol	Reckitt and Colman Products Ltd	Amateur	5595

262 Bioresmethrin

Biosol RTU	Microsol	Professional	5306
Blade	Chemsearch	Professional	4839
Safe Kill RTU	Friendly Systems	Professional	5563

263 Bioresmethrin + Bioallethrin

Days Fly Spray	Day and Sons (Crewe) Ltd	Professional Professional (Animal Husbandry)	6457
Flyclear BHB	Battle, Hayward and Bower Ltd	Professional	6435
Pybuthrin 33 BB	Agrevo UK Limited	Professional	5162
Trilanco Fly Spray	Trilanco	Professional Professional (Animal Husbandry)	6456

264 Bioresmethrin + S-Bioallethrin

Procontrol FIK H20	Agrevo UK Limited	Amateur Professional	6579
Procontrol FIK Super	Agrevo UK Limited	Amateur Professional	6580
Wasp Killer	LC Solutions Ltd	Amateur	6853

265 Boric acid

Baracaf Cockroach Control Sticker	Laboratoire Baracaf Sarl	Professional	3180
Baracaf Cockroach Control Sticker (Domestic)	Laboratoire Baracaf Sarl	Amateur Professional	4964
Boric Acid Concentrate	Rentokil Initial UK Ltd	Professional	4420
Boric Acid Powder	Rentokil Initial UK Ltd	Professional	4373
Cockroach Control	Cockroach Control Services Ltd	Professional	5979
Enviromite	Environetics International Ltd	Amateur Professional	6203
Flea Ban	Animal Care Ltd	Amateur Professional Professional (Animal Husbandry)	6201
Instasective	Instafoam & Fibre Ltd	Professional	5109

Product Name	Marketing Company	Use	HSE No.

265 Boric acid—continued

Killgerm Boric Acid Powder	Killgerm Chemicals Ltd	Professional	4527
Paragon Formula 10 Cockroach Bait	Paragon	Amateur Professional	6438
Pharaohkill	Sabre Pest Control Services	Professional	6565
Roachbuster	Roachbuster UK Ltd	Amateur Professional	5215
Sherley's Flea Busters	Sherley's Ltd	Amateur	6377
VF For Fleas	Veterinarian Formulae Ltd	Amateur Professional Professional (Animal Husbandry)	5973

266 Ceto-stearyl diethoxylate + Oleyl monoethoxylate

Larvex-100	Accotec Ltd	Professional	5393
Larvex-15	Accotec Ltd	Professional	5091

267 Chlorpyrifos

Ant Stop	The Scotts Company (UK) Limited	Amateur	5905
Bob Martin Microshield Household Flea Killing Spray	The Bob Martin Company	Amateur	6015
Chlorpyrifos	Rentokil Initial UK Ltd	Professional	6401
Contra Insect: the vermin bait	Frunol Delicia GmbH	Amateur	6299
Contra-ants bait box	Frunol Delicia GmbH	Amateur	6298
Contra-Insect 480 TEC	Frunol Delicia GmbH	Professional	6224
Duratrol	3M Health Care Ltd	Amateur	5968
D Dursban 4TC	Dow Agrosciences Ltd	Professional	3097
Dursban 4TC	Dow Agrosciences Ltd	Professional	6288
W Dursban Lo	Dowelanco Ltd	Professional	4771
Empire 20	Dowelanco Ltd	Professional	4844
Etisso Ant-ex bait box	Frunol Delicia GmbH	Amateur	6300
Gett	Dow Agrosciences Ltd	Amateur	5863
Killgerm Terminate	Killgerm Chemicals Ltd	Professional	3553
Raid Ant Bait	Johnson Wax Ltd	Amateur	5585
Rentokil Chlorpyrifos Gel	Rentokil Initial UK Ltd	Professional	5862
Swat Gel	Dow Agrosciences Ltd	Professional	6332
Vapona Micro-Tech	Ashe Ltd	Amateur	6591

268 Chlorpyrifos + Cypermethrin

Contra-Insect 200/20 EC	Frunol Delicia GmbH	Professional	6225
Duo 1	Hand Associates Ltd	Professional	6548

269 Chlorpyrifos + Cypermethrin + Pyrethrins

New Tetracide	Killgerm Chemicals Ltd	Professional	3724

Product Name	Marketing Company	Use	HSE No.

270 Chlorpyrifos + Pyrethrins

| Contra-Insect Aerosol | Frunol Delicia GmbH | Amateur | 6372 |
| Contra-Insect Universal | Frunol Delicia GmbH | Amateur | 6405 |

271 Chlorpyrifos + Tetramethrin

| Raid Wasp Nest Destroyer | Johnson Wax Ltd | Amateur | 5597 |

272 Chlorpyrifos-methyl

| D Reldan 50 EC | Dow Agrosciences Ltd | Professional | 4875 |
| W Smite | Agrevo UK Limited | Professional | 5142 |

273 Chlorpyrifos-methyl + Permethrin + Pyrethrins

| W Multispray | Agrevo UK Limited | Professional | 5165 |

274 Cyfluthrin + Pyriproxyfen

| Fleegard Plus | Bayer Plc | Amateur | 6582 |

275 Cyfluthrin + Transfluthrin

| Baygon Insect Spray | Bayer Plc | Amateur | 6546 |

276 Cypermethrin

B and Q New Formula Ant Killer Spray	B & Q Plc	Amateur	5606
Cymperator	Killgerm Chemicals Ltd	Professional	3970
Cyperkill 10	Mitchell Cotts Chemicals Ltd	Professional	4025
Cyperkill 10 WP	Mitchell Cotts Chemicals Ltd	Professional	3649
Cypermethrin 10% EC	Killgerm Chemicals Ltd	Professional	3136
Cypermethrin 10% WP	Killgerm Chemicals Ltd	Professional	3137
Cypermethrin Lacquer	Killgerm Chemicals Ltd	Professional	3164
W Cypermethrin PH-10EC	Zeneca Public Health	Professional	4961
Cytrol Forte WP	Pelgar International Ltd	Professional	6424
Demon 40 WP	Zeneca Public Health	Professional	5457
Doff Ant Killer Spray	Doff Portland Ltd	Amateur	5588
Great Mills Ant Killer Spray	Great Mills (Retail) Ltd	Amateur	5572
Homebase AntKiller Spray	Sainsbury's Homebase House and Garden Centres	Amateur	5573
Major Kil	Anglo Net	Amateur Professional	5874
Maximus RTU	Certified Laboratories	Amateur Professional Professional (Animal Husbandry)	5837
Morgan Ant & Crawling Insect Killer	David Morgan (Nottingham) Ltd	Amateur Professional	6628
Murphy Kil-Ant Ready to Use	Levington Horticulture Ltd	Amateur	5168

Product Name	Marketing Company	Use	HSE No.

276 Cypermethrin—continued

Product Name	Marketing Company	Use	HSE No.
New Siege II	Chemsearch	Amateur Professional	5638
Pure Zap	Pure-Solve Hygiene	Amateur Professional Professional (Animal Husbandry)	6061
Pyrasol C RTU	Microsol	Amateur Professional	5336
Pyrasol CP	Microsol	Professional	5478
Ready Kill	Friendly Systems	Amateur Professional	5566
Rid-Ant	Nehra Cooke's Chemicals Ltd	Amateur Professional	5748
Secto Ant and Crawling Insect Spray	Sinclair Animal and Household Care Ltd	Amateur	5994
Siege II	Chemsearch	Amateur Professional	5251
W Texas Ant Gun	Texas Homecare Ltd	Amateur	5679
Vapona Antpen	Ashe Consumer Products Ltd	Amateur Professional	3592
Vapona Flypen	Ashe Consumer Products Ltd	Amateur	3606
Wilko Ant Killer Spray	Wilkinson Home and Garden Stores	Amateur	5583

277 Cypermethrin + Chlorpyrifos

Product Name	Marketing Company	Use	HSE No.
Contra-Insect 200/20 EC	Frunol Delicia GmbH	Professional	6225
Duo 1	Hand Associates Ltd	Professional	6548

278 Cypermethrin + Methoprene

Product Name	Marketing Company	Use	HSE No.
Killgerm Precor ULV III	Killgerm Chemicals Ltd	Professional	3523

279 Cypermethrin + Permethrin + Tetramethrin + d-Allethrin

Product Name	Marketing Company	Use	HSE No.
Elf Insecticide	Elf Oil UK Ltd	Amateur	6669

280 Cypermethrin + Pyrethrins

Product Name	Marketing Company	Use	HSE No.
Bolt Crawling Insect Killer	Johnson Wax Ltd	Amateur	3609
Raid Cockroach Killer Formula 2	Johnson Wax Ltd	Amateur	4239

281 Cypermethrin + Pyrethrins + Chlorpyrifos

Product Name	Marketing Company	Use	HSE No.
New Tetracide	Killgerm Chemicals Ltd	Professional	3724

282 Cypermethrin + Tetramethrin

Product Name	Marketing Company	Use	HSE No.
Cyperkill Plus WP	Mitchell Cotts Chemicals Ltd	Professional	4855
Cytrol Universal	Pelgar International Ltd	Professional	6675

Product Name	Marketing Company	Use	HSE No.

282 Cypermethrin + Tetramethrin—continued

Cytrol XL	Pelgar International Ltd	Professional	6679
Raid Residual Crawling Insect Killer	Johnson Wax Ltd	Amateur Professional	4584
S C Johnson Wax Raid Ant and Cockroach Killer	Johnson Wax Ltd	Amateur	5216
Shelltox Cockroach and Crawling Insect Killer 3	Temana International Ltd	Amateur Professional	4031
Shelltox Cockroach and Crawling Insect Killer 4	Temana International Ltd	Amateur	4739
Super Shelltox Crawling Insect Killer	Kortman Intradal BV	Amateur	5454
Vapona Ant and Crawling Insect Killer Aerosol	Ashe Consumer Products Ltd	Amateur	5282

283 Cypermethrin + Tetramethrin + d-Allethrin

Bop Flying and Crawling Insect Killer	Robert McBride Ltd - Exports	Amateur Professional	4136
Bop Flying and Crawling Insect Killer (Water Based)	Robert McBride Ltd - Exports	Amateur Professional	4137
Bop Flying Insect Killer	Robert McBride Ltd - Exports	Amateur Professional	4141
Bop Flying Insect Killer (MC)	Robert McBride Ltd - Exports	Amateur Professional	4140
Bop Flying and Crawling Insect Killer (MC)	Robert McBride Ltd - Exports	Amateur Professional	4135

284 Cypermethrin + d-Allethrin

Shelltox Crawling Insect Killer Liquid Spray	Temana International Ltd	Amateur Professional	3565
Vapona Ant and Crawling Insect Spray	Ashe Consumer Products Ltd	Amateur	3593

285 Cyromazine + Permethrin

Staykil Household Flea Spray	Novartis Animal Health UK Ltd	Amateur Professional	5784

286 Deltamethrin

Agrevo Ant Killer	Agrevo UK Limited	Amateur Professional	5877
Ant and Crawling Insect Killer	Doff Portland Ltd	Amateur	6866
Ant Killer Spray	LC Solutions Ltd	Amateur Professional	6843
Ant-off!	Pet and Garden Manufacturing Plc	Amateur Professional	5918
Aqua K-Othrine	Agrevo UK Limited	Professional	6027

Product Name	Marketing Company	Use	HSE No.

286 Deltamethrin—continued

Product Name	Marketing Company	Use	HSE No.
B & Q Ant and Crawling Insect Spray	Miracle Garden Care Ltd	Amateur	5966
B & Q New Ant Killer Spray	B & Q Plc	Amateur Professional	5895
Bio Ant Kill Plus Spray	PBI Home & Garden Ltd	Amateur Professional	6516
Combat Ant and Crawling Insect Spray	Sinclair Animal and Household Care Ltd	Amateur	6837
Crackdown	Agrevo UK Limited	Professional	5097
Crackdown D	Agrevo UK Limited	Professional	6054
Deth-Zap Ant Killer	Gerhardt Pharmaceuticals Ltd	Amateur Professional	6869
Dethlac Insect Lacquer	Gerhardt Pharmaceuticals Ltd	Amateur	6891
Do It All Blitz! Ant Killer	Do It All Ltd	Amateur Professional	6072
Doff New Ant Killer Spray	Doff Portland Ltd	Amateur Professional	5894
Flea-Off!	Pet and Garden Manufacturing Plc	Amateur Professional	5917
Focus New Ant Killer Spray	Focus DIY Ltd	Amateur Professional	5896
Great Mills New Ant Killer Spray	Great Mills (Retail) Ltd	Amateur Professional	5897
Homebase New Ant Killer Spray	Sainsbury's Homebase House and Garden Centres	Professional	5898
Johnson's Home Flea Guard Trigger Spray	Johnson's Veterinary Products Ltd	Amateur	6106
K-Othrine Crawling Insect Powder	Agrevo UK Limited	Amateur	6055
Pro Control CIK Super	Agrevo UK Limited	Amateur	6819
Procontrol CIK Super (P)	Agrevo UK Limited	Professional	6820
S C Johnson Wax Raid Ant Killer Powder	Johnson Wax Ltd	Amateur	6267
Selkil Crawling Insect Killer	Selden Research Ltd	Professional	6008
Vapona Ant and Woodlice Killer Powder	Ashe Consumer Products Ltd	Amateur	6238
Westland Ant Killer RTU	Westland Horticulture	Amateur Professional	6852
Wilko New Ant Killer Spray	Wilkinson Home and Garden Stores	Amateur Professional	5899
Woolworths New Ant Killer Spray	Woolworths Plc	Amateur Professional	5900

287 Deltamethrin + Pyrethrins

Product Name	Marketing Company	Use	HSE No.
Agrevo Pro Control CIK Superfast	Agrevo UK Limited	Amateur	6831
Pro Control CIK Superfast (P)	Agrevo UK Limited	Professional	6836

Product Name	Marketing Company	Use	HSE No.

288 Deltamethrin + s-Bioallethrin

| Crackdown Rapide | Agrevo UK Limited | Professional | 5996 |

289 Diazinon

B and Q Ant Killer lacquer	B & Q Plc	Amateur	4750
Dethlac Insecticidal Lacquer	Gerhardt Pharmaceuticals Ltd	Amateur	4423
Doff Antlak	Doff Portland Ltd	Amateur	3591
Knox Out 2 FM	Rentokil Initial UK Ltd	Professional	4464
Secto Ant and Crawling Insect Lacquer	Sinclair Animal and Household Care Ltd	Amateur	4400
D Secto Kil-A-Line	Sinclair Animal and Household Care Ltd	Amateur	3631
Wilko Ant Killer Lacquer	Wilkinson Home and Garden Stores	Amateur	4965

290 Diazinon + Dichlorvos

| W Vijurrax Spray | Vijusa UK Ltd | Amateur | 5717 |

291 Diazinon + Pyrethrins

| D Ant Gun! | Zeneca Garden Care | Amateur | 4981 |
| D B and Q Antkiller Spray | B & Q Plc | Amateur | 5114 |

292 Dichlorophen + Dichlorvos + Tetramethrin

| W Wintox | Mould Growth Consultants Ltd | Professional | 4236 |

293 Dichlorvos

Betterware Small Space Insect Killer	Betterware UK Ltd	Amateur	5990
Boots Moth Killer	Boots the Chemist Ltd	Amateur	5543
Boots Slow Release Fly and Wasp Killer	Boots the Chemist Ltd	Amateur	5545
Freshways Slow Release Insect Killer	Freshways of York	Amateur	4618
Funnel Trap Insecticidal Strip	Agrisense BCS Ltd	Professional	4292
Globol Small Space Fly and Moth Strip	Globol Chemicals (UK) Ltd	Amateur	4692
Jeyes Expel Moth Killer	Jeyes Group Ltd	Amateur	5504
Jeyes Expel Slow Release Fly and Wasp Killer	Jeyes Group Ltd	Amateur	5505
Jeyes Kontrol Moth Killer	J Sainsbury Plc	Amateur	5546
Jeyes Kontrol Slow Release Fly and Wasp Killer	J Sainsbury Plc	Amateur	5544
Kontrol Kitchen Size Fly Killer	Sinclair Animal and Household Care Ltd	Amateur	5598
Kontrol Slow Release Fly Killer	J Sainsbury Plc	Amateur	5804

413

Product Name	Marketing Company	Use	HSE No.

293 Dichlorvos—continued

Lloyds Supersave Slow Release Fly and Wasp Killer.	Lloyds	Amateur	4920
Murphys Long Lasting Fly and Wasp Killer	Levington Horticulture Ltd	Amateur	6268
Secto Fly Killer Living Room Size	Sinclair Animal and Household Care Ltd	Amateur	4397
Secto Slow Release Fly Killer Kitchen Size	Sinclair Animal and Household Care Ltd	Amateur	4399
Teepol Products Vapona Fly Killer	Teepol Products Ltd	Amateur Professional	3673
Terminator Slow Release Fly Killer	Sanoda Ltd	Amateur	6168
Terminator Small Space Fly and Moth Killer	Sanoda Ltd	Amateur	6167
Vapona Small Space Fly Killer	Ashe Consumer Products Ltd	Amateur	4723

294 Dichlorvos + Diazinon

W Vijurrax Spray	Vijusa UK Ltd	Amateur	5717

295 Dichlorvos + Iodofenphos

D Defest Flea Free	Novartis Animal Health UK Ltd	Amateur	4331
Defest Flea Free	Sherley's Ltd	AmateurI	5715
Nuvan Staykil	Novartis Animal Health UK Ltd	Amateur Professional	4017

296 Dichlorvos + Permethrin + Bioallethrin

Farco Rapid Kill	F A Richard & Company Ltd	Amateur	6152

297 Dichlorvos + Permethrin + S-Bioallethrin

Pif Paf Insecticide	Reckitt and Colman Products Ltd	Amateur Professional	5150

298 Dichlorvos + Permethrin + Tetramethrin

D Laser Insect Killer	Sharpstow International Homecare Products Ltd	Amateur	4910
D Supaswat Insect Killer	Sharpstow International Homecare Products Ltd	Amateur	4630

299 Dichlorvos + Permethrin + Tetramethrin + d-Allethrin

Bop	Robert McBride Ltd - Exports	Amateur	4773

300 Dichlorvos + Tetramethrin

Shelltox Extra FlyKiller	Temana International Ltd	Amateur Professional	3696

Product Name	Marketing Company	Use	HSE No.

300 Dichlorvos + Tetramethrin—continued

| Shelltox FlyKiller | Temana International Ltd | Amateur | 3162 |

301 Dichlorvos + Tetramethrin + Dichlorophen

| W Wintox | Mould Growth Consultants Ltd | Professional | 4236 |

302 Fenitrothion

Antec E-C Kill	Antec International Ltd	Professional Professional (Animal Husbandry)	5468
Demise	Sorex Ltd	Professional	5084
Fenitrothion 47ec	Hand Associates Ltd	Professional	6687
Fenitrothion Dusting Powder	Rentokil Initial UK Ltd	Professional	4372
Fenitrothion Emulsion Concentrate	Rentokil Initial UK Ltd	Professional	4783
Fenitrothion Wettable Powder	Rentokil Initial UK Ltd	Professional	4827
Killgerm Fenitrothion 40 WP	Killgerm Chemicals Ltd	Professional Professional (Animal Husbandry)	4858
Killgerm Fenitrothion 50 EC	Killgerm Chemicals Ltd	Professional	4722
Micromite	Micro-Biologicals Ltd	Professional	4480
Sumithion 20 MC	Pelgar International Ltd	Professional	6209
Sumithion 20% MC	Sumitomo Chemical (UK) Plc	Professional	4905
Sumithion 40 WP	Vetrepharm Ltd	Professional Professional (Animal Husbandry)	6334

303 Fenitrothion + Permethrin + Resmethrin

| Turbair Beetle Killer | Pan Britannica Industries Ltd | Professional | 4648 |

304 Fenitrothion + Permethrin + Tetramethrin

| Connect Insect Killer | Sun Oil Ltd | Amateur | 6109 |
| Motox | HVM International Ltd | Amateur | 5622 |

305 Fenitrothion + Tetramethrin

W Big D Cockroach and Crawling Insect Killer	Domestic Fillers Ltd	Amateur	4098
Big D Cockroach and Crawling Insect Killer	Dylon International Ltd	Amateur	5834
Big D Rocket Fly Killer	Domestic Fillers Ltd	Amateur	3552
Doom Ant and Crawling Insect Killer Aerosol	Napa Products Ltd	Amateur	4753
Killgerm Fenitrothion-Pyrethroid Concentrate	Killgerm Chemicals Ltd	Professional	4623

HSE

INSECTICIDES

Product Name	Marketing Company	Use	HSE No.

305 Fenitrothion + Tetramethrin—continued

Motox Crawling Insect	HVM International Ltd	Amateur	5940
Top Kill Insect Killer	Houdret and Co Ltd	Amateur	6188
Tox Exterminating Fly and Wasp Killer	Keen (World Marketing) Ltd	Amateur	4314

306 Fipronil

Goliath Bait Station	Rhone-Poulenc Rhodic	Professional	6515
Goliath Gel	Rhone-Poulenc Rhodic	Professional	6514
Nexa Cockroach Bait Station	Scotts Celaflor GmbH and Co KG	Amateur	6588

307 Flufenoxuron

Motto	Sorex Ltd	Professional	5970

308 Flufenoxuron + Alphacypermethrin

Tenopa	Sorex Ltd	Professional	6206

309 Hydramethylnon

Faslane	Killgerm Chemicals Ltd	Professional	6380
Faslane	Killgerm Chemicals Ltd	Professional	6471
Maxforce Bait Station	Agrevo UK Limited	Professional	5371
Maxforce Gel	Agrevo UK Limited	Professional	5365
Maxforce Pharaoh's Ant Killer	Agrevo UK Limited	Professional	5082
Maxforce Ultra	Agrevo UK Limited	Professional	6013

310 Hydroprene

W Protrol	Novartis Animal Health UK Ltd	Professional	5333

311 S-Hydroprene

Protrol 9%	Novartis Animal Health UK Ltd	Professional	5909

312 Iodofenphos

Cockroach Bait	Rentokil Initial UK Ltd	Professional	4385
Iodofenphos Granular Bait	Rentokil Initial UK Ltd	Professional	4387
W Nuvanol N 500 Sc	Novartis Animal Health UK Ltd	Professional	4951
Rentokil Iodofenphos Gel	Rentokil Initial UK Ltd	Professional	4359
W Waspex	Rentokil Initial UK Ltd	Professional	4386

313 Iodofenphos + Dichlorvos

D Defest Flea Free	Novartis Animal Health UK Ltd	Amateur	4331
Defest Flea Free	Sherley's Ltd	Amateur	5715

Product Name	Marketing Company	Use	HSE No.

313 Iodofenphos + Dichlorvos—continued

| Nuvan Staykil | Novartis Animal Health UK Ltd | Amateur Professional | 4017 |

314 Lambda-Cyhalothrin

| Demand CS | Zeneca Public Health | Professional | 6287 |
| Icon 2.5 EC | Zeneca Ltd | Professional | 5614 |

315 Lindane

| W Doom Ant and Insect Powder | Napa Products Ltd | Amateur | 4570 |
| Fumite Lindane Pellet No 3 | Hortichem Limited | Professional | 4733 |

316 Lindane + Tetramethrin

| W Doom Flea Killer | Napa Products Ltd | Amateur | 4572 |
| Doom Moth Proofer Aerosol | Napa Products Ltd | Amateur | 4571 |

317 Methoprene

| Biopren BM Ready To Use Pharaoh Ant Killer Bait | Five Star Pest Control | Amateur Professional | 6826 |
| Pharorid | Novartis Animal Health UK Ltd | Professional | 5332 |

318 S-Methoprene

Altosid Liquid Larvicide	Killgerm Chemicals Ltd	Professional	6213
W Dianex	Novartis Animal Health UK Ltd	Professional	5709
Pharorid S	Sandoz Speciality Pest Control Ltd	Professional	5479

319 Methoprene + Cypermethrin

| Killgerm Precor ULV III | Killgerm Chemicals Ltd | Professional | 3523 |

320 Methoprene + Permethrin

Norshield 150	Norbrook Laboratories (GB) Ltd	Amateur	6892
Pedigree and Whiskas Exelpet Household Flea Spray	Thomas's Europe	Amateur	6346
Pedigree Exelpet Household Spray	Thomas's Europe	Amateur	6552
Precor ULV II	Killgerm Chemicals Ltd	Professional	3191
Whiskas Exelpet Household Spray	Thomas's Europe	Amateur	6554

321 S-Methoprene + Permethrin

Acclaim 2000	Sanofi Animal Health Ltd	Amateur	6448
Hartz Rid Flea Control Home Spray	Thomas Cork SM Ltd	Amateur	5835
Room Fogger	Cork International	Amateur	6598
Total Home Guard	Cork International	Amateur	6594

Product Name	Marketing Company	Use	HSE No.

321 S-Methoprene + Permethrin—continued

Vet Kem Acclaim 2000	Sanofi Animal Health Ltd	Amateur	6127
Vet-Kem Pump Spray	Sanofi Animal Health Ltd	Amateur	5475
Zodiac + Flea Spray	Petlove	Amateur	5821
W Zodiac and Pump Spray	Grosvenor Pet Health Ltd	Amateur	5472
Zodiac Maxi Household Flea Spray	Petlove	Amateur	6519

322 Methoprene + Permethrin + Bioallethrin

Arrest	Arnolds Veterinary Products Ltd	Amateur	6389
Bob Martin Home Flea Fogger Plus	The Bob Martin Company	Amateur	6630
Bob Martin Home Flea Spray Plus	The Bob Martin Company	Amateur	6223
R.I.P. Fleas	The Bob Martin Company	Amateur	6207

323 Methoprene + Permethrin + Pyrethrins

Biopren BM Residual Flea Killer Aerosol	Five Star Pest Control	Amateur Professional	6672
Friends Extra Long Lasting Household Flea Spray	Sinclair Animal and Household Care Ltd	Amateur	6685
Johnson's Household Extra Guard Flea and Insect Spray	Johnson's Veterinary Products Ltd	Amateur	6343

324 Methoprene + Permethrin + Tetramethrin

Johnson's Cage and Hutch Insect Spray	Johnson's Veterinary Products Ltd	Amateur	6864
Johnson's New Household Extra Guard Flea and Insect Spray	Johnson's Veterinary Products Ltd	Amateur	6649

325 Methoprene + Pyrethrins

Armitage Pet Care, Pet Bedding & Household Flea Spray	Armitage Brothers Plc	Amateur	6657
Biopren BM 0.68	Five Star Pest Control	Amateur Professional	6618
Canovel Pet Bedding and Household Spray	Smithkline Beecham Animal Health	Amateur	5330
Killgerm Precor RTU	Killgerm Chemicals Ltd	Professional	3165
Killgerm Precor ULV I	Killgerm Chemicals Ltd	Professional	3098
Pedigree Household Anti Flea Spray	Thomas's Europe	Amateur	6553
Pedigree and Whiskas Exelpet Household Anti Flea Spray	Thomas's Europe	Amateur	6347
Protect B Flea Killer Aerosol	Five Star Pest Control	Amateur Professional	6445
Whiskas Household Anti Flea Spray	Thomas's Europe	Amateur	6551

Product Name	Marketing Company	Use	HSE No.

326 S-Methoprene + Pyrethrins

Acclaim	Sanofi Animal Health Ltd	Amateur	5473
Canovel Pet Bedding and Household Spray Plus	Pfizer Animal Health	Amateur	5838
Raid Flea Killer Plus	Johnson Wax Ltd	Amateur	5967
Vet-Kem Acclaim	Sanofi Animal Health Ltd	Amateur	5474
Zodiac + Household Flea Spray	Petlove	Amateur	5822
W Zodiac and Household Flea Spray	Grosvenor Pet Health Ltd	Amateur	5471

327 S-Methoprene + Tetramethrin

Zodiac Household Flea Spray	Petlove	Amateur	6502

328 Naphthalene + Permethrin + Pyrethrins + Benzalkonium chloride

W Roxem D	Horton Hygiene Company	Professional	5107

329 Natamycin + Benzalkonium chloride

D Tymasil	Brocades (Great Britain) Ltd	Amateur	3625

330 Oleyl monoethoxylate + Ceto-stearyl diethoxylate

Larvex-100	Accotec Ltd	Professional	5393
Larvex-15	Accotec Ltd	Professional	5091

331 Permethrin

Amogas Ant Killer	Thornton and Ross Ltd	Amateur	5126
Ant & Crawling Insect Spray	Rentokil Initial UK Ltd	Amateur	6245
Ant Off Powder	Pet and Garden Manufacturing Plc	Amateur	6010
Armitages Good Boy Household Flea Trigger Spray	Armitage Brothers Plc	Amateur	5995
Battles Louse Powder	Battle, Hayward and Bower Ltd	Amateur	6029
Bio New Anti-Ant Duster	PBI Home & Garden Ltd	Amateur Professional	6041
Bio Wasp Nest Destroyer	PBI Home & Garden Ltd	Amateur	6021
Biokill	Bio-Industries Ltd	Professional	6870
Bob Martin Flea Bomb	The Bob Martin Company	Amateur	5891
Bob Martin Flea Kill	The Bob Martin Company	Amateur	5569
Bob Martin Home Flea Spray	The Bob Martin Company	Amateur	3914
Boots Powder Ant Killer	Boots the Chemist Ltd	Amateur	5525
Bug Free	Elite Force UK Ltd	Amateur	6148
Chirton Ant and Insect Killer	Ronson International Ltd	Amateur	3835
Constrain	Historyonics	Amateur	6149
W Coopex 25% EC	Agrevo UK Limited	Professional	5147
Coopex Insect Powder	Agrevo UK Limited	Professional	5052
Coopex WP	Agrevo UK Limited	Professional	5096
Delta Insect Powder	Bennetts Company Ltd	Amateur	4664
Doff New Ant Killer	Doff Portland Ltd	Amateur	6601

HSE
INSECTICIDES

419

Product Name	Marketing Company	Use	HSE No.

331 Permethrin—continued

Product Name	Marketing Company	Use	HSE No.
Dyna-Mite	Stock Nutrition	Amateur Professional Professional (Animal Husbandry)	6411
Flea Off Powder	Pet and Garden Manufacturing Plc	Amateur	6009
Flea Powder	Rentokil Initial UK Ltd	Amateur	5974
W Flee Flea	Flee Flea International	Amateur	5644
Flying Insect Killer Professional	Rentokil Initial UK Ltd	Professional	6216
Fresh-A-Pet Insecticidal Rug and Carpet Freshener	Bilaurand Laboratories Ltd	Amateur	4109
Friends Household Flea Powder	Sinclair Animal and Household Care Ltd	Amateur	6667
Gold Label Kennel and Stable Powder	Stockcare Ltd	Amateur	5699
Good Boy Insecticidal Carpet and Upholstery Powder	Armitage Brothers Plc	Amateur	5984
Hand Anti-Ant Powder	Hand Associates Ltd	Amateur Professional	6848
Hartz Rid Flea Carpet Freshener	Thomas Cork SM Ltd	Amateur	5392
Jeyes Expel Ant Killer Powder	Jeyes Group Ltd	Amateur	5524
Jeyes Kontrol Ant Killer Powder	Jeyes Group Ltd	Amateur	5520
Johnson's Carpet Flea Guard	Johnson's Veterinary Products Ltd	Amateur	4943
Johnson's Household Flea Powder	Johnson's Veterinary Products Ltd	Amateur	5998
Kudos	Agrevo UK Limited	Professional	5263
Lanosol P	Microsol	Professional	5477
Lanosol RTU	Microsol	Professional	5307
Lincoln Lice Control Powder	Battle, Hayward and Bower Ltd	Amateur	5956
Micrapor Environmental Flea Spray	Petlife	Amateur	5444
Mite-Kill	Net-Tex Agricultural Ltd	Amateur	6090
Mosiguard Shield	Medical Advisory Services For Travellers Abroad	Amateur	4515
Muscatrol	Rentokil Initial UK Ltd	Professional	4579
D Nippon Ant Killer Powder	Vitax Ltd	Amateur	4333
Nippon Ant Killer Powder	Vitax Ltd	Amateur	5499
Nippon Woodlice Killer Powder	Vitax Ltd	Amateur	5919
No Fleas On Me	La Compagne Sans Frontiere Ltd	Amateur	6180
Nomad Residex P	Nomad Pharmacy Ltd	Amateur	5177
D One Shot Space Spray	Roussel Uclaf Environmental Health Ltd	Professional	5086
Outright Household Flea Spray	Outright Pet Care Products	Amateur	5623

Product Name	Marketing Company	Use	HSE No.

331 Permethrin—continued

Product Name	Marketing Company	Use	HSE No.
Pedigree Exelpet Anti-Flea Carpet Powder	Thomas's Europe	Amateur	6154
Peripel 10	Agrevo UK Limited	Amateur Professional	5659
Peripel 55	Agrevo UK Limited	Professional	5658
Permasect 0.5 Dust	Mitchell Cotts Chemicals Ltd	Professional	3134
Permasect 10 WP	Mitchell Cotts Chemicals Ltd	Professional	4854
Permasect Powder	Mitchell Cotts Chemicals Ltd	Amateur	5498
Permethrin	Lifesystems Ltd	Amateur	5612
Permethrin Dusting Powder	Rentokil Initial UK Ltd	Professional	4688
Permethrin PH - 25WP	Zeneca Public Health	Professional	4956
Permethrin Wettable Powder	Rentokil Initial UK Ltd	Professional	4581
Permost 0.5% Dust Powder	Hockley International Ltd	Amateur Professional	6184
Permost 25 WP	Hockley International Ltd	Professional Professional (Animal Husbandry)	6081
Pestrin	Hughes and Hughes Ltd	Professional	4499
Pif Paf Crawling Insect Powder	Reckitt and Colman Products Ltd	Amateur	5155
Pynosect 6	Mitchell Cotts Chemicals Ltd	Professional	4528
Pynosect PCO	Mitchell Cotts Chemicals Ltd	Professional	4566
Pynosect PCP	Mitchell Cotts Chemicals Ltd	Professional	5805
Pynosect Powder	Killgerm Chemicals Ltd	Professional	5849
Raid Ant Killer Powder	Johnson Wax Ltd	Amateur Professional	5796
Residex P	Agropharm Ltd	Professional	5170
Residual Powerkill	Yule Catto Consumer Chemicals Ltd	Professional	5180
Secto Extra Strength Insect Killer	Sinclair Animal and Household Care Ltd	Amateur	4329
Secto Flea Free Insecticidal Rug and Carpet Freshener	Sinclair Animal and Household Care Ltd	Amateur	4151
Sherley's Rug-de-Bug	Sherley's Ltd	Amateur	5128
Stapro Insecticide	Stapro Ltd	Professional	4526
Tent Treatment	Ibis Products Ltd	Amateur	6050
Vapona Ant and Crawling Insect Killer Powder	Ashe Consumer Products Ltd	Amateur	5207
Vapona Carpet and Household Flea Powder	Ashe Consumer Products Ltd	Amateur	5650
Vetpet Crusade	Alstoe Ltd	Amateur	6113
Vitax Kontrol Ant Killer Powder	Vitax Ltd	Amateur	5593
Whiskas Exelpet Anti-flea Carpet Powder	Thomas's Europe	Amateur	6153
W Z-Stop Anti-Wasp Strip	Thames Laboratories Ltd	Amateur	3532

Product Name	Marketing Company	Use	HSE No.

332 Permethrin + Bioallethrin

Product Name	Marketing Company	Use	HSE No.
Agrevo Procontrol Crawling Insect Killer	Agrevo UK Limited	Amateur Professional	6116
Agrevo Procontrol Flying Insect Killer	Agrevo UK Limited	Amateur Professional	6117
Agrevo Procontrol Flying Insect Killer Aerosol	Agrevo UK Limited	Professional	6230
Ant Gun! 2	Miracle Garden Care Ltd	Amateur Professional	5629
D Ant Gun! 2	Zeneca Garden Care	Amateur Professional	5184
AntKiller Spray 2	B & Q Plc	Amateur Professional	5185
Baby Bio Houseplant Spray	PBI Home & Garden Ltd	Amateur Professional	5828
Bio Spraydex Ant and Insect Killer	PBI Home & Garden Ltd	Amateur Professional	5829
Crawling Insect and Ant Killer	Rentokil Initial UK Ltd	Amateur Professional	4650
Creepy Crawly Gun!	Miracle Garden Care Ltd	Amateur Professional	5183
Defest II	Sherley's Ltd	Amateur	6002
Farco Flying Insect Killer	F A Richard & Company Ltd	Amateur	6151
Fleegard	Bayer Plc	Amateur	5564
Homebase Ant and Crawling Insect Killer	Sainsbury's Homebase House and Garden Centres	Amateur	5846
Insectrol	Rentokil Initial UK Ltd	Amateur	4568
Insectrol Professional	Rentokil Initial UK Ltd	Professional	4624
Kayo	Nebulous Ltd	Amateur Professional	5881
Nippon Fly Killer Pads	Vitax Ltd	Amateur	5426
Pc Insect Killer	Perycut Chemie Ag	Amateur Professional	4415
Pif Paf Crawling Insect Killer	Reckitt and Colman Products Ltd	Amateur Professional	5153
Pif Paf Flying Insect Killer	Reckitt and Colman Products Ltd	Amateur Professional	5151
Pif Paf Flying Insect Killer Aerosol	Reckitt and Colman Products Ltd	Amateur	5152
Pybuthrin Fly Killer	Agrevo UK Limited	Amateur Professional	5141
Pybuthrin Fly Spray.	Agrevo UK Limited	Amateur	5116
Sainsbury's Crawling Insect and Ant Killer	J Sainsbury Plc	Amateur	5231
W Spraydex Ant & Insect Killer	Pan Britannica Industries Ltd	Amateur	5645
W Spraydex Houseplant Spray	Spraydex Ltd	Amateur	3540
W Spraydex Insect Killer	Spraydex Ltd	Amateur	3834

Product Name	Marketing Company	Use	HSE No.

332 Permethrin + Bioallethrin—continued

Vapona Ant and Crawling Insect Killer	Ashe Consumer Products Ltd	Amateur Professional	5206
Vapona Fly and Wasp Killer	Ashe Consumer Products Ltd	Amateur	5205
D Zap Pest Control	Delsol Ltd	Amateur Professional	5068

333 Permethrin + Bioallethrin + Dichlorvos

Farco Rapid Kill	F A Richard & Company Ltd	Amateur	6152

334 Permethrin + Bioallethrin + Methoprene

Arrest	Arnolds Veterinary Products Ltd	Amateur	6389
Bob Martin Home Flea Fogger Plus	The Bob Martin Company	Amateur	6630
Bob Martin Home Flea Spray Plus	The Bob Martin Company	Amateur	6223
R.I.P. Fleas	The Bob Martin Company	Amateur	6207

335 Permethrin + Cyromazine

Staykil Household Flea Spray	Novartis Animal Health UK Ltd	Amateur Professional	5784

336 Permethrin + Methoprene

Norshield 150	Norbrook Laboratories (GB) Ltd	Amateur	6892
Pedigree and Whiskas Exelpet Household Flea Spray	Thomas's Europe	Amateur	6346
Pedigree Exelpet Household Spray	Thomas's Europe	Amateur	6552
Precor ULV II	Killgerm Chemicals Ltd	Professional	3191
Whiskas Exelpet Household Spray	Thomas's Europe	Amateur	6554

337 Permethrin + Propiconazole

Aqueous Universal	Protim Solignum Ltd	Amateur Professional	5776

338 Permethrin + Pyrethrins

Detia Crawling Insect Spray	K D Pest Control Products	Amateur	3710
Flamil Finale	Flore-Chemie GmbH	Professional	5083
Home Insect Fogger	International Surplus (UK) Ltd	Amateur	6364
D Household Flea Spray	Johnson's Veterinary Products Ltd	Amateur	3908
Insecticidal Aerosol	Aerosols International Ltd	Amateur	5118
Johnson's Household Flea Spray	Johnson's Veterinary Products Ltd	Amateur	5703
Johnson's Pet Housing Spray	Johnson's Veterinary Products Ltd	Amateur	5843

Product Name	Marketing Company	Use	HSE No.

338 Permethrin + Pyrethrins—continued

SC Johnson Wax Raid Ant & Cockroach Killer with Natural Pyrethrum	Johnson Wax Ltd	Amateur	6088
Secto Household Flea Spray	Sinclair Animal and Household Care Ltd	Amateur	4131
X-Gnat M/C (Mosquito Control)	X-Gnat Laboratories Ltd	Professional	6323

339 Permethrin + Pyrethrins + Benzalkonium chloride + Naphthalene

| W Roxem D | Horton Hygiene Company | Professional | 5107 |

340 Permethrin + Pyrethrins + Bioallethrin

| Supabug Crawling Insect Killer | Sharpstow International Homecare Products Ltd | Amateur | 4794 |
| Tyrax | DBC | Amateur | 6829 |

341 Permethrin + Pyrethrins + Chlorpyrifos-methyl

| W Multispray | Agrevo UK Limited | Professional | 5165 |

342 Permethrin + Pyrethrins + Methoprene

Biopren BM Residual Flea Killer Aerosol	Five Star Pest Control	Amateur Professional	6672
Friends Extra Long Lasting Household Flea Spray	Sinclair Animal and Household Care Ltd	Amateur	6685
Johnson's Household Extra Guard Flea and Insect Spray	Johnson's Veterinary Products Ltd	Amateur	6343

343 Permethrin + Pyriproxyfen

| Indorex Spray | Virbac S A | Amateur | 6626 |
| Sergeants Household Patrol | Seven Seas Ltd | Amateur | 6191 |

344 Permethrin + Resmethrin + Fenitrothion

| Turbair Beetle Killer | Pan Britannica Industries Ltd | Professional | 4648 |

345 Permethrin + S-Bioallethrin

Agrevo Procontrol Yellow Flying Insect Killer Aerosol	Agrevo UK Limited	Professional	6232
Aqua Reslin Premium	Agrevo UK Limited	Professional	5172
Aqua Reslin Super	Agrevo UK Limited	Professional	5175
Atta-x Flying Insect Killer	Sanderson Curtis Ltd	Amateur	6348
Bug Wars	Jack Rogers and Co Ltd	Amateur	6361
W Coopex S25% EC	Agrevo UK Limited	Professional	5146
W Imperator Fog S 18/6	Zeneca Public Health	Professional	4960
W Imperator Fog Super S 30/15	Zeneca Public Health	Professional	4955
W Imperator ULV S 6/3	Zeneca Public Health	Professional	4954
W Imperator ULV Super S 10/4	Zeneca Public Health	Professional	4953

Product Name	Marketing Company	Use	HSE No.

345 Permethrin + S-Bioallethrin—continued

W Neopybuthrin Premium	Agrevo UK Limited	Professional	5149
W Neopybuthrin Super	Agrevo UK Limited	Professional	5148
Pif Paf Yellow Flying Insect Killer	Reckitt and Colman Products Ltd	Amateur	5937
Purge	Aztec Aerosols Limited	Amateur	6634
Resigen	Agrevo UK Limited	Professional	5143
Reslin Premium	Agrevo UK Limited	Professional	5174
W Reslin Super	Agrevo UK Limited	Professional	5173

346 Permethrin + S-Bioallethrin + Dichlorvos

Pif Paf Insecticide	Reckitt and Colman Products Ltd	Amateur Professional	5150

347 Permethrin + S-Bioallethrin + Tetramethrin

Agrevo Procontrol Crawling Insect Killer Aerosol	Agrevo UK Limited	Professional	6231
Pif Paf Crawling Insect Killer Aerosol	Reckitt and Colman Products Ltd	Amateur	5595

348 Permethrin + S-Methoprene

Acclaim 2000	Sanofi Animal Health Ltd	Amateur	6448
Hartz Rid Flea Control Home Spray	Thomas Cork Sm Ltd	Amateur	5835
Room Fogger	Cork International	Amateur	6598
Total Home Guard	Cork International	Amateur	6594
Vet Kem Acclaim 2000	Sanofi Animal Health Ltd	Amateur	6127
Vet-Kem Pump Spray	Sanofi Animal Health Ltd	Amateur	5475
Zodiac + Flea Spray	Petlove	Amateur	5821
W Zodiac and Pump Spray	Grosvenor Pet Health Ltd	Amateur	5472
Zodiac Maxi Household Flea Spray	Petlove	Amateur	6519

349 Permethrin + Tetramethrin

Bop Crawling Insect Killer	Robert McBride Ltd - Exports	Amateur Professional	4138
Bop Crawling Insect Killer (MC)	Robert McBride Ltd - Exports	Amateur Professional	4139
Atta-x Cockroach Killer	Sanderson Curtis Ltd	Amateur	6362
W Bolt Ant and Crawling Insect Killer	Johnson Wax Ltd	Amateur	3619
Canned Death	Pearson & Wilkinson	Professional	6112
Combat Fly Killer	Sinclair Animal and Household Care Ltd	Amateur	5808
Corry's Fly and Wasp Killer	Vitax Ltd	Amateur	5322
D-Stroy	Future Developments (Manufacturing) Ltd	Professional	5411
Delta Cockroach and Ant Spray	Bennetts Company Ltd	Amateur	3950
Doom Tropical Strength Fly and Wasp Killer Aerosol	Napa Products Ltd	Amateur	4752

Product Name	Marketing Company	Use	HSE No.

349 Permethrin + Tetramethrin—continued

Product Name	Marketing Company	Use	HSE No.
Dragon FlyKiller	Zeneca Public Health	Amateur	4952
D Fly and Wasp Killer	Dimex Ltd	Professional	5050
Fly and Wasp Killer	Rentokil Initial UK Ltd	Amateur Professional	4643
Fly Free Zone	Helen Dolisznyj's Fly Away	Amateur Professional Professional (Animal Husbandry)	6394
Fly Spray	Merton Cleaning Supplies	Professional	5112
Fly-Kill	Stagchem Ltd	Professional	6202
Hail Flying Insect Killer	Robert McBride Ltd - Exports	Amateur	5568
Insecticide Aerosol	Aerosols International Ltd	Amateur	6567
Insectrol, Fly and Wasp Killer	Rentokil Initial UK Ltd	Amateur	5827
Johnson's New Household Flea Spray	Johnson's Veterinary Products Ltd	Amateur	6623
Johnson's New Pet Housing Spray	Johnson's Veterinary Products Ltd	Amateur	6624
K02 Selkil	Selden Research Ltd	Professional	4828
Nippon Ant and Crawling Insect Killer	Vitax Ltd	Amateur	4414
Nippon Fly Killer Spray	Vitax Ltd	Amateur	4254
Nippon Killaquer Crawling Insect Killer	Arrow Chemicals Ltd	Professional	3694
Nisa Fly and Wasp Killer	Nisa	Amateur Professional	4962
Permost Uni	Hockley International Ltd	Professional Professional (Animal Husbandry)	6399
Purge Insect Destroyer	Aztec Chemicals Ltd	Professional	5048
Pynosect 10	Mitchell Cotts Chemicals Ltd	Professional	3171
Pynosect Extra Fog	Mitchell Cotts Chemicals Ltd	Professional	3172
W Raid Ant and Roach Killer	Johnson Wax Ltd	Amateur	5182
Sactif Flying Insect Killer	Diverseylever Ltd	Professional	4924
Sainsbury's Fly and Wasp Killer	J Sainsbury Plc	Amateur	5335
Sanmex Ant & Crawling Insect Killer	The British Products Sanmex Company Ltd	Amateur	6655
Sanmex Fly and Wasp Killer	The British Products Sanmex Company Ltd	Amateur	4006
Sanmex Supakil Insecticide	The British Products Sanmex Company Ltd	Amateur	4009
Scorpio Fly Spray For Flying Insects	Jangro Ltd	Professional	5049
Secto Fly Killer	Sinclair Animal and Household Care Ltd	Amateur	3862

Product Name	Marketing Company	Use	HSE No.

349 Permethrin + Tetramethrin—continued

Shelltox Insect Killer	Temana International Ltd	Amateur Professional	3697
Shelltox Insect Killer 2	Temana International Ltd	Amateur Professional	4061
Stop Insect Killer	Temana International Ltd	Amateur Professional	3163
Superdrug Fly Killer	Superdrug Stores Plc	Amateur	4185
Zappit	Blp Ltd	Amateur	6181

350 Permethrin + Tetramethrin + Dichlorvos

D Laser Insect Killer	Sharpstow International Homecare Products Ltd	Amateur	4910
D Supaswat Insect Killer	Sharpstow International Homecare Products Ltd	Amateur	4630

351 Permethrin + Tetramethrin + Fenitrothion

Connect Insect Killer	Sun Oil Ltd	Amateur	6109
Motox	HVM International Ltd	Amateur	5622

352 Permethrin + Tetramethrin + Methoprene

Johnson's Cage and Hutch Insect Spray	Johnson's Veterinary Products Ltd	Amateur	6864
Johnson's New Household Extra Guard Flea and Insect Spray	Johnson's Veterinary Products Ltd	Amateur	6649

353 Permethrin + Tetramethrin + d-Allethrin

Hail Plus Crawling Insect Killer	Robert McBride Ltd - Exports	Amateur	5567
New Super Raid Insecticide	Johnson Wax Ltd	Amateur Professional	4583

354 Permethrin + Tetramethrin + d-Allethrin + Cypermethrin

Elf Insecticide	Elf Oil UK Ltd	Amateur	6669

355 Permethrin + Tetramethrin + d-Allethrin + Dichlorvos

Bop	Robert McBride Ltd - Exports	Amateur	4773

356 Permethrin + d-Allethrin

New Vapona Fly and Wasp Killer	Ashe Consumer Products Ltd	Amateur	4201
Raid Outdoor Insectguard	Johnson Wax Ltd	Amateur	5459

357 d-Phenothrin

Aircraft Aerosol Insect Control	Sumitomo Chemical (UK) Plc	Professional	5408
Aircraft Disinfectant	Arrow Chemicals Ltd	Professional	6254

Product Name	Marketing Company	Use	HSE No.

357 d-Phenothrin—continued

D Insecticide Aerosol Single Dose For Aircraft Disinfection	Aerosols International Ltd	Professional	3797
One-Shot Aircraft Aerosol Insect Control	Sumitomo Chemical (UK) Plc	Professional	5407
Phillips Pestkill Karpet Killer	Seven Seas	Amateur	5604
Roebuck Eyot Aircraft Disinfection Spray	Roebuck Eyot Ltd	Professional	6404
Secto Household Dust Mite Control	Sinclair Animal and Household Care Ltd	Amateur	5870
Sergeant's Dust Mite Patrol	Allerayde Ltd	Amateur	6012
Sergeants Car Patrol	Seven Seas Ltd	Amateur	4503
Sergeants Carpet Patrol	Seven Seas Ltd	Amateur	4763
W Sergeants Dust Mite Patrol	Seven Seas Ltd	Amateur	4862
Sergeants Rug Patrol	Seven Seas Ltd	Amateur	4504
Sumithrin 10 Sec	Sumitomo Chemical (UK) Plc	Professional	3762
Sumithrin 10 Sec Carpet Treatment	Sumitomo Chemical (UK) Plc	Amateur	5398
D Wellcome Multishot Aircraft Aerosol	The Wellcome Foundation Ltd	Professional	3990
D Wellcome One-Shot Aircraft Aerosol	The Wellcome Foundation Ltd	Professional	3989
Willo-Zawb	Willows Francis Ltd	Amateur	3076
Willodorm	Willows Francis Ltd	Amateur	4782

358 d-Phenothrin + Pyrethrins

Pyrbek	Hand Associates Ltd	Professional	6550

359 d-Phenothrin + Pyriproxyfen

Nylar ULV 65	Pelgar International Ltd	Professional	6593

360 d-Phenothrin + Tetramethrin

Flak	Robert McBride Ltd - Exports	Amateur Professional	4143
Ant Destroyer Foam	Rentokil Initial UK Ltd	Amateur	6412
Betterware Foaming Ant & Wasp Nest Destroyer	Betterware UK Ltd	Amateur	6615
W Big D Ant and Crawling Insect Killer	Domestic Fillers Ltd	Amateur	4097
Big D Ant and Crawling Insect Killer	Dylon International Ltd	Amateur	5818
Bio Foaming Wasp Nest Destroyer	PBI Home & Garden Ltd	Amateur	6665
Boots Ant and Crawling Insect Killer	Boots the Chemist Ltd	Amateur	5554
Boots Flea Spray	Boots the Chemist Ltd	Amateur	5522
Bop Arabic	Robert McBride Ltd - Exports	Amateur Professional	4142

Product Name	Marketing Company	Use	HSE No.

360 d-Phenothrin + Tetramethrin—continued

Product Name	Marketing Company	Use	HSE No.
Brimpex ULV 1500	Brimpex Metal Treatments	Professional Professional (Animal Husbandry)	4942
Bug Off	Dimex Ltd	Amateur	6356
Deosan Fly Spray	Diverseylever Ltd	Professional	5246
Envar 6/14 EC	Agrimar (UK) Plc	Professional	4354
Envar 6/14 Oil	Agrimar (UK) Plc	Professional	4355
Flying and Crawling Insect Killer	Betterware UK Ltd	Amateur	6607
Friends Household Flea Spray	Sinclair Animal and Household Care Ltd	Amateur	6664
G.C.P. Flying & Crawling Insect Killer	Greenhill Chemical Products Ltd	Amateur	6303
Insect Killer	Rentokil Initial UK Ltd	Amateur	6244
Jeyes Expel Ant and Crawling Insect Killer	Jeyes Group Ltd	Amateur	5552
Jeyes Expel Flea Killer	Jeyes Group Ltd	Amateur	5523
Jeyes Kontrol Ant and Crawling Insect Killer	Jeyes Group Ltd	Amateur	5553
Jeyes Kontrol Flea Killer	Jeyes Group Ltd	Amateur	5521
Killgerm ULV 500	Killgerm Chemicals Ltd	Professional Professional (Animal Husbandry)	4647
Killoff Aerosol Insect Killer	Ferosan Products	Amateur	3974
Kybosh 2	PBI Home & Garden Ltd	Amateur	6622
Murphy Foaming Wasp Nest Destroyer	Levington Horticulture Ltd	Amateur	6609
New Secto Household Flea Spray	Sinclair Animal and Household Care Ltd	Amateur	6504
Pesguard NS 6/14 EC	Sumitomo Chemical (UK) Plc	Professional	4360
Pesguard OBA F 7305 C	Sumitomo Chemical (UK) Plc	Amateur	4064
Pesguard OBA F 7305 D	Sumitomo Chemical (UK) Plc	Amateur	4065
Pesguard OBA F 7305 E	Sumitomo Chemical (UK) Plc	Amateur	4066
Pesguard OBA F 7305 F	Sumitomo Chemical (UK) Plc	Amateur	4067
Pesguard OBA F 7305 G	Sumitomo Chemical (UK) Plc	Amateur	4063
Pesguard OBA F7305 B	Sumitomo Chemical (UK) Plc	Amateur	4419
Pesguard WBA F-2656	Sumitomo Chemical (UK) Plc	Amateur	4263
Pesguard WBA F-2692	Sumitomo Chemical (UK) Plc	Amateur	4262
Pestkill Household Spray	Seven Seas	Amateur	4938
Goodboy Household Flea Spray	Armitage Brothers Plc	Amateur	4090
Raid Fly Wasp and Plant Insect Killer	Johnson Wax Ltd	Amateur	5432
Rentokil Wasp Nest Destroyer	Rentokil Initial UK Ltd	Amateur	6214
S C Johnson Professional Raid Flying Insect Killer	S C Johnson Professional Ltd	Amateur	6407

Product Name	Marketing Company	Use	HSE No.

360 d-Phenothrin + Tetramethrin—continued

Product Name	Marketing Company	Use	HSE No.
S C Johnson Wax Raid Fly and Wasp Killer	Johnson Wax Ltd	Amateur	5181
Sergeant's Dust Mite Patrol Injector and Spray	Seven Seas Ltd	Amateur Professional	5513
W Sergeant's Dust Mite Patrol Spray	Seven Seas Ltd	Amateur	5041
Sergeant's Household Spray	Seven Seas Ltd	Amateur	6011
W Sergeant's Insect Patrol	Seven Seas Ltd	Amateur	5397
Shelltox Flying Insect Killer 2	Temana International Ltd	Amateur Professional	4032
Shelltox Flying Insect Killer 3	Temana International Ltd	Amateur	4741
Shelltox Super Flying Insect Killer	Temana International Ltd	Amateur	4790
Sorex Fly Spray RTU	Sorex Ltd	Professional	5718
Sorex Super Fly Spray	Sorex Ltd	Amateur	6297
Supaswat Insect Killer	Sharpstow International Homecare Products Ltd	Amateur	5691
Super Fly Spray	Sorex Ltd	Amateur	4468
Super Raid II	Johnson Wax Ltd	Amateur	4015
Super Shelltox Flying Insect Killer	Kortman Intradal BV	Amateur	5440
Terminator	Aerosol Logistics	Amateur	6416
Terminator Fly and Wasp Killer	Sanoda Ltd	Amateur	6150
Terminator Outdoor Insect Killer	Sanoda Ltd	Amateur	6171
Terminator Wasp and Insect Nest Destroyer	Sanoda Ltd	Amateur	6144
Vapona House and Plant Fly Spray	Ashe Consumer Products Ltd	Amateur	5236
Vapona Wasp and Fly Killer	Ashe Consumer Products Ltd	Amateur	5754
Vapona Wasp and Fly Killer Spray	Ashe Consumer Products Ltd	Amateur	5280
Vitapet Flearid Household Spray	Seven Seas Ltd	Amateur	5434
Wasp Destroyer Foam	Rentokil Initial UK Ltd	Amateur	6455
Zappit Fly and Wasp Killer	BLP Ltd	Amateur	6336

361 d-Phenothrin + d-Allethrin

Product Name	Marketing Company	Use	HSE No.
Boots New Improved Fly and Wasp Killer	Boots the Chemist Ltd	Amateur	6432
Fly and Ant Killer	Rentokil Initial UK Ltd	Amateur	5561
Jeyes Crawling Insect Spray	Jeyes Group Ltd	Amateur	5270
Kleeneze New Improved Fly Killer	Kleeneze Ltd	Amateur	6431
Kleenoff Crawling Insect Spray	Jeyes Group Ltd	Amateur	5269
Kwik Insecticide	Hand Associates Ltd	Amateur	4290
Nippon Ready For use Fly Killer Spray	Vitax Ltd	Amateur	4631
Pedigree Bedding Spray	Thomas's Europe	Amateur	5427
Pesguard PS 102	Sumitomo Chemical (UK) Plc	Amateur	5312
Pesguard PS 102A	Sumitomo Chemical (UK) Plc	Amateur	4173
Pesguard PS 102B	Sumitomo Chemical (UK) Plc	Amateur	4174

Product Name	Marketing Company	Use	HSE No.

361 d-Phenothrin + d-Allethrin—continued

Product Name	Marketing Company	Use	HSE No.
Pesguard PS 102C	Sumitomo Chemical (UK) Plc	Amateur	4175
Pesguard PS 102D	Sumitomo Chemical (UK) Plc	Amateur	4176
W Secto Ant and Crawling Insect Spray.	Sinclair Animal and Household Care Ltd	Amateur	5714
W Target Flying Insect Killer	Reckitt and Colman Products Ltd	Amateur	4178
Whiskas Bedding Spray	Thomas's Europe	Amateur	4837
Whiskas Bedding Pest Control Spray	Thomas's Europe	Amateur	4596

362 d-Phenothrin + d-Tetramethrin

Product Name	Marketing Company	Use	HSE No.
Pesguard WBA F2714	Sumitomo Chemical (UK) Plc	Amateur	4418
Vapona Max Concentrated Fly and Wasp Killer	Ashe Consumer Products Ltd	Amateur	6062

363 Pirimiphos-methyl

Product Name	Marketing Company	Use	HSE No.
Actellic 25 EC	Zeneca Public Health	Professional Professional (Animal Husbandry) Professional (Food Storage Practice) Professional (Wood Preservative)	4880
Actellic Dust	Zeneca Public Health	Professional	4881
AntKiller Dust 2	Miracle Garden Care Ltd	Amateur	6384

364 Pirimiphos-methyl + Resmethrin + Tetramethrin

Product Name	Marketing Company	Use	HSE No.
Waspend	Miracle Garden Care Ltd	Amateur Professional	4857

365 Potassium salts of fatty acids

Product Name	Marketing Company	Use	HSE No.
Devcol Liquid Ant Killer	Nehra Cooke's Chemicals Ltd	Amateur	5518
Devcol Wasp Killer	Nehra Cooke's Chemicals Ltd	Amateur	5519
Greenco Ant Killer	Doff Portland Ltd	Amateur	4260
Natural Ant Gun	Doff Portland Ltd	Amateur	4916
Natural Wasp Gun	Doff Portland Ltd	Amateur	5327
Natural Wasp Killer	Doff Portland Ltd	Amateur	5326

366 Prallethrin

Product Name	Marketing Company	Use	HSE No.
Boots Mosquito Killer	Boots the Chemist Ltd	Amateur	6590
Etoc 10 MG Mat	Sumitomo Chemical (UK) Plc	Amateur	5925

431

Product Name	Marketing Company	Use	HSE No.

366 Prallethrin—continued

Etoc Liquid Vaporiser	Sumitomo Chemical (UK) Plc	Amateur	5618
Mosqui-Go liquid	Go Travel Products	Amateur	6654
Tourist Flying Insect Killer	Pik-A-Pak	Amateur	6447
Vape Mat	Travel Companions	Amateur	6666

367 Propiconazole + Permethrin

| Aqueous Universal | Protim Solignum Ltd | Amateur
Professional | 5776 |

368 Propoxur

| Killgerm Propoxur 20 EC | Killgerm Chemicals Ltd | Professional | 4546 |

369 Pyrethrins

Agrevo Procontrol Fly Spray	Agrevo UK Limited	Professional	5154
Aquapy	Agrevo UK Limited	Professional	5799
Aquapy Micro	Agrevo UK Limited	Professional	6296
Bob Martin Natural Flea Killing Pet Bedding Spray	The Bob Martin Company	Amateur	6256
Bob Martin Natural Household Flea Spray	The Bob Martin Company	Amateur	6495
Bolt Dry Flying Insect Killer	Johnson Wax Ltd	Amateur	3581
Boots Pump Action Fly and Wasp Killer	Boots the Chemist Ltd	Amateur	5292
Boots Pump Action Fly and Wasp Killer	Boots the Chemist Ltd	Amateur	5547
Cookes Ant Killer	Nehra Cooke's Chemicals Ltd	Amateur	5972
D Coopers Fly Spray N	Pitman-Moore Ltd	Professional Professional (Animal Husbandry)	5377
Dairy Fly Spray	Battle, Hayward and Bower Ltd	Professional Professional (Animal Husbandry)	5579
Days Farm Fly Spray	Day and Sons (Crewe) Ltd	Professional	4749
Detia Pyrethrum Spray	K D Pest Control Products	Amateur	3695
W Devcol Universal Pest Powder	Devcol Ltd	Amateur	4864
Downland Dairy Fly Spray	Downland Marketing Ltd	Professional Professional (Animal Husbandry)	6032
Drione	Agrevo UK Limited	Professional	5254
Emprasan Pendle Mist Fly Spray	Emprasan (Chemicals) Limited	Professional	6239

Product Name	Marketing Company	Use	HSE No.

369 Pyrethrins—continued

Product Name	Marketing Company	Use	HSE No.
Flamil Finale Super	Flore-Chemie GmbH	Amateur Professional	5615
Fortefog	Agropharm Ltd	Amateur Professional	4820
Fortefog	Agropharm Ltd	Amateur Professional	6486
Globol Shake and Spray Insect Killer	Globol Chemicals (UK) Ltd	Amateur	4883
Growing Success Ant Killer	Growing Success Organics Ltd	Amateur	5831
Growing Success Indoor and Outdoor Ant Killer	Growing Success Organics Ltd	Amateur	6442
Jeyes Expel Pump Action Fly and Wasp Killer	Jeyes Group Ltd	Amateur	5506
Jeyes Flying Insect Spray	Jeyes Group Ltd	Amateur	5248
W Karate Dairy Fly Spray	Lever Industrial Ltd	Professional Professional (Animal Husbandry)	5611
Killgerm Pyrethrum Spray	Killgerm Chemicals Ltd	Professional Professional (Animal Husbandry)	4636
Killgerm ULV 400	Killgerm Chemicals Ltd	Professional Professional (Animal Husbandry)	4838
Kleeneze Fly and Flying Insect Pump Spray	Kleeneze Ltd	Amateur	6122
Kleenoff Flying Insect Spray	Jeyes Group Ltd	Amateur	5247
Konk I	Airguard Control Inc	Professional Professional (Animal Husbandry)	5085
Patriot Flying and Crawling Insect Killer	Agropharm Ltd	Amateur Professional	4959
Patriot Flying and Crawling Insect Killer Aerosol	Agropharm Ltd	Amateur Professional	6166
Prevent	Agropharm Ltd	Amateur	4851
Pybuthrin 2/16	Agrevo UK Limited	Professional	5115
Pybuthrin 33	Agrevo UK Limited	Professional	5106
D Pyrematic Flying Insect Killer	Colmart Marketing Ltd	Professional Professional (Animal Husbandry)	4735
Raid Dry Natural Flying Insect Killer	Johnson Wax Ltd	Amateur	4235
Residex	Agropharm Ltd	Amateur	4811

Product Name	Marketing Company	Use	HSE No.

369 Pyrethrins—continued

Product Name	Marketing Company	Use	HSE No.
Samsung Natural Insect Spray	Samsung Deutschtland GmbH	Amateur Professional	6121
SC Johnson Flying Insect Killer	Johnson Wax Ltd	Amateur	5094
Superdrug Fly and Wasp Killer Pump Spray	Superdrug Stores Plc	Amateur	5171
Swak Natural	Technical Concepts International Ltd	Professional	4982
Trappit Ant Powder	Agrisense BCS Ltd	Amateur	6613
Trilanco Farm Fly Spray	Trilanco	Professional	4558
Trilanco Super Fly Spray	Trilanco	Professional	6824
Vapona Green Arrow Fly and Wasp Killer	Ashe Consumer Products Ltd	Amateur	4460
Vapona House and Plant Fly Aerosol	Ashe Consumer Products Ltd	Amateur	5235

370 Pyrethrins + Bendiocarb

Product Name	Marketing Company	Use	HSE No.
Ficam Plus	Agrevo UK Limited	Professional	4830

371 Pyrethrins + Benzalkonium chloride + Naphthalene + Permethrin

Product Name	Marketing Company	Use	HSE No.
W Roxem D	Horton Hygiene Company	Professional	5107

372 Pyrethrins + Bioallethrin + Permethrin

Product Name	Marketing Company	Use	HSE No.
Supabug Crawling Insect Killer	Sharpstow International Homecare Products Ltd	Amateur	4794
Tyrax	DBC	Amateur	6829

373 Pyrethrins + Chlorpyrifos

Product Name	Marketing Company	Use	HSE No.
Contra-Insect Aerosol	Frunol Delicia GmbH	Amateur	6372
Contra-Insect universal	Frunol Delicia GmbH	Amateur	6405

374 Pyrethrins + Chlorpyrifos + Cypermethrin

Product Name	Marketing Company	Use	HSE No.
New Tetracide	Killgerm Chemicals Ltd	Professional	3724

375 Pyrethrins + Chlorpyrifos-methyl + Permethrin

Product Name	Marketing Company	Use	HSE No.
W Multispray	Agrevo UK Limited	Professional	5165

376 Pyrethrins + Cypermethrin

Product Name	Marketing Company	Use	HSE No.
Bolt Crawling Insect Killer	Johnson Wax Ltd	Amateur	3609
Raid Cockroach Killer Formula 2	Johnson Wax Ltd	Amateur	4239

377 Pyrethrins + Deltamethrin

Product Name	Marketing Company	Use	HSE No.
Agrevo Pro Control Cik Superfast	Agrevo UK Limited	Amateur	6831
Pro Control Cik Superfast (P)	Agrevo UK Limited	Professional	6836

Product Name	Marketing Company	Use	HSE No.

378 Pyrethrins + Diazinon

D Ant Gun!	Zeneca Garden Care	Amateur	4981
D B and Q Antkiller Spray	B & Q Plc	Amateur	5114

379 Pyrethrins + Methoprene

Armitage Pet Care, Pet Bedding & Household Flea Spray	Armitage Brothers Plc	Amateur	6657
Biopren BM 0.68	Five Star Pest Control	Amateur Professional	6618
Canovel Pet Bedding and Household Spray	Smithkline Beecham Animal Health	Amateur	5330
Killgerm Precor RTU	Killgerm Chemicals Ltd	Professional	3165
Killgerm Precor ULV I	Killgerm Chemicals Ltd	Professional	3098
Pedigree Household Anti Flea Spray	Thomas's Europe	Amateur	6553
Pedigree and Whiskas Exelpet Household Anti Flea Spray	Thomas's Europe	Amateur	6347
Protect B Flea Killer Aerosol	Five Star Pest Control	Amateur Professional	6445
Whiskas Household Anti Flea Spray	Thomas's Europe	Amateur	6551

380 Pyrethrins + Methoprene + Permethrin

Biopren BM Residual Flea Killer Aerosol	Five Star Pest Control	Amateur Professional	6672
Friends Extra Long Lasting Household Flea Spray	Sinclair Animal and Household Care Ltd	Amateur	6685
Johnson's Household Extra Guard Flea and Insect Spray	Johnson's Veterinary Products Ltd	Amateur	6343

381 Pyrethrins + Permethrin

Detia Crawling Insect Spray	K D Pest Control Products	Amateur	3710
Flamil Finale	Flore-Chemie GmbH	Professional	5083
Home Insect Fogger	International Surplus (UK) Ltd	Amateur	6364
D Household Flea Spray	Johnson's Veterinary Products Ltd	Amateur	3908
Insecticidal Aerosol	Aerosols International Ltd	Amateur	5118
Johnson's Household Flea Spray	Johnson's Veterinary Products Ltd	Amateur	5703
Johnson's Pet Housing Spray	Johnson's Veterinary Products Ltd	Amateur	5843
SC Johnson Wax Raid Ant & Cockroach Killer with Natural Pyrethrum	Johnson Wax Ltd	Amateur	6088
Secto Household Flea Spray	Sinclair Animal and Household Care Ltd	Amateur	4131

Product Name	Marketing Company	Use	HSE No.

381 Pyrethrins + Permethrin—continued

X-Gnat M/C (Mosquito Control)	X-Gnat Laboratories Ltd	Professional	6323

382 Pyrethrins + Resmethrin

Rentokil Houseplant Insect Killer	Rentokil Initial UK Ltd	Amateur	4433

383 Pyrethrins + S-Methoprene

Acclaim	Sanofi Animal Health Ltd	Amateur	5473
Canovel Pet Bedding and Household Spray Plus	Pfizer Animal Health	Amateur	5838
Raid Flea Killer Plus	Johnson Wax Ltd	Amateur	5967
Vet-Kem Acclaim	Sanofi Animal Health Ltd	Amateur	5474
Zodiac + Household Flea Spray	Petlove	Amateur	5822
W Zodiac and Household Flea Spray	Grosvenor Pet Health Ltd	Amateur	5471

384 Pyrethrins + d-Phenothrin

Pyrbek	Hand Associates Ltd	Professional	6550

385 Pyriproxyfen

Nylar 10 EC	Pelgar International Ltd	Professional	6219
Nylar 100	Pelgar International Ltd	Professional	6234
Nylar 4EW	Pelgar International Ltd	Professional	6373
Sumilarv 10 EC	Sumitomo Chemical (UK) Plc	Professional	5991

386 Pyriproxyfen + Cyfluthrin

Fleegard Plus	Bayer Plc	Amateur	6582

387 Pyriproxyfen + Permethrin

Indorex Spray	Virbac S A	Amateur	6626
Sergeants Household Patrol	Seven Seas Ltd	Amateur	6191

388 Pyriproxyfen + d-Phenothrin

Nylar ULV 65	Pelgar International Ltd	Professional	6593

389 Resmethrin + Fenitrothion + Permethrin

Turbair Beetle Killer	Pan Britannica Industries Ltd	Professional	4648

390 Resmethrin + Pyrethrins

Rentokil Houseplant Insect Killer	Rentokil Initial UK Ltd	Amateur	4433

391 Resmethrin + Tetramethrin

151 Dry Fly and Wasp Killer	SB Ltd	Amateur	5169
Apex Fly-Killer	Apex Industrial Chemicals Ltd	Amateur Professional	4300
Chem-kil	Inter-Chem	Amateur Professional	4266

Product Name	Marketing Company	Use	HSE No.

391 Resmethrin + Tetramethrin—continued

Product Name	Marketing Company	Use	HSE No.
Chirton Dry Fly and Wasp Killer	Ronson International Ltd	Amateur	5287
W Cromessol Insect Killer	Cromessol Company Ltd	Amateur	5067
Cromessol Multi-purpose Insect Killer	Wallace Cameron and Co Ltd	Amateur Professional	6045
Fly and Maggot Killer	Net-Tex Agricultural Ltd	Amateur	6097
W Fly Killer	Kalon Agricultural Division	Amateur	4817
W Hospital Flying Insect Spray	Kalon Chemicals	Amateur	4652
W Janisolve Fly Killer	Johmar Enterprises Company	Amateur	4228
K.O. Fly Spray	Genuine Solutions	Amateur Professional	5879
Momar Fly Killer	Momar	Amateur	6264
W Penetone Fly Killer	Penetone Ltd	Amateur	4072
Pynosect 2	Mitchell Cotts Chemicals Ltd	Professional	4654
Pynosect 4	Mitchell Cotts Chemicals Ltd	Professional	4655
Quantum Hygiene Fly Killer	Quantum Hygiene	Amateur Professional	5066
Rapid Exit Fly Spray	LMA Services	Amateur Professional	5787
Sorex Wasp Nest Destroyer	Sorex Ltd	Professional	6294
Swat	Solvitol Ltd	Amateur Professional	3564
Swat A	Solvitol Ltd	Professional	5051
Swot	Southern Chemical Services	Amateur Professional	6007
W Trilanco Fly Killer	Trilanco	Amateur	4816
Wasp Nest Destroyer	Sorex Ltd	Professional	4410

392 Resmethrin + Tetramethrin + Pirimiphos-methyl

Product Name	Marketing Company	Use	HSE No.
Waspend	Miracle Garden Care Ltd	Amateur Professional	4857

393 Silicon Dioxide(amorphous non coated precipitated)

Product Name	Marketing Company	Use	HSE No.
Rid Insect Powder	Rentokil Initial UK Ltd	Professional	5772

394 Tetramethrin

Product Name	Marketing Company	Use	HSE No.
Big D Fly and Wasp Killer	Dylon International Ltd	Amateur	5422
Bolt Flying Insect Killer	Johnson Wax Ltd	Amateur	3791
Boots Fly and Wasp Killer	Boots the Chemist Ltd	Amateur	5632
Di-Fly	Lever Industrial Ltd	Amateur	4509
Dry Fly Killer	Domestic Fillers Ltd	Amateur	4237
Fly and Wasp Killer	Lloyds	Amateur	5639
Jeyes Expel Fly and Wasp Killer	Jeyes Group Ltd	Amateur	5516
Jeyes Kontrol Fly and Wasp Killer	Jeyes Group Ltd	Amateur	5359
Killgerm Py-Kill W	Killgerm Chemicals Ltd	Professional	4632
Kleeneze Fly Killer	Kleeneze Ltd	Amateur	4225

Product Name	Marketing Company	Use	HSE No.

394 Tetramethrin—continued

Murphy Fly and Wasp Killer	Levington Horticulture Ltd	Amateur	6252
W Odex Fly Spray	Odex Ltd	Amateur	4814
Premiere Fly and Wasp Killer	Premiere Products	Amateur Professional	4516
Provence Fly and Wasp Killer	Haventrail Ltd	Amateur	4382
Pyrakill Flying Insect Spray	Forward Chemicals Ltd	Amateur	5357
Raid Flying Insect Killer	Johnson Wax Ltd	Amateur Professional	4585
Red Can Fly Killer	Yellow Can Company Ltd	Amateur	4508
SC Johnson Raid Fly and Wasp Killer	Johnson Wax Ltd	Amateur	5113
Secto Fly and Wasp Killer	Sinclair Animal and Household Care Ltd	Amateur	6314
St Michael Flyspray	Marks and Spencer Plc	Amateur	4177
Super Raid Insect Killer Formula 2	Johnson Wax Ltd	Amateur	4240

395 Tetramethrin + Alphacypermethrin

| Cytrol Alpha Super 50/50 SE | Pelgar International Ltd | Professional | 6678 |

396 Tetramethrin + Bendiocarb

| Camco Insect Spray | Agrevo UK Limited | Amateur | 4222 |
| Devcol Household Insect Spray | Devcol Ltd | Amateur | 5512 |

397 Tetramethrin + Chlorpyrifos

| Raid Wasp Nest Destroyer | Johnson Wax Ltd | Amateur | 5597 |

398 Tetramethrin + Cypermethrin

Cyperkill Plus WP	Mitchell Cotts Chemicals Ltd	Professional	4855
Cytrol Universal	Pelgar International Ltd	Professional	6675
Cytrol XL	Pelgar International Ltd	Professional	6679
Raid Residual Crawling Insect Killer	Johnson Wax Ltd	Amateur Professional	4584
S C Johnson Wax Raid Ant and Cockroach Killer	Johnson Wax Ltd	Amateur	5216
Shelltox Cockroach and Crawling Insect Killer 3	Temana International Ltd	Amateur Professional	4031
Shelltox Cockroach and Crawling Insect Killer 4	Temana International Ltd	Amateur	4739
Super Shelltox Crawling Insect Killer	Kortman Intradal BV	Amateur	5454
Vapona Ant and Crawling Insect Killer Aerosol	Ashe Consumer Products Ltd	Amateur	5282

399 Tetramethrin + Dichlorophen + Dichlorvos

| W Wintox | Mould Growth Consultants Ltd | Professional | 4236 |

Product Name	Marketing Company	Use	HSE No.

400 Tetramethrin + Dichlorvos

| Shelltox Extra FlyKiller | Temana International Ltd | Amateur Professional | 3696 |
| Shelltox FlyKiller | Temana International Ltd | Amateur | 3162 |

401 Tetramethrin + Dichlorvos + Permethrin

| D Laser Insect Killer | Sharpstow International Homecare Products Ltd | Amateur | 4910 |
| D Supaswat Insect Killer | Sharpstow International Homecare Products Ltd | Amateur | 4630 |

402 Tetramethrin + Fenitrothion

W Big D Cockroach and Crawling Insect Killer	Domestic Fillers Ltd	Amateur	4098
Big D Cockroach and Crawling Insect Killer	Dylon International Ltd	Amateur	5834
Big D Rocket Fly Killer	Domestic Fillers Ltd	Amateur	3552
Doom Ant and Crawling Insect Killer Aerosol	Napa Products Ltd	Amateur	4753
Killgerm Fenitrothion-Pyrethroid Concentrate	Killgerm Chemicals Ltd	Professional	4623
Motox Crawling Insect	HVM International Ltd	Amateur	5940
Top Kill Insect Killer	Houdret and Co Ltd	Amateur	6188
Tox Exterminating Fly and Wasp Killer	Keen (World Marketing) Ltd	Amateur	4314

403 Tetramethrin + Fenitrothion + Permethrin

| Connect Insect Killer | Sun Oil Ltd | Amateur | 6109 |
| Motox | HVM International Ltd | Amateur | 5622 |

404 Tetramethrin + Lindane

| W Doom Flea Killer | Napa Products Ltd | Amateur | 4572 |
| Doom Moth Proofer Aerosol | Napa Products Ltd | Amateur | 4571 |

405 Tetramethrin + Methoprene + Permethrin

| Johnson's Cage and Hutch Insect Spray | Johnson's Veterinary Products Ltd | Amateur | 6864 |
| Johnson's New Household Extra Guard Flea and Insect Spray | Johnson's Veterinary Products Ltd | Amateur | 6649 |

406 Tetramethrin + Permethrin

Bop Crawling Insect Killer	Robert McBride Ltd - Exports	Amateur Professional	4138
Bop Crawling Insect Killer (MC)	Robert McBride Ltd - Exports	Amateur Professional	4139
Atta-X Cockroach Killer	Sanderson Curtis Ltd	Amateur	6362
W Bolt Ant and Crawling Insect Killer	Johnson Wax Ltd	Amateur	3619

Product Name	Marketing Company	Use	HSE No.

406 Tetramethrin + Permethrin—continued

Canned Death	Pearson & Wilkinson	Professional	6112
Combat Fly Killer	Sinclair Animal and Household Care Ltd	Amateur	5808
Corry's Fly and Wasp Killer	Vitax Ltd	Amateur	5322
D-Stroy	Future Developments (Manufacturing) Ltd	Professional	5411
Delta Cockroach and Ant Spray	Bennetts Company Ltd	Amateur	3950
Doom Tropical Strength Fly and Wasp Killer Aerosol	Napa Products Ltd	Amateur	4752
Dragon FlyKiller	Zeneca Public Health	Amateur	4952
D Fly and Wasp Killer	Dimex Ltd	Professional	5050
Fly and Wasp Killer	Rentokil Initial UK Ltd	Amateur Professional	4643
Fly Free Zone	Helen Dolisznyj's Fly Away	Amateur Professional Professional (Animal Husbandry)	6394
Fly Spray	Merton Cleaning Supplies	Professional	5112
Fly-Kill	Stagchem Ltd	Professional	6202
Hail Flying Insect Killer	Robert McBride Ltd - Exports	Amateur	5568
Insecticide Aerosol	Aerosols International Ltd	Amateur	6567
Insectrol, Fly and Wasp Killer	Rentokil Initial UK Ltd	Amateur	5827
Johnson's New Household Flea Spray	Johnson's Veterinary Products Ltd	Amateur	6623
Johnson's New Pet Housing Spray	Johnson's Veterinary Products Ltd	Amateur	6624
K02 Selkil	Selden Research Ltd	Professional	4828
Nippon Ant and Crawling Insect Killer	Vitax Ltd	Amateur	4414
Nippon Fly Killer Spray	Vitax Ltd	Amateur	4254
Nippon Killaquer Crawling Insect Killer	Arrow Chemicals Ltd	Professional	3694
Nisa Fly and Wasp Killer	Nisa	Amateur	4962
Permost UNI	Hockley International Ltd	Professional Professional (Animal Husbandry)	6399
Purge Insect Destroyer	Aztec Chemicals Ltd	Professional	5048
Pynosect 10	Mitchell Cotts Chemicals Ltd	Professional	3171
Pynosect Extra Fog	Mitchell Cotts Chemicals Ltd	Professional	3172
W Raid Ant and Roach Killer	Johnson Wax Ltd	Amateur	5182
Sactif Flying Insect Killer	Diverseylever Ltd	Professional	4924
Sainsbury's Fly and Wasp Killer	J Sainsbury Plc	Amateur	5335
Sanmex Ant & Crawling Insect Killer	The British Products Sanmex Company Ltd	Amateur	6655

Product Name	Marketing Company	Use	HSE No.

406 Tetramethrin + Permethrin—continued

Sanmex Fly and Wasp Killer	The British Products Sanmex Company Ltd	Amateur	4006
Sanmex Supakil Insecticide	The British Products Sanmex Company Ltd	Amateur	4009
Scorpio Fly Spray For Flying Insects	Jangro Ltd	Professional	5049
Secto Fly Killer	Sinclair Animal and Household Care Ltd	Amateur	3862
Shelltox Insect Killer	Temana International Ltd	Amateur Professional	3697
Shelltox Insect Killer 2	Temana International Ltd	Amateur Professional	4061
Stop Insect Killer	Temana International Ltd	Amateur Professional	3163
Superdrug Fly Killer	Superdrug Stores Plc	Amateur	4185
Zappit	Blp Ltd	Amateur	6181

407 Tetramethrin + Permethrin + S-Bioallethrin

Agrevo Procontrol Crawling Insect Killer Aerosol	Agrevo UK Limited	Professional	6231
Pif Paf Crawling Insect Killer Aerosol	Reckitt and Colman Products Ltd	Amateur	5595

408 Tetramethrin + Pirimiphos-methyl + Resmethrin

Waspend	Miracle Garden Care Ltd	Amateur Professional	4857

409 Tetramethrin + Resmethrin

151 Dry Fly and Wasp Killer	SB Ltd	Amateur	5169
Apex Fly-Killer	Apex Industrial Chemicals Ltd	Amateur Professional	4300
Chem-kil	Inter-Chem	Amateur Professional	4266
Chirton Dry Fly and Wasp Killer	Ronson International Ltd	Amateur	5287
W Cromessol Insect Killer	Cromessol Company Ltd	Amateur	5067
Cromessol Multi-Purpose Insect Killer	Wallace Cameron and Co Ltd	Amateur Professional	6045
Fly and Maggot Killer	Net-tex Agricultural Ltd	Amateur	6097
W Fly Killer	Kalon Agricultural Division	Amateur	4817
W Hospital Flying Insect Spray	Kalon Chemicals	Amateur	4652
W Janisolve Fly Killer	Johmar Enterprises Company	Amateur	4228
K.O. Fly Spray	Genuine Solutions	Amateur Professional	5879
Momar Fly Killer	Momar	Amateur	6264
W Penetone Fly Killer	Penetone Ltd	Amateur	4072

Product Name	Marketing Company	Use	HSE No.

409 Tetramethrin + Resmethrin—continued

Pynosect 2	Mitchell Cotts Chemicals Ltd	Professional	4654
Pynosect 4	Mitchell Cotts Chemicals Ltd	Professional	4655
Quantum Hygiene Fly Killer	Quantum Hygiene	Amateur Professional	5066
Rapid Exit Fly Spray	LMA Services	Amateur Professional	5787
Sorex Wasp Nest Destroyer	Sorex Ltd	Professional	6294
Swat	Solvitol Ltd	Amateur Professional	3564
Swat A	Solvitol Ltd	Professional	5051
Swot	Southern Chemical Services	Amateur Professional	6007
W Trilanco Fly Killer	Trilanco	Amateur	4816
Wasp Nest Destroyer	Sorex Ltd	Professional	4410

410 Tetramethrin + S-Bioallethrin

Vapona Wasp Killer Aerosol	Ashe Consumer Products Ltd	Amateur	6063

411 Tetramethrin + S-Methoprene

Zodiac Household Flea Spray	Petlove	Amateur	6502

412 Tetramethrin + d-Allethrin

Boots Dry Fly and Wasp Killer	Boots the Chemist Ltd	Amateur	4037
W Boots Dry Fly and Wasp Killer 3	Boots the Chemist Ltd	Amateur	4241
Fly, Wasp and Mosquito Killer	Rentokil Initial UK Ltd	Amateur	4882
New Vapona Fly Killer Dry Formulation	Ashe Consumer Products Ltd	Amateur	4202
Raid Fly and Wasp Killer	Johnson Wax Ltd	Amateur	3000
Sainsbury's Fly, Wasp and Mosquito Killer	J Sainsbury Plc	Amateur	5164

413 Tetramethrin + d-Allethrin + Cypermethrin

Bop Flying and Crawling Insect Killer	Robert McBride Ltd - Exports	Amateur Professional	4136
Bop Flying and Crawling Insect Killer (Water Based)	Robert McBride Ltd - Exports	Amateur Professional	4137
Bop Flying Insect Killer	Robert McBride Ltd - Exports	Amateur Professional	4141
Bop Flying Insect Killer (MC)	Robert McBride Ltd - Exports	Amateur Professional	4140
Bop Flying and Crawling Insect Killer (MC)	Robert McBride Ltd - Exports	Amateur Professional	4135

414 Tetramethrin + d-Allethrin + Cypermethrin + Permethrin

Elf Insecticide	Elf Oil UK Ltd	Amateur	6669

Product Name	Marketing Company	Use	HSE No.

415 Tetramethrin + d-Allethrin + Dichlorvos + Permethrin

Bop	Robert Mcbride Ltd - Exports	Amateur	4773

416 Tetramethrin + d-Allethrin + Permethrin

Hail Plus Crawling Insect Killer	Robert Mcbride Ltd - Exports	Amateur	5567
New Super Raid Insecticide	Johnson Wax Ltd	Amateur Professional	4583

417 Tetramethrin + d-Phenothrin

Flak	Robert Mcbride Ltd - Exports	Amateur Professional	4143
Ant Destroyer Foam	Rentokil Initial UK Ltd	Amateur	6412
Betterware Foaming Ant & Wasp Nest Destroyer	Betterware UK Ltd	Amateur	6615
W Big D Ant and Crawling Insect Killer	Domestic Fillers Ltd	Amateur	4097
Big D Ant and Crawling Insect Killer	Dylon International Ltd	Amateur	5818
Bio Foaming Wasp Nest Destroyer	PBI Home & Garden Ltd	Amateur	6665
Boots Ant and Crawling Insect Killer	Boots the Chemist Ltd	Amateur	5554
Boots Flea Spray	Boots the Chemist Ltd	Amateur	5522
Bop Arabic	Robert Mcbride Ltd - Exports	Amateur Professional	4142
Brimpex ULV 1500	Brimpex Metal Treatments	Professional Professional (Animal Husbandry)	4942
Bug Off	Dimex Ltd	Amateur	6356
Deosan Fly Spray	Diverseylever Ltd	Professional	5246
Envar 6/14 EC	Agrimar (UK) Plc	Professional	4354
Envar 6/14 Oil	Agrimar (UK) Plc	Professional	4355
Flying and Crawling Insect Killer	Betterware UK Ltd	Amateur	6607
Friends Household Flea Spray	Sinclair Animal and Household Care Ltd	Amateur	6664
G.C.P. Flying & Crawling Insect Killer	Greenhill Chemical Products Ltd	Amateur	6303
Goodboy Household Flea Spray	Armitage Brothers Plc	Amateur	4090
Insect Killer	Rentokil Initial UK Ltd	Amateur	6244
Jeyes Expel Ant and Crawling Insect Killer	Jeyes Group Ltd	Amateur	5552
Jeyes Expel Flea Killer	Jeyes Group Ltd	Amateur	5523
Jeyes Kontrol Ant and Crawling Insect Killer	Jeyes Group Ltd	Amateur	5553
Jeyes Kontrol Flea Killer	Jeyes Group Ltd	Amateur	5521

Product Name	Marketing Company	Use	HSE No.

417 Tetramethrin + d-Phenothrin—continued

Product Name	Marketing Company	Use	HSE No.
Killgerm ULV 500	Killgerm Chemicals Ltd	Professional Professional (Animal Husbandry)	4647
Killoff Aerosol Insect Killer	Ferosan Products	Amateur	3974
Kybosh 2	PBI Home & Garden Ltd	Amateur	6622
Murphy Foaming Wasp Nest Destroyer	Levington Horticulture Ltd	Amateur	6609
New Secto Household Flea Spray	Sinclair Animal and Household Care Ltd	Amateur	6504
Pesguard NS 6/14 EC	Sumitomo Chemical (UK) Plc	Professional	4360
Pesguard OBA F 7305 C	Sumitomo Chemical (UK) Plc	Amateur	4064
Pesguard OBA F 7305 D	Sumitomo Chemical (UK) Plc	Amateur	4065
Pesguard OBA F 7305 E	Sumitomo Chemical (UK) Plc	Amateur	4066
Pesguard OBA F 7305 F	Sumitomo Chemical (UK) Plc	Amateur	4067
Pesguard OBA F 7305 G	Sumitomo Chemical (UK) Plc	Amateur	4063
Pesguard OBA F7305 B	Sumitomo Chemical (UK) Plc	Amateur	4419
Pesguard WBA F-2656	Sumitomo Chemical (UK) Plc	Amateur	4263
Pesguard WBA F-2692	Sumitomo Chemical (UK) Plc	Amateur	4262
Pestkill Household Spray	Seven Seas	Amateur	4938
Raid Fly Wasp and Plant Insect Killer	Johnson Wax Ltd	Amateur	5432
Rentokil Wasp Nest Destroyer	Rentokil Initial UK Ltd	Amateur	6214
S C Johnson Professional Raid Flying Insect Killer	S C Johnson Professional Ltd	Amateur	6407
S C Johnson Wax Raid Fly and Wasp Killer	Johnson Wax Ltd	Amateur	5181
Sergeant's Dust Mite Patrol Injector and Spray	Seven Seas Ltd	Amateur Professional	5513
W Sergeant's Dust Mite Patrol Spray	Seven Seas Ltd	Amateur	5041
Sergeant's Household Spray	Seven Seas Ltd	Amateur	6011
W Sergeant's Insect Patrol	Seven Seas Ltd	Amateur	5397
Shelltox Flying Insect Killer 2	Temana International Ltd	Amateur Professional	4032
Shelltox Flying Insect Killer 3	Temana International Ltd	Amateur	4741
Shelltox Super Flying Insect Killer	Temana International Ltd	Amateur	4790
Sorex Fly Spray RTU	Sorex Ltd	Professional	5718
Sorex Super Fly Spray	Sorex Ltd	Amateur	6297
Supaswat Insect Killer	Sharpstow International Homecare Products Ltd	Amateur	5691
Super Fly Spray	Sorex Ltd	Amateur	4468
Super Raid II	Johnson Wax Ltd	Amateur	4015
Super Shelltox Flying Insect Killer	Kortman Intradal BV	Amateur	5440
Terminator	Aerosol Logistics	Amateur	6416
Terminator Fly and Wasp Killer	Sanoda Ltd	Amateur	6150
Terminator Outdoor Insect Killer	Sanoda Ltd	Amateur	6171

Product Name	Marketing Company	Use	HSE No.

417 Tetramethrin + d-Phenothrin—continued

Product Name	Marketing Company	Use	HSE No.
Terminator Wasp and Insect Nest Destroyer	Sanoda Ltd	Amateur	6144
Vapona House and Plant Fly Spray	Ashe Consumer Products Ltd	Amateur	5236
Vapona Wasp and Fly Killer	Ashe Consumer Products Ltd	Amateur	5754
Vapona Wasp and Fly Killer Spray	Ashe Consumer Products Ltd	Amateur	5280
Vitapet Flearid Household Spray	Seven Seas Ltd	Amateur Professional	5434
Wasp Destroyer Foam	Rentokil Initial UK Ltd	Amateur	6455
Zappit Fly and Wasp Killer	BLP Ltd	Amateur	6336

418 d-Tetramethrin + d-Phenothrin

Product Name	Marketing Company	Use	HSE No.
Pesguard WBA F2714	Sumitomo Chemical (UK) Plc	Amateur	4418
Vapona Max Concentrated Fly and Wasp Killer	Ashe Consumer Products Ltd	Amateur	6062

419 Transfluthrin

Product Name	Marketing Company	Use	HSE No.
Baygon Moth Paper	Bayer Plc	Amateur	6228

420 Transfluthrin + Cyfluthrin

Product Name	Marketing Company	Use	HSE No.
Baygon Insect Spray	Bayer Plc	Amateur	6546

421 Trichlorphon

Product Name	Marketing Company	Use	HSE No.
Ant Killer	Rentokil Initial UK Ltd	Amateur	5906
Boots Ant Trap	Boots the Chemist Ltd	Amateur	5542
Detia Ant Bait	K D Pest Control Products	Amateur	3803
Jeyes Expel Ant trap	Jeyes Group Ltd	Amateur	5503
Murphy Kil-Ant Trap	Levington Horticulture Ltd	Amateur	6258
Vapona Ant Bait	Ashe Consumer Products Ltd	Amateur	6092
Vapona Ant Trap	Ashe Consumer Products Ltd	Amateur	3804

HSE INSECTICIDES

5
SURFACE BIOCIDES

Product Name	Marketing Company	Use	HSE No.

Surface Biocides

422 Alkylaryltrimethyl ammonium chloride

Abicide 82	Langlow Products Ltd	Amateur Professional	4470
Albany Fungicidal Wash	C Brewer and Sons Ltd	Amateur Professional	4471
Anti-Mould Solution	Macpherson Paints Ltd	Amateur Amateur (Wood Preservative) Professional Professional (Wood Preservative)	4873
Beeline Fungicidal Wash	Ward Bekker Ltd	Amateur Professional	4224
Fungishield Sterilising Solution Concentrate GS36	Glixtone Ltd	Amateur Professional	4283
Fungishield Sterilising Solution GS37	Glixtone Ltd	Amateur Professional	4284
ICI Decorator 1st Fungicidal Wash	ICI Paints	Amateur Professional	6479
Kibes Sterilising Solution Concentrate GS 36	Kibes (UK) Insulation Ltd	Amateur Professional	5493
Mould Cleaner	J H Woodman Ltd	Amateur	4291

423 Alkylaryltrimethyl ammonium chloride + Boric acid

Kleeneze Anti-mould Spray	Kleeneze Ltd	Amateur Professional	4599

424 Alkylaryltrimethyl ammonium chloride + Disodium octaborate

Boracol 10RH Surface Biocide	Advanced Chemical Specialities Ltd	Professional	4911
Remtox Borocol 10 RH Masonry Biocide	Remtox (Chemicals) Ltd	Professional	4100

425 Azaconazole + Permethrin

Fongix SE Total Treatment for Wood	Liberon Waxes Ltd	Amateur Amateur (Wood Preservative) Professional Professional (Wood Preservative)	5610

Product Name	Marketing Company	Use	HSE No.

425 Azaconazole + Permethrin—continued

D Fongix SE Total Treatment for Wood	V33 SA	Amateur Amateur (Wood Preservative)	4288
Woodworm Killer and Rot Treatment	Liberon Waxes Ltd	Amateur Amateur (Wood Preservative) Professional Professional (Wood Preservative)	4826

426 Benzalkonium bromide

X Moss	Thames Valley Specialist Products Ltd	Amateur Professional	6839

427 Benzalkonium chloride

A-Zygo 3 Sterilising Solution	Premier Condensation and Mould Control Ltd	Professional	5368
Albio B084	Millcliff Ltd	Professional	6059
Algae Remover	Akzo Nobel Woodcare	Amateur Professional	5337
Algaecide	Premkem Ltd	Professional	5842
B and Q Value Fungicide	B & Q Plc	Amateur	6834
Beaver Fungicidal Solution	Philip Johnstone Group Ltd	Amateur Professional	4928
W Bedclear	Fargro Ltd	Professional	5179
Bio-Kil Dentolite Solution	Remtox-Silexine Ltd	Amateur Professional	4593
Bio-Natura's Green Algae Remover	Bio-Natura Ltd	Amateur Professional	6503
BN Algae Remover	Nehra Cooke's Chemicals Ltd	Amateur Professional	5381
W BN Mosskiller	Morgan Nehra Holdings Ltd	Amateur Professional	5080
W Conc Quat	Semitec Ltd	Professional	4904
D Cuprinol Cuprotect Exterior Fungicidal Wash	Cuprinol Ltd	Amateur Professional	4610
D Cuprinol Cuprotect Fungicidal Patio Cleaner	Cuprinol Ltd	Amateur Professional	4606
Cuprinol Exterior Fungicide	Cuprinol Ltd	Amateur Professional	5526
Cuprinol Garden Furniture Algae Killer	Cuprinol Ltd	Amateur Professional	5922

Product Name	Marketing Company	Use	HSE No.

427 Benzalkonium chloride—continued

Product Name	Marketing Company	Use	HSE No.
Cuprinol Garden Timber Algae Killer	Cuprinol Ltd	Amateur Professional	5921
Cuprinol Greenhouse Algae Killer	Cuprinol Ltd	Amateur Professional	5939
Cuprinol Path and Patio Cleaner	Cuprinol Ltd	Amateur Professional	5527
Cuprotect Exterior Fungicide	Cuprinol Ltd	Amateur Professional	4165
Cuprotect Patio Cleaner	Cuprinol Ltd	Amateur Professional	4164
Doff Path & Patio Cleaner	Doff Portland Ltd	Amateur	6305
Doff Path and Patio Cleaner Concentrate	Doff Portland Ltd	Amateur	6248
FEB Fungicide	FEB MBT	Amateur	6833
FMB 451-5 Quat (50%)	Midland Europe	Professional Professional (Wood Preservative) Industrial Industrial (Wood Preservative)	5734
Fungi-Shield Sterilising Solution Concentrate GS36	Glixtone Ltd	Amateur Professional	6292
Fungi-Shield Sterilising Solution GS37	Glixtone Ltd	Amateur Professional	6291
D Gloquat RP	Rhodia Limited	Professional Professional (Wood Preservative)	4815
Granocryl Fungicidal Wash	Kalon Group Plc	Amateur Professional	5797
Green N Clean Algae Killer Concentrate	Phostrogen Ltd	Amateur Professional	6835
Green N Clean Algae Killer Ready to Use	Phostrogen Ltd	Amateur Professional	6830
Homebase Mould Cleaner	Sainsbury's Homebase House and Garden Centres	Amateur Professional	5707
Homebase Weathercoat Fungicidal Wash	Sainsbury's Homebase House and Garden Centres	Amateur Professional	4860
Howes Olympic Algaecide	Killgerm Chemicals Ltd	Professional	4845
Interior Mould Remover Wipes	Cadismark Household Products Ltd	Amateur	5695
Jewson Fungicidal Wash	Jewson Ltd	Amateur Professional	6886

Product Name	Marketing Company	Use	HSE No.

427 Benzalkonium chloride—continued

Product Name	Marketing Company	Use	HSE No.
Leyland Sterilisation Wash	Leyland Paint Company Ltd	Professional	4751
W LPL Biocidal Wash	Liquid Plastics Ltd	Professional	4479
LPL Biowash	Liquid Plastics Ltd	Professional	5853
Mould Killer	Cadismark Household Products Ltd	Amateur	5047
Palace Fungicidal Wash	Palace Chemicals Ltd	Amateur Professional	6253
Paramos	Chemsearch	Professional	4524
Path and Patio Cleaner	Otter Nurseries Ltd	Amateur	6060
Remtox Remwash Extra	Remtox (Chemicals) Ltd	Professional	5178
Remtox-Silexine Scrub out Black Mould/Refill	Remtox-Silexine Ltd	Amateur Professional	4810
Rhodaquat RP50	Rhodia Limited	Professional Professional (Wood Preservative)	5762
Rid Moss	Friendly Systems	Professional	6222
Sigma Fungicidal Solution	Sigma Coatings Ltd	Amateur Professional	6564
Snowcem Algicide	Snowcem	Amateur Professional	4806
Sovereign Dentolite Solution Concentrate	Sovereign Chemical Industries Ltd	Amateur Professional	6354
D Square Deal Fungicidal Solution	Texas Homecare Ltd	Amateur Professional	4644
Texas Fungicidal Wash	Texas Homecare Ltd	Amateur Professional	5745
Thompson's Instant Mould and Moss Killer	Ronseal Ltd	Amateur	6208
Thompson's Patio and Paving Cleaner	Ronseal Ltd	Amateur	5985
Travis Perkins Fungicidal Wash	Travis Perkins Trading Company Ltd	Amateur Professional	6619
Triflex Sterilising Solution Concentrate	Triflex (UK) Ltd	Amateur Professional	6295
Wickes Fungicidal Wash	Wickes Building Supplies Ltd	Amateur Professional	6039

428 Benzalkonium chloride + 2-Phenylphenol

Product Name	Marketing Company	Use	HSE No.
Anti Fungus Wash	Craig & Rose Plc	Professional	6293
C/S Clean R833	Construction Specialties (UK) Ltd	Amateur Professional	6865
W Chelwash	Chelec Ltd	Amateur	3096
Crown Fungicidal Wall Solution	Crown Chemicals Ltd	Professional	5868
Deadly Nightshade	Andura Textured Masonry Coatings Ltd	Professional	4895

Product Name	Marketing Company	Use	HSE No.

428 Benzalkonium chloride + 2-Phenylphenol—continued

Product Name	Marketing Company	Use	HSE No.
D Earnshaws Fungal Wash	E Earnshaw and Company (1965) Ltd	Amateur	4256
Fungicidal Algaecidal Bacteriacidal Wash	Philip Johnstone Group Ltd	Amateur Professional	4216
D Fungicidal Solution FL.2	Manders Paints Ltd	Amateur Professional	4245
W Fungicidal Wash	C W Wastnage Ltd	Amateur	3072
Fungicidal Wash	Everbuild Building Products Ltd	Amateur Professional	6000
W Fungicidal Wash Solution	Johnstone's Paints Plc	Amateur Professional	3776
Fungiguard	Spaldings Ltd	Professional	5582
Houseplan Fungi Wash	Ruberoid Building Products Ltd	Amateur	6374
Johnstone's Fungicidal Wash	Johnstone's Paints Plc	Amateur Professional	5678
Larsen Concentrated Algicide 2	Larsen Manufacturing Ltd	Amateur Professional	6606
Micro FWS	Lectros International Ltd	Professional	6135
Micro Masonry Biocide	Grip-Fast Systems Ltd	Professional	6444
Microguard FWS	Permagard Products Ltd	Professional	5485
Microstel	Bedec Products Ltd	Professional	4198
Mouldcheck Sterilizer 2	Biotech Environmental (UK) Ltd	Amateur Professional	6385
Nitromors Mould Remover	Henkel Home Improvements and Adhesive Products	Amateur	5507
P.J. Fungicidal Solution	Philip Johnstone Group Ltd	Amateur Professional	6318
Permacide Masonry Fungicide	Triton Perma Industries Ltd	Professional	5570
Permarock Fungicidal Wash	Permarock Products Ltd	Professional	5193
POB Mould Treatment	Darcy Industries Ltd	Amateur	4076
Ruberoid Fungicidal Wash Solution	Ruberoid Building Products Ltd	Amateur	6375
Spencer Stormcote Fungicidal Treatment	Spencer Coatings Ltd	Professional	4155
Tetra Concentrated Mould Cleaner	Tetrosyl (Building Products) Ltd	Amateur Professional	4206
Vallance Pro-build Extra Strong Moss and Mould Killer	Vallance Ltd	Amateur Professional	5761
Z.144 Fungicidal Wash	Crosbie Coatings Ltd	Amateur Professional	3003

429 Benzalkonium chloride + Boric acid

Product Name	Marketing Company	Use	HSE No.
Barrettine Premier Multicide Fluid	Barrettine Products Ltd	Amateur Amateur (Wood Preservative)	6312

429 Benzalkonium chloride + Boric acid—continued

Microguard Mouldicidal Wood Preserver	Permagard Products Ltd	Amateur Amateur (Wood Preservative) Professional Professional (Wood Preservative) Industrial Industrial (Wood Preservative)	5100

430 Benzalkonium chloride + Carbendazim + Dialkyldimethyl ammonium chloride

Rentokil Mould Cure Spray	Rentokil Initial UK Ltd	Amateur	4227

431 Benzalkonium chloride + Disodium octaborate

Boracol B8.5 RH Mouldicide/Wood Preservative	Advanced Chemical Specialities Ltd	Amateur Amateur (Wood Preservative) Professional Professional (Wood Preservative)	4809
Boron Biocide 10	Philip Johnstone Group Ltd	Professional Professional (Wood Preservative)	6485
Brunosol Boron 10	Philip Johnstone Group Ltd	Professional Professional (Wood Preservative)	6488
Cuprinol Fungicidal Spray	Cuprinol Ltd	Amateur	4221
Cuprinol Mould Killer	Cuprinol Ltd	Amateur Professional	4163
D Cuprinol No More Mould Fungicidal Spray	Cuprinol Ltd	Amateur	4162
Deepflow 11 Inorganic Boron Masonry Biocide.	Safeguard Chemicals Ltd	Professional	5528
Lectros Boracol 10 RH	Lectros International Ltd	Professional	6499
Permabor Boracol 10RH	Permagard Products Ltd	Professional	6358
Polycell 3 in 1 Mould Killer	Polycell Products Ltd	Amateur Professional	6157

Product Name	Marketing Company	Use	HSE No.

431 Benzalkonium chloride + Disodium octaborate—continued

Product Name	Marketing Company	Use	HSE No.
Probor	Safeguard Chemicals Ltd	Professional Professional (Wood Preservative)	6266
Probor 10	Safeguard Chemicals Ltd	Professional	6595
Probor 50	Safeguard Chemicals Ltd	Professional Professional (Wood Preservative)	5596
Safeguard Boracol 10 RH	Safeguard Chemicals Ltd	Professional	6397
Tribor 10	Triton Chemical Manufacturing Co Ltd	Professional	6286
Weathershield Multi-surface Fungicidal Wash	Imperial Chemical Industries Plc	Amateur Professional	6823

432 Benzalkonium chloride + Naphthalene + Permethrin + Pyrethrins

Product Name	Marketing Company	Use	HSE No.
W Roxem D	Horton Hygiene Company	Professional	5107

433 Boric acid

Product Name	Marketing Company	Use	HSE No.
Rentokil Dry Rot Fluid (E) for Bonded Warehouses	Rentokil Initial UK Ltd	Professional Professional (Wood Preservative)	3986

434 Boric acid + Alkylaryltrimethyl ammonium chloride

Product Name	Marketing Company	Use	HSE No.
Kleeneze Anti-Mould Spray	Kleeneze Ltd	Amateur Professional	4599

435 Boric acid + Benzalkonium chloride

Product Name	Marketing Company	Use	HSE No.
Barrettine Premier Multicide Fluid	Barrettine Products Ltd	Amateur Amateur (Wood Preservative)	6312
Microguard Mouldicidal Wood Preserver	Permagard Products Ltd	Amateur Amateur (Wood Preservative) Professional Professional (Wood Preservative) Industrial Industrial (Wood Preservative)	5100

Product Name	Marketing Company	Use	HSE No.

436 Boric acid + Cypermethrin + Propiconazole

Rentokil Total Wood Treatment	Rentokil Initial UK Ltd	Amateur	6053
		Amateur	
		(Wood	
		Preservative)	
		Professional	
		Professional	
		(Wood	
		Preservative)	

437 Boric acid + Disodium tetraborate

Control Fluid SB	Rentokil Initial UK Ltd	Professional	6018
		Professional	
		(Wood	
		Preservative)	

438 Carbendazim + Dialkyldimethyl ammonium chloride + Benzalkonium chloride

| Rentokil Mould Cure Spray | Rentokil Initial UK Ltd | Amateur | 4227 |

439 Carbendazim + Dichlorophenyl dimethylurea + 2-Octyl-2h-isothiazolin-3-one

| Biocheck A.C.C. | Mould Growth Consultants Ltd | Amateur | 6571 |
| | | Professional | |

440 Cypermethrin + Propiconazole + Boric acid

Rentokil Total Wood Treatment	Rentokil Initial UK Ltd	Amateur	6053
		Amateur	
		(Wood	
		Preservative)	
		Professional	
		Professional	
		(Wood	
		Preservative)	

441 Dialkyldimethyl ammonium chloride

Biophen Barrier	Mycobio Consultants	Amateur	6048
		Professional	
Halophane Bonding Solution	Mould Growth Consultants Ltd	Professional	3929
Halophen BM1165L	Mould Growth Consultants Ltd	Professional	3930
HG Green Slime Remover	HG International B.V.	Amateur	6086
Lichenite	Mould Growth Consultants Ltd	Professional	3936

Product Name	Marketing Company	Use	HSE No.

441 Dialkyldimethyl ammonium chloride—continued

Product Name	Marketing Company	Use	HSE No.
Mosscheck	Biotech Environmental (UK) Ltd	Amateur Professional	4870
Mouldcheck Barrier 2	Biotech Environmental (UK) Ltd	Amateur Professional	4856
Mouldkill	Biotech Environmental (UK) Ltd	Amateur Professional	5192
Remwash Masonry Steriliser	Remtox (Chemicals) Ltd	Amateur Professional	6363
RLT Clearmould Spray	Mould Growth Consultants Ltd	Amateur Professional	4824
RLT Halophen	Mould Growth Consultants Ltd	Amateur Professional	4081
RLT Halophen DS	Mould Growth Consultants Ltd	Amateur Professional	4483
Sovereign Masonry Sterilising Wash	Sovereign Chemical Industries Ltd	Amateur Professional	6360
X Thallo	Thames Valley Specialist Products Ltd	Amateur Professional	6841

442 Dialkyldimethyl ammonium chloride + Benzalkonium chloride + Carbendazim

Product Name	Marketing Company	Use	HSE No.
Rentokil Mould Cure Spray	Rentokil Initial UK Ltd	Amateur	4227

443 Dichlorophen

Product Name	Marketing Company	Use	HSE No.
Bioclean Sterilizer	Mycobio Consultants	Amateur Professional	6047
W Denso Mouldshield Biocidal Cleanser	Winn and Coales (Denso) Ltd	Amateur Professional	4731
W Denso Mouldshield Surface Biocide	Winn and Coales (Denso) Ltd	Professional	4732
En-Tout-Cas Moss Killer	En-Tout-Cas Sports Equipment Ltd	Amateur Professional	5855
Halodec	Mould Growth Consultants Ltd	Professional	4454
W Halophane No.1 Aerosol	Mould Growth Consultants Ltd	Professional	3815
W Halophane No.3 Aerosol	Mould Growth Consultants Ltd	Professional	3816
Mouldcheck Spray	Biotech Environmental (UK) Ltd	Amateur	4093
Mouldcheck Sterilizer	Biotech Environmental (UK) Ltd	Amateur Professional	4505
Nipacide DP30	Nipa Laboratories Ltd	Professional	4925
Panacide M21	Coalite Chemicals Division	Professional	4923
Panaclean 736	Coalite Chemicals Division	Professional	5075
SP.153 Sterilising Detergent Wash	Mason Coatings Plc	Professional	3920

Product Name	Marketing Company	Use	HSE No.

443 Dichlorophen—continued

| SP.154 Fungicidal and Bactericidal Treatment | Mason Coatings Plc | Professional | 3919 |

444 Dichlorophenyl dimethylurea + 2-Octyl-2h-isothiazolin-3-one + Carbendazim

| Biocheck a.c.c. | Mould Growth Consultants Ltd | Amateur Professional | 6571 |

445 Disodium octaborate

ACS Borotreat 10P	Advanced Chemical Specialities Ltd	Professional Professional (Wood Preservative)	6335
Azygo Boron Preservative PDR	Azygo International Direct Ltd	Professional Professional (Wood Preservative)	6497
Probor DB	Safeguard Chemicals Ltd	Professional Industrial	6673

446 Disodium octaborate + Alkylaryltrimethyl ammonium chloride

| Boracol 10RH Surface Biocide | Advanced Chemical Specialities Ltd | Professional | 4911 |
| Remtox Borocol 10 RH Masonry Biocide | Remtox (Chemicals) Ltd | Professional | 4100 |

447 Disodium octaborate + Benzalkonium chloride

Boracol B8.5 RH Mouldicide/Wood Preservative	Advanced Chemical Specialities Ltd	Amateur Amateur (Wood Preservative) Professional Professional (Wood Preservative)	4809
Boron Biocide 10	Philip Johnstone Group Ltd	Professional Professional (Wood Preservative)	6485
Brunosol Boron 10	Philip Johnstone Group Ltd	Professional Professional (Wood Preservative)	6488
Cuprinol Fungicidal Spray	Cuprinol Ltd	Amateur	4221
Cuprinol Mould Killer	Cuprinol Ltd	Amateur Professional	4163

HSE SURFACE BIOCIDES

Product Name	Marketing Company	Use	HSE No.

447 Disodium octaborate + Benzalkonium chloride—continued

D Cuprinol No More Mould Fungicidal Spray	Cuprinol Ltd	Amateur	4162
Deepflow 11 Inorganic Boron Masonry Biocide.	Safeguard Chemicals Ltd	Professional	5528
Lectros Boracol 10 RH	Lectros International Ltd	Professional	6499
Permabor Boracol 10RH	Permagard Products Ltd	Professional	6358
Polycell 3 in 1 Mould Killer	Polycell Products Ltd	Amateur Professional	6157
Probor	Safeguard Chemicals Ltd	Professional Professional (Wood Preservative)	6266
Probor 10	Safeguard Chemicals Ltd	Professional	6595
Probor 50	Safeguard Chemicals Ltd	Professional Professional (Wood Preservative)	5596
Safeguard Boracol 10 RH	Safeguard Chemicals Ltd	Professional	6397
Tribor 10	Triton Chemical Manufacturing Co Ltd	Professional	6286
Weathershield Mulit-Surface Fungicidal Wash	Imperial Chemical Industries Plc	Amateur Professional	6823

448 Disodium tetraborate + Boric acid

Control Fluid SB	Rentokil Initial UK Ltd	Professional Professional (Wood Preservative)	6018

449 Dodecylamine lactate + Dodecylamine salicylate

W Anti Fungus Wash	Craig & Rose Plc	Professional	4375
W Blackfriar Anti Mould Solution 195/9	E Parsons and Sons Ltd	Amateur	4231
W Durham Bl 730	Durham Chemicals	Professional Professional (Wood Preservative) Industrial Industrial (Wood Preservative)	4744

Product Name	Marketing Company	Use	HSE No.

449 Dodecylamine lactate + Dodecylamine salicylate—continued

Product Name	Marketing Company	Use	HSE No.
W Environmental Woodrot Treatment	SCI (Building Products)	Professional Professional (Wood Preservative) Industrial Industrial (Wood Preservative)	5530
W FEB Fungicide	FEB Ltd	Amateur Professional	3009
W Fungicidal Wash	Polybond Ltd	Amateur Amateur (Wood Preservative) Professional Professional (Wood Preservative)	3601
W Fungicidal Wash	Smyth - Morris Chemicals Ltd	Professional Professional (Wood Preservative)	3004
W Green Range Murosol 20	Cementone Beaver Ltd	Professional	4456
W Hyperion Mould Inhibiting Solution	Hird Hastie Paints Ltd	Professional	4412
W Kingfisher Fungicidal Wall Solution	Kingfisher Chemicals Ltd	Professional	4014
W Larsen Concentrated Algicide	Larsen Manufacturing Ltd	Professional	4968
W M-Tec Biocide	M-Protective Coatings Ltd	Professional	4381
W Macpherson Anti-Mould Solution	Macpherson Paints Ltd	Professional Professional (Wood Preservative)	4791
W Macpherson Antimould Solution (RFU)	Macpherson Paints Ltd	Amateur Professional	5190
W Metalife Fungicidal Wash	Metalife International Ltd	Amateur Amateur (Wood Preservative) Professional Professional (Wood Preservative)	5406

HSE SURFACE BIOCIDES

449 Dodecylamine lactate + Dodecylamine salicylate—continued

Product Name	Marketing Company	Use	HSE No.
W Mouldrid	S and K Maintenance Products	Amateur Amateur (Wood Preservative) Professional Professional (Wood Preservative)	3671
W Mrs Clear Mould Fungicidal Wash	Mr (Polymer Cement Products) Ltd	Professional	3620
W Nubex WDR	Nubex Ltd	Professional Professional (Wood Preservative)	3810
W Palace Mould Remover	Palace Chemicals Ltd	Amateur Professional	5204
W Permadex Masonry Fungicide Concentrate	Permagard Products Ltd	Professional	4007
W Permoglaze Micatex Fungicidal Treatment	Permoglaze Paints Ltd	Professional Professional (Wood Preservative)	5249
W Polycell Mould Cleaner	Polycell Products Ltd	Amateur Professional	3612
W Protim Wall Solution II	Protim Solignum Ltd	Amateur Professional	5006
W Protim Wall Solution II Concentrate	Protim Solignum Ltd	Professional	5007
W Ruberoid Fungicidal Wash	Ruberoid Building Products Ltd	Amateur	5949
W Safeguard Deepwood Surface Biocide Concentrate	Safeguard Chemicals Ltd	Professional	4616
W Safeguard Mould and Moss Killer	Safeguard Chemicals Ltd	Amateur Professional	4607
W Solignum Dry Rot Killer	Protim Solignum Ltd	Amateur Professional	5025
W Solignum Fungicide	Protim Solignum Ltd	Amateur Professional	5026
W Thompson's Interior and Exterior Fungicidal Spray	Ronseal Ltd	Amateur	5626
W Thompson's Interior Mould Killer	Ronseal Ltd	Amateur	5624
W Torkill Fungicidal Solution 'W'	Tor Coatings Ltd	Professional	4246
W Unisil S Silicone Waterproofing Solution	Witham Oil & Paint (Lowestoft) Ltd	Professional	4758
W Unitas Fungicidal Wash-Exterior (Solvent Based)	Witham Oil & Paint (Lowestoft) Ltd	Professional	4759
W Unitas Fungicidal Wash-Interior (Water Based)	Witham Oil & Paint (Lowestoft) Ltd	Professional	4760

Product Name	Marketing Company	Use	HSE No.

449 Dodecylamine lactate + Dodecylamine salicylate—continued

W Vallance Fungicidal Wash	Siroflex Ltd	Amateur	5375
		Professional	
W Weathershield Fungicidal Wash	ICI Paints	Amateur	5134
		Amateur	
		(Wood	
		Preservative)	
		Professional	
		Professional	
		(Wood	
		Preservative)	

450 Dodecylamine salicylate + Dodecylamine lactate

W Anti Fungus Wash	Craig & Rose Plc	Professional	4375
W Blackfriar Anti Mould Solution 195/9	E Parsons and Sons Ltd	Amateur	4231
W Durham BI 730	Durham Chemicals	Professional	4744
		Professional	
		(Wood	
		Preservative)	
		Industrial	
		Industrial	
		(Wood	
		Preservative)	
W Environmental Woodrot Treatment	SCI (Building Products)	Professional	5530
		Professional	
		(Wood	
		Preservative)	
		Industrial	
		Industrial	
		(Wood	
		Preservative)	
W FEB Fungicide	FEB Ltd	Amateur	3009
		Professional	
W Fungicidal Wash	Polybond Ltd	Amateur	3601
		Amateur	
		(Wood	
		Preservative)	
		Professional	
		Professional	
		(Wood	
		Preservative)	
W Fungicidal Wash	Smyth - Morris Chemicals Ltd	Professional	3004
		Professional	
		(Wood	
		Preservative)	
W Green Range Murosol 20	Cementone Beaver Ltd	Professional	4456
W Hyperion Mould Inhibiting Solution	Hird Hastie Paints Ltd	Professional	4412
W Kingfisher Fungicidal Wall Solution	Kingfisher Chemicals Ltd	Professional	4014

Product Name	Marketing Company	Use	HSE No.

450 Dodecylamine salicylate + Dodecylamine lactate—continued

Product Name	Marketing Company	Use	HSE No.
W Larsen Concentrated Algicide	Larsen Manufacturing Ltd	Professional	4968
W M-Tec Biocide	M-Protective Coatings Ltd	Professional	4381
W Macpherson Anti-Mould Solution	Macpherson Paints Ltd	Professional Professional (Wood Preservative)	4791
W Macpherson Antimould Solution (RFU)	Macpherson Paints Ltd	Amateur Professional	5190
W Metalife Fungicidal Wash	Metalife International Ltd	Amateur Amateur (Wood Preservative) Professional (Professional (Wood Preservative)	5406
W Mouldrid	S and K Maintenance Products	Amateur Amateur (Wood Preservative) Professional Professional (Wood Preservative)	3671
W Mrs Clear Mould Fungicidal Wash	Mr (Polymer Cement Products) Ltd	Professional	3620
W Nubex WDR	Nubex Ltd	Professional Professional (Wood Preservative)	3810
W Palace Mould Remover	Palace Chemicals Ltd	Amateur Professional	5204
W Permadex Masonry Fungicide Concentrate	Permagard Products Ltd	Professional	4007
W Permoglaze Micatex Fungicidal Treatment	Permoglaze Paints Ltd	Professional Professional (Wood Preservative)	5249
W Polycell Mould Cleaner	Polycell Products Ltd	Amateur Professional	3612
W Protim Wall Solution II	Protim Solignum Ltd	Amateur Professional	5006
W Protim Wall Solution II Concentrate	Protim Solignum Ltd	Professional	5007
W Ruberoid Fungicidal Wash	Ruberoid Building Products Ltd	Amateur	5949
W Safeguard Deepwood Surface Biocide Concentrate	Safeguard Chemicals Ltd	Professional	4616

Product Name	Marketing Company	Use	HSE No.

450 Dodecylamine salicylate + Dodecylamine lactate—continued

Product Name	Marketing Company	Use	HSE No.
W Safeguard Mould and Moss Killer	Safeguard Chemicals Ltd	Amateur Professional	4607
W Solignum Dry Rot Killer	Protim Solignum Ltd	Amateur Professional	5025
W Solignum Fungicide	Protim Solignum Ltd	Amateur Professional	5026
W Thompson's Interior and Exterior Fungicidal Spray	Ronseal Ltd	Amateur	5626
W Thompson's Interior Mould Killer	Ronseal Ltd	Amateur	5624
W Torkill Fungicidal Solution 'W'	Tor Coatings Ltd	Professional	4246
W Unisil S Silicone Waterproofing Solution	Witham Oil & Paint (Lowestoft) Ltd	Professional	4758
W Unitas Fungicidal Wash-Exterior (Solvent Based)	Witham Oil & Paint (Lowestoft) Ltd	Professional	4759
W Unitas Fungicidal Wash-Interior (Water Based)	Witham Oil & Paint (Lowestoft) Ltd	Professional	4760
W Vallance Fungicidal Wash	Siroflex Ltd	Amateur Professional	5375
W Weathershield Fungicidal Wash	ICI Paints	Amateur Amateur (Wood Preservative) Professional Professional (Wood Preservative)	5134

451 3-Iodo-2-propynyl-n-butyl carbamate

Product Name	Marketing Company	Use	HSE No.
Antel FWS	Antel UK Ltd	Professional	6602
BPC-FS Fungicidal Solution Concentrate	Building Preservation Centre	Professional Professional (Wood Preservative)	6484
CME 30/F Concentrate	Terminix Peter Cox Ltd	Professional Professional (Wood Preservative)	5948
EQ109 Surface Biocide	Enviroquest UK Ltd	Professional	6383
W Fungicidal Wall Solution	Sovereign Chemical Industries Ltd	Professional	5387
KF-18 Masonry Biocide (Microemulsion)	Kingfisher Chemicals Ltd	Professional	6529
Lectro FWS	Lectros International Ltd	Professional	6388
Masonry Biocide 3	Terminix Peter Cox Ltd	Professional	6402
Masonry Sterilant	Deal Direct Ltd	Professional	6247
Micro FWS	Construction Chemicals	Professional	6643
Microtech Biocide 25	Philip Johnstone Group Ltd	Professional	5997

Product Name	Marketing Company	Use	HSE No.

451 3-Iodo-2-propynyl-n-butyl carbamate—continued

Product Name	Marketing Company	Use	HSE No.
Omega Biocide	Restoration (UK) Ltd	Professional Professional (Wood Preservative)	6381
Remtox Dry Rot F. W. S.	Remtox (Chemicals) Ltd	Professional	5188
Remtox Dry Rot Paint	Remtox (Chemicals) Ltd	Amateur Amateur (Wood Preservative) Professional Professional (Wood Preservative)	5187
Remtox Fungicidal Wall Solution RS	Remtox (Chemicals) Ltd	Professional	5484
Remtox Microactive FWS W6	Remtox (Chemicals) Ltd	Professional	5589
Restor-MFC	Restoration (UK) Ltd	Professional Professional (Wood Preservative)	6246
Safeguard Fungicidal Micro Emulsifiable Concentrate	Safeguard Chemicals Ltd	Professional Professional (Wood Preservative)	5648
Sovaq Micro FWS	Sovereign Chemical Industries Ltd	Professional	5800
Sprytech 3	Spry Chemicals Ltd	Professional	6491
Stanhope Biocide 25X	Philip Johnstone Group Ltd	Professional	6309
Trisol 23	Triton Chemical Manufacturing Co Ltd	Professional	6274
UKC-2 Dry Rot Killer	UK Chemicals Ltd	Professional	6378
UKC-2 Dry Rot Killer Concentrate	UK Chemicals Ltd	Professional Professional (Wood Preservative)	6107
Ultra Tech 2000M	Crown Chemicals Ltd	Professional	6164

452 Naphthalene + Permethrin + Pyrethrins + Benzalkonium chloride

Product Name	Marketing Company	Use	HSE No.
W Roxem D	Horton Hygiene Company	Professional	5107

453 2-octyl-2h-isothiazolin-3-one

Product Name	Marketing Company	Use	HSE No.
ACS Antimould Emulsion	Advanced Chemical Specialities Ltd	Amateur Professional	6308
Halostain	Mould Growth Consultants Ltd	Professional	6559

Product Name	Marketing Company	Use	HSE No.

454 2-octyl-2h-isothiazolin-3-one + Carbendazim + Dichlorophenyl dimethylurea

Biocheck A.C.C.	Mould Growth Consultants Ltd	Amateur Professional	6571

455 Permethrin + Azaconazole

Fongix SE Total Treatment for Wood	Liberon Waxes Ltd	Amateur Amateur (Wood Preservative) Professional Professional (Wood Preservative)	5610
D Fongix SE Total Treatment for Wood	V33 SA	Amateur Amateur (Wood Preservative)	4288
Woodworm Killer and Rot Treatment	Liberon Waxes Ltd	Amateur Amateur (Wood Preservative) Professional Professional (Wood Preservative)	4826

456 Permethrin + Pyrethrins + Benzalkonium chloride + Naphthalene

W Roxem D	Horton Hygiene Company	Professional	5107

457 Permethrin + Zinc octoate

Premium Wood Treatment	Rentokil Initial UK Ltd	Amateur Amateur (Wood Preservative) Professional Professional (Wood Preservative)	5350

458 2-Phenylphenol + Benzalkonium chloride

Anti Fungus Wash	Craig & Rose Plc	Professional	6293

Product Name	Marketing Company	Use	HSE No.

458 2-Phenylphenol + Benzalkonium chloride—continued

Product Name	Marketing Company	Use	HSE No.
C/S Clean R833	Construction Specialties (UK) Ltd	Amateur Professional	6865
W Chelwash	Chelec Ltd	Amateur	3096
Crown Fungicidal Wall Solution	Crown Chemicals Ltd	Professional	5868
Deadly Nightshade	Andura Textured Masonry Coatings Ltd	Professional	4895
D Earnshaws Fungal Wash	E Earnshaw and Company (1965) Ltd	Amateur	4256
Fungicidal Algaecidal Bacteriacidal Wash	Philip Johnstone Group Ltd	Amateur Professional	4216
D Fungicidal Solution FL.2	Manders Paints Ltd	Amateur Professional	4245
W Fungicidal Wash	C W Wastnage Ltd	Amateur	3072
Fungicidal Wash	Everbuild Building Products Ltd	Amateur Professional	6000
W Fungicidal Wash Solution	Johnstone's Paints Plc	Amateur Professional	3776
Fungiguard	Spaldings Ltd	Professional	5582
Houseplan Fungi Wash	Ruberoid Building Products Ltd	Amateur	6374
Johnstone's Fungicidal Wash	Johnstone's Paints Plc	Amateur Professional	5678
Larsen Concentrated Algicide 2	Larsen Manufacturing Ltd	Amateur Professional	6606
Micro FWS	Lectros International Ltd	Professional	6135
Micro Masonry Biocide	Grip-Fast Systems Ltd	Professional	6444
Microguard FWS	Permagard Products Ltd	Professional	5485
Microstel	Bedec Products Ltd	Professional	4198
Mouldcheck Sterilizer 2	Biotech Environmental (UK) Ltd	Amateur Professional	6385
Nitromors Mould Remover	Henkel Home Improvements and Adhesive Products	Amateur	5507
P.J. Fungicidal Solution	Philip Johnstone Group Ltd	Amateur Professional	6318
Permacide Masonry Fungicide	Triton Perma Industries Ltd	Professional	5570
Permarock Fungicidal Wash	Permarock Products Ltd	Professional	5193
Pob Mould Treatment	Darcy Industries Ltd	Amateur	4076
Ruberoid Fungicidal Wash Solution	Ruberoid Building Products Ltd	Amateur	6375
Spencer Stormcote Fungicidal Treatment	Spencer Coatings Ltd	Professional	4155
Tetra Concentrated Mould Cleaner	Tetrosyl (Building Products) Ltd	Amateur Professional	4206
Vallance Pro-Build Extra Strong Moss and Mould Killer	Vallance Ltd	Amateur Professional	5761
Z.144 Fungicidal Wash	Crosbie Coatings Ltd	Amateur Professional	3003

Product Name	Marketing Company	Use	HSE No.

459 Propiconazole

Trisol 22	Triton Chemical Manufacturing Co Ltd	Professional	6111
Wocosen 100 SL	Janssen Pharmaceutica NV	Professional Professional (Wood Preservative) Industrial Industrial (Wood Preservative)	5743

460 Propiconazole + Boric acid + Cypermethrin

Rentokil Total Wood Treatment	Rentokil Initial UK Ltd	Amateur Amateur (Wood Preservative) Professional Professional (Wood Preservative)	6053

461 Pyrethrins + Benzalkonium chloride + Naphthalene + Permethrin

W Roxem D	Horton Hygiene Company	Professional	5107

462 Sodium 2-phenylphenoxide

Aqueous Fungicidal Irrigation Fluid	Palace Chemicals Ltd	Amateur Professional	3794
Aqueous Fungicidal Irrigation Fluid Concentrate	Palace Chemicals Ltd	Professional	3793
Brunosol Concentrate	Philip Johnstone Group Ltd	Professional Professional (Wood Preservative)	4472
D Coaltec 50	Coalite Chemicals Division	Professional Professional (Wood Preservative) Industrial Industrial (Wood Preservative)	4761

Product Name	Marketing Company	Use	HSE No.

462 Sodium 2-phenylphenoxide—continued

Product Name	Marketing Company	Use	HSE No.
D Coaltec 50TD	Coalite Chemicals Division	Amateur Amateur (Wood Preservative) Professional Professional (Wood Preservative) Industrial Industrial (Wood Preservative)	4762
Cuprinol Dry Rot Killer S for Brickwork and Masonry	Cuprinol Ltd	Amateur Professional	4720
Fungicidal Wash	Laybond Products Ltd	Amateur Professional	5089
Green Range Fungicidal Concentrate	Philip Johnstone Group Ltd	Professional	3689
Mar-Cide	Marcher Chemicals Ltd	Professional Professional (Wood Preservative) Industrial Industrial (Wood Preservative)	5631
D PC-D	Terminix Peter Cox Ltd	Professional	4405
PC-D	Terminix Peter Cox Ltd	Professional	5719
D PC-D Concentrate	Terminix Peter Cox Ltd	Professional	4404
PC-D Concentrate	Terminix Peter Cox Ltd	Professional	5720
PLA Products Dry Rot Killer	Sealocrete PLA Ltd	Amateur Professional	5339
PLA Products Fungicidal Wash	Sealocrete PLA Ltd	Amateur	5373
Protim Wall Solution Concentrate 5	Protim Solignum Ltd	Professional	6437
D Remtox FWS (Low Odour)	Remtox (Chemicals) Ltd	Professional	3882
Restor F.W.S	Restoration (UK) Ltd	Professional	5961
W Safeguard Fungicidal Wall Solution	Safeguard Chemicals Ltd	Professional	3903
D Saturin E30	British Building and Engineering Appliances Plc	Professional	3902
Solignum Anti Fungi Concentrate	Protim Solignum Ltd	Professional	5018
Solignum Anti-Fungi Concentrate 5	Protim Solignum Ltd	Professional	6436
D Sovereign Fungicidal Wall Solution	Sovereign Chemical Industries Ltd	Professional	3933
D Trisol 21	Triton Chemical Manufacturing Co Ltd	Professional	3789

Product Name	Marketing Company	Use	HSE No.

463 Sodium dichlorophen

Bactdet D	Mould Growth Consultants Ltd	Professional	3809
Fungo	Dax Products Ltd	Amateur	4768
RLT Bactdet	Mould Growth Consultants Ltd	Amateur Professional	3928

464 Sodium hypochlorite

Anti-Moss	Keychem Ltd	Professional	6064
Brolac Fungicidal Solution	Akzo Nobel Decorative Coatings	Amateur	4413
Crown Trade Stronghold Fungicidal Solution	Akzo Nobel Decorative Coatings	Professional	5383
D Laura Ashley Home Fungicidal Wash	Laura Ashley Ltd	Amateur	4192
W Palace Fungicidal Wash	Palace Chemicals Ltd	Amateur	4091
Sandtex Fungicide	Akzo Nobel Decorative Coatings	Amateur Professional	5258
Tetra Construction Fungicidal Wash	Tetrosyl (Building Products) Ltd	Amateur	6250
Tetra Fungicidal Wash	Tetrosyl (Building Products) Ltd	Amateur	6240

465 Sodium pentachlorophenoxide

D Mosgo P	Agrichem Ltd	Professional Professional (Wood Preservative)	4337
W Protim Plug Compound	Protim Solignum Ltd	Professional Professional (Wood Preservative)	5012

466 Zinc octoate

Rentokil Dry Rot and Wet Rot Treatment	Rentokil Initial UK Ltd	Amateur Amateur (Wood Preservative) Professional Professional (Wood Preservative)	4378

Product Name	Marketing Company	Use	HSE No.

467 Zinc octoate + Permethrin

| Premium Wood Treatment | Rentokil Initial UK Ltd | Amateur
Amateur
(Wood
Preservative)
Professional
Professional
(Wood
Preservative) | |

6
WOOD PRESERVATIVES

wood preservatives

468 Acypetacs copper

Product Name	Marketing Company	Use	HSE No.
Cuprinol Low Odour Wood Preserver Green	Cuprinol Ltd	Amateur Professional Industrial	5448
W Cuprinol Wood Preserver Green S	Cuprinol Ltd	Amateur Professional	4697

469 Acypetacs copper + Acypetacs zinc + Permethrin

Product Name	Marketing Company	Use	HSE No.
Cuprinol Low Odour End Cut	Cuprinol Ltd	Amateur Professional Industrial	6114
Protim 800p	Protim Solignum Ltd	Industrial	5702

470 Acypetacs zinc

Product Name	Marketing Company	Use	HSE No.
Cuprinol Garden Shed and Fence Preserver	Cuprinol Ltd	Amateur Professional	5873
Cuprinol Low Odour Wet and Dry Rot Killer For Timber	Cuprinol Ltd	Amateur Professional Industrial	5446
Cuprinol Low Odour Wood Preserver Clear	Cuprinol Ltd	Amateur Professional Industrial	5437
W Cuprinol Wet and Dry Rot Killer for Timber(s)	Cuprinol Ltd	Amateur	3589
Cuprinol Wood Preserver	Cuprinol Ltd	Amateur Professional	4698
W Cuprinol Wood Preserver Clear S	Cuprinol Ltd	Amateur Professional Industrial	4708
Cuprisol F	Cuprinol Ltd	Industrial	3734
Cuprisol XQD	Cuprinol Ltd	Industrial	3733
Exterior Wood Preserver S	Cuprinol Ltd	Amateur Professional	4716
Fungicidal Preservative	Protim Solignum Ltd	Amateur Professional Industrial	5489
Protim FDR 800	Protim Solignum Ltd	Industrial	4963
Protim JP 800	Protim Solignum Ltd	Industrial	4987
Protim Paste 800 F	Protim Solignum Ltd	Professional Industrial	4989
Wickes Exterior Wood Preserver	Wickes Building Supplies Ltd	Amateur Professional	5697
Wickes Wood Preserver Clear	Wickes Building Supplies Ltd	Amateur Professional	5694

Product Name	Marketing Company	Use	HSE No.

471 Acypetacs zinc (equivalent to 3% zinc metal) + Permethrin

Product Name	Marketing Company	Use	HSE No.
Protim 800 CWR	Protim Solignum Ltd	Industrial	4993

472 Acypetacs zinc + Dichlofluanid

Product Name	Marketing Company	Use	HSE No.
Cuprinol Decorative Preserver	Cuprinol Ltd	Amateur Professional Industrial	5343
Cuprinol Decorative Preserver	Cuprinol Ltd	Amateur Professional Industrial	6277
Cuprinol Decorative Preserver Red Cedar	Cuprinol Ltd	Amateur Professional Industrial	5342
D Cuprinol Decorative Wood Preserver	Cuprinol Ltd	Amateur	3156
D Cuprinol Decorative Wood Preserver Red Cedar	Cuprinol Ltd	Amateur Professional	4696
Cuprinol Garden Decking Protector	Cuprinol Ltd	Amateur Professional	6862
D Cuprinol Magnatreat F	Cuprinol Ltd	Industrial	3752
D Cuprinol Magnatreat XQD	Cuprinol Ltd	Industrial	3751
Cuprinol Preservative Base	Cuprinol Ltd	Amateur Professional	4929
Wickes Decorative Preserver Rich Cedar	Wickes Building Supplies Ltd	Amateur Professional Industrial	6420
Wickes Decorative Wood Preserver	Wickes Building Supplies Ltd	Amateur Professional Industrial	6419

473 Acypetacs zinc + Permethrin

Product Name	Marketing Company	Use	HSE No.
W Cuprinol 5 Star Complete Wood Treatment S	Cuprinol Ltd	Amateur Professional	4710
Cuprinol Low Odour 5 Star Complete Wood Treatment	Cuprinol Ltd	Amateur Professional	5445
Cuprisol FN	Cuprinol Ltd	Industrial	3632
Cuprisol WR	Cuprinol Ltd	Industrial	3633
Cuprisol XQD Special	Cuprinol Ltd	Industrial	3732
Cut End	Protim Solignum Ltd	Amateur Professional Industrial	5490
Protim 800	Protim Solignum Ltd	Industrial	4991
Protim 800 C	Protim Solignum Ltd	Industrial	4990
Protim 800 WR	Protim Solignum Ltd	Industrial	4992
Protim Brown CDB	Protim Solignum Ltd	Amateur Professional Industrial	6472

Product Name	Marketing Company	Use	HSE No.

473 Acypetacs zinc + Permethrin—continued

Protim Paste 800	Protim Solignum Ltd	Professional	4986
		Industrial	
D Protim WR 800	Fosroc Ltd	Amateur	4487
		Professional	
		Industrial	

474 Acypetacs zinc + Permethrin + Acypetacs copper

Cuprinol Low Odour End Cut	Cuprinol Ltd	Amateur	6114
		Professional	
		Industrial	
Protim 800P	Protim Solignum Ltd	Industrial	5702

475 Acypetacs zinc + Permethrin + Propiconazole + Tebuconazole

Protim 415 T Special Solvent	Protim Solignum Ltd	Industrial	6077

476 Alkylaryltrimethyl ammonium chloride

Anti-Mould Solution	Macpherson Paints Ltd	Amateur (Surface Biocide)	4873
		Amateur (Wood Preservative)	
		Professional (Surface Biocide)	
		Professional (Wood Preservative)	
Barrettine Timberguard	Barrettine Products Ltd	Amateur	3089
Timberdip	Langlow Products Ltd	Professional	4393
		Industrial	
Timberguard 1 Plus 7 Concentrate	Barrettine Products Ltd	Professional	5211
		Industrial	

477 Alkyltrimethyl ammonium chloride

Sinesto B	Finnmex Co	Industrial	5136

478 Ammonium bifluoride + Sodium dichromate + Sodium fluoride

W Rentex	Rentokil Initial UK Ltd	Industrial	4756

479 Arsenic pentoxide + Chromium trioxide + Copper oxide

Celcure AO	Rentokil Initial UK Ltd	Industrial	4220
Celcure CCA Type C	Rentokil Initial UK Ltd	Industrial	5104
Celcure CCA Type C-40%	Rentokil Initial UK Ltd	Industrial	6028
Celcure CCA Type C-50%	Rentokil Initial UK Ltd	Industrial	6030
Celcure CCA Type C-60%	Rentokil Initial UK Ltd	Industrial	5934

Product Name	Marketing Company	Use	HSE No.

479 Arsenic pentoxide + Chromium trioxide + Copper oxide—continued

Injecta Osmose K33 C50	Injecta Osmose Ltd	Industrial	5889
Injecta Osmose K33 C58.7	Injecta Osmose Ltd	Industrial	6498
Injecta Osmose K33 C72	Injecta Osmose Ltd	Industrial	5890
Laporte CCA AWPA Type C	Laporte Wood Preservation	Industrial	3179
Laporte CCA Oxide Type 1	Laporte Wood Preservation	Industrial	3177
Laporte CCA Oxide Type 2	Laporte Wood Preservation	Industrial	3178
Osmose K33 C58.7	Protim Solignum Ltd	Industrial	6889
Protim CCA Oxide 50	Protim Solignum Ltd	Industrial	5537
Protim CCA Oxide 58	Protim Solignum Ltd	Industrial	5686
Protim CCA Oxide 72	Protim Solignum Ltd	Industrial	5594
Tanalith 3302	Hickson Timber Products Ltd	Industrial	5098
Tanalith 3313	Hickson Timber Products Ltd	Industrial	4422
Tanalith C3310	Hickson Timber Products Ltd	Industrial	4669
W Tanalith Oxide C3309	Hickson Timber Products Ltd	Industrial	4668
W Tanalith Oxide C3314	Hickson Timber Products Ltd	Industrial	4774
Tecca LQ1	Tecca Ltd	Industrial	5999

480 Arsenic pentoxide + Copper sulphate + Sodium dichromate

Celcure A Concentrate	Rentokil Initial UK Ltd	Industrial	5458
Celcure A Fluid 10	Rentokil Initial UK Ltd	Industrial	3764
Celcure A Fluid 6	Rentokil Initial UK Ltd	Industrial	3155
Celcure A Paste	Rentokil Initial UK Ltd	Industrial	4523
Injecta CCA-C	Injecta Osmose Ltd	Industrial	4447
D Kemira CCA Type BS	Kemira Kemwood AB	Industrial	3879
Kemwood CCA Type BS	Laporte Kemwood AB	Industrial	5534
Laporte CCA Type 1	Laporte Wood Preservation	Industrial	3175
Laporte CCA Type 2	Laporte Wood Preservation	Industrial	3176
Laporte Permawood CCA	Laporte Wood Preservation	Industrial	3529
Protim CCA Salts Type 2	Protim Solignum Ltd	Industrial	4972
Tanalith 3357	Hickson Timber Products Ltd	Industrial	4431
W Tanalith CL (3354)	Hickson Timber Products Ltd	Industrial	4196
Tanalith CP 3353	Hickson Timber Products Ltd	Industrial	4667
Tecca P2	Tecca Ltd	Industrial	3958

481 Azaconazole

Rodewod 50 SL	Janssen Pharmaceutica NV	Amateur Professional Industrial	3839
Safetray SL	Progress Products	Professional	5464

482 Azaconazole + Dichlofluanid

W Xyladecor Matt U 404	NWE Distributors Ltd	Amateur Professional	4445

Product Name	Marketing Company	Use	HSE No.

482 Azaconazole + Dichlofluanid—continued

| W Xylamon Primer Dipping Stain U 415 | Venilia Ltd | Professional | 4444 |

483 Azaconazole + Permethrin

Fongix SE Total Treatment for Wood	Liberon Waxes Ltd	Amateur (Surface Biocide) Amateur (Wood Preservative) Professional (Surface Biocide) Professional (Wood Preservative)	5610
D Fongix SE Total Treatment for Wood	V33 SA	Amateur (Surface Biocide) Amateur (Wood Preservative)	4288
Woodworm Killer and Rot Treatment	Liberon Waxes Ltd	Amateur (Surface Biocide) Amateur (Wood Preservative) Professional (Surface Biocide) Professional (Wood Preservative)	4826
W Xylamon Brown U 101 C	Venilia Ltd	Amateur Professional	4443
W Xylamon Curative U 152 G/H	NWE Distributors Ltd	Professional	4446

484 Benzalkonium bromide

| X Moss | Thames Valley Specialist Products Ltd | Amateur Professional | 6839 |

485 Benzalkonium chloride

| Algae Remover | Akzo Nobel Woodcare | Amateur Professional | 5337 |

Product Name	Marketing Company	Use	HSE No.

485 Benzalkonium chloride—continued

FMB 451-5 Quat (50%)	Midland Europe	Professional (Surface Biocide) Professional (Wood Preservative) Industrial (Wood Preservative)	5734
Glen Wood Care Wood Preservative	Glen Wood Care	Amateur Professional Industrial	4096
D Gloquat RP	Rhodia Limited	Professional (Surface Biocide) Professional (Wood Preservative)	4815
Langlow Timbershield	Langlow Products Ltd	Amateur Professional	4394
Rhodaquat RP50	Rhodia Limited	Professional (Surface Biocide) Professional (Wood Preservative)	5762

486 Benzalkonium chloride + 2-Phenylphenol

| W Chel Dry Rot Killer for Masonry and Brickwork | Chelec Ltd | Professional | 3067 |

487 Benzalkonium chloride + Boric acid

| Barrettine Premier Multicide Fluid | Barrettine Products Ltd | Amateur (Surface Biocide) Amateur (Wood Preservative) | 6312 |

477

Product Name	Marketing Company	Use	HSE No.

487 Benzalkonium chloride + Boric acid—continued

Product Name	Marketing Company	Use	HSE No.
Microguard Mouldicidal Wood Preserver	Permagard Products Ltd	Amateur (Surface Biocide) Amateur (Wood Preservative) Professional (Surface Biocide) Professional (Wood Preservative) Industrial (Wood Preservative)	5100

488 Benzalkonium chloride + Boric acid + Dialkyldimethyl ammonium chloride + Methylene bis(thiocyanate)

Product Name	Marketing Company	Use	HSE No.
Celbrite MT	Rentokil Initial UK Ltd	Industrial	4550

489 Benzalkonium chloride + Copper oxide

Product Name	Marketing Company	Use	HSE No.
Laporte AcQ 1900	Laporte Wood Preservation	Industrial	6487

490 Benzalkonium chloride + Dialkyldimethyl ammonium chloride

Product Name	Marketing Company	Use	HSE No.
ABL Aqueous Wood Preserver Concentrate 1:9	Advanced Bitumens Ltd	Professional Industrial	4199
Aqueous Wood Preserver	Advanced Bitumens Ltd	Amateur Professional Industrial	4727
Celbrite M	Rentokil Initial UK Ltd	Industrial	4535
Celbronze B	Rentokil Initial UK Ltd	Professional Industrial	4549
W Colourfast Protector	Rentokil Initial UK Ltd	Amateur Professional	4609

491 Benzalkonium chloride + Disodium octaborate

Product Name	Marketing Company	Use	HSE No.
Boracol 20 Rh	Advanced Chemical Specialities Ltd	Professional	4019

Product Name	Marketing Company	Use	HSE No.

491 Benzalkonium chloride + Disodium octaborate—continued

Product Name	Marketing Company	Use	HSE No.
Boracol B8.5 Rh Mouldicide/Wood Preservative	Advanced Chemical Specialities Ltd	Amateur (Surface Biocide) Amateur (Wood Preservative) Professional (Surface Biocide) Professional (Wood Preservative)	4809
Boron Biocide 10	Philip Johnstone Group Ltd	Professional (Surface Biocide) Professional (Wood Preservative)	6485
Brunosol Boron 10	Philip Johnstone Group Ltd	Professional (Surface Biocide) Professional (Wood Preservative)	6488
Preservative Gel	Protim Solignum Ltd	Professional	6528
Probor	Safeguard Chemicals Ltd	Professional (Surface Biocide) Professional (Wood Preservative)	6266
Probor 10	Safeguard Chemicals Ltd	Professional	6595
Probor 50	Safeguard Chemicals Ltd	Professional (Surface Biocide) Professional (Wood Preservative)	5596
Weathershield Multi-Surface Fungicidal Wash	Imperial Chemical Industries Plc	Amateur Professional	6823

492 Benzalkonium chloride + Disodium octaborate + 3-Iodo-2-propynyl-n-butyl carbamate

Product Name	Marketing Company	Use	HSE No.
Antiblu Select	Hickson Timber Products Ltd	Industrial	6087

Product Name	Marketing Company	Use	HSE No.

493 Benzalkonium chloride + Lindane

W Biokil	Jaymar Chemicals	Professional	4247

494 Benzalkonium chloride + Permethrin

W Biokil Emulsion	Jaymar Chemicals	Professional	4562

495 Benzalkonium chloride + Permethrin + Tebuconazole

Vacsol Aqua 6105	Hickson Timber Products Ltd	Industrial	5892

496 Benzalkonium chloride + Tebuconazole

Vacsol Aqua 6101	Hickson Timber Products Ltd	Industrial	5557

497 Boric acid

Basilit B 85	NWE Distributors Ltd	Industrial	3813
Celbor	Rentokil Initial UK Ltd	Industrial	4614
Celbor M	Rentokil Initial UK Ltd	Industrial	4912
Control Fluid FB	Rentokil Initial UK Ltd	Professional	5323
Diffusit	Dr Wolman GmbH	Professional	5992
Dricon	Hickson Timber Products Ltd	Industrial	5688
W Dricon	Lambson Ltd	Industrial	5315
D New Cut 'N' Spray	Rentokil Initial UK Ltd	Amateur	4187
PJG Boron Rods	Philip Johnstone Group Ltd	Professional	6616
Pyro-Red	Hoover Treated Wood Products Inc	Industrial	5867
Pyrolith 3505 Ready to use	Hickson Timber Products Ltd	Industrial	3636
Rentokil Dry Rot Fluid (E) for Bonded Warehouses	Rentokil Initial UK Ltd	Professional (Surface Biocide) Professional (Wood Preservative)	3986
Rentokil Wood Preserver	Rentokil Initial UK Ltd	Amateur Professional	6052

498 Boric acid + 2-(thiocyanomethylthio)benzothiazole

Protim Fentex Europa 1 RFU	Protim Solignum Ltd	Industrial	5119
Protim Fentex Europa I	Protim Solignum Ltd	Industrial	4984

499 Boric acid + 2-Phenylphenol

Protim Fentex P	Protim Solignum Ltd	Industrial	6095

500 Boric acid + 3-iodo-2-propynyl-n-butyl carbamate

Protim Fentex I	Protim Solignum Ltd	Industrial	6096

Product Name	Marketing Company	Use	HSE No.

501 Boric acid + Benzalkonium chloride

Barrettine Premier Multicide Fluid	Barrettine Products Ltd	Amateur (Surface Biocide) Amateur (Wood Preservative)	6312
Microguard Mouldicidal Wood Preserver	Permagard Products Ltd	Amateur (Surface Biocide) Amateur (Wood Preservative) Professional (Surface Biocide) Professional (Wood Preservative) Industrial (Wood Preservative)	5100

502 Boric acid + Chromium acetate + Copper sulphate + Sodium dichromate

Celgard CF	Rentokil Initial UK Ltd	Industrial	4608

503 Boric acid + Chromium trioxide + Copper oxide

Tanalith (3419) CBC	Hickson Timber Products Ltd	Industrial	4022

504 Boric acid + Copper carbonate hydroxide

W Tanalith 3487	Hickson Timber Products Ltd	Industrial	5305

505 Boric acid + Copper carbonate hydroxide + Propiconazole

Wolmanit CX-P	Dr Wolman Gmbh	Industrial	6566

506 Boric acid + Copper carbonate hydroxide + Propiconazole + Tebuconazole

Tanalith E (3492)	Hickson Timber Products Ltd	Industrial	6269
Tanalith E (3494)	Hickson Timber Products Ltd	Industrial	6434

507 Boric acid + Copper carbonate hydroxide + Tebuconazole

Ensele 3428	Hickson Timber Products Ltd	Professional Industrial	5878
Tanalith E (3485)	Hickson Timber Products Ltd	Industrial	5562

508 Boric acid + Copper oxide

W Tanalith 3422	Hickson Timber Products Ltd	Industrial	4403

Product Name	Marketing Company	Use	HSE No.

509 Boric acid + Copper sulphate

Ensele 3426	Hickson Timber Products Ltd	Professional Industrial	4364
W Ensele 3427	Hickson Timber Products Ltd	Amateur	5053
Ensele 3429	Hickson Timber Products Ltd	Amateur Professional	6392
Laporte End Grain Treatment	Laporte Wood Preservation	Amateur Professional Industrial	6885
Rentokil ACB	Rentokil Initial UK Ltd	Industrial	5759

510 Boric acid + Copper sulphate + Potassium dichromate

Tecca CCB 1	Tecca Ltd	Industrial	5584

511 Boric acid + Copper sulphate + Sodium dichromate

Laporte CCB	Laporte Wood Preservation	Industrial	5875
Tanalith CBC Paste 3402	Hickson Timber Products Ltd	Industrial	3833

512 Boric acid + Cypermethrin + Propiconazole

Rentokil Total Wood Treatment	Rentokil Initial UK Ltd	Amateur (Surface Biocide) Amateur (Wood Preservative) Professional (surface Biocide) Professional (Wood Preservative)	6053

513 Boric acid + dialkyldimethyl ammonium chloride

Celbor P	Rentokil Initial UK Ltd	Industrial	5923
Celbor P 25% Solution	Rentokil Initial UK Ltd	Industrial	6033
Celbor P 5% Solution	Rentokil Initial UK Ltd	Industrial	6019
Celbor P25	Rentokil Initial UK Ltd	Industrial	6563
Celbor P5	Rentokil Initial UK Ltd	Industrial	6558
Protim E470	Protim Solignum Ltd	Industrial	6386

514 Boric acid + Dialkyldimethyl ammonium chloride + Methylene bis(thiocyanate) + Benzalkonium chloride

Celbrite Mt	Rentokil Initial UK Ltd	Industrial	4550

515 Boric acid + Disodium octaborate

Acs Woodkeeper Paste Preservative	Advanced Chemical Specialities Ltd	Professional	6221

Product Name	Marketing Company	Use	HSE No.

516 Boric acid + Disodium tetraborate

Control Fluid Sb	Rentokil Initial UK Ltd	Professional (Surface Biocide) Professional (Wood Preservative)	6018
Pc-k	Terminix Peter Cox Ltd	Professional	4409
Rentokil Control Paste Sb	Rentokil Initial UK Ltd	Professional	6501

517 Boric acid + Methylene bis(thiocyanate)

Hickson Antiblu 3737	Hickson Timber Products Ltd	Industrial	4745
Hickson Antiblu 3738	Hickson Timber Products Ltd	Industrial	4746

518 Boric acid + Methylene bis(thiocyanate) + 2-(Thiocyanomethylthio)benzothiazole

Fentex Elite	Protim Solignum Ltd	Industrial	5532

519 Boric acid + Permethrin

Activ-8-ii	Restoration (UK) Ltd	Professional Industrial	6170
Micro Dual Purpose 8	Grip-fast Systems Ltd	Professional Industrial	6408
W Microguard Fi Concentrate	Permagard Products Ltd	Professional Industrial	5101
Microguard Fi Concentrate	Permagard Products Ltd	Professional Industrial	5854

520 Boric acid + Permethrin + Zinc octoate

W Celpruf Bzp	Rentokil Initial UK Ltd	Industrial	4190
W Celpruf Bzp WR	Rentokil Initial UK Ltd	Industrial	4191
Cut'n'treat	Rentokil Initial UK Ltd	Amateur Professional	5802
Premium Grade Wood Treatment	Rentokil Initial UK Ltd	Amateur Professional	4193

521 Boric acid + Propiconazole

Celbor Pr	Rentokil Initial UK Ltd	Industrial	6192

522 Boric acid + Sodium 2,4,6-trichlorophenoxide

Protim Fentex Twr Plus	Protim Solignum Ltd	Industrial	6547

523 Boric acid + Zinc octoate

W Celpruf Bz	Rentokil Initial UK Ltd	Industrial	4188
W Celpruf Bz WR	Rentokil Initial UK Ltd	Industrial	4189
Rentokil Dry Rot Paste (d)	Rentokil Initial UK Ltd	Professional	3987

HSE WOOD PRESERVATIVES

483

Product Name	Marketing Company	Use	HSE No.

524 Boric oxide

ACS Boron Rods	Advanced Chemical Specialities Ltd	Professional	6115

525 Carbendazim

Injecta Osmose ABS 33 A	Injecta Osmose Ltd	Industrial	6533
Osmose ABS33A	Protim Solignum Ltd	Industrial	6887
Tanamix C	Hickson Timber Products Ltd	Industrial	6259

526 Carbendazim + Disodium octaborate

Weathershield Aquatech Preservative Basecoat	ICI Paints	Amateur Professional	5757

527 5-Chloro-2-methyl-4-isothiazolin-3-one + 2-Methyl-4-isothiazolin-3-one

Celkil 90	Rentokil Initial UK Ltd	Industrial	5428
Kathon 886f	Rohm & Haas (UK) Ltd	Industrial	5431
Laporte Mould-ex	Laporte Wood Preservation	Industrial	5430
Tanamix 3743	Hickson Timber Products Ltd	Industrial	5429

528 5-Chloro-2-methyl-4-isothiazolin-3-one + Copper sulphate + 2-Methyl-4-isothiazolin-3-one

Injecta Osmose ABS33	Injecta Osmose Ltd	Industrial	5978

529 Chromium acetate + Copper sulphate + Sodium dichromate

Celcure B	Rentokil Initial UK Ltd	Industrial	4541
Celcure O	Rentokil Initial UK Ltd	Industrial	4539

530 Chromium acetate + Copper sulphate + Sodium dichromate + Boric acid

Celgard CF	Rentokil Initial UK Ltd	Industrial	4608

531 Chromium trioxide + Copper oxide

Osmose K55	Protim Solignum Ltd	Industrial	6867

532 Chromium trioxide + Copper oxide + Arsenic pentoxide

Celcure Ao	Rentokil Initial UK Ltd	Industrial	4220
Celcure CCA Type C	Rentokil Initial UK Ltd	Industrial	5104
Celcure CCA Type C-40%	Rentokil Initial UK Ltd	Industrial	6028
Celcure CCA Type C-50%	Rentokil Initial UK Ltd	Industrial	6030
Celcure Cca Type C-60%	Rentokil Initial UK Ltd	Industrial	5934
Injecta Osmose K33 C50	Injecta Osmose Ltd	Industrial	5889
Injecta Osmose K33 C58.7	Injecta Osmose Ltd	Industrial	6498
Injecta Osmose K33 C72	Injecta Osmose Ltd	Industrial	5890
Laporte CCA Awpa Type C	Laporte Wood Preservation	Industrial	3179
Laporte CCA Oxide Type 1	Laporte Wood Preservation	Industrial	3177
Laporte CCA Oxide Type 2	Laporte Wood Preservation	Industrial	3178
Osmose K33 C58.7	Protim Solignum Ltd	Industrial	6889

Product Name	Marketing Company	Use	HSE No.

532 Chromium trioxide + Copper oxide + Arsenic pentoxide—continued

Protim CCA Oxide 50	Protim Solignum Ltd	Industrial	5537
Protim CCA Oxide 58	Protim Solignum Ltd	Industrial	5686
Protim CCA Oxide 72	Protim Solignum Ltd	Industrial	5594
Tanalith 3302	Hickson Timber Products Ltd	Industrial	5098
Tanalith 3313	Hickson Timber Products Ltd	Industrial	4422
Tanalith C3310	Hickson Timber Products Ltd	Industrial	4669
W Tanalith Oxide C3309	Hickson Timber Products Ltd	Industrial	4668
W Tanalith Oxide C3314	Hickson Timber Products Ltd	Industrial	4774
Tecca LQ1	Tecca Ltd	Industrial	5999

533 Chromium trioxide + Copper oxide + Boric acid

Tanalith (3419) Cbc	Hickson Timber Products Ltd	Industrial	4022

534 Chromium trioxide + Copper oxide + Disodium octaborate

D Celcure Cb Paste	Rentokil Initial UK Ltd	Industrial	4537

535 Coal tar creosote

ABL Brown Creosote	Advanced Bitumens Ltd	Professional Industrial	3859
Bartoline Dark and Light Creosote	Bartoline Ltd	Amateur Professional	4459
Bitmac Creosote	Bitmac Ltd	Professional Industrial	5882
D BS 144 Creosote	Lanstar Coatings Ltd	Industrial	5189
BS 144 Creosote	Oakmere Technical Services Ltd	Industrial	3712
D Builders Mate Creosote	Builders Mate Ltd	Amateur Professional	5110
Carbo Creosote	Talke Chemical Company Ltd	Amateur Professional Industrial	4362
W Coal Tar Creosote	Creohaul and Company	Amateur	3898
Coal Tar Creosote	Great Marsh Ltd	Professional Industrial	4574
W Coal Tar Creosote	Hardmans of Hull Ltd	Amateur Professional	4917
Coal Tar Creosote	James of Bedlington Ltd	Amateur Professional Industrial	6031
Coal Tar Creosote	Laybond Products Ltd	Amateur Professional Industrial	4294
W Coal Tar Creosote	William Mathwin and Son (Newcastle) Ltd	Amateur Professional Industrial	4602

Product Name	Marketing Company	Use	HSE No.

535 Coal tar creosote—continued

Product Name	Marketing Company	Use	HSE No.
D Coal Tar Creosote Blend	Laycocks Agricultural Chemists	Amateur Professional	4450
Creosote	Bagnalls Haulage Ltd	Amateur Professional	5975
Creosote	Blanchard Martin and Simmonds Ltd	Amateur	3884
W Creosote	C W Wastnage Ltd	Amateur Professional	5267
W Creosote	G and B Fuels Ltd	Amateur Professional Industrial	4469
Creosote	Great Mills (Retail) Ltd	Amateur Professional	5666
Creosote	Liver Grease Oil and Chemical Co Ltd	Amateur Professional Industrial	4467
Creosote	T K Bird Ltd	Amateur Professional Industrial	4903
Creosote Blend	Laybond Products Ltd	Amateur Professional Industrial	4948
Creosote Blend Mk1 (Medium Dark)	Great Marsh Ltd	Amateur Professional	4588
Creosote Blend Mk3 (Light Golden)	Great Marsh Ltd	Amateur Professional	4590
Creosote Blended Wood Preservative	Rye Oil Ltd	Amateur Professional	3730
Creosote BS 144	Bitmac Ltd	Amateur Professional Industrial	5885
Creosote BS 144 (Type I and Type II)	Coalite Chemicals Division	Industrial	3811
Creosote BS 144 Type III	Coalite Chemicals Division	Amateur Professional Industrial	3812
Creosote BS 144(3)	Great Marsh Ltd	Professional Industrial	5279
D Dark Brown Creosote	Coal Products Ltd	Amateur Professional	3741
Greenhills Creosote	Greenhills (Wessex) Ltd	Amateur Professional Industrial	4205
D Hickson Timbercare 2511 Brown	Hickson Timber Products Ltd	Industrial	4666
Langlow Creosote	Langlow Products Ltd	Amateur Professional	4639

Product Name	Marketing Company	Use	HSE No.

535 Coal tar creosote—continued

Larsen Creosote	Larsen Manufacturing Ltd	Amateur Professional Industrial	4930
D Light Creosote Blend	Bitmac Ltd	Amateur Professional	5806
W Medium Brown Creosote	C W Wastnage Ltd	Amateur Professional	4691
Middletons Creosote	E W Middleton and Sons	Amateur Professional	4441
Middletons Enviro. Shed & Fence Wood Preservative	E W Middleton and Sons	Amateur Professional	4289
Ovoline 275 Golden Creosote	Bretts Oils Ltd	Amateur Professional Industrial	3651
Solignum Dark Brown	Protim Solignum Ltd	Amateur Professional Industrial	5023
Solignum Medium Brown	Protim Solignum Ltd	Amateur Professional Industrial	5022
Wickes Creosote	Wickes Building Supplies Ltd	Amateur Professional	6101
Wilko Creosote	T K Bird Ltd	Amateur Professional Industrial	5043
Wood Preservative Type 3	Great Marsh Ltd	Professional Industrial	5252

536 Copper carbonate hydroxide + Boric acid

W Tanalith 3487	Hickson Timber Products Ltd	Industrial	5305

537 Copper carbonate hydroxide + Propiconazole + Boric acid

Wolmanit CX-P	Dr Wolman Gmbh	Industrial	6566

538 Copper carbonate hydroxide + Propiconazole + Tebuconazole + Boric acid

Tanalith E (3492)	Hickson Timber Products Ltd	Industrial	6269
Tanalith E (3494)	Hickson Timber Products Ltd	Industrial	6434

539 Copper carbonate hydroxide + Tebuconazole + Boric acid

Ensele 3428	Hickson Timber Products Ltd	Professional Industrial	5878
Tanalith E (3485)	Hickson Timber Products Ltd	Industrial	5562

540 Copper naphthenate

W Bio-Kil Cunap Pole Wrap	Bio-Kil Chemicals Ltd	Professional	5341

HSE WOOD PRESERVATIVES

Product Name	Marketing Company	Use	HSE No.

540 Copper naphthenate—continued

Product Name	Marketing Company	Use	HSE No.
Blackfriar Wood Preserver (WP Green)	E Parsons and Sons Ltd	Amateur Professional	6427
Blackfriars Green Wood Preserver	E Parsons and Sons Ltd	Amateur Professional Industrial	4209
Carbo Wood Preservative Green	Talke Chemical Company Ltd	Amateur Professional Industrial	3820
D Copper Naphthenate Solution To BS 5056	Cuprinol Ltd	Amateur Professional	4711
Cromar Wood Preserver Green	Cromar Building Products Ltd	Amateur Professional	6257
Flag Brand Wood Preservative Green	C W Wastnage Ltd	Amateur Professional Industrial	3073
Glen Wood Care Green	Glen Wood Care	Amateur Professional Industrial	4095
Green Plus Wood Preserver	Philip Johnstone Group Ltd	Amateur Professional	5040
Green Wood Preservative	Advanced Bitumens Ltd	Amateur Professional	4728
Green Wood Preservative	Strathbond Ltd	Professional	4660
Langlow Wood Preservative Green	Langlow Products Ltd	Amateur Professional Industrial	4395
Larsen Green Wood Preservative	Larsen Manufacturing Ltd	Amateur Professional Industrial	3708
Laybond Green Wood Preserver	Laybond Products Ltd	Amateur Professional	5372
Leyland Timbrene Green Environmental Formula	Leyland Paint Company Ltd	Amateur Professional	4799
Premier Barrettine Green Wood Preserver	Barrettine Products Ltd	Amateur Professional Industrial	4159
Preservative for Wood Green	Rentokil Initial UK Ltd	Amateur Professional	4613
Protim Green WR	Protim Solignum Ltd	Amateur Professional Industrial	4976
Ronseal's All-purpose Wood Preserver Green	Ronseal Ltd	Amateur Professional	6187
Supergrade Wood Preserver Green	Rentokil Initial UK Ltd	Professional Industrial	3717
Teamac Woodtec Green	Teal and Mackrill Ltd	Amateur Professional	4429

Product Name	Marketing Company	Use	HSE No.

540 Copper naphthenate—continued

Product Name	Marketing Company	Use	HSE No.
Thompson's All Purpose Wood Preserver Green	Ronseal Ltd	Amateur Professional	5989
Timber Preservative Green	Antel UK Ltd	Professional	4847
Timbrene Green Wood Preserver	Kalon Group Plc	Amateur Professional	4967
Wood Preserver Green	Philip Johnstone Group Ltd	Amateur Professional	4927

541 Copper naphthenate + 3-Iodo-2-propynyl-n-butyl carbamate

Product Name	Marketing Company	Use	HSE No.
Osmose Endcoat Green	Injecta Osmose Ltd	Amateur Professional Industrial	6562
Sovereign Green Timber Preservative/Dipcoat	Sovereign Chemical Industries Ltd	Amateur Professional Industrial	6341

542 Copper naphthenate + Disodium octaborate

Product Name	Marketing Company	Use	HSE No.
W Bio-Kil Sr Pole Wrap	Bio-Kil Chemicals Ltd	Professional	5675

543 Copper naphthenate + Permethrin

Product Name	Marketing Company	Use	HSE No.
Protim Green E	Protim Solignum Ltd	Amateur Professional Industrial	4975
Solignum Green	Protim Solignum Ltd	Amateur Professional Industrial	5020

544 Copper naphthenate + Tri(hexylene glycol)biborate

Product Name	Marketing Company	Use	HSE No.
BCR Green Wood Preserver	Building Chemical Research Ltd	Amateur Professional Industrial	6285
Dark Green Wood Preservative	Palace Chemicals Ltd	Amateur Professional Industrial	5243
Kingfisher Wood Preservative	Kingfisher Chemicals Ltd	Professional	4011
Langlow Pale Green Wood Preserver	Langlow Products Ltd	Amateur Professional Industrial	6068
Pale Green Wood Preservative	Palace Chemicals Ltd	Amateur Professional Industrial	5241
Sovereign Timber Preservative Pale Green	Sovereign Chemical Industries Ltd	Professional	3806

Product Name	Marketing Company	Use	HSE No.

545 Copper naphthenate + Zinc octoate

Weathershield Exterior Timber Preservative	ICI Paints	Amateur Professional	4915

546 Copper oxide + Arsenic pentoxide + Chromium trioxide

Celcure AO	Rentokil Initial UK Ltd	Industrial	4220
Celcure CCA Type C	Rentokil Initial UK Ltd	Industrial	5104
Celcure CCA Type C-40%	Rentokil Initial UK Ltd	Industrial	6028
Celcure CCA Type C-50%	Rentokil Initial UK Ltd	Industrial	6030
Celcure CCA Type C-60%	Rentokil Initial UK Ltd	Industrial	5934
Injecta Osmose K33 C50	Injecta Osmose Ltd	Industrial	5889
Injecta Osmose K33 C58.7	Injecta Osmose Ltd	Industrial	6498
Injecta Osmose K33 C72	Injecta Osmose Ltd	Industrial	5890
Laporte CCA Awpa Type C	Laporte Wood Preservation	Industrial	3179
Laporte CCA Oxide Type 1	Laporte Wood Preservation	Industrial	3177
Laporte CCA Oxide Type 2	Laporte Wood Preservation	Industrial	3178
Osmose K33 C58.7	Protim Solignum Ltd	Industrial	6889
Protim CCA Oxide 50	Protim Solignum Ltd	Industrial	5537
Protim CCA Oxide 58	Protim Solignum Ltd	Industrial	5686
Protim CCA Oxide 72	Protim Solignum Ltd	Industrial	5594
Tanalith 3302	Hickson Timber Products Ltd	Industrial	5098
Tanalith 3313	Hickson Timber Products Ltd	Industrial	4422
Tanalith C3310	Hickson Timber Products Ltd	Industrial	4669
W Tanalith Oxide C3309	Hickson Timber Products Ltd	Industrial	4668
W Tanalith Oxide C3314	Hickson Timber Products Ltd	Industrial	4774
Tecca LQ1	Tecca Ltd	Industrial	5999

547 Copper oxide + Benzalkonium chloride

Laporte Acq 1900	Laporte Wood Preservation	Industrial	6487

548 Copper oxide + Boric acid

W Tanalith 3422	Hickson Timber Products Ltd	Industrial	4403

549 Copper oxide + Boric acid + Chromium trioxide

Tanalith (3419) Cbc	Hickson Timber Products Ltd	Industrial	4022

550 Copper oxide + Chromium trioxide

Osmose K55	Protim Solignum Ltd	Industrial	6867

551 Copper oxide + Dialkyldimethyl ammonium chloride

Laporte Acq 2100	Laporte Wood Preservation	Industrial	6490

552 Copper oxide + Disodium octaborate + Chromium trioxide

D Celcure CB Paste	Rentokil Initial UK Ltd	Industrial	4537

Product Name	Marketing Company	Use	HSE No.

553 Copper sulphate + 2-Methyl-4-isothiazolin-3-one + 5-Chloro-2-methyl-4-isothiazolin-3-one

Injecta Osmose ABS33	Injecta Osmose Ltd	Industrial	5978

554 Copper sulphate + Boric acid

Ensele 3426	Hickson Timber Products Ltd	Professional Industrial	4364
W Ensele 3427	Hickson Timber Products Ltd	Amateur	5053
Ensele 3429	Hickson Timber Products Ltd	Amateur Professional	6392
Laporte End Grain Treatment	Laporte Wood Preservation	Amateur Professional Industrial	6885
Rentokil ACB	Rentokil Initial UK Ltd	Industrial	5759

555 Copper sulphate + Potassium dichromate + Boric acid

Tecca CCB 1	Tecca Ltd	Industrial	5584

556 Copper sulphate + Sodium dichromate

Laporte Cut End Preservative	Laporte Wood Preservation	Industrial	3528

557 Copper sulphate + Sodium dichromate + Arsenic pentoxide

Celcure A Concentrate	Rentokil Initial UK Ltd	Industrial	5458
Celcure A Fluid 10	Rentokil Initial UK Ltd	Industrial	3764
Celcure A Fluid 6	Rentokil Initial UK Ltd	Industrial	3155
Celcure A Paste	Rentokil Initial UK Ltd	Industrial	4523
Injecta Cca-c	Injecta Osmose Ltd	Industrial	4447
D Kemira CCA Type Bs	Kemira Kemwood AB	Industrial	3879
Kemwood CCA Type BS	Laporte Kemwood AB	Industrial	5534
Laporte CCA Type 1	Laporte Wood Preservation	Industrial	3175
Laporte CCA Type 2	Laporte Wood Preservation	Industrial	3176
Laporte Permawood CCA	Laporte Wood Preservation	Industrial	3529
Protim CCA Salts Type 2	Protim Solignum Ltd	Industrial	4972
Tanalith 3357	Hickson Timber Products Ltd	Industrial	4431
W Tanalith CL (3354)	Hickson Timber Products Ltd	Industrial	4196
Tanalith CP 3353	Hickson Timber Products Ltd	Industrial	4667
Tecca P2	Tecca Ltd	Industrial	3958

558 Copper sulphate + Sodium dichromate + Boric acid

Laporte CCB	Laporte Wood Preservation	Industrial	5875
Tanalith CBC Paste 3402	Hickson Timber Products Ltd	Industrial	3833

559 Copper sulphate + Sodium dichromate + Boric acid + Chromium acetate

Celgard CF	Rentokil Initial UK Ltd	Industrial	4608

HSE
WOOD PRESERVATIVES

Product Name	Marketing Company	Use	HSE No.

560 Copper sulphate + Sodium dichromate + Chromium acetate

Celcure B	Rentokil Initial UK Ltd	Industrial	4541
Celcure O	Rentokil Initial UK Ltd	Industrial	4539

561 Creosote

D Arborsan 3 Creosote	Lanstar Coatings Ltd	Industrial	5159
D Arborsan 4 Creosote	Lanstar Coatings Ltd	Amateur Professional Industrial	5160
D Arborsan 6 Creosote	Lanstar Coatings Ltd	Amateur Professional Industrial	5158
B and Q Creosote	B & Q Plc	Amateur Professional	5093
Barrettine Creosote	Barrettine Products Ltd	Amateur Professional Industrial	3146
Chelec Creosote	Chelec Ltd	Amateur Professional	5412
Cindu Creosote	Cindu Chemicals BV	Amateur Professional Industrial	6001
D Co-op Creosote	Co-operative Wholesale Society Ltd	Amateur Professional	5073
Creosote	Great Marsh Ltd	Professional Industrial	4575
W Creosote	Hilbre Building Chemicals Ltd	Amateur Professional	5301
Creosote	Langlow Products Ltd	Amateur Professional	4034
Creosote	Palace Chemicals Ltd	Amateur Professional Industrial	3841
Creosote	PMC Holdings Ltd	Amateur Professional Industrial	5495
W Creosote	Suffolk & Essex Supplies Ltd	Amateur Professional	3141
Creosote (Light Brown and Dark Brown)	Sealocrete PLA Ltd	Amateur Professional	5121
Creosote 131C	Great Marsh Ltd	Professional Industrial	4576
Creosote Blend	Langlow Products Ltd	Amateur Professional	4392
Creosote Coke Oven Oil	Great Marsh Ltd	Professional	4586
W Creosote Emulsion	Oakmere Technical Services Ltd	Industrial	5345

Product Name	Marketing Company	Use	HSE No.

561 Creosote—continued

Creosote Mk2	Great Marsh Ltd	Professional Industrial	4589
DIY Time Creosote	Nurdin and Peacock	Amateur Professional	5314
W Golden Creosote	C W Wastnage Ltd	Amateur Professional	4690
Homebase Creosote	Sainsbury's Homebase House and Garden Centres	Amateur Professional	5054
Jewson Creosote	Jewson Ltd	Amateur Professional	5072
Kalon Creosote	Kalon Group Plc	Amateur Professional	5496
Laybond Creosote Blend	Laybond Products Ltd	Amateur Professional Industrial	6211
Leyland Creosote	Leyland Paint Company Ltd	Amateur Professional	5074
Maxim Creosote	S J Dixon	Amateur Professional	5848
D McDougall Rose Hi-life Creosote	Mcdougall Rose Ltd	Amateur Professional	4308
Nitromors Creosote	Kalon Group Plc	Amateur Professional	5129
Oakmere Creosote Type 2	Oakmere Technical Services Ltd	Amateur Professional Industrial	4390
W Signpost Creosote	Macpherson Paints Ltd	Amateur Professional	4507
Solignum Fencing Fluid	Protim Solignum Ltd	Industrial	5016
Solignum Gold	Protim Solignum Ltd	Amateur Professional Industrial	5024
Southdown Creosote	C Brewer and Sons Ltd	Amateur Professional	4310
Texas Creosote	Texas Homecare Ltd	Amateur Professional	5092
Travis Perkins Creosote	Travis Perkins Trading Company Ltd	Amateur Professional	4941
Valspar Creosote	Akzo Coatings Plc	Amateur Professional	5469
Woodman Creosote and Light Brown Creosote	J H Woodman Ltd	Amateur Professional Industrial	5120

562 Cyfluthrin

Bayer WPC-5	Bayer Plc	Professional	6534

Product Name	Marketing Company	Use	HSE No.

563 Cypermethrin

Product Name	Marketing Company	Use	HSE No.
Cementone Woodworm Killer	Philip Johnstone Group Ltd	Amateur	3850
Crown Micro Insecticide	Crown Chemicals Ltd	Professional	6413
Crown Woodworm Concentrate	Crown Chemicals Ltd	Professional	5470
Devatern 0.5 L	Sorex Ltd	Amateur	3874
Devatern 1.0 L	Sorex Ltd	Professional Industrial	3876
Devatern Ec	Sorex Ltd	Professional Industrial	3875
Green Range Woodworm Killer	Philip Johnstone Group Ltd	Professional	3711
Green Range Woodworm Killer AQ	Philip Johnstone Group Ltd	Professional	3688
Hickson Antiborer 3767	Hickson Timber Products Ltd	Industrial	3583
Killgerm Woodworm Killer	Killgerm Chemicals Ltd	Professional	3126
Lignosol P	Microsol	Professional	5832
Microtech Woodworm Killer AQ	Philip Johnstone Group Ltd	Professional	5452
Nubex Emulsion Concentrate C (Low Odour)	Philip Johnstone Group Ltd	Professional	5268
Palace Microfine Insecticide Concentrate	Palace Chemicals Ltd	Amateur Professional Industrial	5550
W PC-H/3	Terminix Peter Cox Ltd	Professional	4408
PCX-12/P	Terminix Peter Cox Ltd	Professional	4407
PCX-12/P Concentrate	Terminix Peter Cox Ltd	Professional	4406
D Protim Woodworm Killer C	Fosroc Ltd	Amateur Professional Industrial	3778
Water Based Woodworm Killer	Philip Johnstone Group Ltd	Amateur Professional	4874
Waterbased Woodworm Killer 5X	Philip Johnstone Group Ltd	Amateur Professional	5861
Woodworm Killer	Oakmere Technical Services Ltd	Amateur Professional Industrial	4371
Woodworm Killer	Palace Chemicals Ltd	Amateur Professional	3769

564 Cypermethrin + 2-Phenylphenol

Product Name	Marketing Company	Use	HSE No.
Laybond Timber Protector	Laybond Products Ltd	Amateur Professional	5369

565 Cypermethrin + 3-Iodo-2-propynyl-n-butyl carbamate

Product Name	Marketing Company	Use	HSE No.
Microtech Dual Purpose AQ	Philip Johnstone Group Ltd	Professional	5453
D Water Based Wood Preserver (Concentrate)	Cementone Beaver Ltd	Amateur	5515
Water Based Wood Preserver (Concentrate)	Philip Johnstone Group Ltd	Amateur Professional	5630

Product Name	Marketing Company	Use	HSE No.

565 Cypermethrin + 3-Iodo-2-propynyl-n-butyl carbamate—continued

Water-based Wood Preserver	Philip Johnstone Group Ltd	Amateur Professional	4902

566 Cypermethrin + Dichlofluanid

Universal Wood Preservative	Oakmere Technical Services Ltd	Amateur Professional Industrial	4170

567 Cypermethrin + Propiconazole

Celpruf B	Rentokil Initial UK Ltd	Industrial	5935
Cuprinol Quick Drying Clear Preserver	Cuprinol Ltd	Amateur Professional Industrial	6014
W Cuprinol Wood Defender	Cuprinol Ltd	Amateur Professional Industrial	5798
Wocosen 100 SL-C	Janssen Pharmaceutica NV	Professional Industrial	5807

568 Cypermethrin + Propiconazole + Boric acid

Rentokil Total Wood Treatment	Rentokil Initial UK Ltd	Amateur (Surface Biocide) Amateur (Wood Preservative) Professional (surface Biocide) Professional (Wood Preservative)	6053

569 Cypermethrin + Tebuconazole

Celpruf TZC	Rentokil Initial UK Ltd	Industrial	5722

570 Cypermethrin + Tri(hexylene glycol)biborate

BCR Universal Wood Preserver	Building Chemical Research Ltd	Amateur Professional	6290
Cementone Multiplus	Philip Johnstone Group Ltd	Amateur Professional	3849
Crown Fungicide Insecticide	Crown Chemicals Ltd	Professional	6003
Crown Fungicide Insecticide Concentrate	Crown Chemicals Ltd	Professional	5463
Crown Micro Dual Purpose	Crown Chemicals Ltd	Professional	6414

Product Name	Marketing Company	Use	HSE No.

570 Cypermethrin + Tri(hexylene glycol)biborate—continued

Product Name	Marketing Company	Use	HSE No.
Devatern Wood Preserver	Sorex Ltd	Professional Industrial	4073
Green Range Dual Purpose Aq	Philip Johnstone Group Ltd	Professional	3725
Green Range Wykamol Plus	Philip Johnstone Group Ltd	Professional	3731
Killgerm Wood Protector	Killgerm Chemicals Ltd	Professional	3127
Langlow Clear Wood Preserver	Langlow Products Ltd	Amateur Professional Industrial	3802
Nubex Emulsion Concentrate CB (Low Odour)	Philip Johnstone Group Ltd	Professional	5328
Nubex Woodworm All Purpose CB	Philip Johnstone Group Ltd	Professional	3189
Palace Microfine Fungicide Insecticide (Concentrate)	Palace Chemicals Ltd	Professional Industrial	5421
W PC-H/4	Terminix Peter Cox Ltd	Amateur Professional	3740
PCX-122	Terminix Peter Cox Ltd	Professional	3999
PCX-122 Concentrate	Terminix Peter Cox Ltd	Professional	4000

571 Cypermethrin + Zinc versatate

Product Name	Marketing Company	Use	HSE No.
Dipsar G R	Philip Johnstone Group Ltd	Industrial	4449
Everclear	Injecta Osmose Ltd	Industrial	6058

572 Deltamethrin

Product Name	Marketing Company	Use	HSE No.
K-otec AL	Agrevo UK Ltd	Amateur Professional	6570
K-otek EC	Agrevo UK Ltd	Professional	6568
K-otek SL	Agrevo UK Ltd	Professional	6569

573 Dialkyldimethyl ammonium chloride

Product Name	Marketing Company	Use	HSE No.
X Thallo	Thames Valley Specialist Products Ltd	Amateur Professional	6841

574 Dialkyldimethyl ammonium chloride + 3-Iodo-2-propynyl-n-butyl carbamate

Product Name	Marketing Company	Use	HSE No.
Hickson NP-1	Hickson Timber Products Ltd	Industrial	5401

575 Dialkyldimethyl ammonium chloride + Benzalkonium chloride

Product Name	Marketing Company	Use	HSE No.
ABL Aqueous Wood Preserver Concentrate 1:9	Advanced Bitumens Ltd	Professional Industrial	4199
Aqueous Wood Preserver	Advanced Bitumens Ltd	Amateur Professional Industrial	4727
Celbrite M	Rentokil Initial UK Ltd	Industrial	4535
Celbronze B	Rentokil Initial UK Ltd	Professional Industrial	4549

Product Name	Marketing Company	Use	HSE No.

575 Dialkyldimethyl ammonium chloride + Benzalkonium chloride—continued

Product Name	Marketing Company	Use	HSE No.
W Colourfast Protector	Rentokil Initial UK Ltd	Amateur Professional	4609

576 Dialkyldimethyl ammonium chloride + Boric acid

Celbor P	Rentokil Initial UK Ltd	Industrial	5923
Celbor P 25% Solution	Rentokil Initial UK Ltd	Industrial	6033
Celbor P 5% Solution	Rentokil Initial UK Ltd	Industrial	6019
Celbor P25	Rentokil Initial UK Ltd	Industrial	6563
Celbor P5	Rentokil Initial UK Ltd	Industrial	6558
Protim E470	Protim Solignum Ltd	Industrial	6386

577 Dialkyldimethyl ammonium chloride + Copper oxide

Laporte ACQ 2100	Laporte Wood Preservation	Industrial	6490

578 Dialkyldimethyl ammonium chloride + Disodium octaborate

Ecogel	Restoration (UK) Ltd	Professional	6847
Probor 20 Brushable Gel	Safeguard Chemicals Ltd	Professional	6422

579 Dialkyldimethyl ammonium chloride + Methylene bis(thiocyanate) + Benzalkonium chloride + Boric acid

Celbrite MT	Rentokil Initial UK Ltd	Industrial	4550

580 Dialkyldimethyl ammonium chloride + Permethrin + Propiconazole

Traditional Preservative Oil	Oakmasters of Sussex	Amateur Professional Industrial	6023

581 Dichlofluanid

Cedarwood Protector	Rentokil Initial UK Ltd	Amateur Professional Industrial	4587
Cedarwood Special	Cuprinol Ltd	Amateur Professional Industrial	4729
Cromar Wood Preserver	Cromar Building Products Ltd	Amateur Professional Industrial	6243
Cuprinol Hardwood Basecoat	Cuprinol Ltd	Amateur Professional Industrial	5015
Cuprinol Hardwood Basecoat Meranti	Cuprinol Ltd	Industrial	4707
Cuprinol Preservative-wood Hardener	Cuprinol Ltd	Amateur Professional	3590
W Dualprime F	Cuprinol Ltd	Industrial	4706

Product Name	Marketing Company	Use	HSE No.

581 Dichlofluanid—continued

Flexarb Timber Coating	Macpherson Paints Ltd	Amateur Professional	4861
Forsham's Re-treat	Forsham Cottage Arks	Amateur	6281
Hardwood Protector	Rentokil Initial UK Ltd	Amateur Professional Industrial	4580
Impra-Color	Impra Systems Ltd	Amateur Professional	5403
Intertox	International Coatings Ltd	Amateur	5217
Langlow Wood Preservative	Langlow Products Ltd	Amateur Professional Industrial	5548
Langlow Wood Preserver Formulation A	Langlow Products Ltd	Amateur Professional Industrial	3753
Lister Teak Dressing	Lister Lutyens Company Ltd	Amateur Professional	4628
New Formula Cedarwood	Cuprinol Ltd	Amateur Professional Industrial	4699
D Permalene Satin Wood Stain	C Stabler Ltd	Amateur	3644
Regency Garden Buildings Wood Preserver	Regency Garden Buildings	Amateur Professional Industrial	6530
RWN Building Products Wood Preserver	RWN Building Products	Amateur Professional	6523
Sigmalife Impregnant	Philip Johnstone Group Ltd	Professional	6204
Timber Preservative	Philip Johnstone Group Ltd	Amateur Professional	6317
Timberlife	Palace Chemicals Ltd	Amateur Professional	4501
Timberlife Extra	Palace Chemicals Ltd	Amateur Professional	5352
Wood Preservative	Langlow Products Ltd	Amateur Professional Industrial	4600

582 Dichlofluanid + Acypetacs zinc

Cuprinol Decorative Preserver	Cuprinol Ltd	Amateur Professional Industrial	5343
Cuprinol Decorative Preserver	Cuprinol Ltd	Amateur Professional Industrial	6277

Product Name	Marketing Company	Use	HSE No.

582 Dichlofluanid + Acypetacs zinc—continued

Cuprinol Decorative Preserver Red Cedar	Cuprinol Ltd	Amateur Professional Industrial	5342
D Cuprinol Decorative Wood Preserver	Cuprinol Ltd	Amateur	3156
D Cuprinol Decorative Wood Preserver Red Cedar	Cuprinol Ltd	Amateur Professional	4696
Cuprinol Garden Decking Protector	Cuprinol Ltd	Amateur Professional	6862
D Cuprinol Magnatreat F	Cuprinol Ltd	Industrial	3752
D Cuprinol Magnatreat XQD	Cuprinol Ltd	Industrial	3751
Cuprinol Preservative Base	Cuprinol Ltd	Amateur Professional	4929
Wickes Decorative Preserver Rich Cedar	Wickes Building Supplies Ltd	Amateur Professional Industrial	6420
Wickes Decorative Wood Preserver	Wickes Building Supplies Ltd	Amateur Professional Industrial	6419

583 Dichlofluanid + Azaconazole

W Xyladecor Matt U 404	Nwe Distributors Ltd	Amateur Professional	4445
W Xylamon Primer Dipping Stain U 415	Venilia Ltd	Professional	4444

584 Dichlofluanid + Cypermethrin

Universal Wood Preservative	Oakmere Technical Services Ltd	Amateur Professional Industrial	4170

585 Dichlofluanid + Permethrin + Tebuconazole

Impra-Holzschutzgrund (Primer)	Impra Systems Ltd	Professional Industrial	5735

586 Dichlofluanid + Permethrin + Tri(hexylene glycol)biborate

Blackfriar Wood Preserver WP Gold Star Clear	E Parsons and Sons Ltd	Amateur Professional	6417
Blackfriars Gold Star Clear	E Parsons and Sons Ltd	Amateur Professional Industrial	4149
Leyland Timbrene Supreme Environmental Formula	Leyland Paint Company Ltd	Amateur Professional	4800
W Nitromors Timbrene Supreme Environmental Formula	Henkel Home Improvements and Adhesive Products	Amateur Professional Industrial	4107

HSE WOOD PRESERVATIVES

Product Name	Marketing Company	Use	HSE No.

586 Dichlofluanid + Permethrin + Tri(hexylene glycol)biborate—continued

Premier Barrettine New Universal Fluid D	Barrettine Products Ltd	Amateur Professional Industrial	4004
Timbrene Supreme	Kalon Group Plc	Amateur Professional	4995

587 Dichlofluanid + Permethrin + Zinc octoate

OS Color Wood Stain and Preservative	Ostermann and Scheiwe GmbH	Amateur	5531

588 Dichlofluanid + Tri(hexylene glycol)biborate

ABL Universal Woodworm Killer DB	Advanced Bitumens Ltd	Amateur Professional Industrial	3858
Blackfriar Wood Preserver	E Parsons and Sons Ltd	Amateur Professional	6443
Blackfriars New Wood Preserver	E Parsons and Sons Ltd	Amateur Professional Industrial	4150
Double Action Timber Preservative for Doors	International Paint Ltd	Amateur	3707
Double Action Wood Preservative	International Paint Ltd	Amateur	3706
W Nitromors Timbrene Clear Environmental Formula	Henkel Home Improvements and Adhesive Products	Amateur Professional Industrial	4106
Premier Barrettine New Wood Preserver	Barrettine Products Ltd	Amateur Professional Industrial	4038
W Ranch Preservative	International Paint Ltd	Amateur	4085
Ranch Preservative	Plascon International	Amateur	6085
Sadolin New Base	Sadolin UK Ltd	Amateur Professional Industrial	4341
Ultrabond Woodworm Killer	Philip Johnstone Group Ltd	Amateur Professional	4344

589 Dichlofluanid + Tri(hexylene glycol)biborate + Zinc naphthenate

Weathershield Exterior Preservative Basecoat	ICI Paints	Amateur Professional	5245
Weathershield Preservative Primer	ICI Paints	Amateur Professional	5244

590 Dichlofluanid + Tri(hexylene glycol)biborate + Zinc octoate

Timber Preservative Clear TFP7	Johnstone's Paints Plc	Amateur Professional	4016

Product Name	Marketing Company	Use	HSE No.

591 Dichlofluanid + Tributyltin oxide

Product Name	Marketing Company	Use	HSE No.
W Celpruf CP Special	Rentokil Initial UK Ltd	Industrial	4582
Hickson Woodex	Hickson Timber Products Ltd	Industrial	5405
Protim Cedar	Protim Solignum Ltd	Industrial	5042
D Protim R Coloured Plus	Fosroc Ltd	Industrial	4391
Protim Solignum Softwood Basestain CS	Protim Solignum Ltd	Industrial	5647
W Wood Preservative AA 155/00	Becker Acroma Kemira Ltd	Industrial	5308
W Wood Preservative AA 155/03	Becker Acroma Kemira Ltd	Industrial	5310
W Wood Preservative AA155	Becker Acroma Kemira Ltd	Industrial	5311

592 Dichlofluanid + Zinc naphthenate

Product Name	Marketing Company	Use	HSE No.
Bradite Timber Preservative	Bradite Ltd	Amateur	6091
Masterstroke Wood Preserver	Akzo Nobel Decorative Coatings	Amateur	5689
Sikkens Wood Preserver	Akzo Nobel Woodcare	Amateur Professional	6365

593 Dichlofluanid + Zinc octoate

Product Name	Marketing Company	Use	HSE No.
Exterior Preservative Primer	Windeck Paints Ltd	Amateur	3857
Timbercare Microporous Exterior Preservative	Manders Paints Ltd	Amateur Professional	5157
W Wilko Exterior System Preservative Primer	Wilkinson Home and Garden Stores	Amateur	3686

594 Disodium octaborate

Product Name	Marketing Company	Use	HSE No.
ACS Borate Rod	Advanced Chemical Specialities Ltd	Amateur Professional	6545
ACS Boron Wood Preservative Paste	Advanced Chemical Specialities Ltd	Professional	6093
ACS Borotreat 10P	Advanced Chemical Specialities Ltd	Professional (Surface Biocide) Professional (Wood Preservative)	6335
Advance Guard	Injecta Osmose Ltd	Industrial	6674
Azygo Boron Preservative PDR	Azygo International Direct Ltd	Professional (Surface Biocide) Professional (Wood Preservative)	6497
Bio-Kil Boron Paste	Bio-Kil Chemicals Ltd	Professional Industrial	3866
Biokil Timbor Rods	Remtox-Silexine Ltd	Amateur Professional	4502

Product Name	Marketing Company	Use	HSE No.

594 Disodium octaborate—continued

Product Name	Marketing Company	Use	HSE No.
Boracol 20	Advanced Chemical Specialities Ltd	Professional	4018
Boracol 20	Advanced Chemical Specialities Ltd	Professional	6339
Boracol B40	Advanced Chemical Specialities Ltd	Professional	4788
Borax Wood Preservative Clear Odourless	Auro Organic Paints Supplies Ltd	Professional Industrial	6271
Boron Gel 40	Philip Johnstone Group Ltd	Professional	6460
Celgard FP	Rentokil Initial UK Ltd	Industrial	4757
Channelwood Boracol B40 Wood Preservative Gel	Advanced Chemical Specialities Ltd	Professional	6094
Dry Pin	Window Care Systems Ltd	Professional Industrial	5144
Ecobar II	Advanced Chemical Specialities Ltd	Professional	6838
Ensele 3430	Hickson Timber Products Ltd	Amateur Professional	4671
Lectrobor Gel	Lectros International Ltd	Professional	6441
Lectros Boracol 20	Lectros International Ltd	Professional	6493
Lectros Boracol B40 Gel	Lectros International Ltd	Professional	6492
Lectros Lectrobor 11	Lectros International Ltd	Professional	6898
Permabor Boracol 20	Permagard Products Ltd	Professional	6357
Permabor Boracol 40 Paste	Permagard Products Ltd	Professional	6359
Permadip 9 Concentrate	Permagard Products Ltd	Professional Industrial	4160
Probor DB	Safeguard Chemicals Ltd	Professional Industrial	6673
Protek 9 Star Wood Protection	Protek Products (Sun Europa) Ltd	Amateur Professional Industrial	3104
Protek Double 9 Star Wood Protection	Protek Products (Sun Europa) Ltd	Amateur Professional Industrial	3101
Protek Shedstar	Protek Products (Sun Europa) Ltd	Amateur Professional Industrial	3103
Protek Wood Protection Fencegrade	Ashby Timber Treatments	Amateur	3018
Protek Wood Protection Fencegrade 9:1	Ashby Timber Treatments	Amateur	3019
Protek Wood Protection Shed Grade	Ashby Timber Treatments	Amateur	3020
Protek Woodstar	Protek Products (Sun Europa) Ltd	Amateur Professional Industrial	3102

Product Name	Marketing Company	Use	HSE No.

594 Disodium octaborate—continued

Product Name	Marketing Company	Use	HSE No.
Protim B10	Protim Solignum Ltd	Amateur Professional	4971
Protim E480	Protim Solignum Ltd	Industrial	6425
Protim Solignum D.O.T.	Protim Solignum Ltd	Industrial	6429
Pyro-Black	Timber Treatments	Industrial	5866
Remtox Borocol 20 Wood Preservative	Remtox (Chemicals) Ltd	Professional	4094
Remtox-Silexine B40 Paste Wood Preservative	Remtox (Chemicals) Ltd	Professional	4161
Remtox-Silexine Boron Rods	Remtox-Silexine Ltd	Professional	5560
Ronseal Wet Rot Prevention Tablets	Ronseal Ltd	Amateur Professional	6627
Ronseal Wood Preservative Tablets	Ronseal Ltd	Amateur Professional	5271
Safeguard Antiflame 4050 Wd Wood Preservative	Safeguard Chemicals Ltd	Professional	5656
Safeguard Boracol 20	Safeguard Chemicals Ltd	Professional	6393
Safeguard Boracol 40 Paste	Safeguard Chemicals Ltd	Professional	6400
Sovereign Timbor Rod	Sovereign Chemical Industries Ltd	Amateur Professional	4627
Super Andy Man Wood Protection	AMB Products Ltd	Amateur Professional	3535
Tim-bor	Borax Europe Ltd	Professional Industrial	5687
Tim-bor	Borax Europe Ltd	Professional Industrial	6270
Timbertex PI	Marcher Chemicals Ltd	Professional Industrial	3611
D Timbor	Borax Europe Ltd	Industrial	3621
Timbor Paste	Rentokil Initial UK Ltd	Industrial	5402
Tribor 20	Triton Chemical Manufacturing Co Ltd	Professional	6183
Tribor 20	Triton Chemical Manufacturing Co Ltd	Professional	6242
Tribor Gel	Triton Chemical Manufacturing Co Ltd	Professional	6099
Woodtreat Boron 40	Philip Johnstone Group Ltd	Professional	6461

595 Disodium octaborate + 3-Iodo-2-propynyl-n-butyl carbamate + Benzalkonium chloride

Product Name	Marketing Company	Use	HSE No.
Antiblu Select	Hickson Timber Products Ltd	Industrial	6087

596 Disodium octaborate + Benzalkonium chloride

Product Name	Marketing Company	Use	HSE No.
Boracol 20 Rh	Advanced Chemical Specialities Ltd	Professional	4019

Product Name	Marketing Company	Use	HSE No.

596 Disodium octaborate + Benzalkonium chloride—continued

Product Name	Marketing Company	Use	HSE No.
Boracol B8.5 Rh Mouldicide/Wood Preservative	Advanced Chemical Specialities Ltd	Amateur (Surface Biocide) Amateur (Wood Preservative) Professional (Surface Biocide) Professional (Wood Preservative)	4809
Boron Biocide 10	Philip Johnstone Group Ltd	Professional (Surface Biocide) Professional (Wood Preservative)	6485
Brunosol Boron 10	Philip Johnstone Group Ltd	Professional (Surface Biocide) Professional (Wood Preservative)	6488
Preservative Gel	Protim Solignum Ltd	Professional	6528
Probor	Safeguard Chemicals Ltd	Professional (Surface Biocide) Professional (Wood Preservative)	6266
Probor 10	Safeguard Chemicals Ltd	Professional	6595
Probor 50	Safeguard Chemicals Ltd	Professional (Surface Biocide) Professional (Wood Preservative)	5596
Weathershield Multi-surface Fungicidal Wash	Imperial Chemical Industries Plc	Amateur Professional	6823

597 Disodium octaborate + Boric acid

Product Name	Marketing Company	Use	HSE No.
ACS Woodkeeper Paste Preservative	Advanced Chemical Specialities Ltd	Professional	6221

Product Name	Marketing Company	Use	HSE No.

598 Disodium octaborate + Carbendazim

Weathershield Aquatech Preservative Basecoat	ICI Paints	Amateur Professional	5757

599 Disodium octaborate + Chromium trioxide + Copper oxide

D Celcure CB Paste	Rentokil Initial UK Ltd	Industrial	4537

600 Disodium octaborate + Copper naphthenate

W Bio-Kil SR Pole Wrap	Bio-Kil Chemicals Ltd	Professional	5675

601 Disodium octaborate + Dialkyldimethyl ammonium chloride

Ecogel	Restoration (UK) Ltd	Professional	6847
Probor 20 Brushable Gel	Safeguard Chemicals Ltd	Professional	6422

602 Disodium octaborate + Methylene bis(thiocyanate) + 2-(Thiocyanomethylthio)benzothiazole

Timbertex Pd	Marcher Chemicals Ltd	Industrial	5876

603 Disodium octaborate + Permethrin + Propiconazole + Tebuconazole

Protim E410	Protim Solignum Ltd	Industrial	6218
Protim E410(i)	Protim Solignum Ltd	Industrial	6145
Protim E415	Protim Solignum Ltd	Industrial	5908
Protim E415 (i)	Protim Solignum Ltd	Industrial	6017

604 Disodium octaborate + Propiconazole + Tebuconazole

Protim B610	Protim Solignum Ltd	Industrial	6366

605 Disodium octaborate + Sodium pentachlorophenoxide

W Gallwey Abs	Protim Solignum Ltd	Industrial	5014
W Protim Fentex Green Concentrate	Protim Solignum Ltd	Industrial	5122
W Protim Fentex Green Rfu	Protim Solignum Ltd	Industrial	5123
W Protim Fentex M	Protim Solignum Ltd	Industrial	4973
W Protim Kleen II	Protim Solignum Ltd	Industrial	5203

606 Disodium tetraborate + Boric acid

Control Fluid Sb	Rentokil Initial UK Ltd	Professional (Surface Biocide) Professional (Wood Preservative)	6018
PC-K	Terminix Peter Cox Ltd	Professional	4409
Rentokil Control Paste SB	Rentokil Initial UK Ltd	Professional	6501

Product Name	Marketing Company	Use	HSE No.

607 Dodecylamine lactate + Dodecylamine salicylate

Product Name	Marketing Company	Use	HSE No.
W Durham Bl 730	Durham Chemicals	Professional (Surface Biocide) Professional (Wood Preservative) Industrial (Wood Preservative)	4744
W Environmental Woodrot Treatment	SCI (Building Products)	Professional (Surface Biocide) Professional (Wood Preservative) Industrial (Wood Preservative)	5530
W Fungicidal Wash	Polybond Ltd	Amateur (Surface Biocide) Amateur (Wood Preservative) Professional (Surface Biocide) Professional (Wood Preservative)	3601
W Fungicidal Wash	Smyth - Morris Chemicals Ltd	Professional (Surface Biocide) Professional (Wood Preservative)	3004
W Macpherson Anti-Mould Solution	Macpherson Paints Ltd	Professional (Surface Biocide) Professional (Wood Preservative)	4791

Product Name	Marketing Company	Use	HSE No.

607 Dodecylamine lactate + Dodecylamine salicylate—continued

Product Name	Marketing Company	Use	HSE No.
W Metalife Fungicidal Wash	Metalife International Ltd	Amateur (Surface Biocide) Amateur (Wood Preservative) Professional (Surface Biocide) Professional (Wood Preservative)	5406
W Mouldrid	S and K Maintenance Products	Amateur (Surface Biocide) Amateur (Wood Preservative) Professional (Surface Biocide) Professional (Wood Preservative)	3671
W Nubex WDR	Nubex Ltd	Professional (Surface Biocide) Professional (Wood Preservative)	3810
W Permoglaze Micatex Fungicidal Treatment	Permoglaze Paints Ltd	Professional (Surface Biocide) Professional (Wood Preservative)	5249

HSE WOOD PRESERVATIVES

Product Name	Marketing Company	Use	HSE No.

607 Dodecylamine lactate + Dodecylamine salicylate—continued

Product Name	Marketing Company	Use	HSE No.
W Weathershield Fungicidal Wash	ICI Paints	Amateur (Surface Biocide) Amateur (Wood Preservative) Professional (Surface Biocide) Professional (Wood Preservative)	5134

608 Dodecylamine lactate + Dodecylamine salicylate + Permethrin

Product Name	Marketing Company	Use	HSE No.
W Kingfisher Timber Paste	Kingfisher Chemicals Ltd	Professional	4327
W Kingfisher Timberpaste	Kingfisher Chemicals Ltd	Professional	5931
W Safeguard Deepwood Paste	Safeguard Chemicals Ltd	Professional	5758

609 Dodecylamine salicylate + Dodecylamine lactate

Product Name	Marketing Company	Use	HSE No.
W Durham Bi 730	Durham Chemicals	Professional (Surface Biocide) Professional (Wood Preservative) Industrial (Wood Preservative)	4744
W Environmental Woodrot Treatment	SCI (Building Products)	Professional (Surface Biocide) Professional (Wood Preservative) Industrial (Wood Preservative)	5530

Product Name	Marketing Company	Use	HSE No.

609 Dodecylamine salicylate + Dodecylamine lactate—continued

Product Name	Marketing Company	Use	HSE No.
W Fungicidal Wash	Polybond Ltd	Amateur (Surface Biocide) Amateur (Wood Preservative) Professional (Surface Biocide) Professional (Wood Preservative)	3601
W Fungicidal Wash	Smyth - Morris Chemicals Ltd	Professional (Surface Biocide) Professional (Wood Preservative)	3004
W MacPherson Anti-Mould Solution	Macpherson Paints Ltd	Professional (Surface Biocide) Professional (Wood Preservative)	4791
W Metalife Fungicidal Wash	Metalife International Ltd	Amateur (Surface Biocide) Amateur (Wood Preservative) Professional (Surface Biocide) Professional (Wood Preservative)	5406

HSE WOOD PRESERVATIVES

Product Name	Marketing Company	Use	HSE No.

609 Dodecylamine salicylate + Dodecylamine lactate—continued

Product Name	Marketing Company	Use	HSE No.
W Mouldrid	S and K Maintenance Products	Amateur (Surface Biocide) Amateur (Wood Preservative) Professional (Surface Biocide) Professional (Wood Preservative)	3671
W Nubex WDR	Nubex Ltd	Professional (Surface Biocide) Professional (Wood Preservative) Industrial (Wood Preservative)	3810
W Permoglaze Micatex Fungicidal Treatment	Permoglaze Paints Ltd	Professional (Surface Biocide)	5249
W Weathershield Fungicidal Wash	ICI Paints	Amateur (Surface Biocide) Amateur (Wood Preservative) Professional (Surface Biocide) Professional (Wood Preservative)	5134

610 Dodecylamine salicylate + Permethrin + Dodecylamine lactate

Product Name	Marketing Company	Use	HSE No.
W Kingfisher Timber Paste	Kingfisher Chemicals Ltd	Professional	4327
W Kingfisher Timberpaste	Kingfisher Chemicals Ltd	Professional	5931
W Safeguard Deepwood Paste	Safeguard Chemicals Ltd	Professional	5758

611 Flufenoxuron

Product Name	Marketing Company	Use	HSE No.
M8 Super Concentrate	Remtox (Chemicals) Ltd	Professional	6507
Sovaq Flx I	Sovereign Chemical Industries Ltd	Professional	6510

Product Name	Marketing Company	Use	HSE No.

612 Flufenoxuron + Propiconazole + 3-Iodo-2-propynyl-n-butyl carbamate

Product Name	Marketing Company	Use	HSE No.
Remtox M9 Dual Purpose Super Concentrate	Remtox (Chemicals) Ltd	Professional	6508
Sovaq Flx F/I	Sovereign Chemical Industries Ltd	Professional	6509
Sovereign Fungicide/Insecticide	Sovereign Chemical Industries Ltd	Professional	6828

613 3-Iodo-2-propynyl-n-butyl carbamate

Product Name	Marketing Company	Use	HSE No.
Bio-Kil Board Preservative	Bio-Kil Chemicals Ltd	Professional Industrial	4899
BPC-FS Fungicidal Solution Concentrate	Building Preservation Centre	Professional (Surface Biocide) Professional (Wood Preservative)	6484
CME 30/F Concentrate	Terminix Peter Cox Ltd	Professional (Surface Biocide) Professional (Wood Preservative)	5948
Conductive Pilt 80 RFU	PPG Industries (UK) Ltd	Industrial	4500
Crysolite Wood Preservative	Crysolite Protective Coatings	Amateur Professional	5847
Deepkill F	Sovereign Chemical Industries Ltd	Professional Industrial	5250
Omega Biocide	Restoration (UK) Ltd	Professional (Surface Biocide) Professional (Wood Preservative)	6381
Osmose Endcoat Brown	Injecta Osmose Ltd	Amateur Professional Industrial	6561
Osmose Endcoat Clear	Injecta Osmose Ltd	Amateur Professional Industrial	6560
Pilt 80 RFU	PPG Industries (UK) Ltd	Industrial	4877
Pilt NF4 RFU	PPG Industries (UK) Ltd	Industrial	5541
Polyphase Water Based Concentrate	Troy UK	Professional	3944
Polyphase Water Based Ready for Use	Troy UK	Amateur Professional	3945
Protim FDR 250	Protim Solignum Ltd	Industrial	5095
Protim JP 250	Protim Solignum Ltd	Industrial	5105

Product Name	Marketing Company	Use	HSE No.

613 3-Iodo-2-propynyl-n-butyl carbamate—continued

Product Name	Marketing Company	Use	HSE No.
Remecology Spirit Based Fungicide R5	Remtox (Chemicals) Ltd	Professional Industrial	4803
Remtox AQ Fungicide R7	Remtox (Chemicals) Ltd	Professional Industrial	5410
Remtox Dry Rot Paint	Remtox (Chemicals) Ltd	Amateur (Surface Biocide) Amateur (Wood Preservative) Professional (Surface Biocide) Professional (Wood Preservative)	5187
Remtox Fungicide Microemulsion M7	Remtox (Chemicals) Ltd	Professional Industrial	5419
Remtox Fungicide Paste K6	Remtox (Chemicals) Ltd	Professional Industrial	5400
Remtox Microactive Fungicide W7	Remtox (Chemicals) Ltd	Professional	5590
Restor-MFC	Restoration (UK) Ltd	Professional (Surface Biocide) Professional (Wood Preservative)	6246
Sadolin Shed and Fence Preserver	Akzo Nobel Woodcare	Amateur	6821
Safeguard Fungicidal Micro Emulsifiable Concentrate	Safeguard Chemicals Ltd	Professional (Surface Biocide) Professional (Wood Preservative)	5648
D Sov AQ Micro F	Sovereign Chemical Industries Ltd	Professional Industrial	5223
Sovac F	Sovereign Chemical Industries Ltd	Amateur Professional Industrial	5233
W Sovac FWR	Sovereign Chemical Industries Ltd	Professional Industrial	5344
W Sovaq Micro F	Sovereign Chemical Industries Ltd	Professional Industrial	5668
Sovaq Micro F	Sovereign Chemical Industries Ltd	Professional Industrial	5824
W Sovereign AQF	Sovereign Chemical Industries Ltd	Professional Industrial	5229

Product Name	Marketing Company	Use	HSE No.

613 3-Iodo-2-propynyl-n-butyl carbamate—continued

Sovereign Coloured Wood Preservatives	Sovereign Chemical Industries Ltd	Professional Industrial	6337
Timber Water Repellent	Terminix Peter Cox Ltd	Professional	6621
UKC-2 Dry Rot Killer Concentrate	UK Chemicals Ltd	Professional (Surface Biocide) Professional (Wood Preservative)	6107

614 3-Iodo-2-propynyl-n-butyl carbamate + Benzalkonium chloride + Disodium octaborate

Antiblu Select	Hickson Timber Products Ltd	Industrial	6087

615 3-Iodo-2-propynyl-n-butyl carbamate + Boric acid

Protim Fentex I	Protim Solignum Ltd	Industrial	6096

616 3-Iodo-2-propynyl-n-butyl carbamate + Copper naphthenate

Osmose Endcoat Green	Injecta Osmose Ltd	Amateur Professional Industrial	6562
Sovereign Green Timber Preservative/Dipcoat	Sovereign Chemical Industries Ltd	Amateur Professional Industrial	6341

617 3-Iodo-2-propynyl-n-butyl carbamate + Cypermethrin

Microtech Dual Purpose AQ	Philip Johnstone Group Ltd	Professional	5453
D Water Based Wood Preserver (Concentrate)	Cementone Beaver Ltd	Amateur	5515
Water Based Wood Preserver (Concentrate)	Philip Johnstone Group Ltd	Amateur Professional	5630
Water-based Wood Preserver	Philip Johnstone Group Ltd	Amateur Professional	4902

618 3-Iodo-2-propynyl-n-butyl carbamate + Dialkyldimethyl ammonium chloride

Hickson NP-1	Hickson Timber Products Ltd	Industrial	5401

619 3-Iodo-2-propynyl-n-butyl carbamate + Flufenoxuron + Propiconazole

Remtox M9 Dual Purpose Super Concentrate	Remtox (Chemicals) Ltd	Professional	6508
Sovaq FLX F/I	Sovereign Chemical Industries Ltd	Professional	6509
Sovereign Fungicide/insecticide	Sovereign Chemical Industries Ltd	Professional	6828

HSE
WOOD PRESERVATIVES

Product Name	Marketing Company	Use	HSE No.

620 3-Iodo-2-propynyl-n-butyl carbamate + Permethrin

Product Name	Marketing Company	Use	HSE No.
Brunol Dual 8	Philip Johnstone Group Ltd	Professional	5857
Brunol PP	Philip Johnstone Group Ltd	Professional Industrial	5414
Brunol SPI	Philip Johnstone Group Ltd	Professional Industrial	5415
W Deepkill	Sovereign Chemical Industries Ltd	Professional Industrial	5238
Deepkill	Sovereign Chemical Industries Ltd	Professional Industrial	6084
D Fungicide/Insecticide Microemulsion M9	Remtox (Chemicals) Ltd	Professional Industrial	5711
Microtreat Dual	Philip Johnstone Group Ltd	Professional	5858
Profilan - Primer	Impra Systems Ltd	Amateur Professional Industrial	6511
Protim 250	Protim Solignum Ltd	Industrial	5070
Protim 250 WR	Protim Solignum Ltd	Industrial	5071
Protim Universal AQ 250	Protim Solignum Ltd	Professional	5844
Remtox AQ Fungicide/Insecticide R9	Remtox (Chemicals) Ltd	Professional Industrial	5370
Remtox Dual Purpose Paste K9	Remtox (Chemicals) Ltd	Professional Industrial	5399
Remtox Dual Purpose Paste K9	Remtox (Chemicals) Ltd	Professional Industrial	6182
D Remtox Fungicide/Insecticide Microemulsion M9	Remtox (Chemicals) Ltd	Professional Industrial	5360
Remtox Fungicide/Insecticide Microemulsion M9	Remtox (Chemicals) Ltd	Professional Industrial	5774
Remtox Spirit Based F/I K7	Remtox (Chemicals) Ltd	Professional Industrial	5391
D Sadolin Base No 561-2611	Sadolin UK Ltd	Amateur	4330
Sadolin Wood Preserver	Akzo Nobel Woodcare	Amateur Professional	5649
D Sov AQ Micro F/I	Sovereign Chemical Industries Ltd	Professional Industrial	5222
Sovac F/I	Sovereign Chemical Industries Ltd	Professional Industrial	5232
Sovac F/I	Sovereign Chemical Industries Ltd	Professional Industrial	6262
W Sovac F/I WR	Sovereign Chemical Industries Ltd	Professional Industrial	5300
D Sovaq Micro F/I	Sovereign Chemical Industries Ltd	Professional Industrial	5669
W Sovereign AQ F/I	Sovereign Chemical Industries Ltd	Professional Industrial	5221
Sovereign Sovaq Micro F/I	Sovereign Chemical Industries Ltd	Professional Industrial	5756

Product Name	Marketing Company	Use	HSE No.

620 3-Iodo-2-propynyl-n-butyl carbamate + Permethrin—continued

Tripaste	Triton Chemical Manufacturing Co Ltd	Professional	6251
Tripaste PP	Triton Chemical Manufacturing Co Ltd	Amateur Professional	4595
Woodtreat 25	Philip Johnstone Group Ltd	Professional	5413

621 3-Iodo-2-propynyl-n-butyl carbamate + Permethrin + Propiconazole

EQ159 RTU Microemulsion Timber Fluid	Enviroquest UK Ltd	Professional	6832
Timber-gel Ready to Use Microemulsion Timber Fluid	Terminix Peter Cox Ltd	Professional	6840

622 3-Iodo-2-propynyl-n-butyl carbamate + Permethrin + Propiconazole + Tebuconazole

Vacsol 2718 2:1 Conc	Hickson Timber Products Ltd	Industrial	5884
Vacsol 2719 RTU	Hickson Timber Products Ltd	Industrial	5883
Vacsol Aqua 6107	Hickson Timber Products Ltd	Industrial	6082
Vacsol Azure 2722	Hickson Timber Products Ltd	Industrial	6462

623 3-Iodo-2-propynyl-n-butyl carbamate + Propiconazole

EQ158 Timber Fungicide	Enviroquest UK Ltd	Professional	6859
Sadolin Quick Drying Wood Preserver	Akzo Nobel Woodcare	Amateur	6617

624 3-Iodo-2-propynyl-n-butyl carbamate + Propiconazole + Tebuconazole

Protim 418 J	Protim Solignum Ltd	Industrial	6046
Protim FDR418J	Protim Solignum Ltd	Industrial	6467
Vacsol Azure 2720	Hickson Timber Products Ltd	Industrial	6344
Vacsol Azure 2721	Hickson Timber Products Ltd	Industrial	5820

625 Lindane + Benzalkonium chloride

W Biokil	Jaymar Chemicals	Professional	4247

626 Lindane + Pentachlorophenol

W Supergrade Wood Preserver	Rentokil Initial UK Ltd	Professional	4641
W Supergrade Wood Preserver Black	Rentokil Initial UK Ltd	Professional Industrial	3713
W Supergrade Wood Preserver Brown	Rentokil Initial UK Ltd	Professional Industrial	3715
W Supergrade Wood Preserver Clear	Rentokil Initial UK Ltd	Professional Industrial	3716
W Woodtreat	Stanhope Chemical Products Ltd	Professional	4475

Product Name	Marketing Company	Use	HSE No.

627 Lindane + Pentachlorophenol + Tributyltin oxide

W Celpruf PK	Rentokil Initial UK Ltd	Industrial	4540
W Celpruf PK WR	Rentokil Initial UK Ltd	Industrial	4538
W Lar-Vac 100	Larsen Manufacturing Ltd	Industrial	3723
W Protim 80	Protim Solignum Ltd	Industrial	5029
W Protim 80 C	Protim Solignum Ltd	Industrial	5036
W Protim 80 CWR	Protim Solignum Ltd	Industrial	5031
W Protim 80 WR	Protim Solignum Ltd	Industrial	5030
W Protim Brown	Protim Solignum Ltd	Industrial	5000

628 Lindane + Pentachlorophenyl laurate

W Brunol ATP	Stanhope Chemical Products Ltd	Professional	4498
W Mystox BTL	Stanhope Chemical Products Ltd	Professional	4425
W Mystox BTV	Stanhope Chemical Products Ltd	Professional	4424

629 Lindane + Tributyltin oxide

W Cedasol 2320	Hickson Timber Products Ltd	Industrial	4365
W Dipsar	Cementone Beaver Ltd	Industrial	3604
W Hickson Timbercare WRQD	Hickson Timber Products Ltd	Industrial	3594
W Lar-Vac 400	Larsen Manufacturing Ltd	Industrial	5782
Protim 210	Protim Solignum Ltd	Industrial	5033
Protim 210 C	Protim Solignum Ltd	Industrial	5038
Protim 210 CWR	Protim Solignum Ltd	Industrial	5035
Protim 210 WR	Protim Solignum Ltd	Industrial	5039
Protim 215	Protim Solignum Ltd	Industrial	6016
Protim 23 WR	Protim Solignum Ltd	Industrial	5037
Protim 90	Protim Solignum Ltd	Industrial	5276
W Protim Paste	Protim Solignum Ltd	Professional	4979
W Vacsol Mwr Concentrate 2203	Hickson Timber Products Ltd	Industrial	3170
D Vacsol P 2304 RTU	Hickson Timber Products Ltd	Industrial	4597
Vacsol WR Ready to use 2116	Hickson Timber Products Ltd	Industrial	3168

630 2-Methyl-4-isothiazolin-3-one + 5-chloro-2-methyl-4-isothiazolin-3-one

Celkil 90	Rentokil Initial UK Ltd	Industrial	5428
Kathon 886F	Rohm & Haas (UK) Ltd	Industrial	5431
Laporte Mould-Ex	Laporte Wood Preservation	Industrial	5430
Tanamix 3743	Hickson Timber Products Ltd	Industrial	5429

631 2-Methyl-4-isothiazolin-3-one + 5-chloro-2-methyl-4-isothiazolin-3-one + Copper sulphate

Injecta Osmose Abs33	Injecta Osmose Ltd	Industrial	5978

Product Name	Marketing Company	Use	HSE No.

632 Methylene bis(thiocyanate)

D Timbercol Preservative (mbt) Concentrate	Timbercol Industries Ltd	Industrial	3892
Timbertone T10 Preservative	Timbertone	Industrial	4118

633 Methylene bis(thiocyanate) + 2-(Thiocyanomethylthio)benzothiazole

Busan 1009	Buckman Laboratories SA	Industrial	5256
Celbrite Tc	Rentokil Initial UK Ltd	Industrial	3921
Hickson Antiblu 3739	Hickson Timber Products Ltd	Industrial	4841
Injecta Osmose S-11	Injecta Osmose Ltd	Industrial	5957
Laporte Antisapstain	Laporte Wood Preservation	Industrial	5977
Mect	Buckman Laboratories SA	Industrial	5255
Protim Stainguard	Protim Solignum Ltd	Industrial	5009

634 Methylene bis(thiocyanate) + 2-(Thiocyanomethylthio)benzothiazole + Boric acid

Fentex Elite	Protim Solignum Ltd	Industrial	5532

635 Methylene bis(thiocyanate) + 2-(Thiocyanomethylthio)benzothiazole + Disodium octaborate

Timbertex PD	Marcher Chemicals Ltd	Industrial	5876

636 Methylene bis(thiocyanate) + Benzalkonium chloride + Boric acid + Dialkyldimethyl ammonium chloride

Celbrite MT	Rentokil Initial UK Ltd	Industrial	4550

637 Methylene bis(thiocyanate) + Boric acid

Hickson Antiblu 3737	Hickson Timber Products Ltd	Industrial	4745
Hickson Antiblu 3738	Hickson Timber Products Ltd	Industrial	4746

638 Oxine-copper

D Mitrol PQ 8	Kemira Kemwood AB	Industrial	4023
Mitrol PQ 8	Laporte Kemwood AB	Industrial	5533

639 Pentachlorophenol

W Protim GC	Protim Solignum Ltd	Industrial	5202

640 Pentachlorophenol + Lindane

W Supergrade Wood Preserver	Rentokil Initial UK Ltd	Professional	4641
W Supergrade Wood Preserver Black	Rentokil Initial UK Ltd	Professional Industrial	3713
W Supergrade Wood Preserver Brown	Rentokil Initial UK Ltd	Professional Industrial	3715
W Supergrade Wood Preserver Clear	Rentokil Initial UK Ltd	Professional Industrial	3716

HSE WOOD PRESERVATIVES

Product Name	Marketing Company	Use	HSE No.

640 Pentachlorophenol + Lindane—continued

W Woodtreat	Stanhope Chemical Products Ltd	Professional	4475

641 Pentachlorophenol + Tributyltin oxide

W Celpruf JP	Rentokil Initial UK Ltd	Industrial	4548
W Celpruf JP WR	Rentokil Initial UK Ltd	Industrial	4543
W Protim FDR-H	Protim Solignum Ltd	Industrial	4985
D Protim Joinery Lining	Protim Solignum Ltd	Professional	4977
W Protim JP	Protim Solignum Ltd	Industrial	4978
W Protim R Clear	Protim Solignum Ltd	Industrial	5013

642 Pentachlorophenol + Tributyltin oxide + Lindane

W Celpruf PK	Rentokil Initial UK Ltd	Industrial	4540
W Celpruf PK WR	Rentokil Initial UK Ltd	Industrial	4538
W Lar-Vac 100	Larsen Manufacturing Ltd	Industrial	3723
W Protim 80	Protim Solignum Ltd	Industrial	5029
W Protim 80 C	Protim Solignum Ltd	Industrial	5036
W Protim 80 CWR	Protim Solignum Ltd	Industrial	5031
W Protim 80 WR	Protim Solignum Ltd	Industrial	5030
W Protim Brown	Protim Solignum Ltd	Industrial	5000

643 Pentachlorophenyl laurate + Lindane

W Brunol ATP	Stanhope Chemical Products Ltd	Professional	4498
W Mystox BTL	Stanhope Chemical Products Ltd	Professional	4425
W Mystox BTV	Stanhope Chemical Products Ltd	Professional	4424

644 Permethrin

Activ-8-I	Restoration (UK) Ltd	Professional Industrial	6169
Antel Woodworm Killer (P) Concentrate (Water Dilutable)	Antel UK Ltd	Professional	4044
W Barrettine New Woodworm Fluid	Barrettine Products Ltd	Amateur Professional Industrial	3622
Blackfriar Woodworm Killer	E Parsons and Sons Ltd	Amateur Professional Industrial	5814
BPC-8	Safeguard Chemicals Ltd	Professional Industrial	6282
Brunol Insecticidal 8	Philip Johnstone Group Ltd	Professional	5856
Brunol PC	Philip Johnstone Group Ltd	Professional	3863
Brunol PY	Philip Johnstone Group Ltd	Professional	4476
CME 20/P Concentrate	Terminix Peter Cox Ltd	Professional	5785

Product Name	Marketing Company	Use	HSE No.

644 Permethrin—continued

Colron Woodworm Killer	Ronseal Ltd	Amateur	6324
Crown Micro Woodworm	Crown Chemicals Ltd	Professional	6894
W Cuprinol Insecticidal Emulsion Concentrate	Cuprinol Ltd	Professional	4709
Cuprinol Low Odour Woodworm Killer	Cuprinol Ltd	Amateur Professional Industrial	5449
D Cuprinol Woodworm Killer S	Cuprinol Ltd	Amateur Professional	4693
D Cuprinol Woodworm Killer S (aerosol)	Cuprinol Ltd	Amateur	4700
Deal Direct Woodworm Treatment Concentrate	Deal Direct Ltd	Professional Industrial	6249
Deepkill I	Sovereign Chemical Industries Ltd	Professional Industrial	5239
W Deepwood "Clear" Insecticide Emulsion Concentrate	Safeguard Chemicals Ltd	Professional Industrial	4950
Deepwood 8 Micro Emulsifiable Insecticide Concentrate	Safeguard Chemicals Ltd	Professional Industrial	5664
Deepwood Standard Emulsion Concentrate	Safeguard Chemicals Ltd	Professional	5665
Environmental Timber Treatment	SCI (Building Products)	Professional	5451
Environmental Timber Treatment Paste	SCI (Building Products)	Professional	5529
Hickson Antiborer 3768	Hickson Timber Products Ltd	Industrial	3149
Impra-Sanol	Impra Systems Ltd	Amateur Professional	4325
Insecticide 1	Terminix Peter Cox Ltd	Professional	6390
Kenwood Micro Emulsion Woodworm Treatment (Concentrate)	Kenwood Damp Proofing Plc	Professional Industrial	6220
Kenwood Micro Emulsion Woodworm Treatment (Concentrate)	Kenwood Damp Proofing Plc	Professional	6313
Kenwood Woodworm Treatment Concentrate	Kenwood Damp Proofing Plc	Professional Industrial	6103
KF8 Insecticide Microemulsion Concentrate	Kingfisher Chemicals Ltd	Professional	6827
Kingfisher KF-8 Micro Emulsion Insecticide Concentrate	Kingfisher Chemicals Ltd	Professional Industrial	5786
Lectro 8WW	Lectros International Ltd	Professional	6398
Lectros 8P	Lectros International Ltd	Professional	6818
Leyland Timbrene Woodworm Killer	Leyland Paint Company Ltd	Amateur Professional	4825
Low Odour Woodworm Killer	Cuprinol Ltd	Amateur Professional	5435

Product Name	Marketing Company	Use	HSE No.

644 Permethrin—continued

Product Name	Marketing Company	Use	HSE No.
M8 Micro Emulsion Aqueous Insecticide Super Concentrate	Remtox (Chemicals) Ltd	Professional	6273
Micro 8 WW	Lectros International Ltd	Professional Industrial	6132
Micro Woodworm 8	Grip-fast Systems Ltd	Professional Industrial	6415
Micro-Emulsion Woodworm Treatment Concentrate	Kenwood Damp Proofing Plc	Professional	6387
W Microguard Permethrin Concentrate	Permagard Products Ltd	Professional Industrial	5102
Microguard Permethrin Concentrate	Permagard Products Ltd	Professional Industrial	5741
Microguard Woodworm Fluid	Permagard Products Ltd	Amateur Professional Industrial	5103
Microtech Woodworm Killer 25X	Philip Johnstone Group Ltd	Professional	5888
Microtreat Insecticide	Philip Johnstone Group Ltd	Professional	5859
Omega	Restoration (UK) Ltd	Professional Industrial	6379
Palace Microfine 8 -Insecticide Concentrate	Palace Chemicals Ltd	Professional Industrial	5809
Palace Woodworm Killer 8 (Microemulsion)	Palace Chemicals Ltd	Professional	6629
Permethrin WW Conc	Permagard Products Ltd	Professional	4020
Premier Barrettine New Woodworm Killer	Barrettine Products Ltd	Amateur Professional Industrial	4005
Protect and Shine Spray	Rentokil Initial UK Ltd	Amateur	6123
Protim 600WR	Protim Solignum Ltd	Industrial	5833
Protim AQ	Protim Solignum Ltd	Industrial	5201
Protim Aquachem-Insecticidal Emulsion P	Protim Solignum Ltd	Amateur Professional	5161
Protim Insecticidal AQ8	Protim Solignum Ltd	Professional	6526
Protim Insecticidal Emulsion 8	Protim Solignum Ltd	Professional Industrial	5701
Protim Insecticidal Emulsion P	Protim Solignum Ltd	Professional Industrial	5008
Protim Paste P	Protim Solignum Ltd	Amateur Professional	4980
Protim Woodworm Killer P	Protim Solignum Ltd	Amateur Professional Industrial	5011
Remecology Insecticide R8	Remtox (Chemicals) Ltd	Professional	4233
Remecology Spirit Based Insecticide R6	Remtox (Chemicals) Ltd	Professional	4836
D Remtox Insecticide Microemulsion M8	Remtox (Chemicals) Ltd	Professional Industrial	5325

Product Name	Marketing Company	Use	HSE No.

644 Permethrin—continued

Product Name	Marketing Company	Use	HSE No.
W Remtox Insecticide Microemulsion M8	Remtox (Chemicals) Ltd	Professional Industrial	5710
Remtox Insecticide Microemulsion M8	Remtox (Chemicals) Ltd	Professional Industrial	5775
Remtox Insecticide Paste R4	Remtox (Chemicals) Ltd	Professional Industrial	5389
Remtox Microactive Insecticide W8	Remtox (Chemicals) Ltd	Professional	5555
Rentokil Classic Wax Polish	Rentokil Initial UK Ltd	Amateur	3954
Rentokil Woodworm Killer	Rentokil Initial UK Ltd	Amateur Professional	4208
Rentokil Woodworm Treatment	Rentokil Initial UK Ltd	Amateur Professional	3911
Restor-8-Insecticide Concentrate	Restoration (UK) Ltd	Professional Industrial	5962
Safeguard Deepwood I Insecticide Emulsion Concentrate	Safeguard Chemicals Ltd	Professional	5304
Safeguard Deepwood IV Woodworm Killer	Safeguard Chemicals Ltd	Professional	3946
Solignum Insecticidal AQ8	Protim Solignum Ltd	Professional	6527
Solignum Woodworm Killer	Protim Solignum Ltd	Amateur Professional Industrial	5488
Solignum Woodworm Killer Concentrate	Protim Solignum Ltd	Professional Industrial	5466
W Sov Aq Micro I	Sovereign Chemical Industries Ltd	Professional Industrial	5220
W Sovac I	Sovereign Chemical Industries Ltd	Professional Industrial	5230
D Sovaq Micro I	Sovereign Chemical Industries Ltd	Professional Industrial	5663
Sovaq Micro I	Sovereign Chemical Industries Ltd	Professional	6272
W Sovereign Aqueous Insecticide 2	Sovereign Chemical Industries Ltd	Professional	4620
Sovereign Sovaq Micro I	Sovereign Chemical Industries Ltd	Professional Industrial	5755
Sprytech 1	Spry Chemicals Ltd	Professional Industrial	6489
Timbrene Woodworm Killer	Kalon Group Plc	Amateur Professional	4996
Trimethrin 20S	Triton Chemical Manufacturing Co Ltd	Amateur Professional	4621
W Trimethrin 2AQ	Triton Chemical Manufacturing Co Ltd	Professional	4625
W Trimethrin 3AQ	Triton Chemical Manufacturing Co Ltd	Professional	4626

Product Name	Marketing Company	Use	HSE No.

644 Permethrin—continued

Product Name	Marketing Company	Use	HSE No.
W Trimethrin 3OS	Triton Chemical Manufacturing Co Ltd	Amateur Professional	4622
Tritec	Triton Chemical Manufacturing Co Ltd	Professional	5424
Tritec 120	Triton Chemical Manufacturing Co Ltd	Professional	6189
Tritec 121	Triton Chemical Manufacturing Co Ltd	Professional	6541
Tritec 8	Triton Chemical Manufacturing Co Ltd	Professional Industrial	5801
Triton 120	Triton Chemical Manufacturing Co Ltd	Professional	6539
Triton Woodworm Killer	Triton Chemical Manufacturing Co Ltd	Amateur	6226
UKC-8 Woodworm Killer Concentrate	UK Chemicals Ltd	Professional	6108
UKC-8 Woodworm Killer Microemulsion Concentrate	UK Chemicals Ltd	Professional	6376
Ultra Tech 2000 I	Crown Chemicals Ltd	Professional	6143
Water Based Woodworm Fluid	Barrettine Products Ltd	Amateur Professional	6311
Wickes Woodworm Killer	Wickes Building Supplies Ltd	Amateur Professional	5692
Woodworm 8 Micro Emulsifiable Concentrate	Crown Chemicals Ltd	Professional Industrial	5819
W Woodworm Fluid (B) EC	Rentokil Initial UK Ltd	Professional	4213
Woodworm Fluid (B) for Bonded Warehouses	Rentokil Initial UK Ltd	Professional	4725
Woodworm Fluid B	Rentokil Initial UK Ltd	Professional	4743
Woodworm Fluid FP	Rentokil Initial UK Ltd	Professional	4577
Woodworm Fluid Z Emulsion Concentrate	Rentokil Initial UK Ltd	Professional	4264
Woodworm Killer	Langlow Products Ltd	Amateur Professional Industrial	3691
Woodworm Killer (grade P)	Barrettine Products Ltd	Amateur Professional Industrial	5213
Woodworm Killer P8	Antel UK Ltd	Professional	6583
Woodworm Paste B	Rentokil Initial UK Ltd	Professional	4578
D Woodworm Roof Void Paste	Rentokil Initial UK Ltd	Amateur	3765
Woodworm Treatment Spray	Rentokil Initial UK Ltd	Amateur	5619

645 Permethrin + 2-Phenylphenol

Product Name	Marketing Company	Use	HSE No.
Brunol Opa	Philip Johnstone Group Ltd	Professional	4473
Cromar Dry Rot and Woodworm Killer	Cromar Building Products Ltd	Amateur Professional	6421

Product Name	Marketing Company	Use	HSE No.

646 Permethrin + 3-Iodo-2-propynyl-n-butyl carbamate

Product Name	Marketing Company	Use	HSE No.
Brunol Dual 8	Philip Johnstone Group Ltd	Professional	5857
Brunol PP	Philip Johnstone Group Ltd	Professional Industrial	5414
Brunol SP1	Philip Johnstone Group Ltd	Professional Industrial	5415
W Deepkill	Sovereign Chemical Industries Ltd	Professional Industrial	5238
Deepkill	Sovereign Chemical Industries Ltd	Professional Industrial	6084
D Fungicide/insecticide Microemulsion M9	Remtox (Chemicals) Ltd	Professional Industrial	5711
Microtreat Dual	Philip Johnstone Group Ltd	Professional	5858
Profilan - Primer	Impra Systems Ltd	Amateur Professional Industrial	6511
Protim 250	Protim Solignum Ltd	Industrial	5070
Protim 250 WR	Protim Solignum Ltd	Industrial	5071
Protim Universal Aq 250	Protim Solignum Ltd	Professional	5844
Remtox AQ Fungicide/Insecticide R9	Remtox (Chemicals) Ltd	Professional Industrial	5370
Remtox Dual Purpose Paste K9	Remtox (Chemicals) Ltd	Professional Industrial	5399
Remtox Dual Purpose Paste K9	Remtox (Chemicals) Ltd	Professional Industrial	6182
D Remtox Fungicide/Insecticide Microemulsion M9	Remtox (Chemicals) Ltd	Professional Industrial	5360
Remtox Fungicide/Insecticide Microemulsion M9	Remtox (Chemicals) Ltd	Professional Industrial	5774
Remtox Spirit Based F/I K7	Remtox (Chemicals) Ltd	Professional Industrial	5391
D Sadolin Base No 561-2611	Sadolin UK Ltd	Amateur	4330
Sadolin Wood Preserver	Akzo Nobel Woodcare	Amateur Professional	5649
D Sov Aq Micro F/I	Sovereign Chemical Industries Ltd	Professional Industrial	5222
Sovac F/I	Sovereign Chemical Industries Ltd	Professional Industrial	5232
Sovac F/I	Sovereign Chemical Industries Ltd	Professional Industrial	6262
W Sovac F/I WR	Sovereign Chemical Industries Ltd	Professional Industrial	5300
D Sovaq Micro F/I	Sovereign Chemical Industries Ltd	Professional Industrial	5669
W Sovereign Aq F/I	Sovereign Chemical Industries Ltd	Professional Industrial	5221
Sovereign Sovaq Micro F/I	Sovereign Chemical Industries Ltd	Professional Industrial	5756

Product Name	Marketing Company	Use	HSE No.

646 Permethrin + 3-Iodo-2-propynyl-n-butyl carbamate—continued

Tripaste	Triton Chemical Manufacturing Co Ltd	Professional	6251
Tripaste PP	Triton Chemical Manufacturing Co Ltd	Amateur Professional	4595
Woodtreat 25	Philip Johnstone Group Ltd	Professional	5413

647 Permethrin + Acypetacs copper + Acypetacs zinc

Cuprinol Low Odour End Cut	Cuprinol Ltd	Amateur Professional Industrial	6114
Protim 800P	Protim Solignum Ltd	Industrial	5702

648 Permethrin + Acypetacs zinc

W Cuprinol 5 Star Complete Wood Treatment S	Cuprinol Ltd	Amateur Professional	4710
Cuprinol Low Odour 5 Star Complete Wood Treatment	Cuprinol Ltd	Amateur Professional	5445
Cuprisol FN	Cuprinol Ltd	Industrial	3632
Cuprisol WR	Cuprinol Ltd	Industrial	3633
Cuprisol XQD Special	Cuprinol Ltd	Industrial	3732
Cut End	Protim Solignum Ltd	Amateur Professional Industrial	5490
Protim 800	Protim Solignum Ltd	Industrial	4991
Protim 800 C	Protim Solignum Ltd	Industrial	4990
Protim 800 WR	Protim Solignum Ltd	Industrial	4992
Protim Brown CDB	Protim Solignum Ltd	Amateur Professional Industrial	6472
Protim Paste 800	Protim Solignum Ltd	Professional Industrial	4986
D Protim WR 800	Fosroc Ltd	Amateur Professional Industrial	4487

649 Permethrin + Acypetacs zinc (equivalent to 3% zinc metal)

Protim 800 CWR	Protim Solignum Ltd	Industrial	4993

Product Name	Marketing Company	Use	HSE No.

650 Permethrin + Azaconazole

Product Name	Marketing Company	Use	HSE No.
Fongix SE Total Treatment for Wood	Liberon Waxes Ltd	Amateur (Surface Biocide) Amateur (Wood Preservative) Professional (Surface Biocide) Professional (Wood Preservative)	5610
D Fongix SE Total Treatment for Wood	V33 SA	Amateur (Surface Biocide) Amateur (Wood Preservative)	4288
Woodworm Killer and Rot Treatment	Liberon Waxes Ltd	Amateur (Surface Biocide) Amateur (Wood Preservative) Professional (Surface Biocide) Professional (Wood Preservative)	4826
W Xylamon Brown U 101 C	Venilia Ltd	Amateur Professional	4443
W Xylamon Curative U 152 G/H	NWE Distributors Ltd	Professional	4446

651 Permethrin + Benzalkonium chloride

Product Name	Marketing Company	Use	HSE No.
W Biokil Emulsion	Jaymar Chemicals	Professional	4562

652 Permethrin + Boric acid

Product Name	Marketing Company	Use	HSE No.
Activ-8-II	Restoration (UK) Ltd	Professional Industrial	6170
Micro Dual Purpose 8	Grip-Fast Systems Ltd	Professional Industrial	6408
W Microguard FI Concentrate	Permagard Products Ltd	Professional Industrial	5101
Microguard FI Concentrate	Permagard Products Ltd	Professional Industrial	5854

Product Name	Marketing Company	Use	HSE No.

653 Permethrin + Copper naphthenate

Protim Green E	Protim Solignum Ltd	Amateur Professional Industrial	4975
Solignum Green	Protim Solignum Ltd	Amateur Professional Industrial	5020

654 Permethrin + Dodecylamine lactate + Dodecylamine salicylate

W Kingfisher Timber Paste	Kingfisher Chemicals Ltd	Professional	4327
W Kingfisher Timberpaste	Kingfisher Chemicals Ltd	Professional	5931
W Safeguard Deepwood Paste	Safeguard Chemicals Ltd	Professional	5758

655 Permethrin + Propiconazole

Aqueous Universal	Protim Solignum Ltd	Amateur Professional	5776
Celpruf WB11	Rentokil Initial UK Ltd	Industrial	6682
EQ116 Dual Purpose FI	Enviroquest UK Ltd	Professional	6555
Micro 8FI	Construction Chemicals	Professional	6644
Palace Dual Purpose FI8 (Microemulsion)	Palace Chemicals Ltd	Professional	6646
Protim 340 WR	Protim Solignum Ltd	Industrial	5509
Protim 340E	Protim Solignum Ltd	Industrial	5511
Tritec 120 Plus	Triton Chemical Manufacturing Co Ltd	Professional	6289
Tritec 120 Plus	Triton Chemical Manufacturing Co Ltd	Professional	6537
Tritec 121 Plus	Triton Chemical Manufacturing Co Ltd	Professional	6538
Ultra Tech 2000 D	Crown Chemicals Ltd	Professional	6304
Vacsele Aqua 6106	Hickson Timber Products Ltd	Amateur Professional Industrial	5893
W Wocosen 12 OL	Janssen Pharmaceutica NV	Amateur Professional Industrial	5404
Wocosen S-P	Janssen Pharmaceutica NV	Amateur Professional Industrial	5871
Wolsit KD 20	BASF (UK)	Industrial	6307

656 Permethrin + Propiconazole + 3-Iodo-2-propynyl-n-butyl carbamate

EQ159 RTU Microemulsion Timber Fluid	Enviroquest UK Ltd	Professional	6832
Timber-gel Ready to Use Microemulsion Timber Fluid	Terminix Peter Cox Ltd	Professional	6840

657 Permethrin + Propiconazole + Dialkyldimethyl ammonium chloride

Traditional Preservative Oil	Oakmasters of Sussex	Amateur	6023
		Professional	
		Industrial	

658 Permethrin + Propiconazole + Tebuconazole

Protim 415	Protim Solignum Ltd	Industrial	5815
Protim 415T	Protim Solignum Ltd	Industrial	5880
Protim 415TWR	Protim Solignum Ltd	Industrial	5916
Protim 418	Protim Solignum Ltd	Industrial	6466
Protim 418WR	Protim Solignum Ltd	Industrial	6842
Vacsol 2627 Azure	Hickson Timber Products Ltd	Industrial	5558
Vacsol Aqua 6108	Hickson Timber Products Ltd	Industrial	6083
Vacsol Aqua 6109	Hickson Timber Products Ltd	Industrial	6430

659 Permethrin + Propiconazole + Tebuconazole + 3-Iodo-2-propynyl-n-butyl carbamate

Vacsol 2718 2:1 Conc	Hickson Timber Products Ltd	Industrial	5884
Vacsol 2719 RTU	Hickson Timber Products Ltd	Industrial	5883
Vacsol Aqua 6107	Hickson Timber Products Ltd	Industrial	6082
Vacsol Azure 2722	Hickson Timber Products Ltd	Industrial	6462

660 Permethrin + Propiconazole + Tebuconazole + Acypetacs zinc

Protim 415 T Special Solvent	Protim Solignum Ltd	Industrial	6077

661 Permethrin + Propiconazole + Tebuconazole + Disodium octaborate

Protim E410	Protim Solignum Ltd	Industrial	6218
Protim E410(i)	Protim Solignum Ltd	Industrial	6145
Protim E415	Protim Solignum Ltd	Industrial	5908
Protim E415 (i)	Protim Solignum Ltd	Industrial	6017

662 Permethrin + Tebuconazole

Brunol Atp New	Philip Johnstone Group Ltd	Professional	5620
		Industrial	

663 Permethrin + Tebuconazole + Benzalkonium chloride

Vacsol Aqua 6105	Hickson Timber Products Ltd	Industrial	5892

664 Permethrin + Tebuconazole + Dichlofluanid

Impra-holzschutzgrund (primer)	Impra Systems Ltd	Professional	5735
		Industrial	

665 Permethrin + Tri(hexylene glycol)biborate

Antel Dual Purpose (BP) Concentrate (Water Dilutable)	Antel UK Ltd	Professional	4046

Product Name	Marketing Company	Use	HSE No.

665 Permethrin + Tri(hexylene glycol)biborate—continued

Product Name	Marketing Company	Use	HSE No.
W Barrettine New Universal Fluid	Barrettine Products Ltd	Amateur Professional Industrial	3623
BPC8-DP	Building Preservation Centre	Professional	6306
Brunol PBO	Philip Johnstone Group Ltd	Professional	4474
Brunol Special P	Philip Johnstone Group Ltd	Professional	3864
W Chel-wood Preserver/woodworm Dry Rot Killer BP	Chelec Ltd	Amateur Professional	4897
Crown Micro Dual Purpose F/I	Crown Chemicals Ltd	Professional	6895
Deepwood FI Dual Purpose Emulsion Concentrate	Safeguard Chemicals Ltd	Professional Industrial	5087
Dual 8	Deal Direct Ltd	Professional	6255
W Ecology Fungicide Insecticide Aqueous (Concentrate) Wood Preservative Dual Purpose	Kingfisher Chemicals Ltd	Professional	4039
Ecology Fungicide Insecticide Aqueous (Concentrate) Wood Preservative Dual Purpose	Kingfisher Chemicals Ltd	Professional	5930
W insecticide Fungicide Wood Preservative	Kingfisher Chemicals Ltd	Professional	4048
Insecticide Fungicide Wood Preservative	Kingfisher Chemicals Ltd	Professional	5933
KF-8 Dual Purpose Micro Emulsion Concentrate	Kingfisher Chemicals Ltd	Professional Industrial	6075
Kingfisher Timberpaste	Kingfisher Chemicals Ltd	Professional	6686
Larsen Wood Preservative Clear 2 and Brown 2	Larsen Manufacturing Ltd	Amateur Professional Industrial	4287
Larsen Woodworm Killer 2	Larsen Manufacturing Ltd	Amateur Professional Industrial	4285
Mar-Kil S	Marcher Chemicals Ltd	Professional Industrial	5234
Mar-Kil W	Marcher Chemicals Ltd	Professional Industrial	5219
Micron 8 Dual Purpose Emulsion Concentrate	Safeguard Chemicals Ltd	Professional Industrial	5851
Omega Plus	Restoration (UK) Ltd	Professional	6382
Perma AQ Dual Purpose	Triton Perma Industries Ltd	Professional	5617
Permapaste PB	Triton Perma Industries Ltd	Professional	5616
Permethrin F and I Concentrate	Permagard Products Ltd	Professional	4008
Protim CB	Protim Solignum Ltd	Professional Industrial	5864
Protim CDB	Protim Solignum Ltd	Amateur Professional Industrial	4999

Product Name	Marketing Company	Use	HSE No.

665 Permethrin + Tri(hexylene glycol)biborate—continued

Product Name	Marketing Company	Use	HSE No.
Protim WR 260	Protim Solignum Ltd	Amateur Professional Industrial	4997
W Remecology Spirit Based K7	Remtox (Chemicals) Ltd	Professional	4248
Restor-8-Dual Purpose Concentrate	Restoration (UK) Ltd	Professional Industrial	5963
Restor-paste	Restoration (UK) Ltd	Professional	6079
W Safeguard Deepwood II Dual Purpose Emulsion Concentrate	Safeguard Chemicals Ltd	Professional	5303
Safeguard Deepwood II Dual Purpose Emulsion Concentrate	Safeguard Chemicals Ltd	Professional	5850
Safeguard Deepwood III Timber Treatment	Safeguard Chemicals Ltd	Amateur	3937
W Safeguard Deepwood Paste	Safeguard Chemicals Ltd	Professional	4110
Sandhill Aqueous Fungicide - Insecticide	Sandhill Building Products	Professional	5950
Sandhill Fungicide - Insecticide	Sandhill Building Products	Amateur	5952
D Saturin E10	British Building and Engineering Appliances Plc	Professional	3896
D Saturin E5	British Building and Engineering Appliances Plc	Amateur	3897
Solignum Remedial Concentrate	Protim Solignum Ltd	Professional	5021
D Solignum Remedial Fluid PB	Protim Solignum Ltd	Professional	5017
Solignum Remedial PB	Protim Solignum Ltd	Amateur Professional Industrial	5607
Sovereign Aqueous Fungicide/ Insecticide 2	Sovereign Chemical Industries Ltd	Professional Industrial	4619
Sovereign Insecticide/Fungicide 2	Sovereign Chemical Industries Ltd	Professional Industrial	4649
Sprytech 2	Spry Chemicals Ltd	Professional Industrial	6500
Timber Paste	Grip-fast Systems Ltd	Professional	6409
W Trimethrin AQ Plus	Triton Chemical Manufacturing Company Ltd	Professional	3522
Trimethrin OS Plus	Triton Chemical Manufacturing Company Ltd	Amateur Professional	3521
Tripaste PB	Triton Chemical Manufacturing Company Ltd	Professional	6080
Tritec Plus	Triton Chemical Manufacturing Company Ltd	Professional	5425

Product Name	Marketing Company	Use	HSE No.

665 Permethrin + Tri(hexylene glycol)biborate—continued

W Universal Fluid (grade Pb)	Barrettine Products Ltd	Amateur Professional Industrial	5212
Universal Wood Preserver	Langlow Products Ltd	Amateur Professional Industrial	3692
Woodworm Killer	Sealocrete Pla Ltd	Amateur Professional	5028

666 Permethrin + Tri(hexylene glycol)biborate + Dichlofluanid

Blackfriar Wood Preserver WP Gold Star Clear	E Parsons and Sons Ltd	Amateur Professional	6417
Blackfriars Gold Star Clear	E Parsons and Sons Ltd	Amateur Professional Industrial	4149
Leyland Timbrene Supreme Environmental Formula	Leyland Paint Company Ltd	Amateur Professional	4800
W Nitromors Timbrene Supreme Environmental Formula	Henkel Home Improvements and Adhesive Products	Amateur Professional Industrial	4107
Premier Barrettine New Universal Fluid D	Barrettine Products Ltd	Amateur Professional Industrial	4004
Timbrene Supreme	Kalon Group Plc	Amateur Professional	4995

667 Permethrin + Tributyltin naphthenate

Celpruf Tnm	Rentokil Initial UK Ltd	Industrial	4704
Celpruf Tnm WR	Rentokil Initial UK Ltd	Industrial	4703
Protim 230	Protim Solignum Ltd	Industrial	5044
Protim 230 C	Protim Solignum Ltd	Industrial	5781
Protim 230 CWR	Protim Solignum Ltd	Industrial	5778
Protim 230 WR	Protim Solignum Ltd	Industrial	5045
Vacsol 2622/2623 WR	Hickson Timber Products Ltd	Industrial	4645
Vacsol 2625/2626 WR 2:1 Concentrate	Hickson Timber Products Ltd	Industrial	5329
Vacsol 2713/2714	Hickson Timber Products Ltd	Industrial	4808

668 Permethrin + Zinc octoate

D Cuprinol Difusol S	Cuprinol Ltd	Professional	3750
Green Range Timber Treatment Paste	Philip Johnstone Group Ltd	Professional	4651
Larvac 300	Larsen Manufacturing Ltd	Industrial	4461
Palace Timbertreat Ecology	Palace Chemicals Ltd	Professional	4931

Product Name	Marketing Company	Use	HSE No.

668 Permethrin + Zinc octoate—continued

Product Name	Marketing Company	Use	HSE No.
Premium Wood Treatment	Rentokil Initial UK Ltd	Amateur Amateur (Surface Biocide) Professional Professional (Surface Biocide)	5350
W Rentokil Dual Purpose Fluid	Rentokil Initial UK Ltd	Professional	4074
Rentokil Wood Preservative	Rentokil Initial UK Ltd	Amateur Professional	4195
Ronseal's Universal Wood Treatment	Ronseal Ltd	Amateur Professional	6186
Solignum Colourless	Protim Solignum Ltd	Amateur Professional Industrial	5549
Solignum Wood Preservative Paste	Protim Solignum Ltd	Professional	5001
Thompson's Universal Wood Treatment	Ronseal Ltd	Amateur Professional	5986
Wood Preservative Clear	Rentokil Initial UK Ltd	Amateur Professional	4077
Woodtreat BP	Philip Johnstone Group Ltd	Professional	3085

669 Permethrin + Zinc octoate + Boric acid

Product Name	Marketing Company	Use	HSE No.
W Celpruf BZP	Rentokil Initial UK Ltd	Industrial	4190
W Celpruf BZP WR	Rentokil Initial UK Ltd	Industrial	4191
Cut'n'treat	Rentokil Initial UK Ltd	Amateur Professional	5802
Premium Grade Wood Treatment	Rentokil Initial UK Ltd	Amateur Professional	4193

670 Permethrin + Zinc octoate + Dichlofluanid

Product Name	Marketing Company	Use	HSE No.
OS Color Wood Stain and Preservative	Ostermann and Scheiwe Gmbh	Amateur	5531

671 Permethrin + Zinc versatate

Product Name	Marketing Company	Use	HSE No.
Cedasol Ready To Use (2306)	Hickson Timber Products Ltd	Industrial	4117
Celpruf ZOP	Rentokil Initial UK Ltd	Industrial	4681
Celpruf ZOP WR	Rentokil Initial UK Ltd	Industrial	4682
W Imersol 2410	Hickson Timber Products Ltd	Industrial	4674
Protim 220	Protim Solignum Ltd	Industrial	5198
Protim 220 CWR	Protim Solignum Ltd	Amateur Professional Industrial	5196
Protim 220 WR	Protim Solignum Ltd	Industrial	5195
Protim 220C	Protim Solignum Ltd	Industrial	5780

Product Name	Marketing Company	Use	HSE No.
671 Permethrin + Zinc versatate—continued			
Protim Paste 220	Protim Solignum Ltd	Professional Industrial	4983
D Protim WR 220	Fosroc Ltd	Amateur	4113
Solignum Universal	Protim Solignum Ltd	Amateur Professional Industrial	5019
W Vacsele P2312	Hickson Timber Products Ltd	Professional Industrial	4363
Vacsol (2:1 Conc) 2711/2712	Hickson Timber Products Ltd	Industrial	4685
Vacsol 2709/2710	Hickson Timber Products Ltd	Industrial	4678
D Vacsol P (2310)	Hickson Timber Products Ltd	Industrial	4377
Vacsol WR (2:1 Conc) 2614/2615	Hickson Timber Products Ltd	Industrial	4679
Vacsol WR 2612/2613	Hickson Timber Products Ltd	Industrial	4676
672 2-Phenylphenol			
W Basiment 560	Venilia Ltd	Professional Industrial	4411
Laybond Wood Preserver	Laybond Products Ltd	Amateur Professional	5382
Preventol OF	Bayer Plc	Professional Industrial	5200
Protim Joinery Lining 280	Protim Solignum Ltd	Professional	5135
673 2-Phenylphenol + Benzalkonium chloride			
W Chel Dry Rot Killer for Masonry and Brickwork	Chelec Ltd	Professional	3067
674 2-Phenylphenol + Boric acid			
Protim Fentex P	Protim Solignum Ltd	Industrial	6095
675 2-Phenylphenol + Cypermethrin			
Laybond Timber Protector	Laybond Products Ltd	Amateur Professional	5369
676 2-Phenylphenol + Permethrin			
Brunol Opa	Philip Johnstone Group Ltd	Professional	4473
Cromar Dry Rot and Woodworm Killer	Cromar Building Products Ltd	Amateur Professional	6421

Product Name	Marketing Company	Use	HSE No.

677 Pirimiphos-methyl

Actellic 25 EC	Zeneca Public Health	Professional	4880
		Professional (Animal Husbandry)	
		Professional (Food Storage Practice)	
		Professional (Insecticide)	

678 Potassium 2-phenylphenoxide

D Timbercol Preservative Concentrate	Timbercol Industries Ltd	Industrial	3868

679 Potassium dichromate + Boric acid + Copper sulphate

Tecca CCCB 1	Tecca Ltd	Industrial	5584

680 Propiconazole

CME 25/F Concentrate	Terminix Peter Cox Ltd	Professional	6136
Fungicide 2	Terminix Peter Cox Ltd	Professional	6391
Impralan Grund G 200	Impra Systems Ltd	Professional Industrial	6620
Impralan-Grund G 100	Impra Systems Ltd	Professional Industrial	6549
Safetray P	Progress Products	Professional	6275
Timber Fungicide	Timberwise Preservation Scot	Professional	6641
Timber Treatment Plus	Alexander Rose Ltd	Amateur Professional	6648
Trisol 120	Triton Chemical Manufacturing Co Ltd	Professional	6190
Trisol 120	Triton Chemical Manufacturing Co Ltd	Professional	6540
Triton Timber Rot Killer	Triton Chemical Manufacturing Co Ltd	Amateur	6227
Ultra Tech 2000 F	Crown Chemicals Ltd	Professional	6142
Wocosen 100 SL	Janssen Pharmaceutica Nv	Professional	5743
		Professional (Surface Biocide)	
		Industrial	
Wocosen S	Janssen Pharmaceutica Nv	Amateur	5394
		Professional	
		Industrial	

681 Propiconazole + 3-Iodo-2-propynyl-n-butyl carbamate

Eq158 Timber Fungicide	Enviroquest UK Ltd	Professional	6859

Product Name	Marketing Company	Use	HSE No.

681 Propiconazole + 3-Iodo-2-propynyl-n-butyl carbamate—continued

Sadolin Quick Drying Wood Preserver	Akzo Nobel Woodcare	Amateur	6617

682 Propiconazole + 3-Iodo-2-propynyl-n-butyl carbamate + Flufenoxuron

Remtox M9 Dual Purpose Super Concentrate	Remtox (Chemicals) Ltd	Professional	6508
Sovaq FLX F/I	Sovereign Chemical Industries Ltd	Professional	6509
Sovereign Fungicide/Insecticide	Sovereign Chemical Industries Ltd	Professional	6828

683 Propiconazole + 3-Iodo-2-propynyl-n-butyl carbamate + Permethrin

EQ159 RTU Microemulsion Timber Fluid	Enviroquest UK Ltd	Professional	6832
Timber-gel Ready To Use Microemulsion Timber Fluid	Terminix Peter Cox Ltd	Professional	6840

684 Propiconazole + Boric acid

Celbor PR	Rentokil Initial UK Ltd	Industrial	6192

685 Propiconazole + Boric acid + Copper carbonate hydroxide

Wolmanit CX-P	Dr Wolman GmbH	Industrial	6566

686 Propiconazole + Boric acid + Cypermethrin

Rentokil Total Wood Treatment	Rentokil Initial UK Ltd	Amateur Amateur (Surface Biocide) Professional (Surface Biocide)	6053

687 Propiconazole + Cypermethrin

Celpruf B	Rentokil Initial UK Ltd	Industrial	5935
Cuprinol Quick Drying Clear Preserver	Cuprinol Ltd	Amateur Professional Industrial	6014
W Cuprinol Wood Defender	Cuprinol Ltd	Amateur Professional Industrial	5798
Wocosen 100 SL-C	Janssen Pharmaceutica Nv	Professional Industrial	5807

Product Name	Marketing Company	Use	HSE No.

688 Propiconazole + Dialkyldimethyl ammonium chloride + Permethrin

Traditional Preservative Oil	Oakmasters of Sussex	Amateur Professional Industrial	6023

689 Propiconazole + Permethrin

Aqueous Universal	Protim Solignum Ltd	Amateur Professional	5776
Celpruf WB11	Rentokil Initial UK Ltd	Industrial	6682
EQ116 Dual Purpose FI	Enviroquest UK Ltd	Professional	6555
Micro 8FI	Construction Chemicals	Professional	6644
Palace Dual Purpose FI8 (Microemulsion)	Palace Chemicals Ltd	Professional	6646
Protim 340 WR	Protim Solignum Ltd	Industrial	5509
Protim 340E	Protim Solignum Ltd	Industrial	5511
Tritec 120 Plus	Triton Chemical Manufacturing Co Ltd	Professional	6289
Tritec 120 Plus	Triton Chemical Manufacturing Co Ltd	Professional	6537
Tritec 121 Plus	Triton Chemical Manufacturing Co Ltd	Professional	6538
Ultra Tech 2000 D	Crown Chemicals Ltd	Professional	6304
Vacsele Aqua 6106	Hickson Timber Products Ltd	Amateur Professional Industrial	5893
W Wocosen 12 OL	Janssen Pharmaceutica NV	Amateur Professional Industrial	5404
Wocosen S-P	Janssen Pharmaceutica NV	Amateur Professional Industrial	5871
Wolsit KD 20	Basf (UK)	Industrial	6307

690 Propiconazole + Tebuconazole

Protim FDR418	Protim Solignum Ltd	Industrial	6465

691 Propiconazole + Tebuconazole + 3-Iodo-2-propynyl-n-butyl carbamate

Protim 418 J	Protim Solignum Ltd	Industrial	6046
Protim FDR418J	Protim Solignum Ltd	Industrial	6467
Vacsol Azure 2720	Hickson Timber Products Ltd	Industrial	6344
Vacsol Azure 2721	Hickson Timber Products Ltd	Industrial	5820

692 Propiconazole + Tebuconazole + 3-Iodo-2-propynyl-n-butyl carbamate + Permethrin

Vacsol 2718 2:1 Conc	Hickson Timber Products Ltd	Industrial	5884
Vacsol 2719 RTU	Hickson Timber Products Ltd	Industrial	5883
Vacsol Aqua 6107	Hickson Timber Products Ltd	Industrial	6082

Product Name	Marketing Company	Use	HSE No.

692 Propiconazole + Tebuconazole + 3-Iodo-2-propynyl-n-butyl carbamate + Permethrin—continued

Vacsol Azure 2722	Hickson Timber Products Ltd	Industrial	6462

693 Propiconazole + Tebuconazole + Acypetacs zinc + Permethrin

Protim 415 T Special Solvent	Protim Solignum Ltd	Industrial	6077

694 Propiconazole + Tebuconazole + Boric acid + Copper carbonate hydroxide

Tanalith E (3492)	Hickson Timber Products Ltd	Industrial	6269
Tanalith E (3494)	Hickson Timber Products Ltd	Industrial	6434

695 Propiconazole + Tebuconazole + Disodium octaborate

Protim B610	Protim Solignum Ltd	Industrial	6366

696 Propiconazole + Tebuconazole + Disodium octaborate + Permethrin

Protim E410	Protim Solignum Ltd	Industrial	6218
Protim E410(i)	Protim Solignum Ltd	Industrial	6145
Protim E415	Protim Solignum Ltd	Industrial	5908
Protim E415 (i)	Protim Solignum Ltd	Industrial	6017

697 Propiconazole + Tebuconazole + Permethrin

Protim 415	Protim Solignum Ltd	Industrial	5815
Protim 415T	Protim Solignum Ltd	Industrial	5880
Protim 415TWR	Protim Solignum Ltd	Industrial	5916
Protim 418	Protim Solignum Ltd	Industrial	6466
Protim 418WR	Protim Solignum Ltd	Industrial	6842
Vacsol 2627 Azure	Hickson Timber Products Ltd	Industrial	5558
Vacsol Aqua 6108	Hickson Timber Products Ltd	Industrial	6083
Vacsol Aqua 6109	Hickson Timber Products Ltd	Industrial	6430

698 Sodium 2,4,6-trichlorophenoxide + Boric acid

Protim Fentex TWR Plus	Protim Solignum Ltd	Industrial	6547

699 Sodium 2-phenylphenoxide

Aquaguard Wood Preserver	Langlow Products Ltd	Industrial	5936
Brunosol Concentrate	Philip Johnstone Group Ltd	Professional	4472
		Professional (Surface Biocide)	
D Coaltec 50	Coalite Chemicals Division	Professional	4761
		Professional (Surface Biocide)	
		Industrial	

Product Name	Marketing Company	Use	HSE No.

699 Sodium 2-phenylphenoxide—continued

Product Name	Marketing Company	Use	HSE No.
D Coaltec 50TD	Coalite Chemicals Division	Amateur	4762
		Amateur (Surface Biocide)	
		Professional	
		Professional (Surface Biocide)	
		Industrial	
W Fence Protector	Barrettine Products Ltd	Amateur	5208
		Professional	
		Industrial	
Mar-Cide	Marcher Chemicals Ltd	Professional	5631
		Professional (Surface Biocide)	
		Industrial	
Timbertek	Catomance Plc	Amateur	5920
Timbertek Conc	Advanced Bitumens Ltd	Industrial	5904
Timbertex PI/2	Marcher Chemicals Ltd	Professional	5628
		Industrial	
D Valspar Shed and Fence Preservative	Macpherson Paints Ltd	Amateur	4865
		Professional	
D Valspar Timberguard Shed and Fence Preservative	Macpherson Paints Ltd	Amateur	4867
		Professional	
W.S.P. Wood Treatment	Barrettine Products Ltd	Amateur	5209
		Professional	
		Industrial	

700 Sodium dichromate + Arsenic pentoxide + Copper sulphate

Product Name	Marketing Company	Use	HSE No.
Celcure A Concentrate	Rentokil Initial UK Ltd	Industrial	5458
Celcure A Fluid 10	Rentokil Initial UK Ltd	Industrial	3764
Celcure A Fluid 6	Rentokil Initial UK Ltd	Industrial	3155
Celcure A Paste	Rentokil Initial UK Ltd	Industrial	4523
Injecta CCA-C	Injecta Osmose Ltd	Industrial	4447
D Kemira CCA Type BS	Kemira Kemwood AB	Industrial	3879
Kemwood CCA Type BS	Laporte Kemwood AB	Industrial	5534
Laporte CCA Type 1	Laporte Wood Preservation	Industrial	3175
Laporte CCA Type 2	Laporte Wood Preservation	Industrial	3176
Laporte Permawood CCA	Laporte Wood Preservation	Industrial	3529
Protim CCA Salts Type 2	Protim Solignum Ltd	Industrial	4972
Tanalith 3357	Hickson Timber Products Ltd	Industrial	4431
W Tanalith CL (3354)	Hickson Timber Products Ltd	Industrial	4196
Tanalith CP 3353	Hickson Timber Products Ltd	Industrial	4667
Tecca P2	Tecca Ltd	Industrial	3958

HSE WOOD PRESERVATIVES

Product Name	Marketing Company	Use	HSE No.

701 Sodium dichromate + Boric acid + Chromium acetate + Copper sulphate

Celgard CF	Rentokil Initial UK Ltd	Industrial	4608

702 Sodium dichromate + Boric acid + Copper sulphate

Laporte CCB	Laporte Wood Preservation	Industrial	5875
Tanalith CBC Paste 3402	Hickson Timber Products Ltd	Industrial	3833

703 Sodium dichromate + Chromium acetate + Copper sulphate

Celcure B	Rentokil Initial UK Ltd	Industrial	4541
Celcure O	Rentokil Initial UK Ltd	Industrial	4539

704 Sodium dichromate + Copper sulphate

Laporte Cut End Preservative	Laporte Wood Preservation	Industrial	3528

705 Sodium dichromate + Sodium fluoride + Ammonium bifluoride

W Rentex	Rentokil Initial UK Ltd	Industrial	4756

706 Sodium fluoride + Ammonium bifluoride + Sodium dichromate

W Rentex	Rentokil Initial UK Ltd	Industrial	4756

707 Sodium pentachlorophenoxide

D Mosgo P	Agrichem Ltd	Professional Professional (Surface Biocide)	4337
W Protim Fentex WR	Protim Solignum Ltd	Industrial	4974
W Protim Panelguard	Protim Solignum Ltd	Industrial	5010
W Protim Plug Compound	Protim Solignum Ltd	Professional Professional (Surface Biocide)	5012

708 Sodium pentachlorophenoxide + Disodium octaborate

W Gallwey ABS	Protim Solignum Ltd	Industrial	5014
W Protim Fentex Green Concentrate	Protim Solignum Ltd	Industrial	5122
W Protim Fentex Green RFU	Protim Solignum Ltd	Industrial	5123
W Protim Fentex M	Protim Solignum Ltd	Industrial	4973
W Protim Kleen II	Protim Solignum Ltd	Industrial	5203

709 Tebuconazole

Bayer WPC 2-1.5	Bayer Plc	Professional Industrial	5501
Bayer WPC 2-25	Bayer Plc	Professional Industrial	5500
Bayer WPC 2-25-SB	Bayer Plc	Professional Industrial	5502

Product Name	*Marketing Company*	*Use*	*HSE No.*

710 Tebuconazole + 3-Iodo-2-propynyl-n-butyl carbamate + Permethrin + Propiconazole

Vacsol 2718 2:1 Conc	Hickson Timber Products Ltd	Industrial	5884
Vacsol 2719 RTU	Hickson Timber Products Ltd	Industrial	5883
Vacsol Aqua 6107	Hickson Timber Products Ltd	Industrial	6082
Vacsol Azure 2722	Hickson Timber Products Ltd	Industrial	6462

711 Tebuconazole + 3-Iodo-2-propynyl-n-butyl carbamate + Propiconazole

Protim 418 J	Protim Solignum Ltd	Industrial	6046
Protim FDR418J	Protim Solignum Ltd	Industrial	6467
Vacsol Azure 2720	Hickson Timber Products Ltd	Industrial	6344
Vacsol Azure 2721	Hickson Timber Products Ltd	Industrial	5820

712 Tebuconazole + Acypetacs zinc + Permethrin + Propiconazole

Protim 415 T Special Solvent	Protim Solignum Ltd	Industrial	6077

713 Tebuconazole + Benzalkonium chloride

Vacsol Aqua 6101	Hickson Timber Products Ltd	Industrial	5557

714 Tebuconazole + Benzalkonium chloride + Permethrin

Vacsol Aqua 6105	Hickson Timber Products Ltd	Industrial	5892

715 Tebuconazole + Boric acid + Copper carbonate hydroxide

Ensele 3428	Hickson Timber Products Ltd	Professional Industrial	5878
Tanalith E (3485)	Hickson Timber Products Ltd	Industrial	5562

716 Tebuconazole + Boric acid + Copper carbonate hydroxide + Propiconazole

Tanalith E (3492)	Hickson Timber Products Ltd	Industrial	6269
Tanalith E (3494)	Hickson Timber Products Ltd	Industrial	6434

717 Tebuconazole + Cypermethrin

Celpruf TZC	Rentokil Initial UK Ltd	Industrial	5722

718 Tebuconazole + Dichlofluanid + Permethrin

Impra-Holzschutzgrund (Primer)	Impra Systems Ltd	Professional Industrial	5735

719 Tebuconazole + Disodium octaborate + Permethrin + Propiconazole

Protim E410	Protim Solignum Ltd	Industrial	6218
Protim E410(i)	Protim Solignum Ltd	Industrial	6145
Protim E415	Protim Solignum Ltd	Industrial	5908
Protim E415 (i)	Protim Solignum Ltd	Industrial	6017

720 Tebuconazole + Disodium octaborate + Propiconazole

Protim B610	Protim Solignum Ltd	Industrial	6366

HSE
WOOD PRESERVATIVES

Product Name	Marketing Company	Use	HSE No.

721 Tebuconazole + Permethrin

Brunol ATP New	Philip Johnstone Group Ltd	Professional Industrial	5620

722 Tebuconazole + Permethrin + Propiconazole

Protim 415	Protim Solignum Ltd	Industrial	5815
Protim 415T	Protim Solignum Ltd	Industrial	5880
Protim 415TWR	Protim Solignum Ltd	Industrial	5916
Protim 418	Protim Solignum Ltd	Industrial	6466
Protim 418WR	Protim Solignum Ltd	Industrial	6842
Vacsol 2627 Azure	Hickson Timber Products Ltd	Industrial	5558
Vacsol Aqua 6108	Hickson Timber Products Ltd	Industrial	6083
Vacsol Aqua 6109	Hickson Timber Products Ltd	Industrial	6430

723 Tebuconazole + Propiconazole

Protim FDR418	Protim Solignum Ltd	Industrial	6465

724 2-(Thiocyanomethylthio)benzothiazole

Bayer WPC-3	Bayer Plc	Industrial	4969
Bayer WPC-4	Bayer Plc	Industrial	4970
Busan 1111	Buckman Laboratories SA	Industrial	4764
Crysolite Glaramara Concentrated Timber Treatment	Crysolite Protective Coatings	Industrial	5443
Fentex NP-UF	Protim Solignum Ltd	Industrial	5069
Paneltone	Rentokil Initial UK Ltd	Industrial	4253

725 2-(Thiocyanomethylthio)benzothiazole + Boric acid

Protim Fentex Europa 1 RFU	Protim Solignum Ltd	Industrial	5119
Protim Fentex Europa I	Protim Solignum Ltd	Industrial	4984

726 2-(Thiocyanomethylthio)benzothiazole + Boric acid + Methylene bis(thiocyanate)

Fentex Elite	Protim Solignum Ltd	Industrial	5532

727 2-(Thiocyanomethylthio)benzothiazole + Disodium octaborate + Methylene bis(thiocyanate)

Timbertex PD	Marcher Chemicals Ltd	Industrial	5876

728 2-(Thiocyanomethylthio)benzothiazole + Methylene bis(thiocyanate)

Busan 1009	Buckman Laboratories SA	Industrial	5256
Celbrite TC	Rentokil Initial UK Ltd	Industrial	3921
Hickson Antiblu 3739	Hickson Timber Products Ltd	Industrial	4841
Injecta Osmose S-11	Injecta Osmose Ltd	Industrial	5957
Laporte Antisapstain	Laporte Wood Preservation	Industrial	5977
Mect	Buckman Laboratories SA	Industrial	5255
Protim Stainguard	Protim Solignum Ltd	Industrial	5009

Product Name	Marketing Company	Use	HSE No.

729 Tri(hexylene Glycol)biborate

Product Name	Marketing Company	Use	HSE No.
ABL Wood Preservative (B)	Advanced Bitumens Ltd	Amateur Professional Industrial	3743
W Barrettine New Preserver	Barrettine Products Ltd	Amateur Professional Industrial	3624
BCR Wood Preserver	Building Chemical Research Ltd	Amateur Professional Industrial	6315
Brown Wood Preservative	Palace Chemicals Ltd	Amateur Professional Industrial	5242
Clear Wood Preservative	Palace Chemicals Ltd	Amateur Professional Industrial	5240
Crown Timber Preservative	Crown Chemicals Ltd	Amateur Professional	6882
D Green Range Fungicidal Solvent	Cementone Beaver Ltd	Professional	3827
Kingfisher Wood Preservative	Kingfisher Chemicals Ltd	Professional	4012
Langlow Wood Preserver	Langlow Products Ltd	Amateur Professional Industrial	3690
Langlow Wood Preserver Formulation B	Langlow Products Ltd	Amateur Professional Industrial	3754
Larsen Joinery Grade 2	Larsen Manufacturing Ltd	Professional Industrial	4286
Palace Base Coat Wood Preservative	Palace Chemicals Ltd	Amateur Professional Industrial	5351
Palace Microfine Fungicide Concentrate	Palace Chemicals Ltd	Professional Industrial	5551
W PC-F/B	Terminix Peter Cox Ltd	Amateur Professional	3739
W PC-XJ/2	Terminix Peter Cox Ltd	Amateur Professional	3737
Protim 260 F	Protim Solignum Ltd	Amateur Professional Industrial	4998
D Protim Injection Fluid	Fosroc Ltd	Amateur	4204
Safeguard Deepwood Fungicide	Safeguard Chemicals Ltd	Amateur Professional	4183
Sandhill Wood Preservative Clear	Sandhill Building Products	Amateur Professional	5951
W Sovereign AQ/FT	Sovereign Chemical Industries Ltd	Professional Industrial	5433

Product Name	Marketing Company	Use	HSE No.

729 Tri(hexylene Glycol)biborate—continued

Product Name	Marketing Company	Use	HSE No.
Sovereign Coloured Wood Preservative	Sovereign Chemical Industries Ltd	Professional Industrial	6165
Sovereign Dipcoat	Sovereign Chemical Industries Ltd	Professional Industrial	5783
Sovereign Timber Preservative	Sovereign Chemical Industries Ltd	Professional	3807
Timber Preservative With Repellent	Antel UK Ltd	Professional	4848

730 Tri(hexylene glycol)biborate + Copper naphthenate

Product Name	Marketing Company	Use	HSE No.
BCR Green Wood Preserver	Building Chemical Research Ltd	Amateur Professional Industrial	6285
Dark Green Wood Preservative	Palace Chemicals Ltd	Amateur Professional Industrial	5243
Kingfisher Wood Preservative	Kingfisher Chemicals Ltd	Professional	4011
Langlow Pale Green Wood Preserver	Langlow Products Ltd	Amateur Professional Industrial	6068
Pale Green Wood Preservative	Palace Chemicals Ltd	Amateur Professional Industrial	5241
Sovereign Timber Preservative Pale Green	Sovereign Chemical Industries Ltd	Professional	3806

731 Tri(hexylene glycol)biborate + Cypermethrin

Product Name	Marketing Company	Use	HSE No.
BCR Universal Wood Preserver	Building Chemical Research Ltd	Amateur Professional	6290
Cementone Multiplus	Philip Johnstone Group Ltd	Amateur Professional	3849
Crown Fungicide Insecticide	Crown Chemicals Ltd	Professional	6003
Crown Fungicide Insecticide Concentrate	Crown Chemicals Ltd	Professional	5463
Crown Micro Dual Purpose	Crown Chemicals Ltd	Professional	6414
Devatern Wood Preserver	Sorex Ltd	Professional Industrial	4073
Green Range Dual Purpose AQ	Philip Johnstone Group Ltd	Professional	3725
Green Range Wykamol Plus	Philip Johnstone Group Ltd	Professional	3731
Killgerm Wood Protector	Killgerm Chemicals Ltd	Professional	3127
Langlow Clear Wood Preserver	Langlow Products Ltd	Amateur Professional Industrial	3802
Nubex Emulsion Concentrate CB (Low Odour)	Philip Johnstone Group Ltd	Professional	5328
Nubex Woodworm All Purpose CB	Philip Johnstone Group Ltd	Professional	3189

Product Name	Marketing Company	Use	HSE No.

731 Tri(hexylene glycol)biborate + Cypermethrin—continued

Palace Microfine Fungicide Insecticide (Concentrate)	Palace Chemicals Ltd	Professional Industrial	5421
W PC-H/4	Terminix Peter Cox Ltd	Amateur Professional	3740
PCX-122	Terminix Peter Cox Ltd	Professional	3999
PCX-122 Concentrate	Terminix Peter Cox Ltd	Professional	4000

732 Tri(hexylene glycol)biborate + Dichlofluanid

ABL Universal Woodworm Killer DB	Advanced Bitumens Ltd	Amateur Professional Industrial	3858
Blackfriar Wood Preserver	E Parsons and Sons Ltd	Amateur Professional	6443
Blackfriars New Wood Preserver	E Parsons and Sons Ltd	Amateur Professional Industrial	4150
Double Action Timber Preservative for Doors	International Paint Ltd	Amateur	3707
Double Action Wood Preservative	International Paint Ltd	Amateur	3706
W Nitromors Timbrene Clear Environmental Formula	Henkel Home Improvements and Adhesive Products	Amateur Professional Industrial	4106
Premier Barrettine New Wood Preserver	Barrettine Products Ltd	Amateur Professional Industrial	4038
W Ranch Preservative	International Paint Ltd	Amateur	4085
Ranch Preservative	Plascon International	Amateur	6085
Sadolin New Base	Sadolin UK Ltd	Amateur Professional Industrial	4341
Ultrabond Woodworm Killer	Philip Johnstone Group Ltd	Amateur Professional	4344

733 Tri(hexylene glycol)biborate + Dichlofluanid + Permethrin

Blackfriar Wood Preserver WP Gold Star Clear	E Parsons and Sons Ltd	Amateur Professional	6417
Blackfriars Gold Star Clear	E Parsons and Sons Ltd	Amateur Professional Industrial	4149
Leyland Timbrene Supreme Environmental Formula	Leyland Paint Company Ltd	Amateur Professional	4800
W Nitromors Timbrene Supreme Environmental Formula	Henkel Home Improvements and Adhesive Products	Amateur Professional Industrial	4107

HSE
WOOD PRESERVATIVES

Product Name	Marketing Company	Use	HSE No.

733 Tri(hexylene glycol)biborate + Dichlofluanid + Permethrin—continued

Product Name	Marketing Company	Use	HSE No.
Premier Barrettine New Universal Fluid D	Barrettine Products Ltd	Amateur Professional Industrial	4004
Timbrene Supreme	Kalon Group Plc	Amateur Professional	4995

734 Tri(hexylene glycol)biborate + Permethrin

Product Name	Marketing Company	Use	HSE No.
Antel Dual Purpose (BP) Concentrate (Water Dilutable)	Antel UK Ltd	Professional	4046
W Barrettine New Universal Fluid	Barrettine Products Ltd	Amateur Professional Industrial	3623
BPC8-DP	Building Preservation Centre	Professional	6306
Brunol PBO	Philip Johnstone Group Ltd	Professional	4474
Brunol Special P	Philip Johnstone Group Ltd	Professional	3864
W Chel-wood Preserver/Woodworm Dry Rot Killer BP	Chelec Ltd	Amateur Professional	4897
Crown Micro Dual Purpose F/i	Crown Chemicals Ltd	Professional	6895
Deepwood Fl Dual Purpose Emulsion Concentrate	Safeguard Chemicals Ltd	Professional Industrial	5087
Dual 8	Deal Direct Ltd	Professional	6255
W Ecology Fungicide Insecticide Aqueous (Concentrate) Wood Preservative Dual Purpose	Kingfisher Chemicals Ltd	Professional	4039
Ecology Fungicide Insecticide Aqueous (Concentrate) Wood Preservative Dual Purpose	Kingfisher Chemicals Ltd	Professional	5930
W insecticide Fungicide Wood Preservative	Kingfisher Chemicals Ltd	Professional	4048
Insecticide Fungicide Wood Preservative	Kingfisher Chemicals Ltd	Professional	5933
KF-8 Dual Purpose Micro Emulsion Concentrate	Kingfisher Chemicals Ltd	Professional Industrial	6075
Kingfisher Timberpaste	Kingfisher Chemicals Ltd	Professional	6686
Larsen Wood Preservative Clear 2 and Brown 2	Larsen Manufacturing Ltd	Amateur Professional Industrial	4287
Larsen Woodworm Killer 2	Larsen Manufacturing Ltd	Amateur Professional Industrial	4285
Mar-Kil S	Marcher Chemicals Ltd	Professional Industrial	5234
Mar-Kil W	Marcher Chemicals Ltd	Professional Industrial	5219
Micron 8 Dual Purpose Emulsion Concentrate	Safeguard Chemicals Ltd	Professional Industrial	5851
Omega Plus	Restoration (UK) Ltd	Professional	6382

Product Name	Marketing Company	Use	HSE No.

734 Tri(hexylene glycol)biborate + Permethrin—continued

Product Name	Marketing Company	Use	HSE No.
Perma AQ Dual Purpose	Triton Perma Industries Ltd	Professional	5617
Permapaste PB	Triton Perma Industries Ltd	Professional	5616
Permethrin F and I Concentrate	Permagard Products Ltd	Professional	4008
Protim CB	Protim Solignum Ltd	Professional Industrial	5864
Protim CDB	Protim Solignum Ltd	Amateur Professional Industrial	4999
Protim WR 260	Protim Solignum Ltd	Amateur Professional Industrial	4997
W Remecology Spirit Based K7	Remtox (Chemicals) Ltd	Professional	4248
Restor-8-dual Purpose Concentrate	Restoration (UK) Ltd	Professional Industrial	5963
Restor-Paste	Restoration (UK) Ltd	Professional	6079
W Safeguard Deepwood II Dual Purpose Emulsion Concentrate	Safeguard Chemicals Ltd	Professional	5303
Safeguard Deepwood II Dual Purpose Emulsion Concentrate	Safeguard Chemicals Ltd	Professional	5850
Safeguard Deepwood III Timber Treatment	Safeguard Chemicals Ltd	Amateur	3937
W Safeguard Deepwood Paste	Safeguard Chemicals Ltd	Professional	4110
Sandhill Aqueous Fungicide - Insecticide	Sandhill Building Products	Professional	5950
Sandhill Fungicide - Insecticide	Sandhill Building Products	Amateur	5952
D Saturin E10	British Building and Engineering Appliances Plc	Professional	3896
D Saturin E5	British Building and Engineering Appliances Plc	Amateur	3897
Solignum Remedial Concentrate	Protim Solignum Ltd	Professional	5021
D Solignum Remedial Fluid PB	Protim Solignum Ltd	Professional	5017
Solignum Remedial PB	Protim Solignum Ltd	Amateur Professional Industrial	5607
Sovereign Aqueous Fungicide/ Insecticide 2	Sovereign Chemical Industries Ltd	Professional Industrial	4619
Sovereign Insecticide/Fungicide 2	Sovereign Chemical Industries Ltd	Professional	4649
Sprytech 2	Spry Chemicals Ltd	Professional Industrial	6500
Timber Paste	Grip-fast Systems Ltd	Professional	6409
W Trimethrin AQ Plus	Triton Chemical Manufacturing Co Ltd	Professional	3522
Trimethrin OS Plus	Triton Chemical Manufacturing Co Ltd	Amateur Professional	3521

Product Name	Marketing Company	Use	HSE No.

734 Tri(hexylene glycol)biborate + Permethrin—continued

Tripaste PB	Triton Chemical Manufacturing Co Ltd	Professional	6080
Tritec Plus	Triton Chemical Manufacturing Co Ltd	Professional	5425
W Universal Fluid (Grade PB)	Barrettine Products Ltd	Amateur Professional Industrial	5212
Universal Wood Preserver	Langlow Products Ltd	Amateur Professional Industrial	3692
Woodworm Killer	Sealocrete Pla Ltd	Amateur Professional	5028

735 Tri(hexylene glycol)biborate + Zinc naphthenate + Dichlofluanid

Weathershield Exterior Preservative Basecoat	ICI Paints	Amateur Professional	5245
Weathershield Preservative Primer	ICI Paints	Amateur Professional	5244

736 Tri(hexylene glycol)biborate + Zinc octoate + Dichlofluanid

Timber Preservative Clear TFP7	Johnstone's Paints Plc	Amateur Professional	4016

737 Tributyltin naphthenate

Celpruf Primer TN	Rentokil Initial UK Ltd	Industrial	5378
Celpruf TN	Rentokil Initial UK Ltd	Industrial	4701
Protim FDR 230	Protim Solignum Ltd	Industrial	5046
Protim JP 230	Protim Solignum Ltd	Industrial	5779
Vacsol 2652/2653 JWR	Hickson Timber Products Ltd	Industrial	4646
Vacsol 2746/2747 J	Hickson Timber Products Ltd	Industrial	4807

738 Tributyltin naphthenate + Permethrin

Celpruf TNM	Rentokil Initial UK Ltd	Industrial	4704
Celpruf TNM WR	Rentokil Initial UK Ltd	Industrial	4703
Protim 230	Protim Solignum Ltd	Industrial	5044
Protim 230 C	Protim Solignum Ltd	Industrial	5781
Protim 230 CWR	Protim Solignum Ltd	Industrial	5778
Protim 230 WR	Protim Solignum Ltd	Industrial	5045
Vacsol 2622/2623 WR	Hickson Timber Products Ltd	Industrial	4645
Vacsol 2625/2626 WR 2:1 Concentrate	Hickson Timber Products Ltd	Industrial	5329
Vacsol 2713/2714	Hickson Timber Products Ltd	Industrial	4808

739 Tributyltin oxide

W Celpruf Primer	Rentokil Initial UK Ltd	Industrial	3796
Protim 215 PP	Protim Solignum Ltd	Industrial	5034

Product Name	Marketing Company	Use	HSE No.

739 Tributyltin oxide—continued

Protim FDR 2125C	Protim Solignum Ltd	Industrial	6146
Protim FDR 215C	Protim Solignum Ltd	Industrial	6126
Protim FDR210	Protim Solignum Ltd	Industrial	5278
Protim FDR215	Protim Solignum Ltd	Industrial	6322
Protim JP 210	Protim Solignum Ltd	Industrial	5032
D Protim R Coloured	Protim Solignum Ltd	Industrial	5005
Vacsol 2243	Hickson Timber Products Ltd	Industrial	6215
Vacsol 2244	Hickson Timber Products Ltd	Industrial	6351
Vacsol J RTU	Hickson Timber Products Ltd	Industrial	3152
Vacsol JWR Concentrate	Hickson Timber Products Ltd	Industrial	3153
Vacsol JWR RTU	Hickson Timber Products Ltd	Industrial	3154
D Vacsol P Ready to Use (2334)	Hickson Timber Products Ltd	Industrial	4203
W Wood Preservative AA 155/01	Becker Acroma Kemira Ltd	Industrial	5309

740 Tributyltin oxide + Dichlofluanid

W Celpruf CP Special	Rentokil Initial UK Ltd	Industrial	4582
Hickson Woodex	Hickson Timber Products Ltd	Industrial	5405
Protim Cedar	Protim Solignum Ltd	Industrial	5042
D Protim R Coloured Plus	Fosroc Ltd	Industrial	4391
Protim Solignum Softwood Basestain CS	Protim Solignum Ltd	Industrial	5647
W Wood Preservative AA 155/00	Becker Acroma Kemira Ltd	Industrial	5308
W Wood Preservative AA 155/03	Becker Acroma Kemira Ltd	Industrial	5310
W Wood Preservative AA155	Becker Acroma Kemira Ltd	Industrial	5311

741 Tributyltin oxide + Lindane

W Cedasol 2320	Hickson Timber Products Ltd	Industrial	4365
W Dipsar	Cementone Beaver Ltd	Industrial	3604
W Hickson Timbercare WRQD	Hickson Timber Products Ltd	Industrial	3594
W Lar-Vac 400	Larsen Manufacturing Ltd	Industrial	5782
Protim 210	Protim Solignum Ltd	Industrial	5033
Protim 210 C	Protim Solignum Ltd	Industrial	5038
Protim 210 CWR	Protim Solignum Ltd	Industrial	5035
Protim 210 WR	Protim Solignum Ltd	Industrial	5039
Protim 215	Protim Solignum Ltd	Industrial	6016
Protim 23 WR	Protim Solignum Ltd	Industrial	5037
Protim 90	Protim Solignum Ltd	Industrial	5276
W Protim Paste	Protim Solignum Ltd	Professional	4979
W Vacsol MWR Concentrate 2203	Hickson Timber Products Ltd	Industrial	3170
D Vacsol P 2304 RTU	Hickson Timber Products Ltd	Industrial	4597
Vacsol WR Ready to use 2116	Hickson Timber Products Ltd	Industrial	3168

742 Tributyltin oxide + Lindane + Pentachlorophenol

W Celpruf PK	Rentokil Initial UK Ltd	Industrial	4540
W Celpruf PK WR	Rentokil Initial UK Ltd	Industrial	4538
W Lar-Vac 100	Larsen Manufacturing Ltd	Industrial	3723

Product Name	Marketing Company	Use	HSE No.

742 Tributyltin oxide + Lindane + Pentachlorophenol—continued

W Protim 80	Protim Solignum Ltd	Industrial	5029
W Protim 80 C	Protim Solignum Ltd	Industrial	5036
W Protim 80 CWR	Protim Solignum Ltd	Industrial	5031
W Protim 80 WR	Protim Solignum Ltd	Industrial	5030
W Protim Brown	Protim Solignum Ltd	Industrial	5000

743 Tributyltin oxide + Pentachlorophenol

W Celpruf JP	Rentokil Initial UK Ltd	Industrial	4548
W Celpruf JP WR	Rentokil Initial UK Ltd	Industrial	4543
W Protim FDR-H	Protim Solignum Ltd	Industrial	4985
D Protim Joinery Lining	Protim Solignum Ltd	Professional	4977
W Protim JP	Protim Solignum Ltd	Industrial	4978
W Protim R Clear	Protim Solignum Ltd	Industrial	5013

744 Zinc naphthenate

Carbo Wood Preservative	Talke Chemical Company Ltd	Amateur Professional Industrial	3821
Spencer Wood Preservative XP	Spencer Coatings Ltd	Amateur Professional	3947
Teamac Wood Preservative	Teal and Mackrill Ltd	Amateur Professional	4428
Wood Preservative	C W Wastnage Ltd	Amateur Professional Industrial	5284

745 Zinc naphthenate + Dichlofluanid

Bradite Timber Preservative	Bradite Ltd	Amateur	6091
Masterstroke Wood Preserver	Akzo Nobel Decorative Coatings	Amateur	5689
Sikkens Wood Preserver	Akzo Nobel Woodcare	Amateur Professional	6365

746 Zinc naphthenate + Dichlofluanid + Tri(hexylene glycol)biborate

Weathershield Exterior Preservative Basecoat	ICI Paints	Amateur Professional	5245
Weathershield Preservative Primer	ICI Paints	Amateur Professional	5244

747 Zinc octoate

B and Q Clear Wood Preserver	B & Q Plc	Amateur	3013
B and Q Exterior Wood Preservative	B & Q Plc	Amateur	4705
B and Q Formula Wood Preserver	B & Q Plc	Amateur	3014
D Co-op Exterior Wood Preserver	Co-operative Wholesale Society Ltd	Amateur	3640

Product Name	Marketing Company	Use	HSE No.
747 Zinc octoate—continued			
Crown Timber Preservative	Crown Chemicals Ltd	Professional	5442
Homebase Weathercoat Preservative Primer	Sainsbury's Homebase House and Garden Centres	Amateur Professional	5902
Johnstone's Preservative Clear	Kalon Group Plc	Amateur Professional	6319
Leyland Timbrene Environmental Formula	Leyland Paint Company Ltd	Amateur Professional	4801
Leyland Universal Preservative Base	Kalon Group Plc	Amateur Professional	3760
W Nitromors Timbrene Environmental Formula Wood Preservative	Henkel Home Improvements and Adhesive Products	Amateur Professional Industrial	4108
PLA Wood Preserver	Sealocrete PLA Ltd	Amateur Professional Industrial	5340
Premier Pro-Tec(s) Environmental Formula Wood Preservative	Premier Decorative Products	Amateur Professional Industrial	4689
Rentokil Dry Rot and Wet Rot Treatment	Rentokil Initial UK Ltd	Amateur Amateur (Surface Biocide) Professional Professional (Surface Biocide)	4378
Ronseal Low Odour Wood Preserver	Ronseal Ltd	Amateur	4919
Ronseal's All-purpose Wood Preserver	Ronseal Ltd	Amateur Professional	6185
Square Deal Deep Protection Wood Preserver	Texas Homecare Ltd	Amateur Professional	4640
Thompson's All Purpose Wood Preserver	Ronseal Ltd	Amateur Professional	5988
Timber Preservative	Antel UK Ltd	Professional	4846
Timbrene Wood Preserver	Kalon Group Plc	Amateur Professional	4966
Trade Ronseal Low Odour Wood Preserver	Ronseal Ltd	Professional Industrial	4932
748 Zinc octoate + Boric acid			
W Celpruf BZ	Rentokil Initial UK Ltd	Industrial	4188
W Celpruf BZ WR	Rentokil Initial UK Ltd	Industrial	4189
Rentokil Dry Rot Paste (D)	Rentokil Initial UK Ltd	Professional	3987

Product Name	Marketing Company	Use	HSE No.

749 Zinc octoate + Boric acid + Permethrin

W Celpruf BZP	Rentokil Initial UK Ltd	Industrial	4190
W Celpruf BZP WR	Rentokil Initial UK Ltd	Industrial	4191
Cut'n'treat	Rentokil Initial UK Ltd	Amateur Professional	5802
Premium Grade Wood Treatment	Rentokil Initial UK Ltd	Amateur Professional	4193

750 Zinc octoate + Copper naphthenate

Weathershield Exterior Timber Preservative	ICI Paints	Amateur Professional	4915

751 Zinc octoate + Dichlofluanid

Exterior Preservative Primer	Windeck Paints Ltd	Amateur	3857
Timbercare Microporous Exterior Preservative	Manders Paints Ltd	Amateur Professional	5157
W Wilko Exterior System Preservative Primer	Wilkinson Home and Garden Stores	Amateur	3686

752 Zinc octoate + Dichlofluanid + Permethrin

OS Color Wood Stain and Preservative	Ostermann and Scheiwe GmbH	Amateur	5531

753 Zinc octoate + Dichlofluanid + Tri(hexylene glycol)biborate

Timber Preservative Clear TFP7	Johnstone's Paints Plc	Amateur Professional	4016

754 Zinc octoate + Permethrin

D Cuprinol Difusol S	Cuprinol Ltd	Professional	3750
Green Range Timber Treatment Paste	Philip Johnstone Group Ltd	Professional	4651
Larvac 300	Larsen Manufacturing Ltd	Industrial	4461
Palace Timbertreat Ecology	Palace Chemicals Ltd	Professional	4931
Premium Wood Treatment	Rentokil Initial UK Ltd	Amateur Amateur (Surface Biocide) Professional Professional (Surface Biocide)	5350
W Rentokil Dual Purpose Fluid	Rentokil Initial UK Ltd	Professional	4074
Rentokil Wood Preservative	Rentokil Initial UK Ltd	Amateur Professional	4195
Ronseal's Universal Wood Treatment	Ronseal Ltd	Amateur Professional	6186

Product Name	Marketing Company	Use	HSE No.

754 Zinc octoate + Permethrin—continued

Product Name	Marketing Company	Use	HSE No.
Solignum Colourless	Protim Solignum Ltd	Amateur Professional Industrial	5549
Solignum Wood Preservative Paste	Protim Solignum Ltd	Professional	5001
Thompson's Universal Wood Treatment	Ronseal Ltd	Amateur Professional	5986
Wood Preservative Clear	Rentokil Initial UK Ltd	Amateur Professional	4077
Woodtreat BP	Philip Johnstone Group Ltd	Professional	3085

755 Zinc versatate

Product Name	Marketing Company	Use	HSE No.
Celpruf Primer ZV	Rentokil Initial UK Ltd	Professional Industrial	5379
Celpruf ZO	Rentokil Initial UK Ltd	Industrial	4683
Celpruf ZO WR	Rentokil Initial UK Ltd	Industrial	4684
Protim FDR 220	Protim Solignum Ltd	Industrial	5777
Protim JP 220	Protim Solignum Ltd	Industrial	5197
Protim Paste 220 F	Protim Solignum Ltd	Professional Industrial	4988
Vacsol J (2:1 Conc) 2744/2745	Hickson Timber Products Ltd	Industrial	4687
Vacsol J 2742/2743	Hickson Timber Products Ltd	Industrial	4686
Vacsol JWR (2:1 Conc) 2642/2643	Hickson Timber Products Ltd	Industrial	4677
Vacsol JWR 2640/2641	Hickson Timber Products Ltd	Industrial	4680
W Vacsol P Ready to use (2335)	Hickson Timber Products Ltd	Industrial	4116

756 Zinc versatate + Cypermethrin

Product Name	Marketing Company	Use	HSE No.
Dipsar G R	Philip Johnstone Group Ltd	Industrial	4449
Everclear	Injecta Osmose Ltd	Industrial	6058

757 Zinc versatate + Permethrin

Product Name	Marketing Company	Use	HSE No.
Cedasol Ready to use (2306)	Hickson Timber Products Ltd	Industrial	4117
Celpruf ZOP	Rentokil Initial UK Ltd	Industrial	4681
Celpruf ZOP WR	Rentokil Initial UK Ltd	Industrial	4682
W Imersol 2410	Hickson Timber Products Ltd	Industrial	4674
Protim 220	Protim Solignum Ltd	Industrial	5198
Protim 220 CWR	Protim Solignum Ltd	Amateur Professional Industrial	5196
Protim 220 WR	Protim Solignum Ltd	Industrial	5195
Protim 220C	Protim Solignum Ltd	Industrial	5780
Protim Paste 220	Protim Solignum Ltd	Professional Industrial	4983
D Protim WR 220	Fosroc Ltd	Amateur	4113
Solignum Universal	Protim Solignum Ltd	Amateur Professional Industrial	5019

Product Name	Marketing Company	Use	HSE No.

757 Zinc versatate + Permethrin—continued

W Vacsele P2312	Hickson Timber Products Ltd	Professional Industrial	4363
Vacsol (2:1 Conc) 2711/2712	Hickson Timber Products Ltd	Industrial	4685
Vacsol 2709/2710	Hickson Timber Products Ltd	Industrial	4678
D Vacsol P (2310)	Hickson Timber Products Ltd	Industrial	4377
Vacsol WR (2:1 Conc) 2614/2615	Hickson Timber Products Ltd	Industrial	4679
Vacsol WR 2612/2613	Hickson Timber Products Ltd	Industrial	4676

7
WOOD TREATMENT

Wood Treatment

758 Boric acid + Sodium 2,4,6-trichlorophenoxide

Product Name	Marketing Company	Use	HSE No.
Protim Fentex T	Protim Solignum Ltd	Industrial	6367
Protim Fentex TG	Protim Solignum Ltd	Industrial	6440
Protim Fentex TP	Protim Solignum Ltd	Industrial	6370
Protim Fentex TPS	Protim Solignum Ltd	Industrial	6371
Protim Fentex TW	Protim Solignum Ltd	Industrial	6368
Protim Fentex TW	Protim Solignum Ltd	Industrial	6876

759 Carbendazim + Dichlorophenyl dimethylurea + 2-Octyl-2h-isothiazolin-3-one

Product Name	Marketing Company	Use	HSE No.
Cuprinol Ducksback	Cuprinol Ltd	Amateur Professional	6327
Cuprinol Garden Decking Seal	Cuprinol Ltd	Amateur Professional	6858
Cuprinol Garden Shades	Cuprinol Ltd	Amateur Professional	6326

760 Dichlorophenyl dimethylurea + 2-octyl-2h-isothiazolin-3-one + Carbendazim

Product Name	Marketing Company	Use	HSE No.
Cuprinol Ducksback	Cuprinol Ltd	Amateur Professional	6327
Cuprinol Garden Decking Seal	Cuprinol Ltd	Amateur Professional	6858
Cuprinol Garden Shades	Cuprinol Ltd	Amateur Professional	6326

761 3-iodo-2-propynyl-n-butyl carbamate

Product Name	Marketing Company	Use	HSE No.
ACS Dry Rot Paint	Advanced Chemical Specialities Ltd	Amateur Amateur (Biocidal Paint) Professional Professional (Biocidal Paint)	6310
Lectros Dry Rot Paint	Advanced Chemical Specialities Ltd	Amateur Amateur (Biocidal Paint) Professional Professional (Biocidal Paint)	6520

Product Name	Marketing Company	Use	HSE No.

762 2-octyl-2h-isothiazolin-3-one + Carbendazim + Dichlorophenyl dimethylurea

Cuprinol Ducksback	Cuprinol Ltd	Amateur Professional	6327
Cuprinol Garden Decking Seal	Cuprinol Ltd	Amateur Professional	6858
Cuprinol Garden Shades	Cuprinol Ltd	Amateur Professional	6326

763 Sodium 2,4,6-trichlorophenoxide + Boric acid

Protim Fentex T	Protim Solignum Ltd	Industrial	6367
Protim Fentex TG	Protim Solignum Ltd	Industrial	6440
Protim Fentex TP	Protim Solignum Ltd	Industrial	6370
Protim Fentex TPS	Protim Solignum Ltd	Industrial	6371
Protim Fentex TW	Protim Solignum Ltd	Industrial	6368
Protim Fentex TW	Protim Solignum Ltd	Industrial	6876

HSE
WOOD TREATMENT

8

HSE PRODUCT NAME INDEX

HSE
PRODUCT INDEX

563

Micron CSC 200 Series   *16 114*
Micron CSC 300 Series   *18 79*
Micron CSC Extra   *29 99*
Micron Extra   *23 71*
Micron Optima   *46 188*
Micron Plus Antifouling   *29 99*
Microstel   *428 458*
Microtech Biocide 25   *451*
Microtech Dual Purpose AQ   *565 617*
Microtech Woodworm Killer 25X   *644*
Microtech Woodworm Killer aq   *563*
Microtreat Dual   *620 646*
Microtreat Insecticide   *644*
Middletons Creosote   *535*
Middletons Enviro. Shed & Fence Wood
   Preservative   *535*
Miricoat A.F. Coating   *4*
Mite-Kill   *331*
Mitrol PQ 8   *638*
Momar Fly Killer   *391 409*
Morgan Ant & Crawling Insect Killer   *276*
Mosgo P   *465 707*
Mosiguard Shield   *331*
Moskil Mosquito Repellent   *228*
Mosqui-Go Electric   *248*
Mosqui-Go Liquid   *366*
Mosqui-Go Tablets   *366*
Mosquito Killer Travel Pack   *248*
Mosquito Repellent   *248*
Mosscheck   *441*
Mostyn Chlorpyrifos 25 EC   *267*
Moth Repellent   *221*
Mothaks   *222 224*
Motox   *304 351 403*
Motox Crawling Insect   *305 402*
Motto   *307*
Mould Cleaner   *422*
Mould Killer   *427*
Mouldcheck Barrier 2   *441*
Mouldcheck Spray   *443*
Mouldcheck Sterilizer   *443*
Mouldcheck Sterilizer 2   *428 458*
Mouldkill   *441*
Mouldrid   *449 450 607 609*
MPX   *54 120*
Mrs Clear Mould Fungicidal Wash   *449 450*
Multispray   *273 341 375*
Murphy Fly and Wasp Killer   *394*
Murphy Foaming Wasp Nest Destroyer   *360 417*
Murphy Kil-Ant Powder   *242*
Murphy Kil-Ant Ready To Use   *276*
Murphy Kil-Ant Trap   *421*
Murphys Long Lasting Fly and Wasp Killer   *293*
Muscatrol   *331*
Mystox BTL   *628 643*
Mystox BTV   *628 643*

Natural Ant Gun   *365*
Natural Wasp Gun   *365*
Natural Wasp Killer   *365*
Neopybuthrin Premium   *257 345*
Neopybuthrin Super   *257 345*
Net-Guard   *23 71*
Netrex AF   *11*
New Cut 'N' Spray   *497*
New Formula Cedarwood   *581*
New Secto Household Flea Spray   *360 417*
New Siege II   *276*
New Super Raid Insecticide   *234 353 416*
New Tetracide   *269 281 374*
New Vapona Fly and Wasp Killer   *233 356*
New Vapona Fly Killer Dry Formulation   *235 412*
Nexa Cockroach Bait Station   *306*
Nipacide DP30   *443*
Nippon Ant and Crawling Insect Killer   *349 406*
Nippon Ant Killer Powder   *331*
Nippon Fly Killer Pads   *256 332*
Nippon Fly Killer Spray   *349 406*
Nippon Flying Insect Killer Tablets   *248*
Nippon Killaquer Crawling Insect Killer   *349 406*
Nippon Ready For Use Fly Killer Spray   *236 361*
Nippon Woodlice Killer Powder   *331*
Nisa Fly and Wasp Killer   *349 406*
Nitromors Creosote   *561*
Nitromors Mould Remover   *428 458*
Nitromors Timbrene Clear Environmental
   Formula   *588 732*
Nitromors Timbrene Environmental Formula Wood
   Preservative   *747*
Nitromors Timbrene Supreme Environmental
   Formula   *586 666 733*
No Fleas On Me   *331*
No Foul-Zo   *181*
Noa-Noa Rame   *11 29 99*
Nomad Residex P   *331*
Nordrift Antifouling   *11*
Norimp 2000 Black   *11*
Norshield 150   *320 336*
Nu Wave A/F   *44 167 196*
Nu Wave A/F Flat Bottom   *36 108 155*
Nu wave A/F Vertical Bottom   *36 108 155*
Nubex Emulsion Concentrate C (Low Odour)   *563*
Nubex Emulsion Concentrate CB (Low
   Odour)   *570 731*
Nubex WDR   *449 450 607 609*
Nubex Woodworm All Purpose CB   *570 731*
Nuvan Staykil   *295 313*
Nuvanol N 500 SC   *312*
Nylar 10 EC   *385*
Nylar 100   *385*
Nylar 4EW   *385*
Nylar ULV 65   *359 388*
Oakmere Creosote Type 2   *561*

HSE PRODUCT INDEX

Odex Fly Spray   *394*
Omega   *644*
Omega Biocide   *451 613*
Omega Plus   *665 734*
One Shot Space Spray   *331*
One-Shot Aircraft Aerosol Insect Control   *357*
OS Color Wood Stain and Preservative   *587 670 752*
Osmose ABS33a   *525*
Osmose Endcoat Brown   *613*
Osmose Endcoat Clear   *613*
Osmose Endcoat Green   *541 616*
Osmose K33 C58.7   *479 532 546*
Osmose K55   *531 550*
Outright Household Flea Spray   *331*
Ovoline 275 Golden Creosote   *535*
P.J. Fungicidal Solution   *428 458*
Palace Base Coat Wood Preservative   *729*
Palace Dual Purpose FI8 (Microemulsion)   *655 689*
Palace Fungicidal Wash   *427 464*
Palace FWS (Microemulsion)   *451*
Palace Microfine 8 -Insecticide Concentrate   *644*
Palace Microfine Fungicide Concentrate   *729*
Palace Microfine Fungicide Insecticide (Concentrate)   *570 731*
Palace Microfine Insecticide Concentrate   *563*
Palace Mould Remover   *449 450*
Palace Timbertreat Ecology   *668 754*
Palace Woodworm Killer 8 (Microemulsion)   *644*
Pale Green Wood Preservative   *544 730*
Panacide M21   *443*
Panaclean 736   *443*
Pandrol Timbershield Rods   *594*
Paneltone   *724*
Paragon Formula 10 Cockroach Bait   *265*
Paramos   *427*
Patente Laxe   *16 114*
Path and Patio Cleaner   *427*
Patriot Flying and Crawling Insect Killer   *369*
Patriot Flying and Crawling Insect Killer Aerosol   *369*
PC Insect Killer   *256 332*
PC-D   *462*
PC-D Concentrate   *462*
PC-F/B   *729*
PC-H/3   *563*
PC-H/4   *570 731*
PC-K   *516 606*
PC-XJ/2   *729*
PCX-12/P   *563*
PCX-12/P Concentrate   *563*
PCX-122   *570 731*
PCX-122 Concentrate   *570 731*
Pedigree Household Anti Flea Spray   *325 379*

Pedigree and Whiskas Exelpet Household Anti Flea Spray   *325 379*
Pedigree and Whiskas Exelpet Household Flea Spray   *320 336*
Pedigree Bedding Spray   *236 361*
Pedigree Exelpet Anti-Flea Carpet Powder   *331*
Pedigree Exelpet Household Spray   *320 336*
Penetone Fly Killer   *391 409*
Penguin Aqualine   *132*
Penguin Non-Stop   *23 71*
Penguin Non-Stop White   *56 72*
Penguin Racing   *11*
Penguin Racing White   *56 72*
Peripel   *225*
Peripel 10   *331*
Peripel 55   *331*
Perma AQ Dual Purpose   *665 734*
Permabor Boracol 10RH   *431 447*
Permabor Boracol 20   *594*
Permabor Boracol 40 Paste   *594*
Permacide Masonry Fungicide   *428 458*
Permadex Masonry Fungicide Concentrate   *449 450*
Permadip 9 Concentrate   *594*
Permalene Satin Wood Stain   *581*
Permapaste PB   *665 734*
Permarock Fungicidal Wash   *428 458*
Permasect 0.5 Dust   *331*
Permasect 10 WP   *331*
Permasect Powder   *331*
Permethrin   *331*
Permethrin Dusting Powder   *331*
Permethrin F and I Concentrate   *665 734*
Permethrin PH - 25WP   *331*
Permethrin Wettable Powder   *331*
Permethrin WW Conc   *644*
Permoglaze Micatex Fungicidal Treatment   *449 450 607 609*
Permost 0.5% Dust Powder   *331*
Permost 25 WP   *331*
Permost UNI   *349 406*
Pesguard NS 6/14 EC   *360 417*
Pesguard OBA F 7305 C   *360 417*
Pesguard OBA F 7305 D   *360 417*
Pesguard OBA F 7305 E   *360 417*
Pesguard OBA F 7305 F   *360 417*
Pesguard OBA F 7305 G   *360 417*
Pesguard OBA F7305 B   *360 417*
Pesguard PS 102   *236 361*
Pesguard PS 102A   *236 361*
Pesguard PS 102B   *236 361*
Pesguard PS 102C   *236 361*
Pesguard PS 102D   *236 361*
Pesguard WBA F-2656   *360 417*
Pesguard WBA F-2692   *360 417*
Pesguard WBA F2714   *362 418*

Protim Brown CDB   *473 648*
Protim CB   *665 734*
Protim CCA Oxide 50   *479 532 546*
Protim CCA Oxide 58   *479 532 546*
Protim CCA Oxide 72   *479 532 546*
Protim CCA Salts Type 2   *480 557 700*
Protim CDB   *665 734*
Protim Cedar   *591 740*
Protim E410   *603 661 696 719*
Protim E410(i)   *603 661 696 719*
Protim E415   *603 661 696 719*
Protim E415 (i)   *603 661 696 719*
Protim E470   *513 576*
Protim E480   *594*
Protim FDR 2125C   *739*
Protim FDR 215C   *739*
Protim FDR 220   *755*
Protim FDR 230   *737*
Protim FDR 250   *613*
Protim FDR 418V   *690 723*
Protim FDR-H   *641 743*
Protim FDR210   *739*
Protim FDR215   *739*
Protim FDR418   *690 723*
Protim FDR418J   *624 691 711*
Protim Fentex Europa 1 RFU   *498 725*
Protim Fentex Europa I   *498 725*
Protim Fentex Green Concentrate   *605 708*
Protim Fentex Green RFU   *605 708*
Protim Fentex I   *500 615*
Protim Fentex M   *605 708*
Protim Fentex P   *499 674*
Protim Fentex T   *758 763*
Protim Fentex TG   *758 763*
Protim Fentex TP   *758 763*
Protim Fentex TPS   *758 763*
Protim Fentex TW   *758 763*
Protim Fentex TWR Plus   *522 698*
Protim Fentex WR   *707*
Protim GC   *639*
Protim Green E   *543 653*
Protim Green WR   *540*
Protim Injection Fluid   *729*
Protim Insecticidal AQ8   *644*
Protim Insecticidal Emulsion 8   *644*
Protim Insecticidal Emulsion P   *644*
Protim Joinery Lining   *641 743*
Protim Joinery Lining 280   *672*
Protim JP   *641 743*
Protim JP 210   *739*
Protim JP 220   *755*
Protim JP 230   *737*
Protim JP 250   *613*
Protim JP 800   *470*
Protim Kleen II   *605 708*

Protim Panelguard   *707*
Protim Paste   *629 741*
Protim Paste 220   *671 757*
Protim Paste 220 F   *755*
Protim Paste 800   *473 648*
Protim Paste 800 F   *470*
Protim Paste P   *644*
Protim Plug Compound   *465 707*
Protim R Clear   *641 743*
Protim R Coloured   *739*
Protim R Coloured Plus   *591 740*
Protim Solignum D.O.T.   *594*
Protim Solignum Softwood Basestain CS   *591 740*
Protim Stainguard   *633 728*
Protim Universal AQ 250   *620 646*
Protim Wall Solution Concentrate 5   *462*
Protim Wall Solution II   *449 450*
Protim Wall Solution II Concentrate   *449 450*
Protim Woodworm Killer C   *563*
Protim Woodworm Killer P   *644*
Protim WR 220   *671 757*
Protim WR 260   *665 734*
Protim WR 800   *473 648*
Protrol   *310*
Protrol 9%   *311*
Provence Fly and Wasp Killer   *394*
Puma Antifouling   *11*
Pure Zap   *276*
Purge   *257 345*
Purge Insect Destroyer   *349 406*
Pybuthrin 2/16   *369*
Pybuthrin 33   *369*
Pybuthrin 33 BB   *250 263*
Pybuthrin Fly Killer   *256 332*
Pybuthrin Fly Spray.   *256 332*
Pynamin Forte 40MG Mat   *228*
Pynamin Forte Mat 120   *228*
Pynosect 10   *349 406*
Pynosect 2   *391 409*
Pynosect 4   *391 409*
Pynosect 6   *331*
Pynosect Extra Fog   *349 406*
Pynosect PCO   *331*
Pynosect PCP   *331*
Pynosect Powder   *331*
Pyrakill Flying Insect Spray   *394*
Pyrasol C RTU   *276*
Pyrasol CP   *276*
Pyrbek   *358 384*
Pyrematic Flying Insect Killer   *369*
Pyro-Black   *594*
Pyro-Red   *497*
Pyrolith 3505 Ready To Use   *497*
Quantum Hygiene Fly Killer   *391 409*
Quicksilver Antifouling   *54 120*

R.I.P. Fleas   *255 322 334*
Rabamarine A/F 2500   *44 167 196*
Rabamarine A/F No 2500 HS   *36 108 155*
Rabamarine A/F No 2500M HS   *36 108 155*
Raffaello 3   *30 88 144*
Raffaello Alloy   *54 120*
Raffaello Plus Tin Free   *16 114*
Raffaello Racing   *61 89 145*
Raid Ant and Roach Killer   *349 406*
Raid Ant Bait   *267*
Raid Ant Killer Powder   *331*
Raid Cockroach Killer Formula 2   *280 376*
Raid Dry Natural Flying Insect Killer   *369*
Raid Flea Killer Plus   *326 383*
Raid Fly and Wasp Killer   *235 412*
Raid Fly Wasp and Plant Insect Killer   *360 417*
Raid Flyguard   *240*
Raid Flying Insect Killer   *394*
Raid Mothproofer   *225*
Raid Outdoor Insectguard   *233 356*
Raid Residual Crawling Insect Killer   *282 398*
Raid Wasp Nest Destroyer   *271 397*
Ranch Preservative   *588 732*
Rapid Exit Fly Spray   *391 409*
Ravax AF   *11*
Ready Kill   *276*
Red Can Fly Killer   *394*
Regency Garden Buildings Wood Preserver   *581*
Reldan 50 EC   *272*
Remecology Insecticide R8   *644*
Remecology Spirit Based Fungicide R5   *613*
Remecology Spirit Based Insecticide R6   *644*
Remecology Spirit Based K7   *665 734*
Remtox AQ Fungicide R7   *613*
Remtox AQ Fungicide/Insecticide R9   *620 646*
Remtox Borocol 10 RH Masonry Biocide   *424 446*
Remtox Borocol 20 Wood Preservative   *594*
Remtox Dry Rot F. W. S.   *451*
Remtox Dry Rot Paint   *451 613*
Remtox Dual Purpose F/1 K7   *612 619 682*
Remtox Dual Purpose Paste K9   *620 646*
Remtox Fungicidal Wall Solution RS   *451*
Remtox Fungicide Microemulsion M7   *613*
Remtox Fungicide Paste K6   *613*
Remtox Fungicide/Insecticide Microemulsion
  M9   *620 646*
Remtox FWS (Low Odour)   *462*
Remtox Insecticide Microemulsion M8   *644*
Remtox Insecticide Paste R4   *644*
Remtox M9 Dual Purpose Super
  Concentrate   *612 619 682*
Remtox Microactive Fungicide W7   *613*
Remtox Microactive FWS W6   *451*
Remtox Microactive Insecticide W8   *644*
Remtox Remwash Extra   *427*
Remtox Spirit Based F/I K7   *620 646*

Remtox-Silexine B40 Paste Wood
  Preservative   *594*
Remtox-Silexine Boron Rods   *594*
Remtox-Silexine Scrub Out Black Mould/
  Refill   *427*
Remwash Masonry Steriliser   *441*
Rentex   *478 705 706*
Rentokil ACB   *509 554*
Rentokil Ant and Insect Killer   *242*
Rentokil Ant and Insect Powder Professional   *242*
Rentokil Chlorpyrifos Gel   *267*
Rentokil Classic Wax Polish   *644*
Rentokil Control Paste SB   *516 606*
Rentokil Dry Rot and Wet Rot Treatment   *466 747*
Rentokil Dry Rot Fluid (E) for Bonded
  Warehouses   *433 497*
Rentokil Dry Rot Paste (D)   *523 748*
Rentokil Dual Purpose Fluid   *668 754*
Rentokil Flying Insect Killer   *248*
Rentokil Houseplant Insect Killer   *382 390*
Rentokil Iodofenphos Gel   *312*
Rentokil Mould Cure Spray   *430 438 442*
Rentokil Total Wood Treatment   *436 440 460 512*
  568 686
Rentokil Wasp Killer   *242*
Rentokil Wasp Nest Destroyer   *360 417*
Rentokil Wood Preservative   *668 754*
Rentokil Wood Preserver   *497*
Rentokil Woodworm Killer   *644*
Rentokil Woodworm Treatment   *644*
Repel Clothing Spray   *226*
Residex   *369*
Residex P   *331*
Residual Powerkill   *331*
Resigen   *257 345*
Reslin Premium   *257 345*
Reslin Super   *257 345*
Restor F.W.S   *462*
Restor-8-Dual Purpose Concentrate   *665 734*
Restor-8-Insecticide Concentrate   *644*
Restor-MFC   *451 613*
Restor-Paste   *665 734*
Rhodaquat RP50   *427 485*
Rid Insect Powder   *393*
Rid Moss   *427*
Rid-Ant   *276*
Rlt Bactdet   *463*
Rlt Clearmould Spray   *441*
Rlt Halophen   *441*
Rlt Halophen DS   *441*
Roachbuster   *265*
Rodewod 50 SL   *481*
Roebuck Eyot Aircraft Disinfection Spray   *357*
Ronseal Low Odour Wood Preserver   *747*
Ronseal Wet Rot Prevention Tablets   *594*
Ronseal Wood Preservative Tablets   *594*

Shelltox Flying Insect Killer 3   *360 417*
Shelltox Flykiller   *300 400*
Shelltox Insect Killer   *349 406*
Shelltox Insect Killer 2   *349 406*
Shelltox Mat 1   *248*
Shelltox Mat 2   *228*
Shelltox Super Flying Insect Killer   *360 417*
Sherley's Flea Busters   *265*
Sherley's Rug-De-Bug   *331*
Shiprite Sailing   *11*
Shiprite Speed   *16 114*
Siege II   *276*
Sigma Fungicidal Solution   *427*
Sigma Pilot Ecol Antifouling   *9 48 194*
Sigmalife Impregnant   *581*
Sigmaplane Ecol Antifouling   *8 17 29 99 113*
Sigmaplane Ecol HA 120 Antifouling   *32 90 119 134*
Sigmaplane HA Antifouling   *44 167 196*
Sigmaplane HB   *44 167 196*
Sigmaplane HB Antifouling   *44 167 196*
Sigmaplane TA Antifouling   *44 167 196*
Signpost Creosote   *561*
Sikkens Wood Preserver   *592 745*
Silexine Anticon   *210 215*
Silexine Fungi-Chek Emulsion   *210 215*
Silexine Fungi-Chek Oil Matt Paint   *210 215*
Sinesto B   *477*
Skeetal Flowable Concentrate   *241*
Slippy Bottom   *29 99*
Slipstream Antifouling   *23 71*
Smite   *272*
Snowcem Algicide   *427*
Soft Antifouling   *29 99*
Solignum Anti Fungi Concentrate   *462*
Solignum Anti-Fungi Concentrate 5   *462*
Solignum Colourless   *668 754*
Solignum Dark Brown   *535*
Solignum Dry Rot Killer   *449 450*
Solignum Fencing Fluid   *561*
Solignum Fungicide   *449 450*
Solignum Gold   *561*
Solignum Green   *543 653*
Solignum Insecticidal AQ8   *644*
Solignum Medium Brown   *535*
Solignum Remedial Concentrate   *665 734*
Solignum Remedial Fluid PB   *665 734*
Solignum Remedial PB   *665 734*
Solignum Universal   *671 757*
Solignum Wood Preservative Paste   *668 754*
Solignum Woodworm Killer   *644*
Solignum Woodworm Killer Concentrate   *644*
Sorex Fly Spray RTU   *360 417*
Sorex Super Fly Spray   *360 417*
Sorex Wasp Nest Destroyer   *391 409*
Southdown Creosote   *561*

Sov AQ Micro F   *613*
Sov AQ Micro F/I   *620 646*
Sov AQ Micro I   *644*
Sovac F   *613*
Sovac F/I   *620 646*
Sovac F/I WR   *620 646*
Sovac FWR   *613*
Sovac I   *644*
Sovaq FLX F/I   *612 619 682*
Sovaq FLX I   *611*
Sovaq Micro F   *613*
Sovaq Micro F/I   *620 646*
Sovaq Micro FWS   *451*
Sovaq Micro I   *644*
Sovereign AQ F/I   *620 646*
Sovereign AQ/FT   *729*
Sovereign AQF   *613*
Sovereign Aqueous Fungicide/Insecticide 2   *665 734*
Sovereign Aqueous Insecticide 2   *644*
Sovereign Coloured Wood Preservative   *729*
Sovereign Coloured Wood Preservatives   *613*
Sovereign Dentolite Solution Concentrate   *427*
Sovereign Dipcoat   *729*
Sovereign Fungicidal Wall Solution   *462*
Sovereign Fungicide/Insecticide   *612 619 682*
Sovereign Green Timber Preservative/ Dipcoat   *541 616*
Sovereign Insecticide/Fungicide 2   *665 734*
Sovereign Masonry Sterilising Wash   *441*
Sovereign Sovaq Micro F/I   *620 646*
Sovereign Sovaq Micro I   *644*
Sovereign Timber Preservative   *729*
Sovereign Timber Preservative Pale Green   *544 730*
Sovereign Timbor Rod   *594*
SP.153 Sterilising Detergent Wash   *443*
SP.154 Fungicidal and Bactericidal Treatment   *443*
Speedclean Antifouling   *11*
Spencer Stormcote Fungicidal Treatment   *428 458*
Spencer Wood Preservative XP   *744*
Spira 'No-Bite' Mosquito Killer   *248*
Spira 'No-Bite' Outdoor Mosquito Coils   *227*
Spraydex Ant & Insect Killer   *256 332*
Spraydex Houseplant Spray   *256 332*
Spraydex Insect Killer   *256 332*
Sprint   *63 84 97*
Sprytech 1   *644*
Sprytech 2   *665 734*
Sprytech 3   *451*
Square Deal Deep Protection Wood Preserver   *747*
Square Deal Fungicidal Solution   *427*
St Michael Flyspray   *394*

Tiger Cruising-A Antifouling   *60 101*
Tiger Tin Free Antifouling   *23 71*
Tiger White   *54 120*
Tigerline Antifouling   *54 120*
Tigerline Boottop   *54 120*
Tim-Bor   *594*
Timber Fungicide   *680*
Timber Paste   *665 734*
Timber Preservative   *581 747*
Timber Preservative Clear TFP7   *590 736 753*
Timber Preservative Green   *540*
Timber Preservative with Repellent   *729*
Timber Treatment Plus   *680*
Timber Water Repellent   *613*
Timber-Gel Ready To Use Microemulsion Timber
    Fluid   *621 656 683*
Timbercare Microporous Exterior
    Preservative   *593 751*
Timbercol Preservative (MBT) Concentrate   *632*
Timbercol Preservative Concentrate   *678*
Timberdip   *476*
Timberguard 1 Plus 7 Concentrate   *476*
Timberlife   *581*
Timberlife Extra   *581*
Timbertek   *699*
Timbertek Conc   *699*
Timbertex PD   *602 635 727*
Timbertex PI   *594*
Timbertex PI/2   *699*
Timbertone T10 Preservative   *632*
Timbor   *594*
Timbor Paste   *594*
Timbrene Green Wood Preserver   *540*
Timbrene Supreme   *586 666 733*
Timbrene Wood Preserver   *747*
Timbrene Woodworm Killer   *644*
Titan FGA   *29 99*
Titan FGA Antifouling   *16 114*
Titan FGA Antifouling White   *54 120*
Titan FGA-C Antifouling   *67 189*
Titan FGA-E Antifouling   *18 79*
Titan Tin Free   *15 148*
Titan Tin Free Antifouling   *11*
Top Kill Insect Killer   *305 402*
Torkill Fungicidal Solution 'W'   *449 450*
Total Home Guard   *321 348*
Tourist Flying Insect Killer   *366*
Tox Exterminating Fly and Wasp Killer   *305 402*
Trade Ronseal Low Odour Wood Preserver   *747*
Traditional Preservative Oil   *580 657 688*
Trappit Ant Powder   *369*
Travis Perkins Creosote   *561*
Travis Perkins Fungicidal Wash   *427*
Trawler   *11*
Trawler Tin-Free   *127 150*
Tribor 10   *431 447*

Tribor 20   *594*
Tribor Gel   *594*
Triflex Sterilising Solution Concentrate   *427*
Trilanco Farm Fly Spray   *369*
Trilanco Fly Killer   *391 409*
Trilanco Fly Spray   *250 263*
Trilanco Super Fly Spray   *369*
Trilux   *56 72*
Trimethrin 2OS   *644*
Trimethrin 2AQ   *644*
Trimethrin 3AQ   *644*
Trimethrin 3OS   *644*
Trimethrin AQ Plus   *665 734*
Trimethrin OS Plus   *665 734*
Tripaste   *620 646*
Tripaste PB   *665 734*
Tripaste PP   *620 646*
Trisol 120   *680*
Trisol 21   *462*
Trisol 22   *459*
Trisol 23   *451*
Tritec   *644*
Tritec 120   *644*
Tritec 120 Plus   *655 689*
Tritec 121   *644*
Tritec 121 Plus   *655 689*
Tritec 8   *644*
Tritec Plus   *665 734*
Triton 120   *644*
Triton Timber Rot Killer   *680*
Triton Woodworm Killer   *644*
Tropical Super Service Antifouling Paint   *11*
TS 10240 Antifouling ADA160 Series   *11*
Turbair Beetle Killer   *303 344 389*
Tymasil   *246 329*
Tyrax   *258 340 372*
UKC-2 Dry Rot Killer   *451*
UKC-2 Dry Rot Killer Concentrate   *451 613*
UKC-8 Woodworm Killer Concentrate   *644*
UKC-8 Woodworm Killer Microemulsion
    Concentrate   *644*
Ultra Tech 2000 D   *655 689*
Ultra Tech 2000 F   *680*
Ultra Tech 2000 I   *644*
Ultra Tech 2000M   *451*
Ultrabond Woodworm Killer   *588 732*
Unisil S Silicone Waterproofing Solution   *449 450*
Unitas Antifouling Paint Black   *20 50*
Unitas Antifouling Paint Chocolate   *11*
Unitas Antifouling Paint Ted   *11*
Unitas Antifouling Paint White   *51*
Unitas Fungicidal Wash-Exterior (Solvent
    Based)   *449 450*
Unitas Fungicidal Wash-Interior (Water
    Based)   *449 450*
Universal Fluid (Grade PB)   *665 734*

HSE
PRODUCT INDEX

Universal Wood Preservative   *566 584*
Universal Wood Preserver   *665 734*
Vacsele Aqua 6106   *655 689*
Vacsele P2312   *671 757*
Vacsol (2:1 Conc) 2711/2712   *671 757*
Vacsol 2243   *739*
Vacsol 2244   *739*
Vacsol 2622/2623 WR   *667 738*
Vacsol 2625/2626 WR 2:1 Concentrate   *667 738*
Vacsol 2627 Azure   *658 697 722*
Vacsol 2652/2653 JWR   *737*
Vacsol 2658 JWR 2:1 Conc.   *624 691 711*
Vacsol 2709/2710   *671 757*
Vacsol 2713/2714   *667 738*
Vacsol 2718 2:1 Conc   *622 659 692 710*
Vacsol 2719 RTU   *622 659 692 710*
Vacsol 2746/2747 J   *737*
Vacsol Aqua 6101   *496 713*
Vacsol Aqua 6105   *495 663 714*
Vacsol Aqua 6107   *622 659 692 710*
Vacsol Aqua 6108   *658 697 722*
Vacsol Aqua 6109   *658 697 722*
Vacsol Azure 2720   *624 691 711*
Vacsol Azure 2721   *624 691 711*
Vacsol Azure 2722   *622 659 692 710*
Vacsol J (2:1 Conc) 2744/2745   *755*
Vacsol J 2742/2743   *755*
Vacsol J RTU   *739*
Vacsol JWR (2:1 Conc) 2642/2643   *755*
Vacsol JWR 2640/2641   *755*
Vacsol JWR Concentrate   *739*
Vacsol JWR RTU   *739*
Vacsol MWR Concentrate 2203   *629 741*
Vacsol P (2310)   *671 757*
Vacsol P 2304 RTU   *629 741*
Vacsol P Ready To Use (2334)   *739*
Vacsol P Ready To Use (2335)   *755*
Vacsol WR (2:1 Conc) 2614/2615   *671 757*
Vacsol WR 2612/2613   *671 757*
Vacsol WR Ready To Use 2116   *629 741*
Vallance Fungicidal Wash   *449 450*
Vallance Pro-Build Extra Strong Moss and Mould
  Killer   *428 458*
Valspar Creosote   *561*
Valspar Shed and Fence Preservative   *699*
Valspar Timberguard Shed and Fence
  Preservative   *699*
Vape Mat   *366*
Vapona Ant and Crawling Insect Killer   *256 332*
Vapona Ant and Crawling Insect Killer
  Aerosol   *282 398*
Vapona Ant and Crawling Insect Killer
  Powder   *331*
Vapona Ant and Crawling Insect Spray   *229 284*
Vapona Ant and Woodlice Killer Powder   *286*
Vapona Ant Bait   *421*

Vapona Ant Trap   *421*
Vapona Antpen   *276*
Vapona Carpet and Household Flea Powder   *331*
Vapona Citronella Candle   *219*
Vapona Fly and Wasp Killer   *256 332*
Vapona Flypen   *276*
Vapona Green Arrow Fly and Wasp Killer   *369*
Vapona House and Plant Fly Aerosol   *369*
Vapona House and Plant Fly Spray   *360 417*
Vapona Max Concentrated Fly and Wasp
  Killer   *362 418*
Vapona Micro-Tech   *267*
Vapona Mothaks   *222 224*
Vapona Plug-In   *248*
Vapona Plug-In Flying Insect Killer   *228*
Vapona Small Space Fly Killer   *293*
Vapona Wasp and Fly Killer   *360 417*
Vapona Wasp and Fly Killer Spray   *360 417*
Vapona Wasp Killer Aerosol   *260 410*
VC 17M EP-Antifouling   *4*
VC 17M Tropicana   *5 112*
VC 17M-EP   *1*
VC Offshore   *16 54 114 120*
VC Offshore Extra 100 Series   *46 188*
VC Offshore SP Antifouling   *46 188*
VC Prop-O-Drev   *54 120*
VC17M   *4*
VC17M-HS   *3 111 184*
Vectobac 12AS   *241*
Vet Kem Acclaim 2000   *321 348*
Vet-Kem Acclaim   *326 383*
Vet-kem Pump Spray   *321 348*
Vetpet Crusade   *331*
VF For Fleas   *265*
Vijurrax Spray   *290 294*
Viniline   *11*
Vinilstop 9926 Red.   *11*
Vinyl Antifouling 2000   *11*
Vitapet Flearid Household Spray   *360 417*
Vitax Kontrol Ant Killer Powder   *331*
W.S.P. Wood Treatment   *699*
Wahl Envoyage Mosquito Killer   *248*
Wall Fungicide   *451*
Wasp Destroyer Foam   *360 417*
Wasp Killer   *251 264*
Wasp Nest Destroyer   *242 391 409*
Wasp Nest Killer Professional   *242*
Waspend   *364 392 408*
Waspex   *312*
Water Based Wood Preserver (Concentrate)   *565*
  617
Water Based Woodworm Fluid   *644*
Water Based Woodworm Killer   *563*
Water-Based Wood Preserver   *565 617*
Waterbased Woodworm Killer 5X   *563*

Weathershield Aquatech Preservative
  Basecoat *526 598*
Weathershield Exterior Preservative
  Basecoat *589 735 746*
Weathershield Exterior Timber Preservative *545
  750*
Weathershield Fungicidal Wash *449 450 607 609*
Weathershield Mulit-Surface Fungicidal
  Wash *431 447 491 596*
Weathershield Preservative Primer *589 735 746*
Wellcome Multishot Aircraft Aerosol *357*
Wellcome One-Shot Aircraft Aerosol *357*
Westland Ant Killer Powder *242*
Westland Ant Killer RTU *286*
Whiskas Bedding Spray *236 361*
Whiskas Household Anti Flea Spray *325 379*
Whiskas Bedding Pest Control Spray *236 361*
Whiskas Exelpet Anti-Flea Carpet Powder *331*
Whiskas Exelpet Household Spray *320 336*
White Tiger Cruising *131 183*
Wickes Creosote *535*
Wickes Decorative Preserver Rich Cedar *472
  582*
Wickes Decorative Wood Preserver *472 582*
Wickes Exterior Wood Preserver *470*
Wickes Fungicidal Wash *427*
Wickes Mouldkill Emulsion *201 205 212*
Wickes Wood Preserver Clear *470*
Wickes Woodworm Killer *644*
Wilko Ant Killer Lacquer *289*
Wilko Ant Killer Spray *276*
Wilko Creosote *535*
Wilko Exterior System Preservative Primer *593
  751*
Wilko New Ant Killer Spray *286*
Willo-Zawb *357*
Willodorm *357*
Wintox *292 301 399*
Wocosen 100 SL *459 680*
Wocosen 100 SL-C *567 687*
Wocosen 12 OL *655 689*
Wocosen S *680*
Wocosen S-P *655 689*
Wolmanit CX-P *505 537 685*
Wolsit KD 20 *655 689*
Wood Preservative *581 744*
Wood Preservative AA 155/00 *591 740*
Wood Preservative AA 155/01 *739*
Wood Preservative AA 155/03 *591 740*
Wood Preservative AA155 *591 740*
Wood Preservative Clear *668 754*
Wood Preservative Type 3 *535*
Wood Preserver Green *540*
Woodlice Killer *242*
Woodman Creosote and Light Brown
  Creosote *561*

Woodtreat *626 640*
Woodtreat 25 *620 646*
Woodtreat Boron 40 *594*
Woodtreat BP *668 754*
Woodworm 8 Micro Emulsifiable
  Concentrate *644*
Woodworm Fluid (B) EC *644*
Woodworm Fluid (B) For Bonded
  Warehouses *644*
Woodworm Fluid B *644*
Woodworm Fluid FP *644*
Woodworm Fluid Z Emulsion Concentrate *644*
Woodworm Killer *563 644 665 734*
Woodworm Killer (Grade P) *644*
Woodworm Killer and Rot Treatment *425 455
  483 650*
Woodworm Killer P8 *644*
Woodworm Paste B *644*
Woodworm Roof Void Paste *644*
Woodworm Treatment Spray *644*
Woolworth Mosquito Killer *248*
Woolworths New Ant Killer Spray *286*
X Moss *426 484*
X Thallo *441 573*
X-Gnat Advanced Insect Repellent - Fabric
  Spray *226*
X-Gnat M/C (Mosquito Control) *338 381*
XM Anti-Fouling C2000 Cruising Self Eroding *16
  114*
XM Anti-Fouling P4000 Hard *16 114*
XM Antifouling HS3000 High Performance Self
  Eroding *16 114*
Xyladecor Matt U 404 *482 583*
Xylamon Brown U 101 C *483 650*
Xylamon Curative U 152 G/H *483 650*
Xylamon Primer Dipping Stain U 415 *482 583*
Z-Stop Anti-Wasp Strip *331*
Z.144 Fungicidal Wash *428 458*
Zap Pest Control *256 332*
Zappit *349 406*
Zappit Fly and Wasp Killer *360 417*
Zodiac + Flea Spray *321 348*
Zodiac + Household Flea Spray *326 383*
Zodiac and Household Flea Spray *326 383*
Zodiac and Pump Spray *321 348*
Zodiac Household Flea Spray *327 411*
Zodiac Maxi Household Flea Spray *321 348*

9

HSE ACTIVE INGREDIENT INDEX

HSE ACTIVE INGREDIENT INDEX

HSE
INGREDIENT

591

Printed In The United Kingdom For The Stationery Office
J0066093 2/00 C25 10170